AF784842

REPRODUCTIVE TOXICOLOGY

REPRODUCTIVE TOXICOLOGY

Editor

DONALD R. MATTISON

Pregnancy Research Branch
National Institute of Child Health and Human Development
National Institutes of Health
Bethesda, Maryland

CIIT LIBRARY 9/19/83

ALAN R. LISS, INC. • NEW YORK

These papers are being printed both as a combined issue of the American Journal of Industrial Medicine, Volume 4, Number 1/2, 1983, with Dr. Donald R. Mattison as Guest Editor, and as Reproductive Toxicology, PCBR Volume 117, edited by Dr. Mattison.

Address all Inquiries to the Publisher
Alan R. Liss, Inc., 150 Fifth Avenue, New York, NY 10011

Copyright © 1983 Alan R. Liss, Inc.

Printed in the United States of America.

Under the conditions stated below the owner of copyright for this book hereby grants permission to users to make photocopy reproductions of any part or all of its contents for personal or internal organizational use, or for personal or internal use of specific clients. This consent is given on the condition that the copier pay the stated per-copy fee through the Copyright Clearance Center, Incorporated, 21 Congress Street, Salem, MA 01970, as listed in the most current issue of "Permissions to Photocopy" (Publisher's Fee List, distributed by CCC, Inc.), for copying beyond that permitted by sections 107 or 108 of the US Copyright Law. This consent does not extend to other kinds of copying, such as copying for general distribution, for advertising or promotional purposes, for creating new collective works, or for resale.

Library of Congress Cataloging in Publication Data

Main entry under title:

Reproductive toxicology.

(Progress in clinical and biological research; v. 117)
Includes bibliographical references and index.
1. Reproduction—Addresses, essays, lectures.
2. Toxicology—Addresses, essays, lectures. I. Mattison, Donald R. II. American journal of industrial medicine. III. Series. [DNLM: 1. Environmental pollutants—Adverse effects. 2. Reproduction—Drug effects. 3. Occupational diseases—Chemically induced. 4. Toxicology. W1 PR668E v. 117 / WA 400 R425]
QP251.R4448 1983 616.6′5 83-927

ISBN 0-8451-0117-X

Contents

Contributors

Gunnar Ahlborg [293]
Department of Occupational Medicine, University Hospital, 701 85 Örebro, Sweden

Olav Axelson [293]
Department of Occupational Medicine, University Hospital, 581 85 Linköping, Sweden

William F. Blazak [5]
Biochemical Genetics Department, Life Sciences Division, SRI International, Menlo Park, CA 94205

W.C. Brogan III [127]
Department of Pharmacology and Toxicology, West Virginia University Medical Center, Morgantown, WV 26506

William H. Bulger [163]
Worcester Foundation for Experimental Biology, Shrewsbury, MA 01545

Ramona M. Chapman [149]
St. John's Regional Oncology Center, Joplin, MO 64801

David C. Cumming [113]
Departments of Obstetrics and Gynaecology and Medicine (Division of Endocrinology), University of Alberta, Edmonton, Alberta T6G 263, Canada

Jurrien Dean [31]
Laboratory of Chemical Biology, National Institute of Arthritis, Diabetes, Digestive and Kidney Diseases, National Institutes of Health, Bethesda, MD 20205

R. Lowry Dobson [175]
Biomedical Sciences Division, Lawrence Livermore National Laboratory, University of California, Livermore, CA 94550

Sergio Fabro [391]
Reproductive Toxicology Center, Columbia Hospital for Women, Georgetown University, Washington, DC 20037

James S. Felton [175]
Biomedical Sciences Division, Lawrence Livermore National Laboratory, University of California, Livermore, CA 94550

Jacob Hart [309]
Department of Preventive and Social Medicine, Tel Aviv University, Sackler School of Medicine, Tel Aviv 69978, Israel

Kari Hemminki [87,293]
Institute of Occupational Health, SF-00290 Helsinki 29, Finland

Peter F. Infante [383]
Office of Carcinogen Identification and Classification, Health Standards Program, OSHA, US Department of Labor, Washington, DC 20201

Dwight T. Janerich [329]
Division of Community Health and Epidemiology, New York State Department of Health, Albany, NY 12237

Marcus A. Klingberg [309]
Department of Preventive and Social Medicine, Tel Aviv University, Sackler School of Medicine, Tel Aviv 69978, Israel

Herman B.W.M. Koëter [81]
Reproduction Toxicology Section, Department of Biological Toxicology, Institute CIVO–Toxicology and Nutrition TNO, 3700 AJ, Zeist, The Netherlands

David Kupfer [163]
Worcester Foundation for Experimental Biology, Shrewsbury, MA 01545

I.P. Lee [135]
Laboratory of Reproductive and Developmental Toxicology, National Institute of Environmental Health Sciences, Research Triangle Park, NC 27709

The bold face number in brackets following each contributor's name indicates the opening page of that author's paper.

Stephen M. Levin [349]
Environmental Sciences Laboratory, Mount Sinai School of Medicine of the City University of New York, New York, NY 10029

Carolyn H. Lingeman [365]
Chemical Pathology Branch, National Institute of Environmental Health Sciences, Bethesda, MD 20205

Donald R. Mattison [1,17,65,191]
Pregnancy Research Branch, NICHD, NIH, Bethesda, MD 20205

Richard K. Miller [205]
Department of Obstetrics-Gynecology and Pharmacology-Toxicology, University of Rochester School of Medicine and Dentistry, Rochester, NY 14642

Marja-Liisa Niemi [293]
Institute of Occupational Health, SF-00290 Helsinki 29, Finland

Maria S. Nightingale [191]
Pregnancy Research Branch, NICHD, NIH, Bethesda, MD 20205

James W. Overstreet [5]
Departments of Human Anatomy and Obstetrics and Gynecology, School of Medicine, University of California at Davis, Davis, CA 95616

Cheri M. Papier [309]
Department of Preventive and Social Medicine, Tel Aviv University, Sackler School of Medicine, Tel Aviv 69978, Israel

Anthony P. Polednak [329]
Division of Community Health and Epidemiology, New York State Department of Health, Albany, NY 12237

Robert W. Rebar [113]
Department of Reproductive Medicine, University of California at San Diego, La Jolla, CA 92093

Kenji Shiromizu [191]
Pregnancy Research Branch, NICHD, NIH, Bethesda, MD 20205

Carol Grace Smith [107]
Department of Pharmacology, Uniformed Services University of Health Sciences, Washington, DC 20814

Marja Sorsa [87]
Institute of Occupational Health, SF-00290 Helsinki 29, Finland

William J. Swartz [51]
Department of Anatomy, Louisiana State University Medical Center, New Orleans, LA 70112-1394

Ken Takizawa [17]
Pregnancy Research Branch, NICHD, NIH, Bethesda, MD 20205

J.A. Thomas [127]
Department of Pharmacology and Toxicology, West Virginia University Medical Center, Morgantown, WV 26506

Theodora A. Tsongas [383]
Office of Carcinogen Identification and Classification, Health Standards Program, OSHA, US Department of Labor, Washington, DC 20201

H. Tuchmann-Duplessis [245]
Faculté de Médecine, Laboratoire d'Embryologie, 75006 Paris, France

Harri Vainio [87]
Institute of Occupational Health, SF-00290 Helsinki 29, Finland

Allen J. Wilcox [285]
Biometry and Risk Assessment Program, National Institute of Environmental Health Sciences, Research Triangle Park, NC 27709

Mary S. Wolff [259]
Environmental Sciences Laboratory, Mount Sinai School of Medicine of the City University of New York, New York, NY 10029

American Journal of Industrial Medicine 4:1–2 (1983)

Preface: Reproductive Toxicology

Donald R. Mattison, MD

Reproductive toxicology holds in its purview the development and function of the male and female reproductive systems, including fertilization, implantation, embryonic and fetal development, parturition, and postnatal adaptation. This broad perspective—encompassing reproductive and developmental biology, toxicology, teratology, and pharmacology as well as epidemiology, occupational and environmental health and medicine—may seem overwhelming to the health scientist interested in learning about (or participating in) this fascinating science. This volume represents an attempt to review and synthesize these diverse sciences. The need for conciseness and the pressures of editorial deadlines have inevitably produced gaps. Fortunately, knowledge and understanding continue to grow and in time this volume will be supplanted by one which is more complete and lucid.

The basic stimulus for this symposium was derived from efforts to synthesize the sciences of reproductive biology and toxicology. These began while I was a Clinical Associate in the National Institute of Child Health and Human Development (NICHD). This synthesis has been fostered by scientists within several of the Institutes as well as several individuals within NICHD. Those individuals have shared unselfishly of their time, knowledge, laboratory space and budgets.

Like other aspects of science, the development of this volume rests on several previous events. Two deserve special acknowledgment. The Scientific Group on Methodologies for the Safety Evaluation of Chemicals (SGOMSEC) of the Scientific Committee on Problems of the Environment (SCOPE)—sponsored by the World Health Organization—convened a Workshop on Methodologies for the Assessment of the Effects of Chemicals on Reproductive Function in Ispra, Italy, in the spring of 1981. The broad perspective of that meeting was incredibly stimulating and strengthened my resolve to attempt this volume. The other stimulus was the International Course on Occupational Hazards and Reproduction, sponsored by the Finnish Institute of Occupational Health and held outside Helsinki in the summer of 1981. This second meeting (coupled with the legendary hospitality of the Finns and the stimulus of the sauna) removed the last impediments to this volume. The opportunity to meet other scientists at both meetings and discuss a broad range of topics concerned with reproductive toxicology was enormously useful for me.

The first section of this volume reviews the biology of the male and female reproductive systems, early embryonic development and genetics. The second section briefly reviews some aspects of toxicology, including the sensitivity of reproductive events to interference by xenobiotic compounds, and methods for monitoring exposures to genotoxicants. The third section focuses on specific forms, sites, or mechanisms of reproductive toxicity. The fourth section reviews aspects of prenatal, perinatal and postnatal

Pregnancy Research Branch, NICHD, NIH, Bethesda, Maryland.

Address reprint request to Dr. Donald R. Mattison, Building 10, Room 8C313, PRB-NICHD-NIH, Bethesda, MD 20205.

0271-3586/83/0401-02-0001$01.00 **© 1983 Alan R. Liss, Inc.**

toxicity. Authors contributing to the fifth section have focused on isssues relating to surveillance for adverse reproductive effects of occupational, environmental or drug exposures. And the final section addresses some of the important and difficult—but frequently ignored—issues relating to the social impact of reproductive toxicity.

I would like to thank Dr. Irving J. Selikoff and the American Journal of Industrial Medicine for the opportunity to present this "Journal Symposium on Reproductive Toxicology." My wife, Margaret, deserves special thanks for tolerating my forgetfulness and my other less desirable characteristics. Ms. Maria S. Nightingale deserves thanks for keeping my laboratory functioning while I attended to the tasks associated with this volume. And finally I would like to thank the participating authors who have spent valuable time in the preparation of manuscripts designed to summarize specific facets of this field. It is hoped that this volume will be a useful and stimulating resource for health scientists interested in the broad discipline of reproductive toxicology.

REPRODUCTIVE BIOLOGY

American Journal of Industrial Medicine 4:5–15 (1983)

The Biology of Human Male Reproduction: An Overview

James W. Overstreet, MD, PhD, and William F. Blazak, PhD

In this overview, the focus is on the biology of human male reproduction. The structure and function of the testis are described, and the endocrine control of spermatogenesis is discussed. The formation of spermatozoa during spermatogenesis and their maturation in the epididymis are reviewed. The physiology of ejaculation and sperm transport in the female tract are discussed. Sperm capacitation, the acrosome reaction, and fertilization of human gametes are also reviewed.

Key words: spermatogenesis, spermatozoa, epididymis, sperm transport, fertilization

STRUCTURAL AND FUNCTIONAL ASPECTS OF THE TESTIS

The mammalian testis consists of two principal functional components: the seminiferous tubules and the interstitial tissue. Spermatozoa are produced within the seminiferous tubules, which are long, extensively convoluted tubular structures whose ends open into the rete testis. In humans, the tubules in each testis are arranged in about 300 lobules, with each lobule containing one to four tubules. The boundary tissue, or lamina propria, of the tubules has a clearly defined structure consisting of multiple layers of adventitial cells. The seminiferous tubules contain germ cells at various stages of maturity and Sertoli cells. The Sertoli cells extend from the basal lamina propria to the lumen of the seminiferous tubule. These cells, which do not divide after puberty, fulfill a number of important functions. These include participation in the formation of the blood–testis barrier, facilitation of release of spermatozoa from the germinal epithelium, synthesis of specific proteins (eg, androgen-binding protein), secretion of fluid into the lumen of the seminiferous tubule, as well as providing a sustentacular function for germ cells during the later stages of spermiogenesis. The Sertoli cells also

Departments of Human Anatomy and Obstetrics and Gynecology, School of Medicine, University of California at Davis, Davis (J.W.O.).
Biochemical Genetics Department, Life Sciences Division, SRI International, Menlo Park, California (W.F.B.).
Address reprint requests to Dr. James W. Overstreet, Department of Human Anatomy, School of Medicine, University of California at Davis, Davis, CA 95616.

0271-3586/83/0401-02-0005$03.50 © 1983 Alan R. Liss, Inc.

serve important functions in the endocrine control of testicular function. The male germ cells will be discussed in some detail in a later section.

The testicular interstitial tissue, which is located exterior to and between the seminiferous tubules, contains blood and lymph vessels, nerves, and the interstitial or Leydig cells. The Leydig cells normally produce testosterone, estradiol and 17-OH progesterone. In normal men, about 7 mg of testosterone are produced per day mainly by the Leydig cells. The functions of testosterone, as well as the role of Leydig cells in the endocrine control of spermatogenesis, are discussed in the following section.

The testis of the adult human male serves two main functions: (1) production of male gametes—spermatozoa, and (2) production of male hormones or androgens, principally testosterone. These two functions of the testis are highly interrelated in that androgen production is necessary for sperm production, and successful male reproduction requires normal sexual behavior and the development of the secondary sexual characteristics (male accessory organs such as the prostate and seminal vesicles, external genitalia, etc), which are also androgen dependent. It is clear that perturbations in either of these testicular functions will result in reduced reproductive capacity.

ENDOCRINE CONTROL OF ADULT TESTICULAR FUNCTION

The endocrine control of testicular function is a remarkably complex subject, which has been extensively reviewed [cf, Steinberger and Steinberger, 1974; Steinberger et al, 1977; Setchell, 1978; diZerega and Sherins, 1981]. Our purpose here is to outline the salient features of this control briefly; the reader is referred to the referenced reviews for a more detailed discussion and introduction to the original literature.

Adult testicular function is largely under the influence of the two gonadotropic hormones, follicle-stimulating hormone (FSH) and luteinizing hormone (LH) [see Table I]. These hormones are glycoproteins that are synthesized and released by the anterior pituitary under the influence of blood-borne gonadotropin-releasing factors from the hypothalamus. Both hormones bind to specific receptors on the membranes of their respective target cells and stimulate cellular metabolism via increases in adenylate cyclase activity, cyclic AMP, and protein kinase activity. LH binds to the Leydig cells and promotes steroidogenesis. The primary product is the androgen testosterone, with smaller amounts of estradiol and 17-OH progesterone also being synthesized. Testosterone is required for normal spermatogenesis, the integrity of the male accessory sex glands, male secondary sexual characteristics, and is also involved in the regulation of LH secretion via a negative feedback system on the hypothalamus and anterior pituitary. Increased concentration of testosterone in the blood will inhibit the release of LH from the anterior pituitary. Estradiol also inhibits the release of LH.

FSH acts on the Sertoli cells within the seminiferous tubules resulting in an increase in general protein synthesis as well as the synthesis of a number of specific proteins. One such protein, androgen-binding protein (ABP), is synthesized by the Sertoli cells in response to FSH and acts to bind testosterone in the interstitial tissues and facilitate its entry into the seminiferous tubules where it is required in high local concentration for normal spermatogenesis. FSH also enhances the LH stimulation of testosterone production by Leydig cells, perhaps by increasing the number of LH receptors. Experimental evidence also indicates that FSH may be required for the maturation of spermatids into spermatozoa (spermiogenesis).

The release of FSH from the anterior pituitary appears to be modulated by the testosterone concentration of the blood, as described for LH. As with LH, estradiol

TABLE I. The Endocrine Control of Adult Testicular Function*

Hormone	Source	Major target(s)	Direct or indirect effect on target(s)
Luteinizing hormone (LH)	Anterior pituitary	Leydig cells	Stimulate steroidogenesis (testosterone production)
Follicle-stimulating hormone (FSH)	Anterior pituitary	Sertoli cells	Stimulate protein synthesis (eg, androgen-binding protein)
		Sertoli and/or germ cells	Maturation of spermatids into spermatozoa (spermiogenesis)
Testosterone	Leydig cells	Male accessory glands	Maintain structure and function
		Hypothalamus, pituitary	Negative feedback control on release of FSH and LH
Estradiol	Leydig cells[a]	Anterior pituitary	Negative feedback control on release of FSH and LH
Inhibin	Sertoli cells	Anterior pituitary	Negative feedback control on release of FSH

*Adapted from Setchell [1978] and diZerega and Sherins [1981].
[a]About 30% of the estradiol is secreted by the testis; the remainder is derived from the peripheral aromatization of testosterone.

also inhibits FSH release. Unlike that of LH, however, a seminiferous tubule protein termed inhibin is also involved in the regulation of FSH. Inhibin is produced by the Sertoli cells and is believed to operate as a negative feedback control of FSH release at the anterior pituitary.

In addition to the gonadotropic regulation of testicular function, recent studies provide evidence for the existence of an intratesticular Leydig cell–Sertoli cell interaction. Sertoli cells are capable of converting androgen precursors into estrogen, and estrogen inhibits testosterone secretion by the Leydig cells. These events can occur independently of LH and suggest a local hormonal control of testicular testosterone production mediated by the Sertoli cells.

SPERMATOGENESIS

Germ Cells of the Seminiferous Epithelium

Aside from spermatozoa, three distinct types of male germ cells can be distinguished in human seminiferous epithelium: spermatogonia, spermatocytes, and spermatids. Spermatocytes are large, diploid cells, which lie against the basal lamina propria within the seminiferous tubules. Clermont [1963, 1966] described three types of spermatogonia in the human testis, type A-dark (Ad), type A-pale (Ap), and type B. Type Ad spermatocytes undergo numerous mitotic divisions and some of these cells differentiate into Ap spermatogonia; Ap spermatogonia likewise proceed through mitotic divisions, eventually differentiating into type B spermatogonia. The type B spermatogonia divide mitotically and eventually become preleptotene spermatocytes. One characteristic feature of male germ cells is that the cytoplasm does not completely separate upon nuclear division. Such incomplete cytokinesis results in a cohort of cells, held together by cytoplasmic bridges, which develop synchronously. The type Ad spermatogonia, therefore, are thought to be the most primitive spermatogonial cells and represent a class of renewing stem cells [Clermont, 1970]. These stem cells continue to provide Ap spermatogonia for entry into the spermatogenic cycle and eventual transformation into spermatozoa.

The cells resulting from the last mitotic division of the type B spermatogonia are a class of primary spermatocytes termed pre-leptotene spermatocytes. The chromatin in these cells undergoes visual organization into thin filaments resulting in leptotene spermatocytes. Chromosome replication has already occurred, and leptotene spermatocytes contain a tetraploid amount of DNA. Homologous chromosomes then synapse, synaptinemal complexes are formed between pairs of chromosomes, and the nucleoli become more visible; primary spermatocytes at this stage are termed zygotene spermatocytes. In the next stage of prophase of the first meiotic division, pachytene, crossing-over occurs, the chromosomal bivalents become more condensed and thicker, and each chromosome divides longitudinally (except at the centromere) into two sister chromatids. The following stage, diplotene, is characterized by the formation of chiasmata between homologous chromosomes, and the chromosomes appear even more condensed. In the final stage of meiotic prophase, diakinesis, the chiasmata are in the process of moving laterally toward the ends of the chromatids, and the chromosomes display maximal condensation. The chiasmata continue to terminalize and the homologous chromosomes then segregate from one another during anaphase of the first meiotic division. This division results in secondary spermatocytes that contain a diploid amount of DNA. After a short interphase, each secondary spermatocyte divides in a mitotic fashion (without interphase chromosome replication, however) to yield two spermatids, each containing a haploid amount of DNA.

The spermatids that result from the second meiotic division are initially small, round cells containing a moderate amount of cytoplasm. These cells then undergo a remarkable transformation into spermatozoa, a process termed spermiogenesis. This transformation includes nuclear condensation, formation of an acrosome, elimination of most of the cytoplasm, development of a tail, and the arrangement of mitochondria into the sperm middle piece. It is during this transformation that the nucleus of the spermatid assumes the shape characteristic of the spermatozoa of the species.

Upon completion of spermiogenesis, the elongated nucleus of the spermatid is embedded in the cytoplasm of a Sertoli cell, with the tail projecting into the lumen of the seminiferous tubule. The spermatid is extruded into the lumen of the tubule largely through the activity of the Sertoli cell. This extrusion process leaves the majority of the cytoplasm of the spermatid embedded in the cytoplasm of the Sertoli cell.

Dynamics of Human Spermatogenesis

Before a type B spermatogonium becomes a spermatozoon, many other spermatogonial cells will have entered the spermatogenic cycle. This results in several superimposed generations of germ cells in any cross-section of a seminiferous tubule. Owing to the regularity of this process, certain germ cells in the seminiferous epithelium are always found in association with one another, and never with certain other types of germ cells. Thus, the various generations of germ cells are not associated at random but form cellular associations of fixed composition. It follows that only a limited number of such cellular associations will exist in various cross-sections of a seminiferous tubule. Each such association of cells is termed a stage of the cycle of the seminiferous epithelium; a cycle of the seminiferous epithelium is formed by the series of successive cellular associations that occurs along a seminiferous tubule between two successive appearances of the same cellular association. In man, Clermont [1963] described six different stages based on nuclear morphology and the topographical arrangement of the cells. The length of time required for one complete cycle of the seminiferous epithelium in man was shown to be 16 days [Heller and Clermont, 1964].

Dear Book or CD Buyer,

Thank you for your purchase from Research Triangle Books. We aim to provide a wide variety of books and CDs, give accurate descriptions, package items well, and ship quickly. Your satisfaction is our goal.

If you have any problems, please contact us via Amazon, and please check that our response is not in your Spam folder. Or you can contact us directly at researchtrianglebooks@gmail.com . You may also contact us directly if you have questions about other items that you are looking for. If you must return an item, please try to reuse the original shipping materials if possible.

Please note: New, sealed, CDs and DVDs cannot be returned after opening them, unless there is a manufacturing problem with them, so please make sure that you have the correct item before opening.

Lastly, we ask that you provide seller feedback when requested by Amazon, or you may do it now by clicking "My Account" and "Leave seller feedback".

Bob and MJ

The spermatogenic cycle extends from the time a type B spermatogonium differentiates into a pre-leptotene spermatocyte until this same cell is released into the lumen of the seminiferous tubule as a spermatozoon. The duration of the spermatogenic cycle, therefore, is the amount of time required for the production of a sperm cell from a pre-leptotene spermatocyte. The duration of the spermatogenic cycle in man was determined by Heller and Clermont [1964] using intratesticular injections of tritiated thymidine, and following the most advanced, labeled cells in testicular biopsies obtained at various times after injection. They concluded that the duration of the human spermatogenic cycle was 74 days, with a possible variation of 4 or 5 days. This represents the time required for about 4.6 cycles of the seminiferous epithelium. The duration of the spermatogenic cycle in other mammalian species is approximately equivalent to this number of cycles of the seminiferous epithelium, but differs in the amount of time required because of differing durations of the cycle of the seminiferous epithelium among species [Ortavant et al, 1977].

Rate of Sperm Production

The rate of production of spermatozoa by men has been assessed in a few studies. By enumerating the spermatozoa in human ejaculates collected every 3 days, mean daily sperm production rates of 95 $\times$ 10^6 to 155 $\times$ 10^6 cells have been reported [Freund, 1962, 1963; Zimmerman et al, 1965]. In a recent report, Amann and Howards [1980] calculated the daily rate of production of spermatozoa in 23 men after enumerating the number of elongated spermatids in homogenates prepared from the testes of these men. The mean number of spermatozoa produced daily by these men was 123 $\pm$ 18 $\times$ 10^6, with a range of 21 to 374 $\times$ 10^6 cells per day. These rates are equivalent to 4.45 $\times$ 10^6 sperm per day per gram of testis. As discussed by Amann and Howards [1980], daily sperm production by humans is considerably below that of other mammalian species, owing mainly to a paucity of germ cells within the seminiferous tubules.

THE EPIDIDYMIS AND SPERM MATURATION

Upon release of spermatozoa from the seminiferous epithelium, sperm pass through the rete testis and enter the epididymis via the ductuli efferentes. The cells at this stage are immature and are virtually immobile, incapable of fertilizing ova, and often bear a remnant of cytoplasm (cytoplasmic droplet) in the neck and middle piece region. Final maturation of the cells occurs during transit through the epididymis, a convoluted duct located on the posterior surface of the testis. The epididymis is divided anatomically into three segments: the caput, corpus, and cauda epididymidis. The epididymis is dependent upon testicular androgen for the maintenance of its structural and functional properties.

The maturation of spermatozoa in the human epididymis has been studied by Bedford et al [1973]. These authors demonstrated a number of maturational changes in human spermatozoa as they progressed from the caput to the cauda epididymidis including (1) loss of the cytoplasmic droplet, (2) structural maturation of the nucleus and tail organelles through the formation of disulfide bonds between protein sulfhydryl groups, (3) acquisition of progressive motility, and (4) changes in the surface of the spermatozoal plasma membrane. The nature of these changes indicated that human spermatozoa, like those of other mammals, gained the capacity to fertilize ova after maturation in the epididymis. In a recent report, Hinrichsen and Blaquier [1980] found that human spermatozoa collected from the cauda epididymidis were capable of undergoing capacitation and entering the ooplasm of the zona pellucida-free hamster ovum in vitro, whereas those collected from the corpus and caput segments of the epididymis failed to penetrate ova. These findings provide further evidence that human spermatozoa gain the capacity to fertilize ova after transit through the epididymis.

In contrast to other mammalian species, the transit time of spermatozoa through the human epididymis is very short and highly variable. Rowley et al [1970] found that the rate of transport varied from 1 to 21 days within an individual. Amann and Howards [1980] identified transit times of 0.72 ± 0.11, 0.71 ± 0.10, and 1.76 ± 0.24 days in the caput, corpus, and cauda segments, respectively, of the human epididymis. This relatively short and variable transit time may be a significant factor in the structural and functional heterogeneity of spermatozoa in the human ejaculate [Bedford et al, 1973].

EJACULATION

The normal sequence of sexual functions in the human male includes erection, emission, and ejaculation. Emission of spermatozoa into the prostatic urethra is effected by the sympathetic hypogastric nerves innervating the epididymis, the vas deferens, and the prostate gland [Hotchkiss, 1970]. Preceding emission, there is an initial movement of spermatozoa from the distal epididymis toward the ampulla of the vas deferens [Mitsuya et al, 1960]. During emission, the vas deferens and its ampulla become narrowed and the ampulla becomes less cavernous. The contents of the ampulla and the prostatic secretions first enter the prostatic urethra. This is followed by contraction of the seminal vesicles during which about one-tenth of their contents is emptied into the urethra. These events of emission are followed by ejaculation, which is a reflex action mediated by afferent impulses in the pundendal nerves and efferent impulses along the same nerves, causing contraction of the pelvic and perineal muscles. As pressure builds in the prostatic urethra, its contents are propelled into the penile urethra. Additional prostatic secretions then enter the prostatic urethra and further contractions of the seminal vesicles occur, the cycle being repeated an average of six additional times [Mitsuya et al, 1960].

MALE ACCESSORY SECRETIONS

The human ejaculate is composed of spermatozoa from the vas deferens and cauda epididymidis, together with accessory fluids from the prostate gland, the seminal vesicles and Cowper's gland [Mann, 1964]. The role of the accessory gland secretions in relation to sperm physiology is poorly understood. One recognized function of the seminal plasma is its buffering effect on the vaginal environment. In women, the vaginal acidity is usually pH 5 or less, but the buffering capacity of the seminal plasma can raise the vaginal pH to more than 7 within 8 sec after ejaculation [Fox et al, 1973].

The semen is usually ejaculated in a specific sequence. The first fraction of the ejaculate characteristically contains the majority of the spermatozoa and the prostatic secretion, whereas the second portion is composed of seminal vesicle secretion and relatively fewer spermatozoa. When these fractions are mixed in vitro or within the vagina, the semen spontaneously coagulates. The clot-like material that is formed liquefies spontaneously within the next 20 min. It has been suggested that the seminal vesicles may provide specific proteins involved in formation of the coagulum and that the prostate may be the source of hydrolytic enzymes required for its liquefaction, but the biochemistry of these processes is very complicated and not well understood [Tauber et al, 1980]. The physiological functions of semen coagulation may include initial protection of spermatozoa in the vaginal environment, since the sperm cells are virtually immotile within the clot. However, motile sperm have been observed swimming freely in the cervical mucus within 90 sec after ejaculation, long before liquefaction of the coagulum [Sobrero and MacLeod, 1962]. Initial entry of spermatozoa into the cervical mucus, before coagulation of the semen, must result from contractions of the uterus and the vagina. This contractile activity may be stimulated by pharmacologically active substances in the seminal plasma [Freund, 1973].

SPERM TRANSPORT IN THE FEMALE TRACT

During human coitus, semen is ejaculated within the posterior vagina and directly into contact with cervical mucus protruding from the external cervical os. A pressure gradient is generated across the cervix during coitus by penile thrusting and contractions of the female viscera [Fox et al, 1970]. These pressures and the coordinated muscular contractions of the uterus and uterine tubes instantaneously transport thousands of spermatozoa into the reproductive tract [Overstreet and Katz, 1981]. A few of these sperm cells may be propelled all the way to the peritoneal cavity [Settlage et al, 1973]. This initial phase of rapid sperm transport may be initiated by reflex uterine contractions mediated by the sympathetic fibers of the autonomic nervous system. Stretch receptors in the vagina and/or pharmacologically active spasmogens in the semen may stimulate the onset of sperm transport, but sexual orgasm does not appear to be required. The importance of rapid sperm transport in women may relate primarily to the accelerated uptake of spermatozoa from the vagina into the more favorable environment of the cervical mucus. These female contractions may transport small amounts of other seminal constituents along with the sperm cells, and it has been suggested that these could serve as local messengers to the upper reproductive tract [Overstreet, 1983].

The coordinated muscular contractions that result in rapid sperm transport are very short lived and are probably restricted to the coital period. Thereafter, spermatozoa swim out of the seminal pool within the vagina and into the cervical mucus. This sperm migration from the vagina to the cervix appears to occur within the first half hour following coitus, and few spermatozoa enter the cervical mucus after this time [Tredway et al, 1978]. The migration of spermatozoa through the uterus and into the upper reproductive tract is thought to involve the intrinsic motility of the sperm cells and has been described as a sequential buildup and release of spermatozoa from sites of storage or "reservoirs" [Overstreet and Katz, 1977]. The cervix appears to be an important sperm reservoir in women. The mucus that fills the cervical canal is penetrable by spermatozoa for only 24 to 48 hr near the time of ovulation. At its peak of quality, the mucus is clear, watery, and copious. Spermatozoa reaching the cervix may swim along microscopic channels in the mucus, which direct them to the cervical epithelium. The complex, folded cervical mucosa may be a reservoir in which spermatozoa aggregate, or they may swim continuously in the mucus until its properties change in the luteal phase of the cycle. Within 10 to 15 min after insemination, sperm numbers in the mucus reach a plateau, after which the size of the cervical sperm population remains fairly constant for 24 hr [Tredway et al, 1978]. Within 48 hr the number of spermatozoa in the cervical mucus has fallen significantly in most cases, but on rare occasions motile sperm can be found in the mucus for 5 days.

Our understanding of the physiology of sperm transport beyond the cervix is based primarily on animal studies. There are a number of sperm transport mechanisms that are common to several mammalian species and may, therefore, operate in humans as well. In many mammals there is a second site for sperm storage in the isthmus of the Fallopian tube adjacent to the uterus [Overstreet, 1983]. Experiments with rabbits have demonstrated a critical importance of the lower oviduct in regulating sperm–ovum interaction. Vigorous spermatozoa entering this region from the uterus are temporarily immobilized. Their upward migration is thereby restricted, and the isthmic reservoir is filled. There is no significant sperm release from this compartment until the periovulatory period that occurs 8 to 12 hr after mating in the rabbit. The final ascent of spermatozoa to the site of fertilization in the tubal ampulla appears to be synchronized with the arrival of ova. The mechanism of sperm escape from the isthmic reservoir involves reactivation of sperm motility as well as changes in the contractility of the Fallopian tube [Overstreet, 1983]. The limited studies of tubal sperm transport in women have yet to identify an isthmic reservoir, but the experimental techniques for isolation of human tubal sperm may not have been sufficiently precise to allow definitive assessment of sperm location within the oviduct.

A tremendous reduction in sperm numbers is accomplished during sperm transport through the female tract. The number of sperm reaching the ovum is less than one of every one hundred thousand ejaculated by the male. This appears to be an effective mechanism for reducing the risk of simultaneous fertilization of an ovum by more than one sperm. It is unknown whether the fertilizing sperm population is a random sample of the ejaculate, or a group of sperm cells selected by the female on the basis of unique and presumably favorable characteristics. Experiments carried out thus far to address this question have been inconclusive, and neither support nor refute the notion that sperm selection may occur during transport through the female tract [Overstreet and Katz, 1977].

SPERM CAPACITATION, THE ACROSOME REACTION, AND FERTILIZATION

The events of fertilization begin with the arrival of spermatozoa and the ovum at the site of fertilization in the ampulla of the Fallopian tube. The freshly ovulated oocyte is surrounded by a thick acellular zona pellucida and several layers of granulosa cells. Together these layers constitute the vestments of the ovum. To reach the surface of the ovum the spermatozoon must penetrate through each of the investing layers. This passage requires a series of precisely timed, coordinated steps that depend on several independent functions of the sperm cell.

Experiments carried out more than 50 years ago demonstrated that when mating of rabbits was progressively advanced toward the time of ovulation, there was a corresponding decrease in litter size. Although it was not appreciated at that time, this was due to the short interval of time the spermatozoa spent in the reproductive tract before the ova lost their viability. About 15 years later, other experiments showed that spermatozoa when injected into the female tract did not immediately fertilize the eggs. These data lead investigators to hypothesize that mature spermatozoa from the epididymis or ejaculate were initially infertile and required a period of incubation in the female, during which the capacity to fertilize was attained. During this interval of sperm capacitation, a series of poorly understood physiological changes were presumed to take place in the spermatozoa [see Overstreet and Katz, 1981].

The term capacitation is now used in reference to the physiological changes in the sperm cell that are required both for penetration of the ovum vestments and for fusion with the vitellus. The capacitation phenomena have been studied in many species including humans, but they remain poorly understood. Capacitation appears to include both alterations of the sperm surface and chemical changes within the sperm cell [Yanagimachi, 1981]. There are two distinct events that follow the completion of capacitation. One of these is the acrosome reaction, which involves a series of fusions between specific sperm membranes [Meizel, 1978]. The openings that are formed in the acrosome allow release of its contents, including hydrolytic enzymes such as hyaluronidase, which may facilitate the passage of spermatozoa through the granulosa cell investments of the ovum [Yanagimachi, 1981]. There also appears to be a relationship between capacitation and fertilization, which is distinct from the acrosome reaction. Unique patterns of sperm motility, which are associated with capacitation, have been observed in many species of mammal. This so-called activated motility may provide additional thrust for penetration of the ovum vestments [Katz and Overstreet, 1980].

Specific receptors on the sperm may be involved in their attachment and binding to the zona pellucida [Yanagimachi, 1981]. Capacitation may be a prerequisite for this initial interaction. There is no agreement at this time on the precise sequence of events undergone by the fertilizing sperm cell, but in some species the binding of spermatozoa to the zona surface may take place before completion of the acrosome reaction.

After completion of the acrosome reaction, the sperm may shed its outer membranes, exposing other lytic enzymes such as acrosin. Acrosin may be bound to the sperm membrane and thus may be directly exposed to the zona surface, digesting a pathway through the zona substance as the sperm penetrates [Yanagimachi, 1981]. The zona pellucida is a thick protective vestment. A significant propulsive thrust of the sperm flagellum is probably required to push the sperm head through the zona substance [Katz and Yanagimachi, 1981], and rigidity of the sperm head may also be important for this shearing action [Bedford, 1974].

After passing through the zona pellucida, the fertilizing spermatozoon resides for only a brief interval in the perivitelline space. Within seconds after initial contact with the oolemma, flagellar activity ceases and fusion with the vitellus begins [Yanagimachi, 1981]. Sperm incorporation into the ooplasm involves functions of the sperm cell, which are distinct from those associated with either zona attachment or zona penetration. In fact, different regions of the sperm head are probably involved in each of these events. Fusion with the vitellus occurs over the mid-region of the sperm head, and phagocytic engulfment of the sperm by the ooplasm follows. Both capacitation and the acrosome reaction appear to be required for sperm fusion with the oolemma, since these events are required for fertilization of ova from which the zona pellucida has been removed [Yanagimachi, 1981].

ACKNOWLEDGMENTS

Dr. Overstreet is supported by National Institutes of Health Research Career Development Award HD00224. The original research described in this paper was supported by research grants from the National Institutes of Health (HD11186 and HD15149 to J. W. Overstreet and HD12971 to D. F. Katz), the Environmental Protection Agency (R009089), and the National Institute for Occupational Safety and Health (OH01148).

REFERENCES

Amann RP, Howards SS (1980): Daily spermatozoal production and epididymal spermatozoal reserves of the human male. J Urol 124:211–215.

Bedford JM (1974): Mechanisms involved in penetration of spermatozoa through the vestments of the mammalian egg. In Coutinho EM, Fuchs F (eds): "Physiology and Genetics of Reproduction." Part B. New York: Plenum Press, pp 55–68.

Bedford JM, Calvin J, Cooper GW (1973): The maturation of spermatozoa in the human epididymis. J Reprod Fertil [Suppl] 18:199–213.

Clermont Y (1963): The cycle of the seminiferous epithelium in man. Am J Anat 112:35–51.

Clermont Y (1966): Renewal of spermatogonia in man. Am J Anat 118:509–524.

Clermont Y (1970): Dynamics of human spermatogenesis. In Rosenberg E, Paulsen CA (eds): "The Human Testis." New York: Plenum Press, pp 47–61.

diZerega GS, Sherins RJ (1981): Endocrine control of adult testicular function. In Burger H, deKrester D (eds): "The Testis." New York: Raven Press, pp 127–140.

Fox CA, Wolff HS, Baker JA (1970): Measurement of intravaginal and intrauterine pressure during human coitus by radiotelemetry. J Reprod Fertil 22:243–251.

Fox CA, Meldrum SJ, Watson BW (1973): Continuous measurement by radiotelemetry of vaginal pH during human coitus. J Reprod Fertil 33:69–75.

Freund M (1962): Interrelationships among the characteristics of human semen and factors affecting semen-specimen quality. J Reprod Fertil 4:143–159.

Freund M (1963): Effect of frequency of emission on semen output and an estimate of daily sperm production in man. J Reprod Fertil 6:269–286.

Freund M (1973): Mechanisms and problems of sperm transport. In Segal SJ, Crozier R, Corfman PA, Condliffe PG (eds): "The Regulation of Mammalian Reproduction." Springfield, Illinois: Charles C. Thomas, pp 352–361.

Heller CG, Clermont Y (1964): Kinetics of the germinal epithelium in man. Recent Progr Horm Res 20: 545–575.

Hinrichsen MJ, Blaquier JA (1980): Evidence supporting the existence of sperm maturation in the human epididymis. J Reprod Fertil 60:291–294.

Hotchkiss RS (1970): Physiology of the male genital system as related to reproduction. In Campbell MF, Harrison JH (eds): "Urology." Third ed. Philadelphia: WB Saunders Co, pp 161–196.

Katz DF, Overstreet JW (1980): Mammalian sperm movement in the secretions of the male and female genital tracts. In Steinberger A, Steinberger E (eds): "Testicular Development, Structure, and Function." New York: Raven Press, pp 481–489.

Katz DF, Yanagimachi R (1981): Movement characteristics of hamster and guinea pig spermatozoa upon attachment to the zona pellucida. J Reprod Fertil 25:785–791.

Mann T (1964): "The Biochemistry of Semen and of the Male Reproductive Tract." New York: John Wiley & Sons, Inc.

Meizel S (1978): The mammalian sperm acrosome reaction, a biochemical approach. In Johnson MH (ed): "Development in Mammals." Vol 3. New York: North-Holland Publishing Company, pp 1–64.

Mitsuya H, Asai J, Suyama K, Vshida T, Hosoe K (1960): Application of x-ray cinematography in urology. I. Mechanism of ejaculation. J Urol 83:86–92.

Ortavant R, Courot M, Hochereau de Reviers MT (1977): Spermatogenesis in domestic mammals. In Cole HH, Cupps PT (eds): "Reproduction in Domestic Animals." Third edition. New York: Academic Press, pp 203–227.

Overstreet JW (1983): Transport of gametes in the reproductive tract of the female mammal. In Hartmann JF (ed): "Mechanism and Control of Animal Fertilization." New York: Academic Press, in press.

Overstreet JW, Katz DF (1977): Sperm transport and selection in the female genital tract. In Johnson MH (ed): "Development in Mammals." Vol 2. Elsevier: Amsterdam, pp 31–65.

Overstreet JW, Katz DF (1981): Sperm transport and capacitation. In Sciarra JJ (ed): "Gynecology and Obstetrics." Vol 5, Chapter 45. New York: Harper and Row Publishers, pp 1–10.

Rowley MJ, Teshima JF, Heller CG (1970): Duration of transit of spermatozoa through the human male ductular system. Fertil Steril 21:390–396.

Setchell BP (1978): "The Mammalian Testis." New York: Cornell University Press.

Settlage DSF, Motoshima M, Tredway DR (1973): Sperm transport from the external cervical os to the fallopian tubes in women: A time and quantitation study. Fertil Steril 24:655–661.

Sobrero AJ, MacLeod J (1962): The immediate post-coital test. Fertil Steril 13:184–189.

Steinberger E, Steinberger A (1974): Hormonal control of testicular function in mammals. In Knobil K, Sawyer WH (eds): "Handbook of Physiology." Vol 4, part 2. Washington, DC: American Physiological Society, pp 325–345.

Steinberger E, Swerdloff RS, Horton R (1977): The control of testicular function. In Greep RO, Koblinsky MA (eds): "Frontiers in Reproduction and Fertility Control." Cambridge, MA: MIT Press, pp 264–292.

Tauber PF, Propping D, Schumacher GFB, Zaneveld LJD (1980): Biochemical aspects of the coagulation and liquefaction of human semen. J Androl 1:280–288.

Tredway DR, Buchanan GC, Drake TS (1978): Comparison of the fractional post coital test and semen analysis. Am J Obstet Gynecol 130:647–652.

Yanagimachi R (1981): Mechanisms of fertilization in mammals. In Mastroianni L Jr, Biggers JD (eds): "Fertilization and Embryonic Development In Vitro." Chapter 5. New York: Plenum Press, pp 81–182.

Zimmerman SJ, Maude MB, Moldawer M (1965): Frequent ejaculation and total sperm count, motility, and form in humans. Fertil Steril 16:342–345.

American Journal of Industrial Medicine 4:17–30 (1983)

Female Reproduction

Ken Takizawa, MD, PhD, and Donald R. Mattison, MD

Reproduction in the female requires a complex sequence of integrated events including ovulation, gamete transport, fertilization, conceptus transport, implantation, fetal support, and finally parturition. Each of these individual reproductive events represents the sum of many systemic, local, cellular, and molecular interactions. All of these processes have varying degrees of susceptibility to interference by xenobiotic compounds. This chapter presents a brief review of some of these mechanisms of female reproductive biology.

Key words: female reproduction, ovulation, fallopian tube, fertilization, conceptus transport, implantation, parturition

INTRODUCTION

The events responsible for female fertility are frustratingly complex. Unfortunately, these processes do not occur in the absence of exogenous influences. Reproduction occurs in the environment we have inherited and modified. For some of us reproduction may occur in an environment free from major pollution or drug ingestion. Others, however, will reside in heavily polluted areas or require the use of single or multiple drugs throughout a part or all of their reproductive life span. In many cases we may not know if these xenobiotic exposures will adversely alter reproduction. This chapter provides an overview of normal female reproductive biology as a guide to understanding female reproductive toxicology.

The sequential mechanisms of ovulation, fertilization, implantation, and maintenance of pregnancy necessary for female reproduction are not completely understood (Fig. 1). Historically, most of our basic knowledge has been acquired piece by piece from subprimate research. However, recently many reproductive biologists have successfully utilized nonhuman primates. The cumulative information gained from sub-

Pregnancy Research Branch, NICHD, NIH, Bethesda, Maryland.
Address reprint requests to Dr. Donald R. Mattison, Building 10 ACRF, Room 8C313, Pregnancy Research Branch, NICHD, NIH, Bethesda, MD 20205.

0271-3586/83/0401-02-0017$04.00 © 1983 Alan R. Liss, Inc.

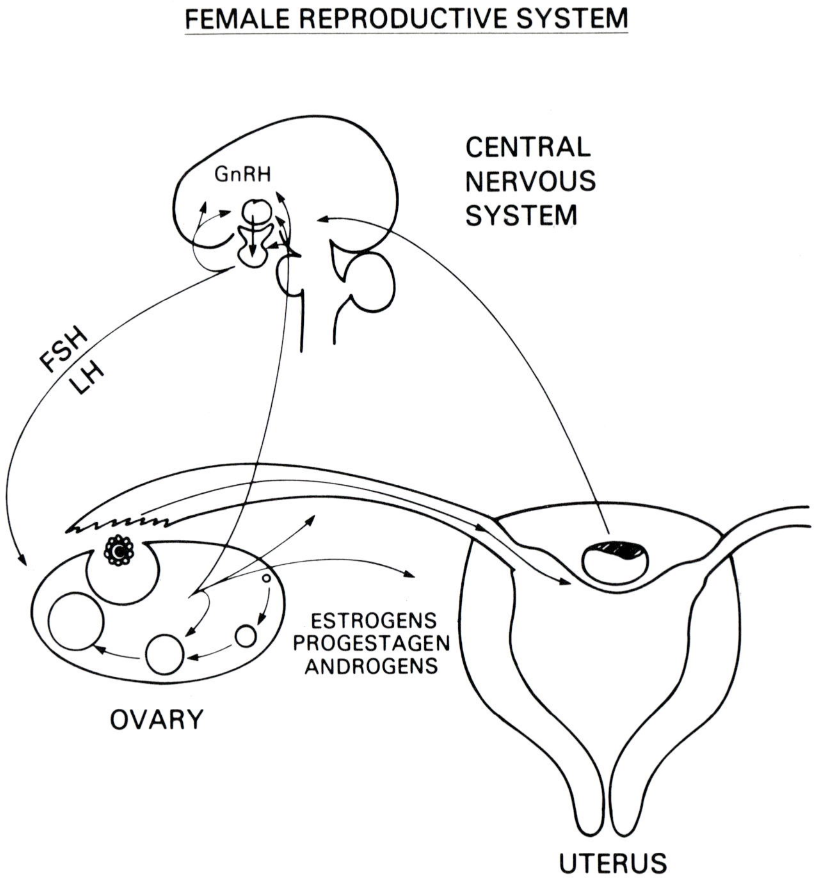

Fig. 1. Schematic representation of the control mechanisms involved in female reproduction. The gonadotropins, follicle-stimulating hormone (FHS), and luteinizing hormone (LH) are released from the pituitary into the circulation in a pulsatile pattern by pulsatile release of gonadotropin-releasing hormone (GnRH). FSH supports the growth of the dominant follicle, which produces increasing quantities of estrogen. Ovulation occurs after the LH surge and is followed by luteinization of the dominant follicle and increasing production of progesterone. (Reproduced with permission from Mattison [in press]).

human primates and subprimates are very useful and applicable to humans. These efforts have produced detailed advances allowing increasing success in the treatment of many reproductive disorders. In the near future the complicated hormonal and neurological control mechanism(s) involved in female reproduction will be elucidated. In this review, we summarize ovarian processes involved in follicle growth and ovulation; the role of the fallopian tube, uterotubal junction and uterus in gamete transport, fertilization, implantation and maintenance of pregnancy, and finally parturition.

THE OVARY

In the human, as well as most primates, ovarian development, oogenesis, and the initial stages of meiosis occur prenatally [Van Wagenen and Simpson, 1965, 1973]. Formation and proliferation of oogonia, stem cells for the oocytes, occur only during the fetal period. Oogonia differentiate into primordial oocytes which enter meiosis before, and become arrested in the prophase of the first meiotic division after formation of the follicle complex. This occurs around the time of birth [Moor and Warnes, 1979]. In this respect, the ovary is different from the testis in that the male gonad contains spermatogonia, stem cells which can proliferate throughout the life of the male, while oogonia are not present after birth.

In mammals, the majority of the follicles are in the "resting pool" and consist of a small oocyte and several granulosa cells surrounded by a basement membrane. The onset of follicle growth or recruitment from the resting pool appears to depend in part on gonadotropin stimulation. Follicle recruitment and growth normally occurs throughout the life of the female; however, in the absence of adult levels of gonadotropins, continued growth and development of the follicle complex cannot be supported and atresia occurs. With the onset of puberty and maturation of the hypothalamic-pituitary-ovarian axis elevated levels of gonadotropins support continued growth and development of the follicle, as well as ovulation and luteinization (Fig. 1).

The mechanism by which a single dominant follicle is selected and nurtured for ovulation and luteinization represents an interesting but poorly understood facet of primate reproduction. Unless conception occurs, luteolysis is inevitable and the ovarian cycle is repeated. Although these processes are not yet completely understood, we would like to focus on follicle maturation, ovulation, luteinization, and luteolysis.

Follicle Recruitment, Growth, and the Selection of the Dominant Follicle

The mechanism by which a single follicle, the "dominant follicle," is selected for ovulation in primates during the first half of cycle is indeed fascinating. Several hypotheses have been proposed to explain selection of the dominant follicle. The first hypothesis postulates the influence of favorable conditions arising from the relationship between the dominant follicle and its vascular supply. A group of primordial follicles which are recruited from the resting pool begin to grow, the follicle complex with the most favorable location in the vascular bed is selected predominantly on the basis of better hormonal and metabolic support [Bahr et al, 1977]. Another hypothesis advanced to explain selection of the dominant follicle focuses on the variability of follicles and individual differences in follicle growth. A group of follicles which respond to gonadotropin stimulation with follicular growth contains one follicle complex which is more sensitive to gonadotropin stimulation. This sensitive follicle is selectively matured and ovulated. At present, both of these hypotheses concerning the mechanism of selection of the dominant follicle are equally plausible.

Once selection of the dominant follicle occurs it assumes control of the remainder of the cycle. As it grows, the dominant follicle secretes increasing quantities of sex steroids, predominantly estrogen, during the follicular phase (Fig. 2). This increase in circulating estrogen decreases follicle-stimulating hormone (FSH) levels through negative feedback in the hypothalamus and pituitary [Ross et al, 1970]. In spite of the decrease in circulating levels of FSH the follicular fluid of the dominant follicle contains increasing quantities of gonadotropins (especially FSH) and sex steroids. As the dominant follicle matures it becomes increasingly sensitive to FSH and luteinizing hormone (LH) ultimately resulting in ovulation following the gonadotropin surge [Goldenberg et al, 1972]. In contrast, the other follicles in the growing cohort lack sufficient FSH for continued growth. These follicles may be desensitized to FSH by elevated estrogen. Whatever the mechanism, these nondominant follicles stop growing

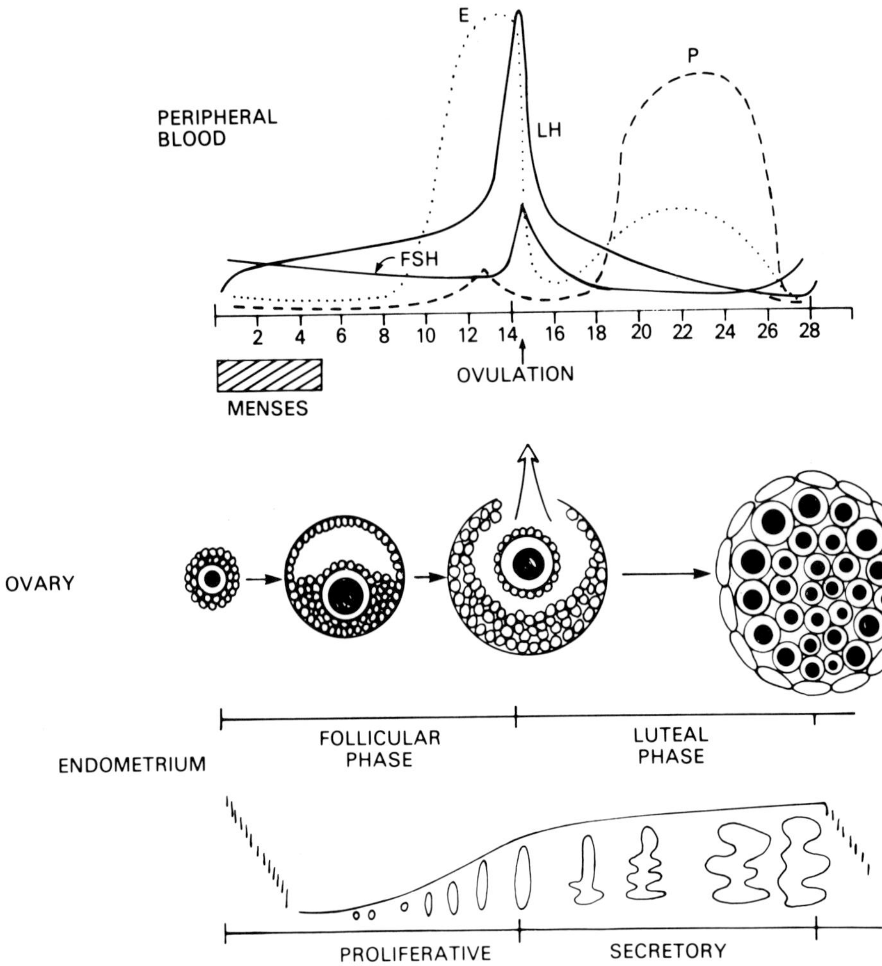

Fig. 2. The endocrine and cellular events involved in the menstrual cycle along the pituitary-ovarian-uterine axis. Under the influence of follicle-stimulating hormone (FSH) the dominant follicle grows and secretes increasing quantities of estrogen (E) into the circulation. E secreted by the dominant follicle feeds back on the hypothalamus and pituitary decreasing FSH secretion. The increasing levels of E also stimulate proliferation of the endometrium. A continued increase in the level of circulating E ultimately leads to the surge of luteinizing hormone, which is followed by ovulation. After ovulation the morphology of the follicle changes with formation of the corpus luteum. The major secretory product of the corpus luteum is progesterone (P). Increasing quantities of P stimulate the release of secretory granules into endometrial glands, proliferation of endometrial stromal cells and spiral arterioles in preparation for implantation. (Reproduced with permission from Mattison [in press]).

and become atretic. Of the more than one million oocytes which exist in the ovaries after birth, less than 500 are ovulated, and considerably fewer are fertilized and result in term births. The advantage of this redundancy is not known, but appears to parallel that observed in the male in which many millions of sperm are ejaculated but only a single sperm fertilizes the egg.

Ovulation

Ovulation or oocyte release results from a complex interaction involving sex steroids, gonadotropins, and prostaglandins. Tonic levels of FSH stimulate follicular complex growth and selection of the dominant follicle (Fig. 2). The increasing secretion of estrogen from the dominant follicle initially suppresses, via negative-feedback mechanisms, the secretion of FSH from the pituitary. Continued secretion of estrogen, however, ultimately triggers a reflex discharge of LH and FSH from the pituitary producing the preovulatory surge [Vande Wiele et al, 1970]. Following the LH surge, LH enters into follicular fluid and participates in preovulatory oocyte maturation [McNatty et al, 1975], from completion of the first meiotic division, producing the first polar body to the start of the second division; preovulatory progesterone secretion by the granulosa cells [Delforge et al, 1972]; and stimulation of ovarian prostaglandin synthesis from arachidonic acid [Chasalow and Pharriss, 1972]. Finally, the Graffian follicle ruptures and the oocyte is released. As the administration of indomethacin, an inhibitor of prostaglandin synthesis, increases the frequency of unruptured luteinized follicles, prostaglandins are thought to mediate ovum release [Grinwich et al, 1972; Maia et al, 1978]. The sequential interaction of FSH in follicle maturation, FSH and LH in ovum maturation, and LH and prostaglandins in ovum expulsion appears to be required for a fertile ovarian cycle.

Luteinization

Following ovulation the morphology of the follicle changes considerably, as does its major secretory product. After the preovulatory gonadotropin surge the basement membrane surrounding the dominant follicle complex breaks down and is penetrated by capillaries. The small compact cuboidal granulosa cells also change, differentiating into larger polygonal luteal cells. During the follicular phase of the cycle the major secretory product was estrogen, following ovulation the major steroid secreted is progesterone (Fig. 2). The corpus luteum is composed of luteal cells and a large component of morphologically distinct thecal cells.

Following fertilization, and even before implantation, the conceptus secretes many molecules, some of which are recognized by mother and are considered to block luteolysis [Heap et al, 1979]. In pigs, production of estrogen by the blastocyst leads to maternal recognition of conception. Estrogen produced by the porcine blastocyst is not only luteotrophic through its effects on the maternal hypothalamic-pituitary axis [Gardner et al, 1963], but also by suppression of prostaglandin synthesis [Frank et al, 1977]. In cows, pregnancy-specific proteins of embryonic origin suppress synthesis of prostaglandins which are responsible for luteolysis [Goding, 1974]. In contrast, in primates, chorionic gonadotropin produced by the early conceptus supports the corpus luteum required for the maintenance of the endometrium early in pregnancy [Knobil, 1973].

Luteolysis

One significant difference between subprimates (rodents, bovine, porcine, and ovine) and primates (monkey, human) may be the mechanism of luteolysis. In subprimates, administration of prostaglandins during the postovulatory period can induce luteolysis [Goding et al, 1972] and inhibitors of prostaglandin synthesis such as indomethacin can prolong the life span of corpus luteum [Horton and Poyser, 1973]. For these reasons prostaglandins are regarded as the substances which control the life span of the corpus luteum. In subprimates, ovulation leads to endogenous prostaglandin synthesis by the uterus. These prostaglandins are then transported to the ovary through the utero-ovarian circulation and produce breakdown of the corpus luteum [McCracken et al, 1972]. Prostaglandin-induced vasoconstriction also aids the process of luteolysis [Novy and Cook, 1973].

In contrast, in primates prostaglandins do not appear to control luteolysis. Luteolysis may be induced by the withdrawal of FSH, LH, and prolactin, which are responsible for maintaining luteal function [Vande Viele et al, 1970]. Alternatively as the life span of the corpus luteum is intrinsically short in primates, luteolysis may not require any special mechanism [Knobil, 1973]. As we have seen, a complex interaction between FSH for follicle maturation, LH for oocyte maturation, ovulation, and luteinization; prolactin for luteinization; and prostaglandins for ovulation and luteolysis is essential for control of the sex cycle, conception, and maintenance of pregnancy in mammals.

THE FALLOPIAN TUBE

The fallopian tube is anatomically separated into fimbria, ampulla, ampullary-isthmic junction (A-I-J), and the isthmus (Fig. 3). Each of these segments of the fallopian tube consists of a mucosal epithelium composed of ciliated and secretory cells, a muscle layer and a serosal or peritoneal surface, and each has a morphologically characteristic appearance. The lumen of each segment of the fallopian tube is bathed with fluid whose composition, consistency and direction of flow appears to vary during the cycle. In primates as well as subprimates, the motility and secretions produced by each

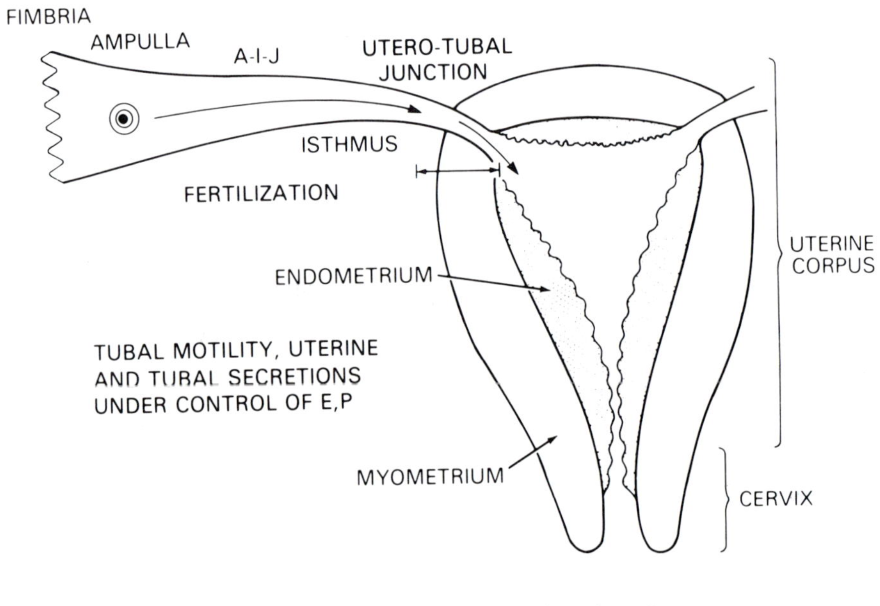

Fig. 3. Schematic anatomy of the fallopian tube and uterus.

segment of the tube is controlled by sex steroids and the autonomic nervous system. Because of this, tubal motility and secretion varies during the menstrual cycle. This joint endocrine and neural control enables the tube to function in gamete transport, fertilization, and maintenance and transport of the conceptus.

The Fimbria

The fimbria is the funnel-shaped entrance of tube near the ovary. The lush, convoluted surface of the fimbria plays an important role in picking up the egg encased in the cumulus mass from the ovarian surface following ovulation and transporting it to the lumen of the tube. As most subprimates, especially rodents, have a closed ovarian bursae which surrounds the ovary partly or completely, ovulation essentially occurs within the tube [Blandau, 1973]. In these species, the egg is transported to the ampulla by the action of cilia present on ciliated epithelial cells in the fimbria [Gaddum-Rosse and Blandau, 1973]. As primates do not have an ovarian bursa, ovum pick-up from the ovarian surface following ovulation requires active muscular contraction of the engorged fimbria [Hafez, 1973]. Once the egg has been picked up and carried into the infundibulum, it is transported into the ampulla chiefly by cilial motion beating away from the ovary.

Ciliated cells of the fimbria appear to play a crucial role in egg transport in mammals. These ciliated cells are under hormonal regulation during the sex cycle. For instance, in the rhesus monkey, fimbrial epithelium loses its cilia and becomes atrophic during the luteal phase. Early in the follicular phase these cells hypertrophy and the cilia regenerate [Brenner, 1969]. At the time of ovulation, the ciliated cells of the fimbria are developed for efficient egg transport.

The Ampulla

Following transport of the egg into the ampulla it must be carried to the ampullary-isthmic junction where fertilization occurs. In rodents, transport to the ampullary-isthmic junction depends mainly on peristaltic muscular contraction rather than cilial beat [Harper, 1961]. In primates, however, ovum transport to the ampullary-isthmic junction appears to depend predominantly on cilial motion toward the uterus, as the muscle layer of the ampulla is very thin [Blandau, 1973]. The hormonal change from follicular phase (estrogen dominant) to the luteal phase (progesterone dominant) produces increased secretion of tubal fluid and the development of cilia. The cumulative effect of cilial beat, muscle contractions, and fluid flow appear to cooperate to transport the egg toward the uterus.

The Ampullary-Isthmic Junction

The fallopian tube is the site of fertilization. The prerequisite condition for fertilization is the timely meeting between an adequately matured oocyte and sperm in the appropriate environment. The ampullary-isthmic junction (A-I-J) apparently satisfies these requirements, as it is the site of fertilization in most mammals. The A-I-J, located between the ampulla and isthmus, has also been called the "physiological stricture" or "physiological adrenergic sphincter" [Brundin and Wirsen, 1964] of the fallopian tube. The function of the A-I-J is retention of the egg for a period between 15 hours and two days following ovulation. In order to function in egg retention, the A-I-J has a rich autonomic innervation, especially receptors for noradrenaline [Brundin, 1965]. During the early postovulatory phase, the synergistic action of estrogen and progesterone on the autonomic nervous system may act differentially on the muscular layer of the fallopian tube to relax the ampulla and contract the isthmus. This contracture of the isthmic muscles coupled with the narrowing lumen due to mucosal edema may produce temporary stricture. During the initial period of residence of the oocyte at the A-I-J, the egg loses the corona radiata cells and meiosis progresses to the second period of arrest in the metaphase stage of the second meiotic division [Whittingham, 1979], waiting the arrival of sperm.

Tubal Fluid

The tubal fluid is composed of a transudate of serum and secretions formed by secretory cells of the tubal epithelium. Because the fimbria of the fallopian tube open into the peritoneal cavity, tubal fluid may also contain peritoneal fluid, as well as follicular fluid and uterine fluid. Fallopian-tube fluid is the medium in which the egg and sperm are nurtured and transported, and may provide the appropriate environment for sperm capacitation, fertilization, and early embryogenesis. In humans, during the follicular phase, secretory cells of the tubal epithelium hypertrophy, with a concomitant increase in glycoprotein synthesis [Hashimoto et al, 1964]. Secretory granules stored in the Golgi apparatus during the follicular phase are evacuated into the tubal fluid during the early postovulatory phase [Hafez, 1973b]; these secretory proteins may play an important role in regulating the compositions of tubal fluid and influence osmolarity, electrolyte composition, pH, oxygen tension, buffering by bicarbonate, enzyme activity, and the content of pyruvate or lactate.

The Isthmus

The isthmic region of the tube contains an abundant muscle layer and is richly supplied with autonomic nerve fibers. As a result of this rich innervation it may take part in the transport of the conceptus, or gametes by muscle contraction. The wall of isthmus is so thick in primates that the exact mechanism of conceptus transport has not been determined. However, it is generally felt that peristaltic contractions following fertilization during the luteal phase control transport of the conceptus into the uterus. During the periovulatory period isthmic contractions are toward the ovary for the transport of sperm [Blandau and Gaddum-Rosse, 1974]. As the critical period for sperm transport through the isthmus is entirely different from the period for conceptus transport, the direction of peristaltic contraction is expected to change. In fact, in rabbit and pig, the direction of contraction is toward the ovary during the preovulatory and early postovulatory periods. During the late postovulatory period (24 to 48 hours after ovulation) the contraction pattern becomes undulatory—backward and forward [Blandau et al, 1977]. In primates muscle contraction and cilial beat may vary with the phase of the cycle so that during the luteal phase transport is toward the uterus.

THE UTEROTUBAL JUNCTION

The uterotubal junction (U-T-J) is anatomically separated into the intramural or intrauterine portion and the extramural portion (Fig. 3). Generally speaking, in litter-bearing species like rodents, as well as species like the pig and horse in which a greater volume of semen is ejaculated into the uterine cavity, the U-T-J is well developed with a large muscular layer, in addition to the fibrous investments of tubal ligaments and the complicated projections of the mucosa and submucosa found in this region of the tube

[Hafez, 1973a]. In these species, during the early postovulatory phase, the U-T-J becomes a physiological sphincter blocking access to the tube by virtue of muscle contraction and mucosal edema, and may regulate access of sperm to the egg, or allow the uterus to function as a sperm reservoir by intermittent opening and closing.

In the human, an annular fold of uterine mucosa at the intramural portion of the U-T-J may also functon as a sphincter. Clinically, the function of this sphincter is thought to be measured as the uterotubal opening pressure (the peak pressure) by the insufflation technique [Hafez and Black, 1969]. With respect to sperm transport in the human, however, the U-T-J is thought to represent only a mechanical barrier. Sperm ejaculated in the vagina reaches the tube within 30 minutes [Rubenstein et al, 1951], and is distributed with decreasing concentrations from the vagina, cervix, uterus, and tube. At present it is thought that the U-T-J does not represent a barrier to egg transport in most mammals.

THE UTERUS

The uterus can be divided anatomically into a body, or corpus, and cervix. The uterine corpus is histologically divided into endometrium and myometrium (Fig. 3). In many species the uterus functions as a simple conduit for passage of sperm after ejaculation, however in some species, especially in rodents, cows, or pigs, the uterus may function as a sperm reservoir. Hormonally dependent alterations in the endometrium during the ovarian cycle prepare the uterus for implantation of the embryo and a mechanical support for the developing conceptus. The uterus undergoes enormous physiological changes in preparation for, and in response to, pregnancy. The endometrium develops glandular and vascular elements which support implantation of the conceptus and development of the placenta. The myometrium, initially quiescent during pregnancy, gradually develops a coordinated contraction pattern which if effective will result in the delivery of the fetus at the end of gestation. In this section we will explore the mechanisms of implantation and the suppression and stimulation of muscle contractions in the myometrium.

The Endometrium

The conceptus implants in the endometrium six to seven days after fertilization in the human and five to six days after fertilization in the rat. Before implantation the embryo develops through the morula and blastula stage, producing several kinds of molecules which signal the maternal organism of the start of pregnancy. Proteins produced by the early conceptus are luteotrophic, and especially in domestic animals, proteins secreted by the early conceptus inhibit prostaglandin synthesis, blocking luteolysis [Goding, 1974]. The sequential production of ovarian steroids and the stimulation from the embryo supporting the corpus luteum produce an endometrium which is receptive for implantation.

During the preovulatory phase of the cycle, estrogen produced by the follicle complex increases uterine blood flow and stimulates endometrial proliferation. Following ovulation the synergistic action of estrogen and progesterone completes the preparation of the endometrium for implantation (Fig. 2). The effects of progesterone on an endometrium previously stimulated with estrogen are pleotropic producing increased protein synthesis, and formation of microvilli [Nilsson, 1962]. The vascularity of the endometrium increases, secretory products of endometrial cells are discharged into the lumen of endometrial glands, and the increase of capillary permeability in the endometrium produces a generalized edema [Noyes et al, 1950]. As a result of this epithelial proliferation and edema when the conceptus (blastula) enters the uterus, the endometrial cavity is almost completely obliterated, with close apposition of microvilli-coated epithelial surfaces [Enders and Schlafke, 1969]. This increases the likelihood that the embryo will make physical contact with the endometrium enhancing the interaction between trophoblast and endometrial cell plasma membranes.

Physiological mechanisms controlling the timing of implantation are not completely understood in primates. It is known, however, that endometrial receptivity for implantation occurs during a limited time span. In rodents, various treatments including administration of anti-LH serum [Laurence and Ichikawa, 1969], tranquilizers [Psychoyos, 1962], or lactation [Rothchild, 1960] can delay implantation. These treatments suppress pituitary gonadotropin secretion either directly or through the hypothalamus, this in turn decreases the production of ovarian steroids which synchronize uterine development and apparently also modulate the timing of implantation. It is thought that endometrial receptivity is a response to circulating estrogen and progesterone and once the endometrial response has occurred (ie, alteration of the endometrial architecture and discharge of secretory vesicles), the endometrium loses its receptivity for implantation [Toft and Gorski, 1966]. As mentioned above, as a result of the precise synergistic action of estrogen and progesterone, the growing embryo synchronized with the uterine environment can implant during a limited period of time, representing a window of endometrial receptivity.

The Myometrium

Successful intrauterine development followed by the timely onset of labor requires two different responses from the uterine myometrium: quiescence and contractility. The mechanism by which the myometrium remains quiescent during pregnancy but generates strong, coordinated contractions at the end of pregnancy are not completely understood. Csapo [1956] has suggested that local and/or circulating concentrations of progesterone (progesterone block theory) play an important role in regulating uterine contractility in rabbits. The rabbit myometrium remains spontaneously active and contractile during pregnancy; however, the high concentration of progesterone elevates the resting membrane potential of the uterine muscle cells, producing hyperpolarization and increasing the threshold for oxytocin-induced contractions. In humans as well as in rabbits, intrauterine-pressure recordings [Csapo, 1970] have demonstrated that the myometrium is not absolutely quiescent during gestation. In women the amplitude of uterine contractions increases gradually, but the frequency of contractions decreases from the 7th to the 36th week of gestation. Recordings throughout the volume of the uterus demonstrate that these contractions are not coordinated like contractions in active labor, but appear to represent intermittent contractions of different regions of the myometrium. Monitoring the electrical activity of the uterus also demonstrates the discoordinate nature of this contractile activity. Systemic treatment of premature labor with progesterone does not inhibit uterine contractions in humans, guinea pigs, or rats suggesting that the progesterone-block theory cannot completely account for the suppression of uterine contractions [Porter and Finn, 1977]. This, however, does not exclude the possibility that progesterone may act locally in some species.

Since prostaglandins have been identified as a smooth-muscle stimulant, they have subsequently been recognized as an endogenous compound likely to initiate uterine muscle contraction during labor [Bygdeman et al, 1968]. In contrast with the mechanism of progesterone which increases the level of the energy-dependent Ca^{2+} binding by the smooth endoplasmic reticulum to result in suppression of uterine contraction, prostaglandins decrease Ca^{2+} binding to induce myometrial contraction [Carsten, 1973]. At present, the rapid catabolism of prostaglandins by fetal membranes seems to be more essential for the suppression of uterine contractions than suppression by progesterone.

The mechanism controlling the onset of myometrial contraction at delivery may be considered to be a reversal of the mechanism controlling uterine relaxation. The increase of activity in the fetal adrenal cortex may alter placental steroid metabolism, producing a decrease in progesterone synthesis and an increase in estrogen synthesis due to a richer supply of C19 precursor from fetal adrenal [Liggins et al, 1972]; promote prostaglandin synthesis and decrease degradation of prostaglandins in the decidual portion of the placenta [Keirse and Turnbill, 1975], which may initiate uterine contractions. Additionally, luteolysis (especially in subprimates) induced by prostaglandins [Pharriss and Wyngarden, 1969] and fetoplacental damage caused by the prostaglandin-induced vasoconstriction and uterine contraction in primates [Bolognese and Corson, 1973] produces an additional decrease in progesterone synthesis, which may act as a further stimulus for uterine contraction. As oxytocin itself does not stimulate muscle contraction but increases prostaglandin production [Vane and Williams, 1973], it is likely to promote uterine contractions through prostaglandin synthesis.

At the present time, it is thought that the mechanism of relaxation of uterine smooth muscle cells during pregnancy is due to both a decrease in prostaglandin synthesis and an increase in prostaglandin degradation. Then, near the end of gestation, when these suppressive effects are removed, effective coordinated uterine contractions are inevitable.

CONCLUSION

The ovary is essential for female reproduction, controlling the hypothalamus, pituitary, and uterus, and allowing conception and maintenance of pregnancy in mammals. In order to perform these functions, the following special roles are periodically assigned to the ovary: selection and maturation of the dominant follicle, ovulation, luteinization, and luteolysis. These ovarian tasks are finely regulated by interactions with FSH for follicle maturation; LH for oocyte maturation, ovulation, and luteinization; prolactin for luteinization; and prostaglandins for ovulation and luteolysis.

The fallopian tube is anatomically separated into fimbria, ampulla, ampullary-isthmic junction, and isthmus, each segment of which has a morphologically characteristic appearance. In addition to the cilial beat of the ciliated cells which hypertrophy at the time of ovulation, the muscle contractions which are neuroendocrinologically controlled can transport the egg and early conceptus toward the uterus. Tubal fluid in which the gametes are nurtured and transported provides the appropriate environment for sperm capacitation, fertilization, and embryogenesis during the early postovulatory period.

In litter-bearing species like rodents as well as species like the pig and horse, the uterotubal junction may regulate access of sperm to the egg or allow the uterus to function as a sperm reservoir by intermittent opening and closing access from the uterus to the tube during the early postovulatory phase.

If conception occurs, macromolecules secreted by the embryo are recognized by the maternal organism and the production of ovarian steroids is maintained, supporting the endometrium and establishing a uterine environment receptive for implantation. As a result of the synergistic action of estrogen and progesterone, the growing embryo synchronized with the uterine environment can implant during a limited period of time.

As prostaglandins are endogenous compounds which are thought to induce uterine muscle contractions, it seems reasonable to assume that the mechanism of myometrial relaxation is chiefly due to a decrease in local and/or circulating prostaglandins. At the end of gestation the fetus may also trigger an increase in maternal prostaglandin synthesis as well as a decrease in catabolism of prostaglandins and may inhibit progesterone synthesis in the placenta. Each of these events individually, or in concert, may produce the onset of coordinated uterine contractions which occur with increased frequency and duration.

REFERENCES

Bahr JM, Ross GT, Nalbandov AV (1977): Hormonal regulation of the development, maturation and ovulation of the ovarian follicle. In Greep RO, Koblinsky MA (eds): "Frontiers in Reproduction and Fertility Control, Part 2." Cambridge, Massachusetts: The MIT Press, pp 40–43.

Blandau RJ (1973): Gamete transport in the female mammal. In Greep RO, Astwood EB (eds): "Handbook of Physiology, Section 7; Endocrinology, Volume II; Female Reproductive System, Part 2." Baltimore: Williams & Wilkins, pp 153–163.

Blandau RJ, Gaddum-Rosse P (1974): Mechanism of sperm transport in pig oviducts. Fertil Steril 25:61–67.

Blandau RJ, Brackett B, Brenner RM, Boling JL, Broderson SH, Hamner C, Mastroianni L (1977): The oviduct. In Greep RO, Koblinsky MA (eds): "Frontiers in Reproduction and Fertility Control, Part 2." Cambridge, Massachusetts: The MIT Press, pp 132–145.

Bolognese RJ, Corson SL (1973): Abortion of early pregnancy by the intravaginal administration of prostaglandin F_2-alpha. Am J Obstet Gynecol 117:246–250.

Brenner RM (1969): The biology of oviductal cilia. In Hafez ESE, Blandau RJ (eds): "The Mammalian Oviduct." Chicago: University of Chicago Press, pp 203–229.

Brundin J (1965): Distribution and function of adrenergic nerves in the rabbit fallopian tube. Acta Physiol Scand (Suppl) 259:5–56.

Brundin J, Wirsen C (1964): Adrenergic nerve terminals in the human Fallopian tube by fluorescence microscopy. Acta Physiol Scand 61:505–506.

Bygdeman M, Kwon SU, Mukherjee T, Wiqvist N (1968): Effect of intravenous infusion of prostaglandin E_1 and E_2 on motility of the pregnant human uterus. Am J Obstet Gynecol 102:317–332.

Carsten ME (1973): Prostaglandins and cellular calcium transport in the pregnant human uterus. Am J Obstet Gynecol 117:824–832.

Chasalow FI, Pharriss BB (1972): Luteinizing hormone stimulation of ovarian prostaglandin biosynthesis. Prostaglandins 1:107–117.

Csapo A (1956): Progesterone "block." Am J Anat 98:273–291.

Csapo A (1970): The diagnostic significance of the intrauterine pressure. Obstet Gynecol Survey 25: 403–435.

Delforge JP, Thomas K, Roux F, Carneiro de Siqueira J, Ferin J (1972): Time relationships between granulosa cells growth and luteinization, and plasma luteinizing hormone discharge in humans. I. A morphometric analysis. Fertil Steril 23:1–11.

Enders AC, Schlafke S (1969): Cytological aspects of trophoblast-uterine interaction in early implantation. Am J Anat 125:1–30.

Frank M, Bazeer FW, Thatcher WW, Wilcox CJ (1977): A study of prostaglandin $F_{2}\alpha$ as the luteolysin in swine. III. Effects of estradiol valerate on prostaglandin F progestins, estrone and estradiol concentrations in the utero-ovarian vein of nonpregnant gilts. Prostaglandins 14:1183–1196.

Gaddum-Rosse P, Blandau RJ (1973): In vitro studies of ciliary activity within the oviducts of the rabbit and pig. Am J Anat 136:91–104.

Gardner ML, First NL, Casida LE (1963): Effect of exogenous estrogens on corpus luteum maintenance in gilts. J Anim Sci 22:132–134.

Goding JR (1974): The demonstration that $PGF_{2}\alpha$ is the uterine luteolysin in the ewe. J Reprod Fertil 38: 261–271.

Goding JR, Cumming IA, Charnley WA, Brown JM, Cain MD, Cerini JC, Cerini MED, Findlay JK, O'Shea JD, Pemberton DH (1972): Prostaglandin $F_{2}\alpha$: "The" luteolysin in the mammal ? Hormones and antagonists. Gynecol Invest 2:73–97.

Goldenberg RL, Vaitukaitis JL, Ross GT (1972): Oestrogen and follicle stimulating hormone interactions on follicle growth in rats. Endocrinology 90:1492–1498.

Grinwich DL, Kennedy TG, Armstrong DT (1972): Dissociation of ovulatory and steroidogenic actions of luteinzing hormone in rabbits with indomethacin, an inhibitor of prostaglandin synthesis. Prostaglandin 1:89–96.

Hafez ESE (1973a): Anatomy and physiology of the mammalian uterotubal junction. In Greep RO, Astwood EB (eds): "Handbook of Physiology, Section 7; Endocrinology, Volume II; Female Reproductive System, Part 2." Baltimore: Williams & Wilkins, pp 87–95.

Hafez ESE (1973b): Endocrine control of the structure and function of the mammalian oviduct. In Greep RO, Astwood EB (eds): "Handbook of Physiology, Section 7; Endocrinology, Volume II; Female Reproductive System, Part 2." Baltimore: Williams & Wilkins, pp 97–122.

Hafez ESE, Black DL (1969): The mammalian uterotubal junction. In Hafez ESE, Blandau RJ (eds): "The Mammalian Oviduct." Chicago: University of Chicago Press, pp 85–126.

Harper MJK (1961): The mechanisms involved in the movement of newly ovulated eggs through the ampulla of the rabbit Fallopian tube. J Reprod Fertil 2:522–524.

Hashimoto M, Shimoyama T, Kosaka M, Komori A, Hirasawa T, Yokoyama Y, Kawase N, Nakamura T (1964): Electron microscopic studies on the epithelial cells of the human Fallopian tube (Report III). J Jpn Obstet Gynecol Soc 11:92–100.

Heap RB, Flint AP, Gadsby JE (1979): Role of embryonic signals in the establishment of pregnancy. Br Med Bull 35:129–135.

Horton EW, Poyser NL (1973): Elongation of the estrous cycle in the guinea pig following subcutaneous or intrauterine administration of indomethacine. Br J Pharmacol 49:98–105.

Keirse MJNC, Turnbill AC (1975): Metabolism of prostaglandin $F_{2}\alpha$ within the human uterus in early pregnancy. Br J Obstet Gynaecol 82:142–145.

Knobil E (1973): On the regulation of the primate corpus luteum. Biol Reprod 8:246–258.

Laurence KA, Ichikawa S (1969): The effects of antiserum to bovine luteinizing hormone on the oestrus cycle and early pregnancy in the female rat. Int J Fertil 14:8–15.

Liggins GC, Grieves SA, Kendall JZ, Knox BS (1972): The physiological roles of progesterone, oestradiol-17β and prostaglandin $F_{2}\alpha$ in the control of ovine parturition. J Reprod Fertil (Suppl) 16:85–103.

Maia H, Barbosa I, Continho EM (1978): Inhibition of ovulation in Marmoset monkeys by indomethacin. Fertil Steril 29:565–570.

Mattison DR (1983): Ovarian toxicity: Effects on sexual maturation, reproduction and menopause. In Clarkson T, Nordberg G (eds): "Reproductive and Developmental Toxicity of Metals." New York: Plenum Publishing Co (in press).

McCracken JA, Carlson JC, Glew ME, Goding JR, Baird DT, Green K, Samuelsson B (1972): Prostaglandin F_2-alpha identified as a luteolytic hormone in sheep. Nature New Biol 238:129–134.

McNatty KP, Hunter WM, McNeilly AS, Sawers RS (1975): Changes in the concentrations of pituitary and steroid hormones in the follicular fluid of human Graffian follicles throughout the menstrual cycle. J Endocrinol 64:555–571.

Moor RM, Warnes GM (1979): Regulation of meiosis in mammalian oocytes. Br Med Bull 35:99–103.

Nilsson O (1962): Correlation of structure to function of the luminal cell surface in the uterine epithelium of mouse and man. Z Zellforsch Microsk Anat 56:803–808.

Novy MJ, Cook MJ (1973): Redistribution of blood flow by prostaglandin F_2-alpha in the rabbit ovary. Am J Obstet Gynecol 117:381–385.

Noyes RW, Hertig AT, Rock J (1950): Dating the endometrial biopsy. Fertil Steril 1:3–25.

Pharriss BB, Wyngarden LJ (1969): The effect of prostaglandin $F_{2}\alpha$ on the progestogen content of ovaries from pseudopregnant rats. Proc Soc Exp Biol Med 130:92–94.

Porter DG, Finn CA (1977): The biology of the uterus. In Greep RO, Koblinsky MA (eds): "Frontiers in Reproduction and Fertility Control, Part 2." Cambridge, Massachusetts: The MIT Press, pp 146–156.

Psychoyos A (1962): A study on the hormonal requirements for the ovum implantation in the rat by means of delayed nidation-inducing substances (chlorpromazine, trifluoperazine). J Endocrinol 27:337–343.

Ross GT, Cargille CM, Lipsett MB, Rayford PL, Marshall JR, Strott CA, Rodbard D (1970): Pituitary and gonadal hormones in women during spontaneous and induced ovulatory cycles. Recent Prog Horm Res 26:1–62.

Rothchild I (1960): The corpus luteum-pituitary relationship. The association between the cause of luteotrophin secretion and the cause of follicular quiescence during lactation: The basis for a tentative theory of the corpus luteum pituitary relationship in the rat. Endocrinology 67:9–41.

Rubenstein BB, Straus H, Lazarus ML, Hawking H (1951): Sperm survival in women. Fertil Steril 2:15–19.

Toft D, Gorski J (1966): AS receptor molecule for estrogens; isolation from the rat uterus and preliminary characterization. Proc Natl Acad Sci USA 55:1574–1581.

Vande Wiele RL, Bogumiel J, Dyrenfurth I, Ferin M, Jewelewicz R, Warren ML, Rizkallah T, Mikhail G (1970): Mechanisms regulating the menstrual cycle in women. Recent Prog Horm Res 26:63–103.

Vane JR, Williams KI (1973): Prostaglandin production contributes to the contractions of rat isolated uterus. Br J Pharmacol p 146.

Van Wagenen G, Simpson ME (1965): "Embryology of the Ovary and Testis: Homo Sapiens and Macaca Mulatta." New Haven: Yale University.

Van Wagenen G, Simpson ME (1973): "Postnatal Development of the Ovary in Homo Sapiens and Macaca Mulatta and Induction of Ovulation in the Macaque." New Haven: Yale University.

Whittingham DG (1979): In-vitro fertilization, embryo transfer and storage. Br Med Bull 35:105–111.

American Journal of Industrial Medicine 4:31–49 (1983)

Preimplantation Development: Biology, Genetics, and Mutagenesis

Jurrien Dean, MD

After fertilization, the one-cell zygote grows to the morula stage and then the blastocyst stage prior to its implantation in the uterus. This series of events is under the control of the embryonic genome, although maternal cytoplasmic components may initially play a major role. The study of molecular biology has revealed a great deal concerning the organization and expression of the eukaryotic genome. Particular interest has been focused on the mechanism of control of gene expression. These mechanisms must ultimately control cellular differentiation and the early development of the embryo. The study of reproductive toxicology requires a sound molecular appreciation of gene expression and cell differentiation. Mutations of the genome can dramatically affect the embryo. Although repair mechanisms have evolved to compensate for these mutations, repairs are not uniformly successful and unrepaired mutations can lead to early fetal loss.

Key words: **embryogenesis, development, meiosis, gametogenesis, fertilization, preimplantation embryo, mutagenesis, DNA repair mechanisms**

INTRODUCTION

After ovulation and ejaculation, the mammalian oocyte and sperm meet in the oviduct. Fertilization results in the formation of the one-cell zygote, which develops to the blastocyst stage prior to implantation in the uterine wall. During this preimplantation period, the embryo is particularly susceptible to genetic abnormalities and certain toxic insults. Damage to one or several cells early in mammalian development is often lethal and, when not, can affect a vast number of cells in the adult mammal that trace their lineage to the damaged cell.

The study of modern reproductive toxicology requires a broad understanding of development and a molecular appreciation of cellular biology. In this review, current

Laboratory of Chemical Biology, National Institute of Arthritis, Diabetes, Digestive and Kidney Diseases, National Institutes of Health, Bethesda, Maryland.
Address reprint requests to Dr. Jurrien Dean, Laboratory of Chemical Biology, National Institute of Arthritis, Diabetes, Digestive and Kidney Diseases, Building 10, Room 9N-307, National Institutes of Health, Bethesda, MD 20205.

0271-3586/83/0401-02-0031$05.50 © 1983 Alan R. Liss, Inc.

knowledge will be described of normal mammalian development from fertilization through the preimplantation embryo. Recent data that has been gained in molecular and cellular biology will be emphasized and the implications for gene expression and cell differentiation will be examined. Using this material as background, the mechanism and genetics of mutations, cellular repair systems in mammalian cells, and the consequences of unrepaired mutations in early development will be described.

NORMAL DEVELOPMENT

Mammalian development begins with gametogenesis and results in the production of a fully formed adult animal. Although our ultimate interest is in humans, much of what we have learned about development has been in experimental animals. Many conclusions discussed in this paper are based on such work, and although they are likely to be applicable to humans, this has not as yet been rigorously demonstrated in every case.

Meiosis

During gametogenesis both the oocyte and the sperm precursor cells divide meiotically into mature gametes, each containing a haploid genome (Table I). By the earliest portion of the first meiotic prophase (leptotene stage), the genomic DNA has been replicated by DNA polymerase to produce a tetraploid (4C) cell. Pairs of homologous sister chromatids are joined at a centromere. Unlike mitosis, during the zygotene phase pairs of homologous sister chromatids join to form bivalents (four chromatids). As the bivalents shorten (pachytene phase), breaks may occur in the DNA. Random repair of these breaks allows recombination between the two homologous pairs of sister chromatids. When each homologous pair of chromatids begins to separate (diplotene stage), the recombination sites can be seen as chiasmata. This recombination in large part provides for the genetic diversity of subsequent generations. In the human genome it is

TABLE I. Meiosis

	Oogenesis	Spermatogenesis
First division		
Prophase I		Primary spermatocytes are formed after several mitotic divisions at puberty. The first meiotic division takes 12–14 days and occurs in the testes. Spermatogenesis is an ongoing process with no "meiotic arrest."
leptotene zygotene pachytene diplotene	Occurs during fetal development "Meiotic arrest" until after puberty	
Metaphase I Anaphase I Teleophase I Second division Prophase II	Preovulatory changes	
Metaphase II	Ovulation	Secondary spermatocytes. Meiosis is completed in 24 days and results in four haploid spermatids.
Anaphase II Teleophase II	Fertilization	

estimated that 40–50 recombination events take place during the meiosis of each gamete [Lange et al, 1975].

During the first meiotic metaphase, the nuclear membrane dissolves and the bivalents line up on the meiotic spindle [Petzelt, 1980] to be divided (anaphase) between two daughter cells each containing a diploid (2C) complement of DNA. Each daughter cell goes through a second meiotic division, which is very similar to mitosis with the formation of a spindle in metaphase II and division of identical chromatids to form two daughter cells each with a haploid genome (1C).

Maturation of Gametes

In the human female the primordial germ cells migrate from the wall of the yolk sac to the developing gonads where they mitotically multiply. Between the third and sixth months of human gestation, some of these cells enter the first meiotic prophase, but further development is arrested prior to metaphase I. At five months gestation there are approximately 7 million oocytes in the primate ovary. However, many of these degenerate, so that at birth there are no more than 2 million, and this number dwindles further to 400,000 at menarche. At birth each of the oocytes in "meiotic arrest" is surrounded by granulosa cells which together make up the primordial follicle [Baker and O, 1976]. During meiotic arrest the oocyte remains transcriptionally and translationally active [Mangia and Canipari, 1977] as they accumulate stores for early development. The surrounding granulosa cells transport nutrients into the metabolically active oocyte via gap junctions [Heller et al, 1981].

After menarche, one follicle is selected by an as yet unknown mechanism to mature into a Graffian follicle that will be ovulated during the next menstrual cycle (Fig. 1). Just prior to ovulation, the nuclear membrane disappears (germinal vesicle breakdown, GVBD) and meiosis progresses through the first division to metaphase II, resulting in two asymmetrical daughter cells enclosed by the zona pellucida (Table I). The smaller daughter cell is known as a polar body [Zamboni et al, 1972]. The secondary oocyte (90–110 μm in diameter) with its encompassing cumulus oöphorous is ovulated into the oviduct. The cumulus oöphorous is composed of granulosa cells enmeshed in a matrix of hyaluronic acid, which disperses shortly after ovulation. Sperm may pass through the cumulus, and its penetration into the oocyte's cytoplasm triggers the completion of the oocyte's second meiotic division. A second polar body is extruded and the female pronucleus is formed [Baker and O, 1976]. Some oocytes can take up to 40 years to complete meiosis, and only those that are fertilized undergo two complete meiotic divisions.

The process in the male is quite different. Here, the primordial germ cells migrate to the primitive gonads but do not differentiate into spermatogonia until several years before puberty. At puberty the spermatogonia proliferate, both replacing themselves mitotically and producing primary spermatocytes. These primary spermatocytes undergo two reductive divisions (Table I) and upon the completion of meiosis there are four spermatids per primary spermatocyte, each containing a haploid genome [Clermont, 1972]. This process takes 24 days. The spermatids undergo a maturation process known as spermiogenesis as they pass along the seminiferous tubes of the testes. The round spermatids elongate and form a head containing the nucleus tipped by the acrosome, a midpiece containing mitochondria, and a tail for motility [Fawcett, 1975]. During this process nuclear histones are modified and replaced by protamines rich in disulfide bonds [Balhorn et al, 1977]. The haploid DNA, now a densely packaged struc-

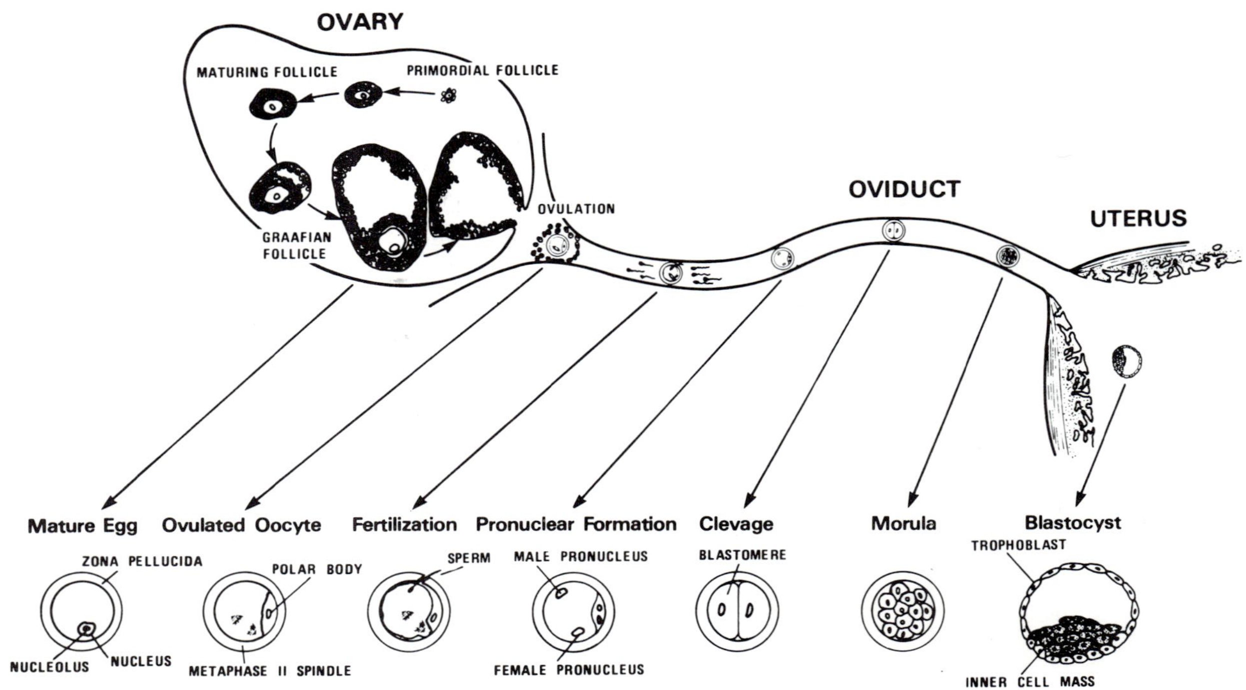

Fig. 1. Mammalian preimplantation development: Diagrammatic representation of oogenesis, ovulation, fertilization and development of the zygote to the preimplantation blastocyst.

ture has little transcriptional and translational activity [Monesi et al, 1978]. The spermatozoa are released into the lumen of the seminiferous tubes and pass into the epididymis where they become fully motile prior to ejaculation [Clermont, 1972].

Sperm-Oocyte Fusion

After ejaculation and capacitation in the female genital tract, the mature spermatozoa encounters the ovulated oocyte usually within 12 hours after ovulation [Rock and Hertig, 1944]. Of the tens of millions of sperm deposited into the vagina, the vast majority die. Probably somewhere between 10 and 100 sperm actually come into close proximity to the oocyte [Austin, 1972]. During its approach to the oocyte the acrosome releases its enzymatic contents (hyaluronidase, acrosin, glycosidases, etc), which are thought to facilitate the sperm's passage through the outer vestments of the egg. However, the in vivo timing and geographical location of the release of the acrosomal enzymes is still a matter of speculation as is their exact biological function [Bedford, 1981].

The nature of the interaction between the sperm and the zona pellucida is not entirely clear, but one or more of the zona proteins may serve as a species-specific sperm receptor [Bleil and Wassarman, 1980]. After penetration through the zona, the sperm fuses its postacrosomal plasma membrane with the egg's plasma membrane and enters into the egg's cytoplasm. The sperm plasma membrane vesiculates and when exposed to the egg's cytoplasm, the sperm nucleus rapidly decondenses [Yanagimachi, 1981] and forms the male pronucleus.

Subsequent sperm are prevented from entering the oocyte by several mechanisms [Wolf, 1981; Gulyas, 1980]. During oocyte maturation small cortical granules (0.3–0.5 μm in diameter) appear and migrate to the periphery coming to rest near the plasma membrane. The fusion of sperm and egg triggers the cortical reaction in which the granules fuse with the egg's plasma membrane releasing their contents into the perivitelline space (between the plasma membrane and the overlying zona pellucida). Concomitant with this reaction, the zona pellucida is structurally modified both by specific proteolytic cleavage and a general hardening reaction, which may be mediated by ovoperoxidase catalyzed cross-linking of tyrosine residues in the zona proteins. Following these changes, sperm do not bind to the zona pellucida. In addition to this relatively rapid block to polyspermy, there is also an apparent rearrangement of the plasma membrane (now made up in part of cortical granule membrane), which, after several hours, aids in preventing further sperm fusion.

The penetration of the sperm into the egg's cytoplasm triggers the completion of second meiotic division of the egg and formation of the female pronucleus. The rapid approach of the two pronuclei is presumably mediated by contractile proteins. Just prior to syngamy, when their membranes merge, each pronucleus duplicates its DNA so that every pair of chromosomes is composed of four sister chromatids. Subsequently, the chromosomes condense on the newly formed mitotic spindle and the one-celled zygote splits into two daughter cells each containing half of the cytoplasm and a diploid nucleus [Austin, 1972].

Preimplantation Embryo

Subsequent cleavages begin slowly but increase in frequency to produce blastomeres. The total volume of the early embryo is constrained by the encircling zona pellucida and remains approximately constant. Thus, with each cleavage the blastomeres becomes smaller, and eventually their size is reduced to that of a normal somatic cell [McLaren, 1972; Sherman, 1979]. During cleavage there is also partitioning of the maternal cytoplasm, which is responsible for the maternal inheritance of mitochondrial genes. To the extent that the partitioning of cytoplasmic components is nonuniform, cleavage may play an important role in subsequent early cell differentiation [Johnson and Ziomek, 1981]. At the 12–16 cell stage, the blastomeres undergo Ca^{++} mediated compaction to form the morula (Latin for mulberry) in which individual cells are not as readily distinguished [McLaren, 1972; Hyafil et al, 1981]. Although the cells appear morphologically similar, biochemical studies suggest that the outer cells are already destined to become the trophoblasts (precursors of the placenta) and the inner cells will form the inner cell mass of the embryo [Ducibella et al, 1977; Handyside and Johnson, 1978]. Furthermore, only cells of the inner cell mass maintain their totipotency as evidenced by nuclear transfer experiments [Illmensee and Hoppe, 1981].

A small fluid-filled cavity gradually forms in the center of the early embryo as it enters into the blastocyst stage. The outer cells flatten against the zona pellucida, and as the embryo passes into the uterus, it escapes entirely from the confines of the zona [McLaren, 1972; Sherman, 1979]. By this time cells of the inner cell mass and the trophectoderm differ in morphology, biochemistry, and cell surface antigens [Handyside and Johnson, 1978; Willison and Stern, 1978; Brulet et al, 1980].

The free-floating human blastocyst implants on the uterine wall approximately seven days after ovulation. At this time the proliferating trophectodermal cells are flattened and the inner cell mass is confined to one pole of the blastocyst. This pole ultimately becomes the site of uterine attachment. The inner cell mass then differentiates into endodermal and ectodermal cells [McLaren, 1972; Sherman, 1979]. Recently, some lines of teratocarcinoma cells have been shown to mimic the differentiation of endodermal cells in the presence of appropriate inducers [Strickland et al, 1980; Hogan et al, 1981]. The study of these model systems may yield greater insight into the mechanism of early development.

NORMAL MOLECULAR AND CELL BIOLOGY

Modern reproductive toxicology requires a detailed knowledge of the molecular and cellular events that control differentiation and development. Many of the recent advances in understanding the eukaryotic genome and its products provide particular insight into the mechanism and effect of mutagenesis on the early preimplantation embryo. It is becoming increasingly clear that the genome is not a static repository of DNA sequences coding for an array of gene products. Rather, it is a dynamic structure which can be modified and rearranged during development.

Gene Organization

A haploid human cell contains 2–3 pg of DNA [Lewin, 1980], the bulk of which is located in the nucleus, although a very small amount (less than 1%) is found in the mitochondria [Lewin, 1980; Clayton, 1982]. The haploid DNA complement contains about 2.5×10^9 base pairs, which in a totally extended form would measure 0.87 m [Lewin, 1980]. However, the DNA is condensed onto nucleosome structures containing DNA and an octamer of small basic proteins called histones (H2A, H2B, H3, and H4) [McGhee and Felsenfeld, 1980]. These nucleosomes can be seen by electron microscopy as "beads on a string" separated by linking segments of DNA [Olins and

Olins, 1974]. The primary nucleosomes are packed into higher order structures and eventually interact with scaffold proteins to form chromatids, each containing a centromere [Paulson and Laemmli, 1977]. During the S or synthesis phase of the cell cycle, the genome is replicated semiconservatively by DNA polymerase acting as a part of a multiunit holoenzyme. The replication is semidiscontinuous in the 5′ to 3′ direction and involves the production of short fragments (Okazaki fragments) on the lagging strand later filled in by DNA polymerase I [Ogawa and Okazaki, 1980]. Because of the huge size of the mammalian genome this process is initiated at thousands of sites, each of which becomes a replicating fork.

Of the 2–3 $\times$ 10^9 base pairs of DNA in the haploid genome approximately 8–10% represents segments of highly repetitive sequences, 20% represents moderately repetitive sequences, and 70% are unique sequences [Britten and Kohne, 1968]. The highly repetitive and moderately repetitive sequences can occur either as stretches of tandem repeats or can be interspersed throughout the genome [Zuker and Lodish, 1981; Singer, 1982]. Some are transcribed; but their functions are unknown, although postulated to affect the regulation of gene expression. The remainder of the genomic DNA is composed of unique sequences, most of which are transcribed into precursor mRNA and could code for more than 1.8 $\times$ 10^6 average-sized (35,000-dalton) proteins. This is far in excess of the estimated 5–10 $\times$ 10^4 proteins in the human body. Thus, many unique DNA sequences that are transcribed also may play a regulatory role [Davidson and Britten, 1979].

The unique sequences are found on all 22 of the somatic chromosomes as well as the X and Y chromosomes. Gradually a human gene map is being constructed using linkage studies, in situ hybridization, and somatic cell hybridization to indicate where on a standard human karyotype unique sequences reside. This localization can be done to the nearest of the 320 bands seen on a stained human karyotype arrested in metaphase [Dutrillaux and Lejeune, 1975]. Each metaphase band encompasses roughly 8 $\times$ 10^6 base pairs or enough DNA to code for 8000 average-sized (35,000-dalton) proteins. Thus, even with the localization of a gene to a particular band, there is still a significant gap between the localization of genes in the chromosome and at the DNA level.

Flow of Information

Although the initiation of transcription of RNA is thought to be a major control point for gene expression in eukaryotic cells, little is known of its molecular mechanism. Once initiated RNA polymerase synthesizes complementary RNA in a 5′ to 3′ direction along the DNA template. Only one of the DNA strands (the sense strand) is transcribed. For the most part the DNA remains packaged in nucleosomes, and although the core histones have not been demonstrated to play a direct regulatory role, variants of H2A and H2B do affect the structure of nucleosomes [Simpson, 1981; Savic et al, 1981]. Only small portions of the genome are transcribed at any moment in time. The molecular signals for the termination of most RNA transcription in mammalian cells are at present poorly understood [Benoist et al, 1980; Breathnach and Chambon, 1981], although a termination sequence has been reported in yeast [Zaret and Sherman, 1982].

RNA polymerase II transcribes unique gene sequences into precursor RNA molecules, which are processed by a series of modifications into the mature mRNAs that are responsible for gene expression. The 5′ end is capped with 7-methylguanosine [Shatkin, 1976]. Intervening sequences (introns) which do not code for proteins are present in most eukaryotic mRNA and are spliced out of the precursor mRNA [Leder et al, 1980]. About 70% of eukaryotic mRNA is polyadenylated on the 3′ end with up to 200 adenosine residues [Lewin, 1980; Breathnach and Chambon, 1981]. Also during transcription, the elongating transcripts can be complexed with proteins to form ribonucleoprotein particles [Martin et al, 1980]. The exact role that these proteins play in RNA processing, export into the cytoplasm, and eventual binding to ribosomes is an area under intense investigation.

Mature mRNA is translated into protein on cytoplasmic ribosomes according to its triplet codons. A 5′ untranslated sequence [Grosveld et al, 1982] precedes the AUG initiation codon that begins the coding sequence. At least seven eukaryotic initiation factors (eIF) play a role in the formation of the 80S initiation complex [Hershey, 1980; Kramer and Hardesty, 1980; Jagus et al, 1981]. One, eIF-2, mediates the binding of Met-$tRNA_f$ to the 40S ribosomal subunit and is important in the control of translation. Many ribosomal complexes can translate a particular mRNA at the same time. Once started, a series of elongation factors [Hershey, 1980] promotes the binding of aminoacyl-tRNAs to ribosomes and the elongation of the polypeptide chain. Protein synthesis stops when one of three termination codons is reached and involves releasing factors, one of which has been characterized [Hershey, 1980]. After release from the ribosome complex, the protein chain is often posttranslationally modified by the removal of one or more amino acid residues, by the formation of disulfide bonds, and by the hydroxylation phosphorylation or glycosylation of specific residues. The proteins are then routed by one of several possible mechanisms [Sabatini et al, 1982; Engleman and Steitz, 1981; Kriel, 1981] to their ultimate destination within the cell.

Control of Gene Expression

There is much evidence that transcription and posttranscriptional modifications of RNA, and translation and degradation of protein are major control points in gene expression. However, structural changes in the genome may also be important. Three mechanisms have been described: rearrangement, transposition, and amplification. Rearrangement of the immunoglobulin genes during development has been demonstrated in man and plays a significant role in the creation of antibody diversity [Leder et al, 1980]. Transposable elements have been shown to account for the changes in the mating substance expressed by the simple eukaryotic yeast cell and may be operative in higher animals as well [Hicks et al, 1979]. Amplification of the rRNA genes occurs in sea urchin and Drosophila melanogaster [Long and Dawid, 1980] and amplication of chorion genes has been detected in Drosophila [Long and Dawid, 1980; Spradling and Mahowald, 1980].

Much attention has been directed at the control of transcription via RNA polymerase II. In contrast to prokaryotic cells, no control proteins have been demonstrated to affect RNA polymerase II; and, indeed, the exact binding site of the polymerase to the DNA is not known. However, certain sequences 5′ to the coding sequence of cloned genes do appear to play an important role in the control of transcription. A consensus sequence 5′TATA$^{T}_{A}$A$^{T}_{A}$3′ (the Goldberg-Hogness or ATA box) is located approximately 30 base pairs upstream from the cap site in a large number of genes [Gannon et al, 1979]. Deletion of this region leads to inappropriate initiation of transcripts and in some cases a decrease in efficiency. Another consensus sequence, the CCAAT box, is located ~80 base pairs upstream from the cap site and appears to be important for the production of a large number of transcription products [Efstatiadis et al, 1980; Benoist et al, 1980].

In addition, certain structural differences have been detected in genes that are transcriptionally active in contrast to those that are not. One difference is that active genes have regions that are highly sensitive to digestion by DNase I [Weintraub and Grondine, 1976; Elgin, 1981], although at least one exception has been noted [Croce et al, 1981]. Another difference is the extent of DNA methylation. Normally, between 50 and 70% of the cytosine residues in the DNA of any given eukaryotic cell are methylated as 5-methylcytosine. Some active genes have been found to be hypomethylated when compared to the same genes in tissues where the gene is not active [Razin and Riggs, 1980]. It is not clear how DNase I sensitivity or methylation might influence transcription.

Cellular Differentiation

Transcriptional control may prove to play the major overall role in cellular differentiation and development. However, in early mammalian development this may not be entirely true. After ovulation, transcription ceases in the mouse oocyte [Wassarman and Letourneau, 1976] and has not been detected until the two-pronuclei stage after fertilization [Clegg and Piko, 1982]. However, during this period of transcriptional inactivity there are quantitative and qualitative changes in gene products [Schultz and Wassarman, 1977]. Most of these changes also occur in unfertilized eggs [Van Blerkom, 1979] as well as in physically and chemically enucleated oocytes [Schultz et al, 1979; Braude et al, 1979]. Although some of these changes can be attributed to post-translational modifications [Van Blerkom, 1981], there appear to be newly translated products as well [Cascio and Wassarman, 1982]. Presumably maternal mRNAs code for these new proteins, and it has been suggested that these mRNAs are masked either as RNA:RNA duplexes or as complex ribonucleoprotein particles (RNP) [Martin et al, 1980; Raff, 1980]. The release of mRNAs from mRNPs may be an important control over developmental events.

Thus, it is likely that control of translation is also an important mechanism of gene control in early development. In addition to the possible sequestration mRNAs, the initiation of protein synthesis (the rate-limiting step in translation) is a major control site for gene expression [Ochoa and de Haro, 1979; Kramer and Hardesty, 1980; Jagus et al, 1981]. For example, the phosphorylation of the smallest subunit of eIF-2 by eIF kinase, which can be induced by a variety of mechanisms, leads to the inhibition of Met-$tRNA_f$ binding to the 40S ribosomal subunit. Alternatively protein synthesis can be inhibited during elongation. Production of interferon by a cell is accompanied by (among other effects) the synthesis of a unique oligonucleotide with 2′–5′ phosphodiester linkage, pppA2′p5′A2′p5′A. This polynucleotide appears to interfere with polypeptide chain elongation, perhaps by activating a nuclease to degrade the mRNA. Interestingly, interferon also appears to inhibit translation by inducing the phosphorylation of eIF-2.

The earliest defined differentiation in the embryo occurs with the formation of the trophectoderm and inner cell mass, which morphologically can be detected in the blastocyst [McLaren, 1972; Sherman, 1979]. The more recent use of cell surface and biochemical markers allow this distinction to be made more rigorously at the morula stage [Ducibella et al, 1977; Handyside and Johnson, 1978; Johnson and Ziomek, 1981] and eventually these changes will be defined at the level of gene expression. Another recent study has compared the developmental potential of the nuclei from the inner cell mass and trophectoderm cells [Illmensee and Hoppe, 1981]. Inner cell mass nuclei when injected into enucleated mouse oocytes supported complete development to produce fertile adults. Nuclei from the trophectoderm were unable to support development much beyond the morula stage.

GENETIC ABNORMALITIES

Radiation and chemical toxins have a wide range of effects on the preimplantation embryo. Alterations in membrane permeability, disruption of the mitotic spindle, and inhibition of enzyme activities (particularly those involved with detoxification of environmental agents) can lead to disruption and cell death. However, mutation of the genome of the early embryo will be the focus of this review. Such mutations can be inherited via the germ cells or can occur in somatic cells during embryogenesis. The nature of these abnormalities varies from specific point mutation in the genome to chromosomal derangement. Some are corrected by self-repair mechanisms; others, although uncorrected, can be compensated for by the embryo and still others result in early embryonic death.

Mechanisms of Point Mutation

Point mutations are those that affect a single nucleotide and include transition, transversion, and frameshift mutations [Goldstein et al, 1974; Roberts, 1978; Lindahl, 1979]. Transition mutations of the DNA involve the switching of a purine for a purine or a pyrimidine for a pyrimidine. They can be induced by incorporation into DNA of base analogues such as 5-bromouracil or 2-aminopurine; by nitrous acid, which deaminates adenine to hypoxanthine and cytosine to uracil; or by hydroxylamine, which modifies cytosine to a derivative having base-pairing properties like thymidine. Less commonly, transversion mutations occur with the substitution of a purine for a pyrimidine or vice versa.

Frameshift mutations are caused by the addition or deletion of a nucleotide that shifts the reading frame of the triplet coden during transcription. Transcription proceeds normally, but the coding is abnormal downstream from the frameshift mutation and can result in a variety of abnormal translation products. Acridine dyes intercalate into the DNA helix and cause unwinding and the spreading apart of the bases. If they are present during DNA replication, an additional nucleotide can be inserted leading to a frameshift mutation. Acridine dyes also increase the frequency of unequal crossing-over during meiosis, which can cause deletions or insertions. Experimentally, deletions can be induced by the hydrolytic loss of purines, the covalent cross-linking of same-strand pyrimidines, or by alkylating and deaminating agents that form bases that cannot pair. Cross-linking of bases and DNA strand breaks induced either by radiation or chemical carcinogens interfere with DNA replication and if the damage is unrepaired can lead to cell death [von Sonntag et al, 1981].

Mechanism of Chromosome Abnormalities

Chromosomal abnormalities are divided into those that affect the number (aneuploidy) and those that affect the structure of chromosomes. Although some of these alterations occur during preimplantation development, most occur earlier during meiosis and gametogenesis. The vast majority of these mutations result in fetal loss before implantation on the uterine wall.

Chromosome number can be affected by nondysjunction in which either an autosome or a sex chromosome does not separate from its sister chromatid during the first or second meiotic division. More rarely, chromosome number can be affected by anaphase lagging in which one chromosome is lost during cell division. Structural anomalies of chromosomes require that one or two breaks occur in the DNA [Bloom, 1972]. Probably most single breaks are repaired by the cell with no resulting syndrome. However, if unrepaired, the genetic material distal to the break can be lost in subsequent cell divisions. Some breaks result in sister chromatid exchange which can be detected by culturing cells in the presence of bromodeoxyuridine [Latt et al, 1980].

If there are two breaks in a chromosome, the intervening material can invert before rejoining occurs and yield a paracentric inversion (if it does not include the centromere) or a pericentric inversion (if it does). Alternatively, the two ends of the intervening material can join in a ring structure with the loss of the genetic material distal to the breaks, or the ends may link end-to-end with one another to form dicentric chromosomes [Bloom, 1972]. If the two breaks occur on two different chromosomes, translocations can occur. Most often this involves the acrocentric chromosomes (13–15 and 21–22).

Repair Mechanisms

Eukaryotic cells have a variety of mechanisms to repair spontaneous, radiation, or chemically induced damage to their DNA [Hanawalt et al, 1979; Lindahl, 1979; Grossman, 1981]. Although excision repair and postreplication repair systems appear particularly important in mammalian cells, other repair mechanisms have been described. Insertase can catalyze the insertion of bases into apurinic or apyrimidinic sites, and photoreactivation can separate intrastrand thymidine dimers [Hanawalt et al, 1979; Hall and Mount, 1981]. Much interest has recently been focused on the inducible SOS repair system in bacteria [Witkin, 1976], and it appears that similar enzymes may also repair DNA in mammalian cells [Hanawalt et al, 1979; Bockstahler, 1981].

More demonstrably useful for repair of point mutations in mammalian cells are the excision-repair mechanisms. Two general systems have been described [Lindahl, 1979] (Fig. 2). The first, nucleotide excision, involves the recognition of an abnormal nucleotide, breakage of the phosphodiester bond, removal and degradation of a patch of DNA containing the abnormal nucleotide, repair synthesis of DNA using the undamaged strand as a template, and finally, ligation of the newly patched DNA. A second system, base excision, utilizes a DNA glycosylase to remove abnormal bases such as uracil and an apurinic/apyrimidinic endonuclease to break the DNA strand prior to repair synthesis and ligation. These excision-repair mechanisms can repair thymidine dimers and replace abnormal bases caused by irradiation and chemical mutagens. Defects in these repair mechanisms have been implicated in human disease [Friedberg et al, 1979].

During DNA replication, the DNA strands separate and without a complementary strand of DNA, excision-repair enzymes cannot repair nucleotide damage. However, postreplication repair mechanisms first described in bacterial cells are also present in mammalian cells [Hanawalt et al, 1979]. The model for this type of repair is based on the formation of thymidine dimers after UV irradiation (Fig. 3). During replication, gaps, which may be hundreds or thousands of nucleotides long, appear opposite the dimers. These gaps are terminated by DNA representing the start of a new Okazaki fragment. The daughter strand containing the gap is switched with the parental strand oriented in the same direction (5′ to 3′ or 3′ to 5′) by sister strand exchange. The gap is then repaired using the intact daughter strand as a template. This process is called daughter-strand gap repair and does not result in the repair of the thymidine dimer

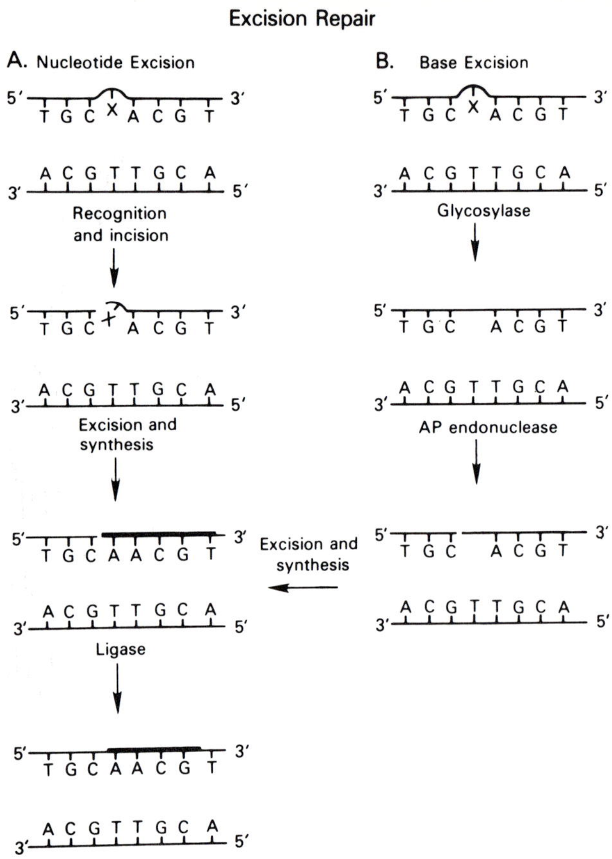

Fig. 2. Excision-repair mechanism: Schematic representation of two types of excision repair. A. Nucleotide excision with subsequent repair (see text). B. Base excision with the production of an apurinic/apyrimidinic (AP) site followed by repair (see text).

per se but does permit continued DNA replication. The thymidine dimer will be diluted out in subsequent cell division or can be repaired by excision repair systems or photoreactivation.

Manifestations of Unrepaired Mutations in the Early Embryo

Mutations can and probably do occur both in coding and noncoding regions of the genome. Many of these mutations will go unnoticed as silent mutations. Their "silence" is most often a reflection of the accuracy and sensitivity of our analytic techniques, which range from morphologic analysis to protein function, or DNA sequence data. If an abnormality is determined by a single genetic locus, it is a monogenic or Mendelian mutation and can be inherited as a recessive or dominant trait. However, the terms "recessive" and "dominant," which refer to phenotypic expression, are becoming less meaningful in the light of present-day knowledge of molecular genetics.

Other abnormalities are inherited as polygenic diseases, a term which is primarily an expression of our ignorance of the molecular mechanism of their inheritance and the effect of the environment on their expression. It appears that more than one genetic locus is responsible for their expression, although the interaction of these various loci is unclear. Examples of polygenic inheritance are seen in some common adult diseases (diabetes, coronary artery disease, hypertension) psychiatric diseases (schizophrenia,

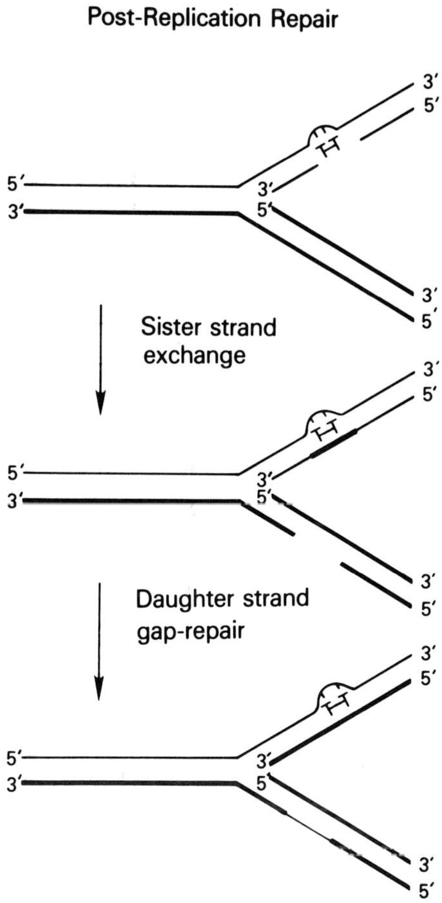

Fig. 3. Postreplication repair: Diagrammatic representation of repair of a gap caused by the presence of a thymidine dimer (T-T) during replication (see text).

depression), and some congenital birth defects (cleft lip or palate, club foot, neural tube defects, congenital heart defects).

The gametes that form the new embryo can carry abnormal genes from the preceding generation that either were inherited in the germ tissue or arose de novo during gametogenesis. The effects of ionizing radiation and chemical mutagens on gametes vary [Newcombe, 1971]. Studies involving large numbers of atomic bomb survivors showed no statistical increase in the number of mutant genes in the offspring [Schull et al, 1981]. Smaller-scale studies on patients who have received medical radiation for testicular cancer or lymphomas support these conclusions [Speiser et al, 1973; Smithers et al, 1973]. Although there is a marked increase in the risk of cancer, the major genetic effect of the radiation is to cause loss of gametes and infertility. Oligospermia is usually reversible, indicating the relative resistance of the spermatogonia stem cells. Premature ovarian failure persists, in that the ovary is unable to regenerate more oocytes. Studies of the effects of chemical mutagens are less clear-cut. For example, epidemiological and genetic data on the unfortunate chemical exposure at Love Canal is still fragmentary, and no firm conclusions can be drawn from it [Maugh, 1982].

As evidenced by unscheduled DNA synthesis, spermatocytes and oocytes are able to repair damage to their DNA induced by UV light, ionizing radiation, and chemical mutagens. Mature spermatids lose this ability. However, in dominant lethal studies in which mouse sperm were mutagenized with X-rays or chemicals before being used to fertilize oocytes, the damaged genome was repaired by components of the egg prior to the first cleavage in the zygote [Pedersen and Mangia, 1978; Generoso et al, 1979].

However, repair is not always accomplished and the oocyte appears to be particularly susceptible to mutagenesis late in oogenesis [Newcombe, 1971]. Genetically abnormal preimplantation embryos often die during early development or fail to implant in the uterine wall. It has been estimated that as many as 45% of all human conceptions are aborted—most during this early period [Carr, 1971]. In addition, the early embryo, although relatively resistant, can undergo mutation by radiation or chemicals [Carr, 1971; Newcombe, 1971; Bloom, 1972]. If mutated somatic cells survive, they will pass the mutation on to their considerable progeny making a mosaic embryo comprising normal and abnormal cells. Such mosaics have been experimentally created in mice [Dewey et al, 1977]. The embryo has a remarkable ability to compensate for the loss of a relatively large percentage of its cells with only a small delay in its development [Willison and Stern, 1975]. However, if cell death is extensive enough the embryo will be spontaneously aborted, often prior to implantation.

FUTURE DIRECTIONS

Two difficulties that plague investigators in mammalian developmental biology are the paucity of biological material and the lack of adequate model systems. However, new tools now available in genetics and molecular biology are able, in part, to offset these drawbacks. These new advances should markedly facilitate the study of toxicology in reproduction and early embryogenesis. Three deserve particular mention.

Genetic Mutants

The best-defined genetic mutants in early mammalian development are those of the mouse T complex, which is associated with suppression of recombination, poor sperm motility, and early embryopathies [Bennett, 1980; Silver, 1981]. It is located on chromosome 17 and occupies approximately 1% of the genome. A series of dominant and recessive mutations have been mapped to this region; thus, each haplotype observed is a composite of alleles at the diverse genetic loci in the T complex. These loci appear to code for cell surface glycoproteins [Cheng and Bennett, 1980] both on the sperm and the early embryo.

Several of the T complex haplotypes have been associated with early embryonic death. One mutant, t^{12}, when homozygous in the early embryo prevents the differentiation of the blastomeres into trophectoderm cells and leads to cell death at the morula stage. A second mutant, t^{w73}, prevents implantation and interferes with the growth of the inner cell mass. Other mutants interfere with later stages of development. These mutants may express abnormal cell surface components and lead to deranged cellular differentiation. Recent advances now permit the establishment of long-term euploid teratocarcinoma cell lines derived directly from early embryos [Evans and Kaufman, 1981; Martin, 1981] and should prove useful in further analysis of these mutants. This may lead to greater understanding of the molecular mechanism of recombination and the function of these cell surface antigens in early embryogenesis.

Teratocarcinoma Cell Lines

A variety of teratocarcinoma cell lines have been developed that can mimic certain aspects of early mammalian development [Martin, 1980]. They can be grown in long-term culture and in relatively large quantities. The cell lines are pluripotent and some euploid cells appear to be totipotent [Stewart and Mintz, 1981]. These cells can be induced to differentiate into various tissues and are thought to closely resemble cells of the early embryo. They have been useful in studies of the mechanism of compaction of the early cleaving embryo into the morula [Hyafil et al, 1981], of the role of cell surface antigens in the differentiation of the inner cell mass into endodermal cells [Strickland et al, 1980], and have been used as a model system to study X-inactivation in the early embryo [McBurney and Strutt, 1980]. Also, because of their ability to form chimeric mice after mixing with early embryonic cells, they are a useful vehicle for the introduction into embryos of new genetic material [Dewey et al, 1977]. This may be done by transfection of teratocarcinoma cells with cloned and isolated DNA sequences [Pellicer et al, 1980], or with SV-40 recombinants containing the appropriate DNA sequence [Linnenbach et al, 1980].

Microinjection

Microinjection of nuclei or of cloned genetic material into the pronucleus is perhaps the most elegant way to introduce new genetic material into mice to study events in reproduction and early development. This technique has been used to study the relative ability of nuclei from cells of the inner cell mass and from trophectoderm cells to sustain development through birth [Illmensee and Hoppe, 1981]. It has been used to introduce heterologous cloned genetic material into the male pronucleus of the one-cell zygote for analysis of embryonic and adult gene expression [Brinster et al, 1981; Wagner et al, 1981; Brinster et al, 1982]. Some new genetic material appears to be integrated into the genome and transmitted to their offspring.

These techniques should lead to important progress in understanding the developmental potential of different cell types and perhaps, the creation of animal models of certain human diseases. The microinjection of well defined portions or normal DNA will permit in vivo analysis of the signals that control transcription and processing of precursor RNA. Injection of mutant DNA will complement these studies as well as lend greater insight into DNA repair mechanism and the effect of specific mutations on development. Injection of RNA will facilitate studies on translational control of gene expression. Potentially such systems can also be used to study in vivo recombination and gene-dosage effects in the growing embryo.

ACKNOWLEDGMENTS

I wish to express my gratitude to Drs Camerini-Otero, A. Dean, I. East, M. Fordis, D. Mattison, and A. N. Schechter for their critical reading of the manuscript. I thank Mrs Laura Barry for her excellent preparation of the manuscript.

REFERENCES

Austin CR (1972): Fertilization. In Austin CR, Short, RV (eds): "Reproduction in Mammals: 1. Germ Cells and Fertilization." Cambridge: Cambridge University Press, pp 102–133.

Baker TG, O WS (1976): Development of the ovary and oogenesis. In MacNaughton MD, Govan ADT (eds): "Clinics in Obstetrics and Gynaecology, Vol 3." London: W.B. Saunders, pp 3–26.

Balhorn R, Gledhill BL, Wyrobek AJ (1977): Mouse sperm chromatin proteins: Quantitative isolation and partial characterization. Biochemistry 16:4074–4080.

Bedford JM (1981): Why mammalian gametes don't mix. Nature 291:286–288.

Benoist C, O'Hare K, Breathnach R, Chambon P (1980): The ovalbumin gene—Sequence of putative control regions. Nucleic Acids Res 8:127–142.

Bennett D (1980): The T-complex in the mouse: An assessment after 50 years of study. Harvey Lect 74: 1–21.

Bleil JD, Wassarman PM (1980): Mammalian sperm-egg interaction: Identification of a glycoprotein in mouse egg zonae pellucidae possessing receptor activity for sperm. Cell 20:873–882.

Bloom AD (1972): Induced chromosomal aberrations in man. Adv Hum Genet 3:99–172.

Bockstahler LE (1981): Induction and enhanced reactivation of mammalian viruses by light. Prog Nucleic Acid Res Mol Biol 26:303–313.

Braude P, Pelham H, Flach G, Lobatto R (1979): Post-transcriptional control in the early mouse embryo. Nature 282:102–105.

Breathnach R, Chambon P (1981): Organization and expression of eucaryotic split genes coding for proteins. Annu Rev Biochem 50:349–383.

Brinster RL, Chen HY, Trumbauer M, Senear AW, Warren R, Palmiter RD (1981): Somatic expression of herpes thymidine kinase in mice following injection of a fusion gene into eggs. Cell 27:223–231.

Brinster RL, Chen HY, Warren R, Sarthy A, Palmiter RD (1982): Regulation of metallothinein—thymidine kinase fusion plasmids injected into mouse eggs. Nature 296:39–44.

Britten RL, Kohne DE (1968): Repeated sequences in DNA. Science 161:529–540.

Brulet P, Babinet C, Kemler R, Jacob F (1980): Monoclonal antibodies against trophectoderm-specific markers during mouse blastocyst formation. Proc Natl Acad Sci USA 77:4113–4117.

Carr DH (1971): Chromosomes and abortion. Adv Hum Genet 2:201–257.

Cascio SM, Wassarman PM (1982): Program of early development in mammal: Post-transcriptional control of a class of proteins synthesized by mouse oocytes and early embryos. Dev Biol 89:397–408.

Cheng CC, Bennett D (1980): Nature of the antigenic determinants of T locus antigens. Cell 19:537–543.

Clayton DA (1982): Replication of animal mitochondrial DNA. Cell 28:693–705.

Clegg KB, Piko L (1982): RNA synthesis and cytoplasmic polyadenylation in the one-cell mouse embryo. Nature 295:342–345.

Clermont Y (1972): Kinetics of spermatogenesis in mammals: Seminiferous epithelium cycle and spermatogonial renewal. Physiol Rev 52:198–236.

Croce CM, Linnebach A, Huebner K, Parnes JR, Margulies DH, Appella E, Seidman JG (1981): Control of expression of histocompatibility antigens (H-2) and β_2-microglobulin in F9 teratocarcinoma stem cells. Proc Natl Acad Sci USA 78:5754–5758.

Davidson EH, Britten RJ (1979): Regulation of gene expression: Possible role of repetitive sequences. Science 204:1052–1059.

Dewey MJ, Martin DW, Jr, Martin GR, Mintz B (1977): Mosaic mice with teratocarcinoma-derived mutant cells deficient in hypoxanthine phosphoribosyltransferase. Proc Natl Acad Sci USA 74: 5564–5568.

Ducibella T, Ukena T, Karnovsky M, Anderson E (1977): Changes in cell surface and cortical cytoplasmic organization during early embryogenesis in the preimplantation mouse embryo. J Cell Biol 74: 153–167.

Dutrillaux B, Lejeune J (1975): New techniques in the study of human chromosomes: Methods and applications. Adv Hum Genet 5:119–156.

Efstatiadis A, Posakony JW, Maniatis T, Lawn RM, O'Connell C, Spritz RA, DeRiel JK, Forget BG, Weissman, SM, Slightom JL, Blechl AE, Smithies O, Baralle FE, Shoulders CC, Proudfoot NJ (1980): The structure and evolution of the human β-globin family. Cell 21:653–668.

Elgin SCR (1981): DNase I—Hypersensitive sites on chromatin. Cell 27:413–415.

Engelman DM, Steitz TA (1981): The spontaneous insertion of proteins into and across membranes: The helical hairpin hypothesis. Cell 23:411–422.

Evans MJ, Kaufman MH (1981): Establishment in culture of pluripotential cells from mouse embryos. Nature 292:154–156.

Fawcett DW (1975): The mammalian spermatozoon. Dev Biol 44:394–436.

Friedberg EC, Ehmann UK, Williams JI (1979): Human diseases associated with defective DNA repair. Adv Radiat Biol 8:85–174.

Gannon F, O'Hare K, Perrin F, LePennec JP, Benoist C, Cochet M, Breathnach R, Royal A, Garapin A, Cami B, Chambon P (1979): Organization and sequences at the 5′ end of a cloned complete ovalbumin gene. Nature 278:428–434.

Generoso WM, Cain KT, Krishna M, Huff SW (1979): Genetic leisons induced by chemicals in spermatozoa and spermatids of mice are repaired in the egg. Proc Natl Acad Sci USA 76:435–437.

Goldstein A, Aronow L, Kalman SM (1974): "Principles of Drug Action." New York: John Wiley & Sons.

Grossman L (1981): Enzymes involved in the repair of damaged DNA. Arch Biochem Biophys 211:511–522.

Grosveld GC, de Boer E, Shewmaker CK, Flavell RA (1982): DNA sequences necessary for transcription of the rabbit β-globin in vivo. Nature 295:120–126.

Gulyas BJ (1980): Cortical granules of mammalian eggs. Int Rev Cytol 63:357–392.

Hall JD, Mount DW (1981): Mechanism of DNA replication and mutagenesis in ultraviolet-irradiated bacteria and mammalian cells. Prog Nucleic Acid Res Mol Biol 25:53–126.

Hanawalt PC, Cooper PK, Ganesan AK, Smith CA (1979): DNA repair in bacteria and mammalian cells. Ann Rev Biochem 48:783–836.

Handyside AH, Johnson MH (1978): Temporal and spatial patterns of the synthesis of tissue-specific polypeptides in the preimplantation mouse embryo. J Embryol Exp Morphol 44:191–199.

Heller DT, Cahill DM, Schultz RM (1981): Biochemical studies of mammalian oogenesis: metabolic cooperativity between granulosa cells and growing mouse oocytes. Dev Biol 84:455–464.

Hershey JWB (1980): The translation machinery: Components and mechanism. In Prescott DM, Goldstein L (eds): "Gene Expression: Translation and the Behavior of Proteins, Cell Biology, Vol 4." New York: Academic Press, pp 1–68.

Hicks J, Strathern JH, Klar AJS (1979): Transposable mating type genes in Saccharomyces cerevisiae. Nature 282:478–483.

Hogan BLM, Taylor A, Adamson E (1981): Cell interactions modulate embryonal carcinoma cell differentiation into parietal or visceral endoderm. Nature 291:235–237.

Ilyafil F, Babinet C, Jacob F (1981): Cell-cell interactions in early embryogenesis: A molecular approach to the role of calcium. Cell 26:447–454.

Illmensee K, Hoppe PC (1981): Nuclear transplantation in Mus musculus: Developmental potential of nuclei from preimplantation embryos. Cell 23:9–18.

Jagus R, Anderson WF, Safer B (1981): The regulation of initiation of mammalian protein synthesis. Prog Nucleic Acid Res Mol Biol 25:127–185.

Johnson MH, Ziomek CA (1981): The foundation of two distinct cell lineages within the mouse morula. Cell 24:71–80.

Kramer G, Hardesty B (1980): Regulation of eukaryotic protein synthesis. In Prescott DM, Goldstein L (eds): "Gene Expression: Translation and the Behavior of Proteins, Cell Biology, Vol 4." New York: Academic Press, pp 69–105.

Kreil G (1981): Transfer of proteins across membranes. Ann Rev Biochem 50:317–348.

Lange K, Page BM, Elston RC)1975): Age trends in human chiasma frequencies and recombination fractions. I. Chiasma frequencies. Am J Hum Genet 27:410–418.

Latt SA, Schreck RR, Loveday KS, Dougherty CP, Schuler CF (1980): Sister chromatid exchange. Adv Hum Genet 10:267–331.

Leder P, Max EE, Seidman JG, Kwan S-P, Scharff M, Nau M, Norman B (1980): Recombination events that activate, diversify, and delete immunoglobulin genes. Cold Spring Harbor Symp Quant Biol 45:859–865.

Lewin B (1980): "Gene Expression, Vol 2, Eucaryotic Chromosomes." New York: John Wiley & Sons.

Lindahl T (1979): DNA glycosylases, endonucleases for apurinic/apyrimidinic sites, and base excision-repair. Prog Nucleic Acid Res Mol Biol 22:135–192.

Linnenbach A, Huebner K, Croce CM (1980): DNA-Transformed murine teratocarcinoma cells: Regulation of expression of simian virus 40 tumor antigen in stem versus differentiated cells. Proc Natl Acad Sci USA 77:4875–4879.

Long EO, Dawid IB (1980): Repeated genes in eukaryotes. Ann Rev Biochem 49:727–764.

Mangia F, Canipari R (1977): Biochemistry and maturation in mammalian oocytes. In Johnson MH (ed): "Development in Mammals, Vol 2." Amsterdam: North-Holland, pp 1–29.

Martin GR (1980): Teratocarcinomas and mammalian embryogenesis. Science 209:768–776.

Martin GR (1981): Isolation of a pluripotent cell line from early mouse embryos cultured in medium conditioned by teratocarcinoma stem cells. Proc Natl Acad Sci USA 78:7634–7638.

Martin TE, Pullman JM, McMullen MD (1980): Structure and function of nuclear and cytoplasmic ribonucleoprotein complexes. In Precott DM, Goldstein L (eds): "Gene Expression: Translation and the Behavior of Proteins. Cell Biology, Vol 4." New York: Academic Press, pp 137–174.

Maugh TH, II (1982): Biological markers for chemical exposure. Science 215:643–647.

McBurney MW, Strutt BJ (1980): Genetic activity of X chromosomes in pluripotent female teratocarcinoma cells and their differentiated progeny. Cell 21:357–364.

McGhee JD, Felsenfeld G (1980): Nucleosome structure. Ann Rev Biochem 49:1115–1156.

McLaren A (1972): The embryo. In Austin CR, Short RV (eds): "Reproduction in Mammals: 2. Embryonic and Fetal Development." Cambridge: Cambridge University Press, pp 1–42.

Monesi V, Geremia R, D'Agostino A, Boitani C (1978): Biochemistry of male germ cell differentiation in mammals: RNA synthesis in meiotic and postmeiotic cells. Curr Top Dev Biol 12:11–36.

Newcombe HB (1971): The genetic effects of ionizing radiations. Adv Genet 16:240–303.

Ochoa S, de Haro C (1979): Regulation of protein synthesis in eukaryotes. Ann Rev Biochem 48:549–580.

Ogawa T, Okazaki T (1980): Discontinuous DNA replication. Ann Rev Biochem 49:421–457.

Olins AL, Olins DE)1974): Spheroid chromatin units (v bodies). Science 183:330–332.

Paulson JR, Laemmli UK (1977): The structure of histone-depleted metaphase chromosomes. Cell 12: 817–828.

Pedersen RA, Mangia F (1978): Ultraviolet-light-induced unscheduled DNA synthesis by resting and growing mouse oocytes. Mutat Res 49:425–429.

Pellicer A, Wagner EF, Kareh AE, Dewey MJ, Reuser AJ, Silverstein S, Axel R, Mintz B (1980): Introduction of a viral thymidine kinase gene and the human β-globin gene into developmentally multipotential mouse teratocarcinoma cells. Proc Natl Acad Sci USA 77:2098–2102.

Petzelt C (1980): Biochemistry of the mitotic spindle. Int Rev Cytol 60:53–92.

Raff RA (1980): Masked messenger RNA and regulation of protein synthesis in eggs and embryos. In Prescott DM, Goldstein L (eds): "Gene Expression: Translation and Behavior of Proteins, Cell Biology, Vol 4." New York: Academic Press, pp 107–136.

Razin A, Riggs AD (1980): DNA methylation and gene function. Science 210:604–610.

Roberts JJ (1978): The repair of DNA modified by cytotoxic, mutagenic and carcinogenic chemicals. Adv Radiat Biol 7:211–436.

Rock J, Hertig AT (1944): Information regarding the time of human ovulation derived from a study of 3 unfertilized and 11 fertilized ova. Am J Obstet Gynecol 47:343–356.

Sabatini DD, Kreibich G, Morimoto T, Adesnik M (1982): Mechanism for the incorporation of proteins in membranes and organelles. J Cell Biol 92:1–22.

Savic A, Richman P, Williamson P, Poccia D (1981): Alterations in chromatin structure during sea urchin embryogenesis. Proc Natl Acad Sci USA 78:3706–3710.

Schull WJ, Otake M, Neel JM (1981): Genetic effects of the atomic bombs: A reappraisal. Science 213: 1220–1227.

Schultz RM, Letourneau GE, Wassarman PM (1979): Meiotic maturation of mouse oocytes in vitro: Protein synthesis in nucleate and anucleate oocyte fragments. J Cell Sci 30:251–264.

Schultz RM, Wassarman PM (1977): Specific changes in the pattern of protein synthesis during meiotic maturation of mammalian oocytes in vitro. Proc Natl Acad Sci USA 74:538–541.

Shatkin AJ (1976): Capping of eucaryotic mRNAs. Cell 9:645–653.

Sherman MI (1979): Developmental biochemistry of preimplantation mammalian embryos. Ann Rev Biochem 48:443–470.

Silver LM (1981): Genetic organization of the mouse t complex. Cell 27:239–240.

Simpson RT (1981): Modulation of nucleosome structure by histone subtypes in sea urchin embryos. Proc Natl Acad Sci USA 78:6803–6807.

Singer MF (1982): SINES and LINES: Highly repeated short and long interspersed sequences in mammalian genomes. Cell 28:433–434.

Smithers DW, Wallace DM, Austin DE (1973): Fertility after unilateral orchidectomy and radiotherapy for patients with malignant tumors of the testes. Br Med J 4:77–79.

Speiser B, Rubin P, Casarett G (1973): Aspermia following lower truncal irradiation in Hodgkin's disease. Cancer 32:692–698.

Spradling AC, Mahowald AP (1980): Amplification of genes for chorion proteins during oogenesis in Drosophila melanogaster. Proc Natl Acad Sci USA 77:1096–1100.

Stewart TA, Mintz B (1981): Successive generations of mice produced from an established culture line of euploid teratocarcinoma cells. Proc Natl Acad Sci USA 78:6314–6318.

Strickland S, Smith KK, Marotti KR (1980): Hormonal induction of differentiation in teratocarcinoma stem cells: Generation of parietal endoderm by retinoic acid and dibutycryl cAMP. Cell 21:347–355.

Van Blerkom J (1979): Molecular differentiation of the rabbit ovum. Dev Biol 72:188–194.

Van Blerkom J (1981): Structural relationship and posttranslational modifications of stage-specific proteins synthesized during early preimplantation development in the mouse. Proc Natl Acad Sci USA 78:7629–7633.

von Sonntag C, Hagen U, Schon-Bopp A, Schulte-Frohlinde D (1981): Radiation-induced strand breaks in DNA: Chemical and enzymatic analysis of end groups and mechanistic aspects. Adv Radiat Biol 9: 109–142.

Wagner TE, Hoppe PC, Jollick JD, Scholl DR, Hodinka RL, Gault JB (1981): Microinjection of a rabbit β-globin gene into zygotes and its subsequent expression in adult mice and their offspring. Proc Natl Acad Sci USA 78:6376–6380.

Wassarman PM, Letourneau GE (1976): RNA synthesis in fully-grown mouse oocytes. Nature 261:73–74.

Weintraub H, Grondine M (1976): Chromosomes subunits in active genes have an altered conformation. Science 193:848–856.

Willison KR, Stern PL (1978): Expression of a Forssman antigenic specificity in the preimplantation mouse embryo. Cell 14:785–793.

Wilson IB, Stern MS (1975): Organization in the pre-implantation embryo. In Balls M, Wild AE (eds): "The Early Development of Mammals." Cambridge: Cambridge University Press, pp 81–95.

Witkin EM (1976): Ultraviolet mutagenesis and inducible DNA repair in Escherichia coli. Bact Rev 40: 869–907.

Wolf DP (1981): The mammalian eggs block to polyspermy. In Mastroianni L, Jr, Biggers JD (eds): "Fertilization and Embryonic Development in vitro." New York: Plenum Press, pp 183–197.

Yanagimachi R (1981): Mechanism of fertilization in mammals. In Mastroianni L, Jr, Biggers JD (eds): "Fertilization and Embryonic Development in vitro." New York: Plenum Press, pp 81–182.

Zamboni L, Thompson RS, Smith DM (1972): Fine morphology of human oocyte maturation in vitro. Biol Reprod 7:425–457.

Zaret KS, Sherman F (1982): DNA sequence required for efficient transcription termination in yeast. Cell 28:563–573.

Zuker C, Lodish HF (1981): Repetitive DNA sequences cotranscribed with developmentally regulated Dictyostelium discoideum mRNAs. Proc Natl Acad Sci USA 78:5386–5390.

American Journal of Industrial Medicine 4:51–61 (1983)

Early Mammalian Embryonic Development

William J. Swartz, PhD

This report examines the major events occurring during the first month of development in the human embryo. Developmental events are detailed, beginning with cleavage of the zygote and ending with the formation of the three primary germ layers. Certain aspects of early human development are compared with that in lower forms of vertebrates to emphasize species differences in these basic embryological processes. A discussion of the formation of identical twins is included, since it has its inception during the first month of development.

Key words: embryo, cleavage, blastocyst, implantation, twinning

INTRODUCTION

There are many reports in the literature dealing with early development of the embryo. However, very few of these deal with the development of the human embryo during the first month following fertilization. Much of the information ascribed to early human development in most of the popular embryology texts has been acquired from studies on subhuman primates, principally monkeys. Such early stages in development are difficult to examine in the human for obvious reasons. With the advent of in vitro fertilization studies and its limited successes by Patrick Steptoe and Robert Edwards in England more attention is being given to these critical periods of development. Most of the events to be discussed here occur prior to the time a prospective mother becomes aware of her pregnancy.

The present report will describe early embryonic development in the human beginning with events occurring following fertilization and continuing to the end of the implantation process. This covers a period lasting until approximately the end of the fourth week of development. It is the intent of the author not only to describe some of the major developmental events occurring during this time but also to correlate aspects of clinical importance associated with these events.

Department of Anatomy, Louisiana State University Medical Center, New Orleans.
Address reprint requests to William J. Swartz, PhD, Department of Anatomy, Louisiana State University Medical Center, 1901 Peridido Street, New Orleans, LA 70112-1394.

0271-3586/83/0401-02-0051$03.50 **© 1983 Alan R. Liss, Inc.**

CLEAVAGE

The stages in embryonic development, of course, begin with fertilization. At fertilization, sex is determined, and the diploid number of chromosomes is restored. Because fertilization is the union of one male and one female sex cell, a mixing of the hereditary complements of these two sex cells occurs, thus providing for species variation. In addition, as a result of fertilization, cleavage is initiated.

Cleavage is the means whereby the unicellular fertilized egg (zygote, which measures about 0.1 mm in all mammalian species) is transformed into a multicellular package as a result of consecutive mitotic divisions. The zygote cleaves (divides) into two cells and then two cells become four, four become eight, etc. Cleavage divisions differ from mitotic divisions in that the former follow very rapidly upon one another, and as a result of these divisions, no cell growth occurs. Cleavage divisions result from mitotic cycles with much shortened interphase periods. Even though cell divisions are occurring during this period, there is no growth of the entire mass; however, replications of chromosomes do occur. With the decrease in size of the cells following each cleavage division, the cells become nearer to the size characteristic of the definitive cells of the organism.

The method of cleavage varies in different animals (Fig. 1). It is correlated with the amount of yolk that is stored in the ovum as food material for growth. Yolk itself is nonliving and inert and plays no *active* part in cleavage, but it does serve as a mechanical impediment to the process. In fertilized ova containing a large amount of yolk (megalecithal), the cleavage is limited to a small area of cytoplasm on the surface of the yolk (Fig. 1). The yolk itself is not divided. The yolk-laden zygote is incompletely and unequally cleaved. This is the case in birds, reptiles, higher fishes, and monotremes (egg-laying mammals). Amphibians possess eggs with a medium amount of yolk (telolecithal). In these animals, cleavage is slowed down as the cleavage plane encounters yolk at the vegetal pole. Before one cleavage division is completed, other cleavage divisions begin. As a result, more cells are found at the animal pole than at the vegetal pole, the latter ones being much larger since they are lagging behind in numbers of cleavage divisions. Cleavage, in this instance, is complete and unequal (Fig. 1). The ova of marsupials and placental mammals, including man, have very little yolk (isolecithal). A large amount of yolk is not necessary in these species since there will be an association with the mother, via the placenta, very early in development whereby the embryo will draw upon the uterine circulation of the mother for its nutrition. The small amount of yolk elicits little resistance to the cleavage process. As a result cleavage is complete and equal and results in cells of equal size (Fig. 1).

In lower vertebrate forms, a clearly defined polarity of the zygote exists. Yolk is concentrated at the vegetal pole and the nucleus and cytoplasm come to lie at the animal pole. Because of the lack of yolk material in the mammalian zygote, no clearly defined polarity exists. The point at which the polar body of the ovum is given off during meiosis is arbitrarily referred to as the animal pole [Corliss, 1976]. The mitotic spindle of the cleavage divisions forms at right angles to an imaginary axis passing through the ovum from animal to vegetal pole. The resulting cells of these cleavage divisions are referred to as blastomeres.

During the cleavage process the zona pellucida, surrounding the ovum, remains intact forcing the blastomeres to dispose themselves within the confines of this membrane (Figs. 1, 2). The zona pellucida acts to hold the cleaving egg together and prevent it from adhering to the walls of the oviduct [McLaren, 1973]. The number of cells pres-

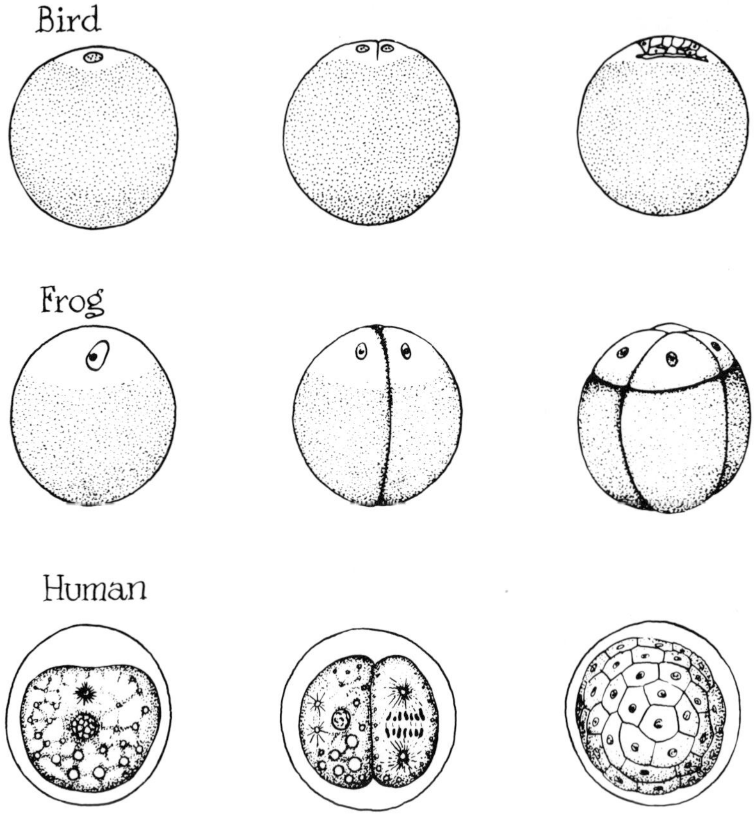

Fig. 1. Cleavage patterns in three vertebrate species possessing different amounts of yolk material

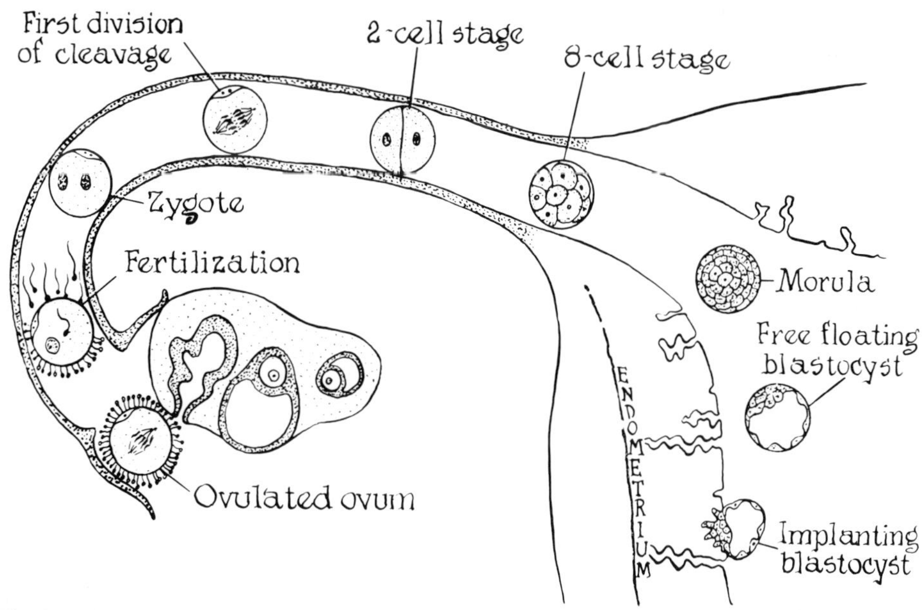

Fig. 2. Schematic representation of events occurring during the first week of embryonic development in the oviduct and uterus. Events depicted are ovulation, fertilization, cleavage, formation of morula and blastocyst, and the beginning of implantation.

ent in the cleaving embryo and the approximate time period of their appearance following ovulation have been determined by Lewis and Hartmann [1933] for the rhesus monkey. It is assumed that human cleavage divisions occur quite similarly. These stages and times of appearance following ovulation are as follows:

1-cell stage—ovulation–24 hours
2-cell stage—24–36 hours
3- to 4-cell stage—36–48 hours
5- to 8-cell stage—48–72 hours
9- to 16-cell stage—72–96 hours

As the cleaving egg passes down the oviduct, its main nutrients come from within itself—from stores in the blastomeres. From studies on the mouse, Tarkowski [1965] suggested that the blastomeres attain the ability to secrete fluid into the blastocoele at the end of the fifth cleavage division. It is possible that the zona pellucida acts as a membrane permitting transfer of certain substances of low molecular weight like amino acids and mucosaccharides resulting from a breakdown of tubal secretions containing mucopolysaccharides and glycogen [Hamilton and Mossman, 1972].

The 16-cell stage embryo takes on the appearance of a mulberry, and the embryonic mass has been ascribed the name morula. This morula is a solid ball of cells all of which are similar in appearance. It is at the morula stage that the embryo enters the uterus approximately two to four days following fertilization (Fig. 2).

During or just after entrance into the uterus, the morula undergoes specific changes. Fluid passes from the uterine cavity through the zona pellucida and into the intercellular spaces between the blastomeres. With an increase in the amount of intercellular fluid present in the morula, the spaces coalesce forming a single cavity known as the blastocoele surrounded by blastomeres. Borland et al [1977] has reported concentration gradients of different inorganic ions in the fluid of the blastocoele indicative of active transport processes occurring here. Uterine proteins have been found in the fluid of rabbit embryos by Beier and Maurer [1975], indicating early transport of such materials from the mother to the embryo prior to implantation. When the blastomeres of the morula begin to be rearranged and organized about a central cavity, the entire structure is referred to as a blastula. After the morula stage the term cleavage is no longer applied to the cellular divisions taking place.

Thus, looking back on the cleavage period the principal effects of this process are 1) a partitioning of the protoplasm of the zygote among the blastomeres and 2) an approximation in size of developing cells to that characteristic of the definitive cells of the organism.

BLASTOCYST FORMATION

With the end of the cleavage period and the beginning of the blastocyst stage, cell division does not cease, but continues unabated. Whereas cleavage was merely an increase in cell number through repeated cell divisions, now processes of regrouping and divergent specializations begin to make their appearance. Up to this time, the embryo has been very independent, but it must prepare itself for acquiring an intimate association with the maternal organism if pregnancy is to continue. The first indication of this is a segregation of cells composing the blastocyst into two distinct cell groups (inner cell

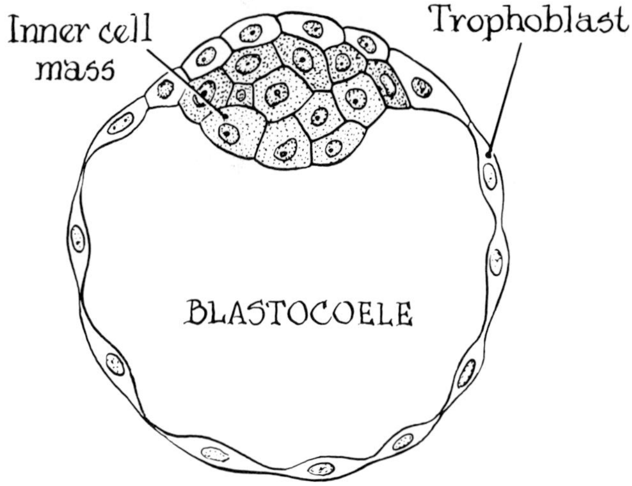

Fig. 3. Diagram of a section through a human blastocyst approximately five days after fertilization showing the basic components of the blastocyst.

mass and trophoblast) surrounding a fluid-filled cavity (blastocoele). Thus, the blastocyst resembles a hollow sphere (Fig. 3). An internal cluster of cells, the inner cell mass, becomes established at one pole of the blastocyst. The other cell group is a layer that constitutes the outer wall of the blastocyst known as the trophoblast. The inner cell mass is primarily concerned with the formation of the embryo proper, while the peripheral trophoblastic layer will give rise to a portion of the placenta and protective embryonic membranes that develop around the embryo.

The blastocoele is initially the result of a functional activity of precociously differentiated trophoblastic cells that are concerned with the passage of fluid with solutes to the inner cell mass. This fluid permits movements of substances within the blastocoele. By its influence on tension within the blastocyst, it is usually considered to play a part in the mechanism of implantation [Hamilton and Mossman, 1972]. The blastocyst arrives in the uterus about five days after fertilization and remains free floating for about two days.

Formation of the blastocyst brings with it a disintegration of the zona pellucida. Up to this point, the zona pellucida has prevented the adherence of the sticky embryonic mass to the uterine tube and uterus.

Although the development and further differentiation of the trophoblast and inner cell mass occur concurrently, they will be presented separately for clarity's sake.

IMPLANTATION

Development of Trophoblast

The period of implantation is defined as that interval between the loss of the zona pellucida around the blastocyst and the complete encirclement of the trophoblast by uterine endometrium [Young et al, 1968]. Human implantation is interstitial in nature since the invading blastocyst comes to lie entirely within the substance of the endometrium.

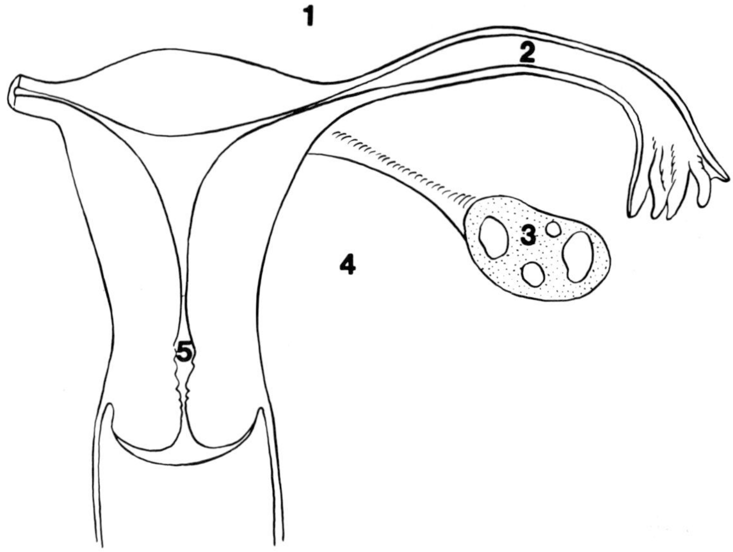

Fig. 4. Diagram indicating abnormal implantation sites in the human. 1, abdomen; 2, uterine tube; 3, ovary; 4, pelvis; 5, near internal os of cervix.

In the human, the blastocyst makes contact with the uterine mucosa via its trophoblast cells and begins to penetrate the endometrium at approximately seven days following fertilization. The penetration and erosion of the uterine mucosa results from the action of proteolytic enzymes produced by the trophoblast [Billington, 1970]. The uterine mucosa also promotes the proteolytic action of the blastocyst. Thus, by the end of the first week of development, the human zygote has passed through the cleavage and blastocyst stages and has begun to invade the endometrium (Fig. 2).

The normal uterine site for implantation is the endometrium of the body of the uterus on the upper part of the anterior or posterior wall near the midline. Abnormal sites for implantation include the abdominal and pelvic cavities, the uterine tube and the ovary (Fig. 4). When implantation occurs within the uterine tube, continued growth of the embryo results in rupture of the oviduct with resultant internal hemorrhage during the second month. Abnormal implantation may also occur within the uterus near the internal os (opening). This becomes clinically important when the placenta develops over the cervical canal (placenta praevia) and may result in severe bleeding in mid-pregnancy.

Most of the information on the early stages of implantation has been obtained from studies on primates other than humans. During the second week of development, the cuboidal trophoblastic cells of the blastocyst begin to differentiate into two morphologically distinct cell types. The first is an inner mononucleated cell layer, the cytotrophoblast, and an outer multinucleated area, the syncytiotrophoblast (syntrophoblast). The highly mitotically active cytotrophoblasts give rise to the syntrophoblast. The cytotrophoblast layer will retain its cellular integrity, while cell boundaries are indistinguishable in the syntrophoblast (Fig. 5). It appears likely that the contact of the trophoblast with the endometrium stimulates the expansion of the trophoblast since in the early stages of implantation that part of the trophoblast still bounding the uterine

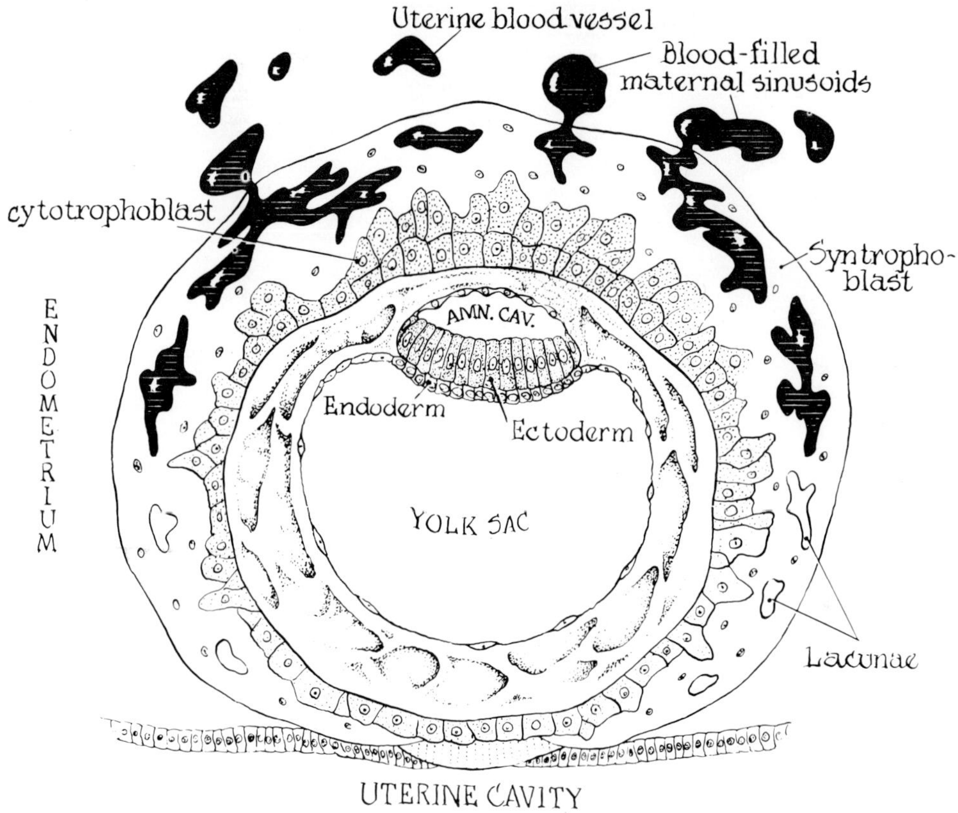

Fig. 5. Diagram of a section through an implanting blastocyst at about ten days after fertilization. Observe the bilaminar germ disc and the appearance of the primitive stages in the uteroplacental circulation.

cavity remains a single layer [Carlson, 1981]. Nourishment of the embryo during the early stages of implantation appears to occur by diffusion of materials from the maternal interstitial fluids or possibly debris from eroded uterine endometrial cells.

The further development of the syntrophoblast is quite complex. At nine days, isolated spaces begin to appear in the amorphous trophoblast area, which eventually coalesce. The syncytial cells continue to penetrate deeper into the endometrium. Maternal endometrial capillaries around the invading organism become dilated and congested forming sinusoids. The syntrophoblast eventually invades the lining of these sinusoids. As a result, maternal blood then seeps directly into the lacunar networks and begins to flow slowly through the lacunar system. This results in a continuity between the lacunae of the syntrophoblast, which is of embryonic origin, and the endometrial sinusoids, which are of maternal origin. This heralds the beginning of an intimate maternal and embryonic association—a placental circulation. By the 12th day, the blastocyst is completely embedded in the uterine stroma.

At 15 days both components of the trophoblast are seen to contribute to the formation of multiple branched projections called villi. During the third week of development, these villi become highly vascularized and become bathed by maternal blood in the lacunae (Fig. 6). It is in these areas that the gaseous and nutrient exchanges between mother and fetus take place. The blood vessels formed in these villi establish

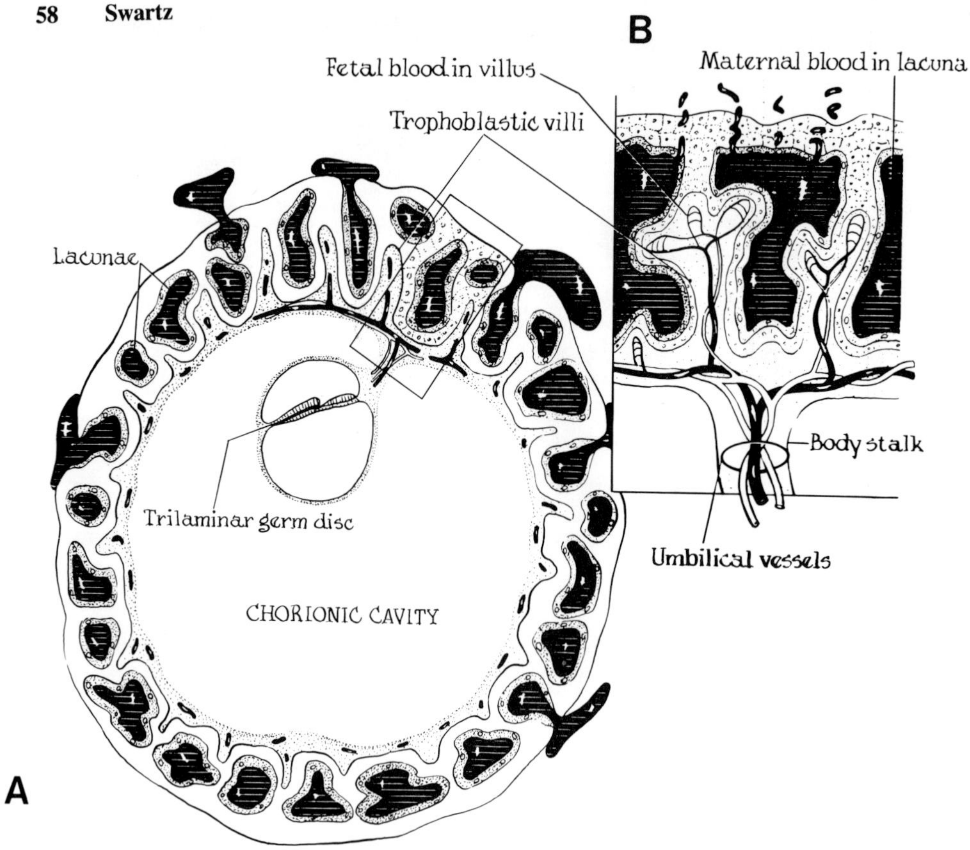

Fig. 6. A. Diagram of a section through an implanted embryo at approximately three weeks postfertilization. Note the trilaminar germ disc connected to the trophoblastic villi by means of the body stalk. B. High-power view of the area outlined in 6A showing the umbilical vessels and the relationship between fetal and maternal blood in the villi.

connections with vessels in the connecting stalk, which in turn make contact with the intraembryonic circulatory system and thereby connect the placenta to the embryo (Fig. 6). Thus, when the heart begins to beat during the fourth week of development, the pathway is present to allow exchange of nutrients between the mother and embryo.

Development of Embryoblast

At first, the cells of the inner cell mass develop much more slowly than the trophoblast cells. Cells on the under surface of the inner cell mass arrange themselves into a single layer facing the blastocoele. These are the endodermal cells. The remaining cells become columnar and are known as the ectoderm. The embryo is now a two-layered structure known as the bilaminar germ disc (Fig. 5).

In the area of the trophoblast adjacent to the ectodermal layer of the embryo, a layer of cells (amnioblasts) appear that become continuous with the embryonic ectoderm at the periphery. The exact origin of these cells is disputed. Hertig [1968] reported their appearance to occur as a result of delamination of the adjacent cytotrophoblast, while Luckett [1978] attributed the origin of the amnion to cells of the inner cell mass.

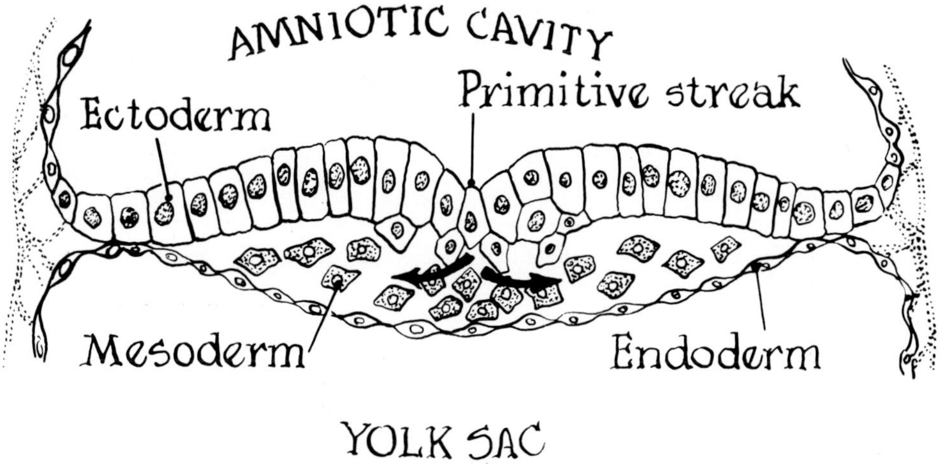

Fig. 7. Cross-sectional diagram through the germ disc of a 16-day embryo showing the delamination of cells from the ectodermal layer and their migration into a middle position between the ectoderm and the endoderm. These cells form the third germ layer called mesoderm.

Spaces then begin to appear between the amnioblasts and the ectoderm. These spaces coalesce and form a space called the amniotic cavity which upon further development will become filled with fluid (Fig. 5, 6). Although this fluid-filled cavity serves as a means of cushioning the developing embryo from mechanical shock, its chief function is to provide a medium in which the embryonic body can be suspended in a freely moving fashion and thus allow for symmetrical development [Mossman, 1937]. During amniotic cavity formation, the endodermal layer becomes involved in the formation of a yolk sac, which is basically nonfunctional in humans because of the paucity of yolk material.

At about the 13th day of gestation, a longitudinal depression (primitive streak) appears in the ectodermal layer of the blastodisc (Fig. 7). The appearance of this streak is the first indication of a craniocaudal axis of the developing embryo. Some of the ectodermal cells migrate in the direction of this primitive streak. Upon nearing the streak they detach themselves from the ectoderm and invaginate between the ectoderm and endoderm. These cells will form the definitive mesodermal layer of the embryo. The embryo is now composed of three layers and is now known as the trilaminar germ disc (Figs. 6, 7). From these three embryonic germ layers all tissues, organs, and organ systems will develop. The formation of these systems is not within the scope of this paper but a listing of what structures develop from which germ layer is apropos (Table I).

TWINNING

A discussion of the development of identical twins is pertinent since it is initiated during the early periods of embryonic development. Identical twins result from the fertilization of one egg by one sperm followed later by a splitting of the embryonic mass. Thus, they are of the same sex. They are morphologically and physiologically identical, they have identical blood characteristics, and they exhibit a tolerance for transplants between themselves. Fraternal twins result from the fertilization of two eggs by two dif-

TABLE I. Major Derivatives of the Three Primary Germ Layers

Ectoderm	Mesoderm	Endoderm
Central nervous system	Muscles	Liver
Peripheral nervous system	Bones	Pancreas
Lens of the eye	Cartilage	Urinary bladder
Adrenal medulla	Blood	Linings of pharynx, esophagus, stomach, intestines
Pituitary gland	Heart and blood vessels	
Epidermis	Spleen	Linings of trachea, bronchi, and lungs
Hair	Adrenal cortex	
Nails	Urogenital system	
Sweat and sebaceous glands		

ferent sperm. Thus, they are two individuals who may differ in many respects, as normal siblings do. Twins occur in about 1% of all pregnancies of which about 30% are of the identical type [Bulmer, 1970].

In most cases, splitting of embryonic material occurs at the early blastocyst stage. The inner cell mass splits evenly and each half of the divided mass develops into a separate individual [Bulmer, 1970]. Identical twinning can also occur slightly later at the bilaminar germ disc stage. A duplication of the primitive streak may occur with the resultant development of one embryo from each primitive streak. In some instances an incomplete duplication or splitting of the primitive streak occurs. This incomplete separation leads to the formation of conjoined or Siamese twins.

An unusual but consistent incidence of this duplication occurs in the nine-banded armadillo. In this species, the blastocyst produces four primitive streaks, probably as a result of a duplication of two originally formed ones, each of which develop into an embryo. Since all four resulting embryos develop from a single zygote, they are identical quadruplets.

CONCLUSION

The early stages of development following fertilization, as discussed here, are critical in that during this period not only is the developing embryo and its mother embarking on the formation of an intimate relationship via the placenta, but also the sources of all adult structures, the three primary germ layers, are being established.

It is essential that an orderly sequence of events occur for development to proceed normally. Little evidence is available regarding the susceptibility of these early developmental events to teratogenic agents. Most investigations of this nature deal with the effects of teratogenic agents during the second and third months of gestation, the time during which differentiation of the organ systems are occurring.

In reality, the chances of embryonic exposure to noxious agents during the first month of development are much greater than at other times. Developmental events such as cleavage, blastocyst formation, and implantation are occurring prior to the future mother even being aware of her pregnancy. Although pregnant women exhibit extreme caution in avoiding exposing themselves to certain diseases, drugs, and chemicals, they exhibit little if any during the first month of pregnancy since they are basically unaware of their condition.

Whether exposure of the embryo to specific toxic agents during the first month following fertilization can, in fact, modify developmental events in such a manner that a congenital abnormality results or whether such exposure can alter development so severely that spontaneous abortion ensues is not clear. To elucidate this problem, much more information is needed regarding the physical changes and metabolic activities of the fertilized zygote as it passes down the oviduct into the uterus and then erodes its way into the maternal endometrium.

ACKNOWLEDGMENTS

This work was supported by grant 1 R01 OH 00835-03 awarded by the National Institute for Occupational Safety and Health. The author would like to express his appreciation to Mr. Donald Alvarado for his artistry and Mrs. Mary Ann Wilde for her typing of the manuscript.

REFERENCES

Beier HM, Maurer RR (1975): Uteroglobin and other proteins in rabbit blastocyst fluid after development in vivo and in vitro. Cell Tissue Res 149:1–10.

Billington WD (1970): Trophoblast. In Phillipp EE, Barnes J, Newton M (eds): "Scientific Foundations of Obstetrics and Gynecology." London: William Heinemann.

Borland RM, Biggers JD, Lechene CP (1977): Studies in the composition and formation of mouse blastocoele fluid using electron probe microanalysis. Dev Biol 55:9–32.

Bulmer MJ (1970): "The Biology of Twinning in Man." Oxford: Clarendon Press.

Carlson BM (1981): "Patten's Foundations of Embryology," Ed. 4. New York: McGraw Hill.

Corliss CE (1976): "Patten's Human Embryology: Elements of Clinical Development." New York: McGraw Hill.

Hamilton WJ, Mossman HW (1972): "Hamilton, Boyd and Mossman's Human Embryology: Prenatal Development of Form and Function." Baltimore: Williams and Wilkins.

Hertig, AT (1968): "Human Trophoblast." Springfield, Ill: Charles C. Thomas.

Lewis WH, Hartmann CG (1933): Early cleavage stages of the egg of the monkey (Macacus rhesus). Contrib Embryol Carnegie Inst 24:187–201.

Luckett WP (1978): Origin and differentiation of the yolk sac and extraembryonic mesoderm in presomite human and rhesus monkey embryos. Am J Anat 152:59–98.

McLaren A (1973): Blastocyst activation. In Segal SJ, Crozier R, Corfman PA, Condliffe PG (eds): "The Regulation of Mammalian Reproduction." Springfield, Ill: Charles C. Thomas, pp 321–328.

Mossman HW (1937): Comparative morphogenesis of the fetal membranes and accessory uterine structures. Carnegie Inst Wash Publ 479, Contrib Embryol 26:129–246.

Tarkowski AJ (1965): From a discussion of a paper by Mulnard JG (1965). Studies of regulation of mouse ova in vitro. In Wolstenholme GEW, O'Connor M (eds): "Preimplantation Stages of Pregnancy (Ciba Foundation Symposium)." London: Churchill, pp 123–144.

Young MP, Whicher JT, Potts DM (1968): The ultrastructure of implantation in the golden hamster (Cricetus auratus). J Embryol Exp Morphol 19:341–345.

APPENDIX I. SUGGESTED READING

Langman J (1981): "Medical Embryology," Ed. 4. Baltimore: Williams and Wilkins.

McLaren A (1972): The embryo. In Austin CR, Short RV (eds): "Embryonic and Fetal Development." Cambridge: University Press, pp 1–42.

Moore KL (1977): "The Developing Human: Clinically Oriented Embryology," Ed. 2. Philadelphia: W.B. Saunders.

Moore KL (1982): "The Developing Human: Clinically Oriented Embryology," Ed. 3. Philadelphia: W.B. Saunders.

O'Rahilly R (1973): Developmental Stages in Human Embryos. Part A: Embryos of the First Three Weeks (Stages 1–9)." Carnegie Inst Wash Publ No. 631.

TOXICOLOGY

American Journal of Industrial Medicine 4:65–79 (1983)

The Mechanisms of Action of Reproductive Toxins

Donald R. Mattison, MD

A general discussion of the mechanisms of action of reproductive toxins is presented. Reproductive toxicants may produce their adverse effects by one of several mechanisms. Some xenobiotics may act directly, either by virtue of chemical reactivity (eg, oocyte destruction by alkylating agents) or by structural similarity to endogenous molecules (eg, hormone antagonists). Other xenobiotics interrupt reproductive processes indirectly either by metabolic processing to a direct-acting toxicant (eg, prohormone or metabolic activation to reactive intermediate) or by altering endocrine homeostasis (eg, increased steroid clearance). Selected examples of compounds demonstrating reproductive toxicity by one or more of these mechanisms are tabulated and an extensive bibliography of reproductive toxins is attached.

Key words: reproductive toxins, metabolic activation, chemical reactivity, hormone agonist, hormone antagonist, endocrine homeostasis

INTRODUCTION

Reproduction is a complex stepwise process that begins with gametogenesis, continues through gamete interaction, implantation, embryonic development, growth, parturition, and postnatal adaptation and is completed with the development and sexual maturation of the newly formed individual. Unfortunately these reproductive processes do not take place in a chemically pristine environment, but rather in an environment increasingly contaminated with the products and by-products of the chemical age in which we live. Certain environmental pollutants are known to be carcinogenic, mutagenic, or toxic to the reproductive system. The vast majority of xenobiotics, however, have not been adequately tested for reproductive toxicity.

Just as reproduction is complex, biological mechanisms underlying toxicology are similarly complex and involve absorption, distribution, metabolism (toxification and/or detoxification), excretion, and repair. The synthesis of these maturing sciences into the nascent science of reproductive toxicology may represent the most difficult and critical challenge facing humankind over the next decade. Before presenting an overview of the mechanisms through which reproductive toxins can act, I will identify several salient issues that illustrate the complexity of reproductive toxicology.

Pregnancy Research Branch, National Institute of Child Health and Human Development, National Institutes of Health, Bethesda, Maryland.
Address reprint requests to Dr. Donald R. Mattison, Bldg. 10, Rm. 8C313, PRB, NICHD, NIH, Bethesda, MD 20205.

0271-3586/83/0401-02-0065$04.50 **© 1983 Alan R. Liss, Inc.**

The issue of species differences in reproductive toxicology is complex and is dependent on mechanisms of hormonal control, anatomy, pharmacokinetics, metabolism, etc. In some instances these species differences are poorly understood. A reproductive toxin in one species may not be toxic in another (including humans) because of differences in reproductive or toxicological mechanisms. The teratogenicity of thalidomide is an instructive example of species susceptability, in that rat and mouse are insensitive, while rabbit, human, and non-human primates are sensitive [Shepard, 1976]. Another example is the difference in sensitivity to oocyte destruction by polycyclic hydrocarbons exhibited by rats and mice [Mattison et al, 1979].

Another pertinent issue in toxicology concerns possible gender differences [Doull, 1975]. This is also crucial in reproductive toxicology because of the differences in anatomy and biological control mechanisms for reproduction in the male and female. Because of the ease of accessibility of gametes and gonads in the male, more xenobiotic compounds have been tested and demonstrated toxic to male reproductive processes than female reproductive processes. At the present time it is not known if this represents an actual gender difference in gonadal or gamete toxicity, or is simply an artifact of experimental designs. For example, more parameters are accessible for scoring sperm (eg: shape, motility, penetrability) than oocytes, which are frequently difficult to obtain.

Another critical issue in reproductive toxicity is the sensitive window, or time frame during which toxicity occurs. This has been demonstrated on many occasions in studies of teratogens and toxins acting on spermatogenesis. Several studies have demonstrated that compounds like galactose or azathioprine are toxic to the developing ovary, but have no effect on the developed ovary [Mattison, 1981; Mattison and Nightingale, 1982]. Similar studies, however, have seldom explored the issue of differential follicular or oocyte sensitivity. A small number of experimental and clinical studies have attempted to explore the effects of age on gonadal sensitivity to chemotherapeutic agents or other xenobiotic compounds [Mattison, 1981;1982].

As reproduction is essential for continued existence of any species, it is clearly necessary to develop an understanding of reproductive toxins and mechanisms of action, site of action, as well as species (especially human) susceptibility.

General aspects of testicular [Lucier et al, 1977; Gomes, 1970] and ovarian toxicity [Mattison, 1980a; Mattison, 1981; Mattison and Nightingale, 1982] have been addressed, and a recent conference was devoted to gonadal toxicity [Dixon, 1978]. General aspects of reproductive toxicology have been explored in several recent reviews [Dixon, 1980; Mattison, 1981], and gonadal metabolism of xenobiotics has also been reviewed [Heinrichs and Juchau, 1980]. An ever-expanding literature is devoted to teratology [Wilson, 1973, 1974; Schardein, 1976; Shepard, 1976; Heinonen et al, 1977]. Reproductive effects of occupational exposures have been considered [Hunt, 1975; American College of Obstetricians and Gynecologists, 1976; Bingham, 1977; Messite and Bond, 1980]. Several reviews have addressed issues of comparative reproductive biology [Greep and Koblinsky, 1977; van Tienhoven, 1968; Benirschke, 1967; Diczfalusy and Standley, 1972].

MECHANISMS OF REPRODUCTIVE TOXICITY

The mechanisms of toxicity can be reduced ultimately to some effect which interrupts the normal functioning of a cell, organ, or organism [Casarett and Doull, 1975].

Toxicants act to interrupt the flow of matter, energy, or information that are necessary for normal functioning of cells, organs, or organisms (Fig. 1). Following exposure to a reproductive toxin, the compound must be distributed to the target organ (gonad, hypothalamus, pituitary, uterus, epididymus, liver, etc) where it exerts its adverse effect. If the compound is metabolized and cleared, no adverse effect will occur. Within the target organ the toxin will interact with a critical cell or subcellular component disrupting an event necessary for normal reproductive function. If this interaction is not repaired, the toxic effect, altered reproduction, will be produced. This toxic effect may be very specific, affecting only a single function of a single cell type or broad and nonspecific, with multiple sites of toxicity within the organism. Within each target, however, this multistep process must be completed before reproductive toxicity occurs.

Reproductive toxins may act directly (Fig. 2), by either virtue of structural similarity to an endogenous compound (ie, hormone, nutrient) or because of chemical reactivity (ie, alkylating agent, denaturant, chelator).

Some reproductive toxins may act indirectly, requiring metabolic processing within the organism or organ before exerting a toxic effect. The metabolite formed may then exert its toxic effect through one of the direct mechanisms of reproductive toxicity described (structural similarity or chemical reactivity). Other indirect-acting reproductive toxins may exert their effects by producing alterations in physiological control mechanisms of the organism (ie, enzyme induction or inhibition).

It is also possible that reproductive toxins may exert adverse effects through more than one mechanism. For example, the halogenated polycyclic hydrocarbons (ie, polychlorinated or polybrominated biphenyls) may act indirectly by induction of microsomal monooxygenases or transferases. These compounds may also act directly by virtue of steroid hormone agonist properties.

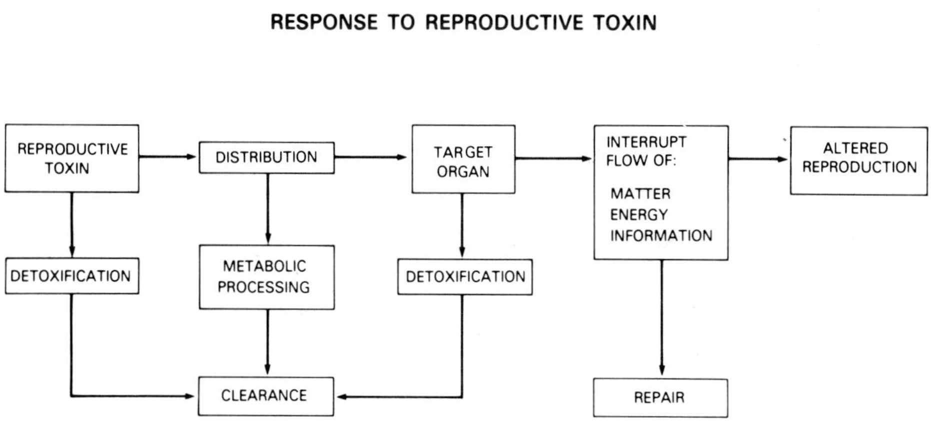

Fig. 1. The response of an organism to a reproductive toxin is complex and involves distribution, detoxification, clearance, and possibly repair before the adverse effect is produced. If the multistep defense network of the organism is unable to block the toxic interruption of the flow of matter, energy, or information necessary for normal reproduction, impaired function will result.

Examples of the reproductive toxins discussed below are presented in Table I, and Figure 2 provides a schematic classification of these mechanisms.

DIRECT-ACTING TOXINS

Structural Similarity

One mechanism of action of direct-acting reproductive toxins results from structural similarity to a biologically important molecule. These compounds may best be considered wolves in sheeps' clothing, in that they mislead an organism. The compounds considered in this category are generally agonists or antagonists of endogenous hormones. Well-known examples of this type of direct-acting "reproductive toxin" are the oral contraceptives, which act predominantly by suppression of gonadotropin secretion. Occupational exposure to oral contraceptives, for example, during formulation, represents an obvious example of a reproductive and endocrine toxin [Harrington et al, 1978]. Occupational or environmental exposure to xenobiotics that are distributed to the hypothalamus or pituitary and act like estrogen, progesterone, or testosterone will similarly inhibit gonadotropin secretion and gonadal function.

Chemical Reactivity

Compounds that are chemically reactive may be nonspecific in their site of action. Because of chemical reactivity, exposure to these compounds generally results from occupational exposure or drug treatment. Examples considered here include alkylating agents, which are found in the chemical industry and used in the treatment of many neoplastic and some nonneoplastic diseases. Because of their reactivity, most of these compounds are toxic, carcinogenic, or mutagenic in other organ systems, and their reproductive toxicity has often been overlooked. This has been the case, until recently, with alkylating agents used in the treatment of human diseases. The reproductive system, however, may be more sensitive to the toxic effects of these compounds than other organ systems. For example, the incidence of sterility following cancer chemotherapy is considerably higher than the incidence of second tumors. Other examples of direct-acting reproductive toxins include boron and metals such as cadmium, lead, and mercury, which have been demonstrated toxic to the developing as well as mature reproductive system [Mattison, 1983].

INDIRECT-ACTING TOXINS

Metabolic Activation

In addition to direct interference with reproductive processes, certain toxins may act indirectly to impair production. Examples of this kind of toxin are those that are metabolized to form a toxic metabolite. The metabolites formed may be chemically reactive or mimic an endogenous molecule.

One of the mechanisms utilized to remove hydrophobic xenobiotics from the body is oxidation by microsomal monooxygenases [Fleischer and Packer, 1978], which increases the polarity of the molecule. The polar metabolite is then conjugated or excreted directly. Unfortunately, some of the metabolites formed in this process are chemically reactive. These reactive metabolites formed in vivo may then interact with

TABLE I. Mechanisms of Action of Reproductive Toxins

Mechanism	Compound	Species	Gender	Selected references
Direct-acting reproductive toxins				
Structural similarity				
	Steroid hormones	Mammalian	M/F	Harrington et al [1978]
	Cimetidine	Primate	M	Winters et al [1978] Van Thiel et al [1979a] Wolf [1979] Eversmann et al [1979] Hall [1976]
	Diethylstilbesterol	Mammalian	M/F	McLachlan et al [1975] Forsberg [1969, 1970, 1972] Vorherr et al [1979] Boylan [1978] Bibbo et al [1975, 1977] Kaufman et al [1977] Haney et al [1979] Herbst et al [1971, 1980] Schmidt et al [1980] Barnes et al [1980] Korach et al [1979]
	Azathioprine 6-mercaptopurine	Mammalian	M/F	Reimers and Sluss [1978] Reimers et al [1980] Githens et al [1965] Gross et al [1977]
	Halogenated polycyclic hydrocarbons	Mammalian	M/F	Kimbrough [1974] Kupfer [1975] Orberg and Kihlstrom [1973] Linder et al [1974] Welch et al [1971] Derr and Decker [1979] Nowicki and Norman [1972] Fahim et al [1968, 1970] Poland et al [1970] Kolmodin et al [1969] Bitman and Cecil [1970]
Chemical reactivity				
	Alkylating agents	Mammalian	M/F	Hemsworth and Jackson [1962, 1963] Jackson et al [1969] Heller et al [1964] Chapmann et al [1979a,b] Sieber and Adamson [1975] Himmelstein-Braw et al [1977, 1978] Hsu et al [1979] Parra et al [1978] Ettendorf et al [1976]

TABLE I. Continued

Mechanism	Compound	Species	Gender	Selected References
				Oishi and Hiraga [1979] Fairley et al [1972] Freeman-Narrod and Narrod [1977]
	Cadmium	Mammalian	M/F	Patanelli [1975] Gunn and Gould [1970] Johnson [1977] Dixon et al [1976] Parizek [1960, 1964] Parizek et al [1968]
	Boron	Primate	M	Krasovski et al [1976] Lee et al [1978] Weir and Fisher [1972] Truhant et al [1964] Dixon et al [1976]
	Lead	Mammalian Primate	M/F	Wide [1980] Jacquet et al [1977]
	Mercury	Mammalian	M/F	Lamperti and Niewenhuis [1976]
Indirect-acting reproductive toxins				
Metabolic activation				
	Diethylstilbesterol	Mammalian	F	Blackburn et al [1976] Rudiger et al [1979] Metzler [1976] Metzler and McLachlan [1978]
	Ethanol	Mammalian	F/M	Van Thiel et al [1979b] Anderson et al [1980] Hinckers [1978] Ouellette et al [1977] Jones and Smith [1975]
	Chlorcyclizine	Mammalian	M	Wong and Hruban [1972]
	Dibromachloropropane	Mammalian	M	Torkelson et al [1961] Whorton et al [1977] Marshall et al [1978] Moody et al [1980] Teramoto et al [1980]
	Polycyclic aromatic hydrocarbons	Mammalian	M/F	Mattison [1980] Mattison and Thorgeirsson [1978, 1979] Wyrobeck and Bruce [1975] Nebert [1981] Felton et al [1978] Basler and Rohrborn [1976]
	Cyclophosphamide	Mammalian	M/F	Koyama et al [1977] Sirus et al [1976] Sobrinho et al [1971] Warne et al [1973] Kumar et al [1972] Rohrborn and Hansmann [1971] Uldall et al [1972] Lentz et al [1977]

Mechanism	Compound	Species	Gender	Selected References
Disrupted homeostasis				
	Salicylazosulfa-pyridine	Mammalian	M	Levi et al [1979] Traub et al [1979] Toth [1979a,b]
	Halogenated polycyclic hydrocarbons	Avian Mammalian	M/F	Kimbrough [1974] Kupfer [1975] Heinrichs et al [1971] Gellert et al [1972, 1974, 1978] Allen et al [1979] Iatropoulos et al [1976] Lucien and McDaniel [1979] Thorpe [1967, 1969] Welch et al [1971] Gellert [1978] Gellert and Heinrich [1975]
	Anticonvulsants	Primate	F/M	Aronson and Grahame-Smith [1981]

cellular macromolecules, just as exogenously administered chemically reactive compounds do. This mechanism results in the reproductive toxicity of cyclophosphamide, dibromochloropropane, and the polycyclic aromatic hydrocarbons.

The ovary and testis have been demonstrated to have microsomal monooxygenases, epoxide hydrases [Sims and Grover, 1974], and transferases responsible for metabolic processing of many xenobiotic compounds [Mattison and Thorgeirsson, 1978, 1979; Dixon and Lee, 1980; Heinrichs and Juchau, 1980].

Enzyme Modification

Other indirect-acting reproductive toxins may induce or inhibit enzyme systems important in reproduction, enhancing or suppressing steroid secretion or clearance. Because reproduction requires hormonal feedback loops for successful control, xenobiotics that alter the rate of steroid synthesis or clearance may alter reproductive processes. This has been demonstrated in rodents treated with several of the halogenated polycyclic hydrocarbon pesticides, including DDT, polychlorinated biphenyls, and polybrominated biphenyls.

Another form of reproductive toxins may enhance rather than impair fertility. One commonly used method for fertility regulation, oral contraceptives, relies on feedback inhibition of gonadotropin secretion. Xenobiotics that stimulate clearance of the estrogenic and/or progestogenic component of the oral contraceptive will decrease circulating levels and increase the probability of ovulation. This effect may be especially crucial in an occupational setting where other xenobiotics are teratogenic or fetotoxic.

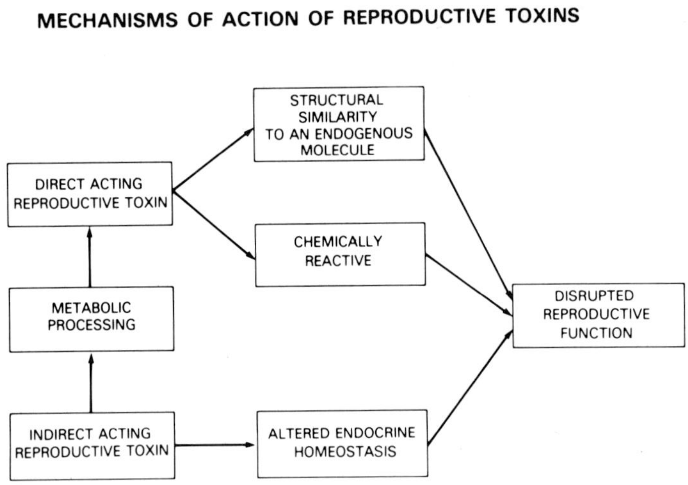

Fig. 2. Reproductive toxins, irrespective of their site(s) of action produce their adverse effect(s) by one or more of these mechanisms.

DETOXIFICATION

Biological organisms have several mechanisms available for responding to toxic stresses. These mechanisms are classified as detoxification mechanisms and exhibit a relatively broad specificity, as expected from the wide range of xenobiotics processed by these pathways [Aitio, 1978; Oesch, 1972]. Impaired detoxification either because of enzyme deficiences in the exposed individual [Speilberg and Gordon, 1981] or because of the nature of the substrate [Moore et al, 1980] will increase the toxicity observed.

Detoxification mechanisms represent an immediate cellular response to a toxic exposure and may involve conjugation with sulfate, glucuronate, or glutathione, or hydrolysis by epoxide hydratase, all acting to decrease the concentration of chemically reactive toxins within the organism, organ, or cell. Other mechanisms of detoxification may involve metabolism of the toxin to a less toxic or more easily excreted (ie, more polar) compound.

The mechanisms used by the cell for detoxification depend in part on the nature of the toxic molecule. Compounds that are chemically reactive, such as alkylating agents, or epoxides are detoxified by epoxide hydratase and/or conjugative systems, whereas nonpolar compounds are first metabolized by monooxygenases to more polar products before conjugation and excretion. Some of the detoxification pathways have been demonstrated in the ovary and testis [Mukhtar et al, 1978; Lee and Dixon, 1978; Heinrichs and Juchau, 1980].

REPAIR

Although the cell has mechanisms for detoxification, in some instances those mechanisms provide inadequate protection resulting in cell or organ damage. When a cell or organ has been damaged by a toxin, it must attempt to repair that damage.

The mechanisms of repair may be as simple as renewed or increased protein synthesis to replace nonfunctioning proteins destroyed by the toxin. More sophisticated (biologically) repair mechanisms appear to have evolved for DNA damage [Lehmann and Bridges, 1977]. These mechanisms involve screening for damage in DNA, as well as several enzymes for excision of the damaged DNA region and subsequent replacement of the region excised.

Recent investigations have suggested that the mature oocyte can repair damage in oocyte DNA before and after fertilization [Pedersen and Manigia, 1978], and sperm DNA after fertilization [Generoso et al, 1979]. However, it is not known if other types of repair mechanisms exist in the oocyte. Studies by Dixon and Lee [1980] have also demonstrated DNA repair capability in the developing sperm.

SUMMARY

Successful reproduction is of crucial importance for all animals and depends upon the genetic, anatomic, and endocrine integrity of the reproductive tract and gametes. I have briefly explored some mechanisms of reproductive toxicity and tabulated selected reproductive toxins in this discussion, but have by no means discussed all known or suspect reproductive toxins. The xenobiotic compounds chosen illustrate the multiple modes and sites of action available to reproductive toxins. Because of the cryptic and periodic nature of human reproduction, other drugs or environmental xenobiotic compounds may be presently impairing human or domestic animal reproduction, but remain undetected.

REFERENCES

Aitio A (1978): "Conjugation Reactions in Drug Biotransformation." New York: Elsevier.

Allen JR, Barsotti DA, Lambrecht LK, Van Miller JP (1979): Reproductive effects of halogenated aromatic hydrocarbons on nonhuman primates. Ann NY Acad Sci 320:419–425.

American College of Obstetricians and Gynecologists (1976): "Guidelines on Pregnancy and Work." Chicago: American College of Obstetrics and Gynecology.

Anderson RA, Willis BR, Oswald C, Reddy JM, Beyler SA, Zaneveld, LJD (1980): Hormonal imbalance and alterations in testicular morphology induced by chronic ingestion of ethanol. Biochem Pharmacol 29:1409–1419.

Aronson JK, Grahame-Smith DG (1981): Clinical pharmacology: Adverse drug interactions. Br Med J 282:288–291.

Barnes AB, Colton T, Gundersen J, Noller KL, Tilley BC, Strama T, Townsend DE, Hatab P, O'Brain PC (1980): Fertility and outcome of pregnancy in women exposed in utero to diethylstilbesterol. N Engl J Med 302:609–613.

Basler A, Rohrborn G (1976): Chromosome aberrations in oocytes of NMRI mice and bone marrow cells of chinese hamsters induced with 3,4-benzpyrene. Mutat Res 38:327–332.

Benirschke KT (1967): "Comparative Aspects of Reproductive Failure." New York: Springer-Verlag.

Bibbo M, Al-Nageeb M, Baccarini I, Gill W, Newton M, Sleeper K, Sonek M, Weid GL (1975): Follow-up study of male and female offspring of DES-treated mothers. A preliminary report. J Reprod Med 15:29–32.

Bibbo M, Gill WB, Azizi F, Blaigh R, Fang VS, Rosenfield RL, Schumacher FG, Sleeper K, Sonek M, Weid GL (1977): Follow-up study of male and female offspring of DES-exposed mothers. Obstet Gynecol 49:1–8.

Bingham E (1977): Proceeding. Washington, DC: Conference on Women and The Workplace, Society for Occupational and Environmental Health.

Bitman J, Cecil HC (1970): Estrogenic activity of DDT analogs and polychlorinated biphenyls. J Agric Food Chem 18:1108–1112.

Blackburn GM, Thompson MH, King HWS (1976): Binding of diethylstilbesterol to deoxyribonucleic acid by rat liver microsomal fractions in vitro and in mouse foetal cells in culture. Biochem J 158:643–646.

Boylan ES (1978): Morphological and functional consequences of prenatal exposure to diethylstilbesterol in the rat. Biol Reprod 19:854–863.

Casarett LJ, Doull J (eds)(1975): "Toxicology. The Basic Science of Poisons." New York: Macmillan.

Chapman RM, Sutcliffe SB, Malpas JS (1979a): Cytotoxic-induced ovarian failure in women with Hodgkin's disease. I. Hormone function. JAMA 242:1877–1881.

Chapman RM, Sutcliffe SB, Malpas JS (1979b): Cytotoxic-induced ovarian failure in women with Hodgkins disease. II. Effects on sexual function. JAMA 242:1882–1884.

Derr SK, Decker J (1979): Alterations of androgenicity in rats exposed to PCB's (Aroclor. 1254). Bull Environ Contam Toxicol 21:43–45.

Diczfalusy E, Standley CC (1972): "The Use of Non-Human Primates in Research in Human Reproduction." Stockholm: WHO, Karolinska Institutet.

Dieckmann WJ, Davis ME, Rynkiewicz SM, Pottinger RE (1953): Does the administration of diethylstilbesterol during pregnancy have therapeutic value. Am J Obstet Gynecol 66:1062–1081.

Dixon RL (1978): Symposium on Target Organ Toxicity: Gonads (Reproduction and Genetic Toxicity). Environ Health Perspect 24:1–127.

Dixon RL (1980): Toxic responses of the reproductive system. In Doull J, Klassen CD, Amdur MO (eds): "Toxicology. The Basic Science of Poisons." New York: Macmillan, pp 332–354.

Dixon RL, Lee IP (1980): Pharmacokinetic and adaptation factors involved in testicular toxicity. Fed Proc 39:66–72.

Dixon RL, Lee IP, Sherins RJ (1976): Methods to assess reproductive effects of environmental chemicals: Studies of Cadmium and Boron administered orally. Environ Health Perspect 13:59–67.

Doull J (1975): Factors influencing toxicology. In Caserett LJ, Doull J (eds): "Toxicology: The Basic Science of Poisons." New York: Macmillan.

Ettendorf JN, West CD, Pitcock JA, Williams DL (1976): Gonadal function, testicular histology, and meiosis following cyclophosphamide therapy in patients with nephrotic syndrome. J Pediatr 88:206–212.

Eversmann T, Landfraf R, Londong W, Von Werder K (1979): Effect of cimetidine on prolactin-secretion and glucose tolerance in men. Horm Metab Res 11:412–413.

Fahim MS, Dement DG, Hall DG, Fahim Z (1970): Induced alterations in hepatic metabolism of androgens in the rat. Am J Obstet Gynecol 107:1085–1091.

Fahim MS, King TM, Hall DG (1968): Induced alterations in the biologic activity of estrogen. Am J Obstet Gynecol 100:171–175.

Fairley KF, Barrie JU, Johnson W (1972): Sterility and testicular atrophy related to cyclophosphamide therapy. Lancet 1:568–569.

Felton JS, Kwan TC, Wuebbles BJ, Dobson RL (1978): Genetic differences in polycyclic aromatic hydrocarbon metabolism and their effects on oocyte killing in developing mice. In Mahlum DD, Sikov MR, Hackett PL, Andrew FD (eds): "Developmental Toxicology of Energy Related Pollutants." D.O.E. Symp Ser 47:1526.

Fleischer S, Packer L (eds)(1978): Biomembranes, Part C: Biological oxidations, microsomal, cytochrome P-450, and other hemoprotein systems. In: "Methods in Enzymology," vol 52. New York, Academic Press.

Forsberg JG (1969): The development of atypical epithelium in the mouse uterine cervix and vaginal fornix after neonatal oestradiol treatment. Br J Exp Pathol 50:187–195.

Forsberg JG (1970): An estradiol mitotic rate inhibiting effect in the mullerian epithelium in neonatal mice. J Exp Zool 175:369–374.

Forsberg JG (1972): Estrogen, vaginal cancer, and vaginal development. Am J Obstet Gynecol 113:83–87.

Freeman-Narrod M, Narrod SA (1977): Chronic toxicity of methotrexate in mice. J Natl Cancer Inst 58:735–739.

Gellert RJ (1978): Uterotrophic activity of polychlorinated biphenyls (PCB) and induction of precocious reproductive aging in neonatally treated female rats. Environ Res 16:123–130.

Gellert RJ, Heinrichs WL (1975): Effects of DDT homologs administered to female rats during the perinatal period. Biol Neonate 26:283–290.

Gellert RJ, Heinrichs WL, Swerdloff RS (1972): DDT homologues: Estrogenlike effects on the vagina, uterus and pituitary of the rat. Endocrinology 91:1095–1100.

Gellert RJ, Heinrichs WL, Swerdloff R (1974): Effects of neonatally administered DDT homologs on reproductive function in male and female rats. Neuroendocrinology 16:84–94.

Gellert RJ, Wallace CA, Weismeier EM, Shuman RM (1978): Topical exposure of neonates to hexachlorophene: Long-standing effects on mating behavior and prostatic development in rats. Toxicol Appl Pharmacol 43:339–349.

Generoso WM, Cain KT, Krishna M, Huff S (1979): Genetic lesions induced by chemicals in spermatozoa and spermatids of mice are repaired in the egg. Proc Natl Acad Sci USA 76:435–437.

Githens JH, Rosenkrantz JG, Tunnock SM (1965): Teratogenic effects of azathioprine (Imuran). J Pediatr 66:959–961.

Gomes WR (1970): Chemical agents affecting testicular function and male fertility. In Johnson AD, Gomes WR, Vandemark NL (eds): "The Testis." New York: Academic Press, pp 483–554.

Gross A, Fein A, Serr DM, Nebel L (1977): The effect of Imuran on implantation and early embryonic development in rats. Obstet Gynecol 50:713–718.

Greep RO, Koblinsky MA (1977): Frontiers in reproduction and fertility control. "A Review of the Reproductive Sciences and Contraceptive Development. Part 2." Cambridge, Massachusetts: MIT Press.

Gunn SA, Gould TC (1970): Cadmium and other mineral elements. In Johnson AD, Gomes WR, Vandemark NL (eds): "The Testis," Vol. III. New York: Academic Press, pp 377–481.

Hall WH (1976): Breast changes in males on cimetidine. N Engl J Med 295:841.

Haney AF, Hammond CB, Soules MR, Creasman WT (1979): Diethylstilbesterol induced upper genital tract abnormalities. Fertil Steril 31:142–146.

Harrington JM, Stein GF, Rivera RO, deMorales AV (1978): Occupational hazards of formulating oral contraceptives—A survey of plant employees. Arch Environ Health 33:12–15.

Heinonen OP, Slone D, Shapiro S (1977): "Birth Defects and Drugs in Pregnancy." Littelton, Massachusetts: Publishing Sciences Group.

Heinrichs WL, Juchau MR (1980): Extrahepatic drug metabolism: The gonads. In Gram TE (ed): "Extrahepatic Metabolism of Drugs and Other Foreign Compounds." New York: SP Medical and Scientific Books, pp 313–332.

Heinrichs WL, Gellert RJ, Bakke JL, Lawrence NL (1971): DDT administered to neonatal rats induces persistent estrus syndrome. Science 173:642–643.

Heller RH, Jones HW, Blanchard M (1964): Production of ovarian dysgenesis in the rat and human by busulphan. Am J Obstet Gynecol 89:414–420.

Hemsworth BN, Jackson H (1962): Effect of busulphan on the developing gonad of the male rat. J Reprod Fertil 5:187–194.

Hemsworth BN, Jackson H (1963): Effect of busulphan on the developing ovary in the rat. J Reprod Fertil 6:229–233.

Herbst AL, Hubby MM, Blough RR, Azizi F (1980): A comparison of pregnancy experience in DES-exposed and DES-unexposed daughters. J Reprod Med 24:62–69.

Herbst AL, Ulfelder H, Poskanser DC (1971): Adenocarcinoma of the vagina. Association of maternal stilbesterol therapy with tumor appearance in young women. N Engl J Med 284:878–881.

Himmelstein-Braw R, Peters H, Faber M (1977): Influence of irradiation and chemotherapy on the ovaries of children with abdominal tumors. Br J Cancer 36:269–275.

Himmelstein-Braw R, Peters H, Faber M (1978): Morphological appearance of the ovaries of leukemic children. Br J Cancer 38:82–87.

Hinckers HJ (1978): The influence of alcohol on the fetus. J Perinat Med 6:3–14.

Hsu AC, Folami AO, Bain J, Rance CP (1979): Gonadal function in males treated with cyclophosphamide for nephrotic syndrome. Fertil Steril 31:173–177.

Hunt VR (1975): "Occupational Health Problems of Pregnant Women." DHEW No. 5A-5304-75.

Iatropoulos MJ, Hobson W, Knauf V, Adams HP (1976): Morphological effects of hexachlorobenzene toxicity in female rhesus monkeys. Toxicol Appl Pharmacol 37:433–444.

Jackson H, Fox BW, Craig AW (1959): The effect of alkylating agents on male rat fertility. Br J Pharmacol 14:149–157.

Jacquet P, Gerber GB, Leonard A, Maes J (1977): Plasma hormone levels in normal and lead-treated preg-

nant mice. Experiencia 33:1375–1377.
Johnson AD (1977): The influence of cadmium on the testis. In Johnson AD, Gomes WR (eds): "The Testis," Vol IV. New York: Academic Press, pp 565–576.
Jones KL, Smith DW (1975): The fetal alcohol syndrome. Teratology 12:1–10.
Kaufman RH, Binder GL, Gray PM, Adam E (1977): Upper genital tract changes associated with exposure in utero to diethylstilbesterol. Am J Obstet Gynecol 128:51–59.
Kimbrough RD (1974): The toxicity of polychlorinated polycyclic compounds and related chemicals. Crit Rev Toxicol 2:445–498.
Kolmodin B, Azarnoff DL, Sjoqvist F (1969): Effect of environmental factors on drug metabolism: Decreased plasma half-life of antipyrine in workers exposed to chlorinated hydrocarbon insecticides. Clin Pharmacol Ther 10:638–642.
Korach KS, Metzler M, McLachlan JA (1979): Diethylstilbesterol metabolites and analogs. New probes for the study of hormone action. J Biol Chem 254:8963–8968.
Koyama H, Wada T, Nishizawa Y, Iwanaga T, Aoki Y, Terasawa T, Kosaki G, Yamamoto T, Wasa A (1977): Cyclophosphamide induced ovarian failure and its therapeutic significance in patients with breast cancer. Cancer 39:1403–1409.
Krasovski GN, Varshavskaya SP, Borisova AF (1976): Toxic and gonadotropic effects of cadmium and boron relative to standards for these substances in drinking water. Environ Health Perspect 13:69–75.
Kumar R, Biggart JD, McEvoy J, McGeown MG (1972): Cyclophosphamide and reproductive function. Lancet 1:1212–1214.
Kupfer D (1975): Effects of pesticides and related compounds on steroid metabolism and function. Crit Rev Toxicol 4:83–124.
Lamperti AA, Niewenhuis R (1976): The effects of mercury on the structure and function of the hypothalamo-pituitary axis in the hamster. Cell Tissue Res 170:315–324.
Lee IP, Dixon RL (1978): Factors influencing reproductive and genetic toxic effects on male gonads. Environ Health Perspect 24:117–127.
Lee IP, Sherins RJ, Dixon RL (1978): Evidence for induction of germinal aplasia in male rats by environmental exposure to boron. Toxicol Appl Pharmacol 45:577–590.
Lehmann AR, Bridges BA (1977): DNA repair, Essays Biochem 13:71–119.
Lentz RD, Bergstein J, Steftes MW, Brown DR, Prem K, Michael AF, Vernier RL (1977): Postpubertal evaluation of gonadal function following cyclophosphamide therapy before and during puberty. J Pediatr 91:385–394.
Levi AJ, Fisher AM, Hughes L, Hendry WF (1979): Male infertility due to sulphasalazine. Lancet 2: 276–278.
Linder RE, Gaines TB, Kimbrough RD (1974): The effect of polychlorinated biphenyls on rat reproduction. Food Cosmet Toxicol 12:63–77.
Lucier BW, McDaniel OS (1979): Developmental toxicology of the halogenated aromatics: Effects on enzyme development. Ann NY Acad Sci 320:449–457.
Lucier GW, Lee IP, Dixon RL (1977): Effects of environmental agents on male reproduction. In Johnson AD, Gomes WR (eds): "The Testis," Vol. III. New York: Academic Press, pp 577–628.
Marshall S, Whorton D, Krauss RM, Palmer WS (1978): Effect of pesticides on testicular function. Urology 11:257–259.
Mattison DR (1979): Difference in sensitivity of rat and mouse primordial oocyte to destruction by polycyclic aromatic hydrocarbons. Chem Biol Interact 28:133–137.
Mattison DR (1980a): How xenobiotic compounds can destroy oocytes. Contemp Obstet Gynecol 15: 157–169.
Mattison DR (1980b): Morphology of oocyte and follicle destruction by polycyclic aromatic hydrocarbons in mice. Toxicol Appl Pharmacol 53:249–259.
Mattison DR (1981): Drugs, xenobiotics and the adolescent: Implications for reproduction. In Soyka LF, Redmond GP (eds): "Drug Metabolism in the Immature Human." New York: Raven Press, pp 129–143.
Mattison DR (1981): Effects of biologically foreign compounds on reproduction. In Abdul-Karim RW (ed): "Drugs During Pregnancy: Clinical Perspectives." Philadelphia: G.F. Stickley, pp 129–143.
Mattison DR (1983): Ovarian toxicity: Effects on sexual maturation, reproduction and menopause. In Clarkson WT, Nordbert G (eds): *Reproductive and Developmental Toxicity of Metals.* Plenum

Press, New York, in press.
Mattison DR, Nightingale MS (1982): Prepubertal ovarian toxicity. In Hunt VR, Smith MK, Worth D (eds): "Banbury Report 11; Environmental Factors in Human Growth and Development." Cold Spring Harbor Laboratory, pp 395–409.
Mattison DR, Thorgeirsson SS (1978): Gonadal aryl hydrocarbon hydroxylase in rats and mice. Cancer Res 38:1368–1373.
Mattison DR, Thorgeirsson SS (1979): Ovarian aryl hydrocarbon hydroxylase activity and primordial oocyte toxicity of polycyclic aromatic hydrocarbon in mice. Cancer Res 39:3471–3475.
Mattison DR, West DM, Menard RA (1979): Differences in benzo(a)pyrene metabolic profile in rat and mouse ovary. Biochem Pharmacol 28:2101–2104.
McLachlan JA, Newbold RR, Bullock B (1975): Reproductive tract lesions in male mice exposed prenatally to diethylstilbesterol. Science 190:991–992.
Messite J, Bond MB (1980): Reproductive toxicology and occupational exposure. In Zenz C (ed): "Developments in Occupational Medicine." Chicago: Year Book Medical Publishers, pp 59–129.
Metzler M (1976): Metabolic activation of carcinogenic diethylstilbesterol in rodents and humans. J Toxicol Environ Health [Suppl] 1:21–35.
Metzler M, McLachlan JA (1978): Peroxidase-mediated oxidation, a possible pathway for metabolic activation of diethylstilbestrol. Biochem Biophys Res Commun 85:874–884.
Moody DE, Head B, Smuckler EA (1980): Reduction in hepatic microsomal cytochromes P-450 and b5 in rats exposed to 1,2-dibromo-3-chloropropane and carbon tetrachloride: Enhancement of effect by pretreatment with phenobarbital. J Environ Pathol Toxicol 3:177–190.
Moore CJ, McCulsky GA, Fischer SM, Macleod MC, Slaga TJ, Selkirk JK (1980): Metabolism of carcinogenic 2-hydroxybenzo(a)pyrene in rodent and human cells. Carcinogenesis 1:979–987.
Mukhtar H, Philpot RM, Bend JR (1978): The postnatal development of microsomal epoxide hydrose, cytosolic glutathione-S-transferase, and mitochondrial and microsomal cytochrome P-450 in adrenals and ovaries in female rats. Drug Metab Dispos 6:577–583.
Nebert DW (1981): Birth defects and the potential role of genetic differences in drug metabolism. In Bloom AD, James LS (eds): "Birth Defects Original Article Series," Vol XVII. New York: Alan R. Liss, pp 51–70.
Nowicki HG, Norman AW (1972): Enhanced hepatic metabolism of testosterone 4-androstene-3,17-dione, and estradiol-17β in chickens pretreated with DDT or PCB. Steroids 19:85–97.
Oesch F (1972): Mammalian epoxide hydrase: Inducible enzymes catalyzing the inactivation of carcinogenic and cyctotoxic metabolic derived from aromatic and olefinic compounds. Xenobiotics 3:305–340.
Oishi S, Hiraga K (1979): Effect of phthalic acid esters on gonadal function in male rats. Bull Environ Contam Toxicol 21:65–67.
Orberg J, Kihlstrom JE (1973): Effects of long term feeding of polychlorinated biphenyls (PCB, Clophen A 60) on the length of the oestrous cycle and on the frequency of implanted ova in the mouse. Environ Res 6:176–179.
Ouellette EM, Rosett HL, Rosman NP, Weiner L (1977): Adverse effects on offspring of maternal alcohol abuse during pregnancy. N Engl J Med 297:528–530.
Parizek J (1960): Sterilization of the male by cadmium salts. J Reprod Fertil 1:294–309.
Parizek J (1964): Vascular changes at sites of oestrogen biosynthesis produced by parenteral injection of cadmium salts: The destruction of the placenta by cadmium salts. J Reprod Fertil 7:263–265.
Parizek J, Ostadalova I, Benes I, Pitha J (1968): The effect of a subcutaneous injection of cadmium salts on the ovaries of adult rats in persistent oestrus. J Reprod Fertil 17:559–562.
Parra A, Santos D, Cervantes C, Sojo I, Carranzo A, Cortes-Gallegos V (1978): Plasma gonadotropins and gonadal steroids in children treated with cyclophosphamide. J Pediatr 92:117–124.
Patanelli DJ (1975): Suppression of fertility in the male. In Hamilton DW, Greep RO (eds): "Handbook of Physiology. Section 7: Endocrinology, vol 5, The Male Reproductive System." Washington, DC: American Physiological Society, pp. 245–258.
Pedersen RA, Manigia F (1978): Ultraviolet light induced unscheduled DNA synthesis by resting and growing mouse oocytes. Mutat Res 49:425–429.
Poland A, Smith D, Kuntzman R, Jacobson M, Conney AH (1970): Effect of intensive occupational exposure to DDT on phenylbutazone and cortisol metabolism in human subjects. Clin Pharmacol Ther 11:724–732.

Reimers TJ, Sluss PM (1978): 6-Mercaptopurine treatment of pregnant mice: Effects on second and third generations. Science 201:65–67.

Reimers TJ, Sluss PM, Goodwin J, Seidel GE (1980): Bi-generated effects of 6-mercaptopurine on reproduction in mice. Biol Reprod 22:367–375.

Rohrborn G, Hansmann I (1971): Induced chromosome aberrations in unfertilized oocytes of mice. Humangenetik 13:184–198.

Rudiger HW, Haenisch F, Metzler M, Oesch F, Glatt HR (1979): Metabolites of diethylstilbesterol induce sister chromatid exchange in cultured human fibroblasts. Nature 281:392–394.

Schardein JL (1976): "Drugs as Teratogens." Cleveland, Ohio: CRC Press.

Schmidt G, Fowler WC, Talbert LM, Edelman DA (1980): Reproductive history of women exposed to diethylstilbesterol in utero. Fertil Steril 33:21–24.

Shepard TH (1976): "Catalog of Teratogenic Agents." 2nd ed. The Johns Hopkins University Press. Baltimore, Maryland.

Sieber SM, Adamson RH (1975): Toxicity of antineoplastic agents in man: Chromosomal aberrations, antifertility effects, congenital malformations and carcinogenic potential. Adv Cancer Res 22:57–155.

Sims P, Grover PL (1974): Epoxides in polycyclic aromatic hydrocarbon metabolism and carcinogenesis. Adv Cancer Res 20:165–274.

Sirus ES, Leventhal BG, Vaitukaitis JL (1976): Effects of childhood leukemia and chemotherapy on puberty and reproductive function in girls. N Engl J Med 294:1143–1147.

Sobrinho LG, Levine RA, De Conti RC (1971): Amenorrhea in patients with Hodgkin's disease treated with antineoplastic agents. Am J Obstet Gynecol 109:135–139.

Speilberg SP, Gordon GB (1981): Glutathione synthetase-deficient lymphocytes and acetaminophen toxicity. Clin Pharmacol Ther 29:51–55.

Teramoto S, Saito R, Aoyama H, Shirasu Y (1980): Dominant lethal mutation induced by male rats by 1,2-dibromo-3-chloropropane (DBCP). Mutat Res 77:71–81.

Thorpe E (1967): Some pathological effects of hexachlorophene in the rat. J Comp Pathol 77:137–143.

Thorpe E (1969): Some toxic effects of hexachlorophene in sheep. J Comp Pathol 79:167–171.

Torkelson TR, Sadek SE, Rowe VK, Kodama JK, Anderson HH, Loquvam GS, Hine CH (1961): Toxicologic investigation of 1,2-dibromo-3-chloropropane. Toxicol Appl Pharmacol 3:545–559.

Toth A (1979a): Reversible toxic effect of salicylazosulfapyridine on semen quality. Fertil Steril 31:538–540.

Toth A (1979b): Male infertility due to sulphasalazine. Lancet 2:904.

Traub AI, Thompson W, Carville J (1979): Male infertility due to sulphasalazine. Lancet 2:639–640.

Truhant R, Phu-Lich H, Loisillier F (1964): Sur les' effects de l'ingestin repetee de petites doses de derives du bore sur les fonctions de reproduction du rat. CHR Acad Sci 258:5099–5102.

Uldall PR, Kerr DNS, Tacchi D (1972): Sterility and cyclophosphamide. Lancet 1:693–694.

Van Thiel DH, Gravaler JS, Cobb CF, Sherins RJ, Lester R (1979a): Alcohol-induced testicular atrophy in the adult male rat. Endocrinology 105:888–895.

Van Thiel DH, Gavaler JS, Smith WI, Paul G (1979b): Hypothalamic-pituitary-gonadal dysfunction in men using cimetidine. N Engl J Med 300:1012–1015.

van Tienhoven A (1968): "Reproductive Physiology of Vertebrates." Philadelphia: W.B. Saunders Co.

Vorherr H, Messer RH, Vorherr UF, Jordan SW, Kornfeld M (1979): Teratogenesis and carcinogenesis in rat offspring after transplacental and transmammary exposure to diethylstilbesterol. Biochem Pharmacol 28:1865–1877.

Warne GL, Fairley KF, Hobbs JB, Martin FIR (1973): Cyclophosphamide induced ovarian failure. N Engl J Med 289:1159–1162.

Weir RJ, Fisher RS (1972): Toxicologic studies on borax and boric Acid. Toxicol Appl Pharmacol 23: 351–364.

Welch RM, Levin W, Conney AH (1969): Estrogenic action of DDT and its analogs. Toxicol Appl Pharmacol 14:358–367.

Welch RM, Levin W, Kuntzman, Jacobson M, Conney AH (1971): Effect of halogenated hydrocarbon insecticides on the metabolism and uterotropic action of estrogens in rats and mice. Toxicol Appl Pharmacol 19:234–246, 1971.

Whorton D, Krauss RM, Marshall S, Milby TH (1977): Infertility in male pesticide workers. Lancet 2: 1259:1261.

Wide M (1980): Interference of lead with implantation in the mouse: Effect of exogenous estradiol and progesterone. Teratology 21:187–191.
Wilson JG (1973): "Environment and Birth Defects." New York: Academic Press.
Wilson JG (1974): Factors Determining the teratogenicity of drugs. Annu Rev Pharmacol Toxicol 14:205–217.
Winters SJ, Banks JL, Loriaux DL (1979): Cimetidine is an antiandrogen in the rat. Gastroenterology 76:504–508.
Wolf MM (1979): Impotence on cimetidine treatment. N Engl J Med 300:94.
Wong TW, Hruban Z (1972): Testicular degeneration and necrosis induced by chlorcyclizine. Lab Invest 26:278–289.
Wyrobeck AJ, Bruce WR (1975): Chemical induction of sperm abnormalities in mice. Proc Natl Acad Sci USA 72:4425–4429.

American Journal of Industrial Medicine 4:81–86 (1983)

Relevance of Parameters Related to Fertility and Reproduction in Toxicity Testing

Herman B.W.M. Koëter, BSc

Toxicity data of 37 compounds tested in both a subchronic study and one or more studies focused on reproduction were evaluated to obtain information on the significance of fertility and reproduction parameters in setting no-effect levels (NEL). It appeared that a) the test system aimed at reproduction and development was more sensitive than the subchronic study for 35% of the selected compounds and was equally sensitive for another 35%; b) for 65% of these compounds parameters related to reproduction and development were at least codecisive in determining the minimal-effect level (MEL); c) for 21% of all compounds, parameters related to reproduction were the most sensitive and therefore decisive for the MEL. It was concluded that reproduction in its broadest sense is a very sensitive tool in toxicity testing. The possibilities of either applying a simple screen of fertility and reproduction at an early stage of toxicity testing or incorporating fertility and reproduction parameters in the current subchronic study are discussed.

Key words: Reproduction, fertility, toxicity, no-effect level, comparative study

INTRODUCTION

The reproductive faculty can be considered a prime criterion for life and in consequence a major aspect of the biology of any organism. Therefore, since reproductive impairments of one kind or another are frequent and widespread [Council on Environmental Quality, 1981], it is astonishing that safety testing programs for different categories of chemicals do not always include parameters related to fertility, reproduction, or prenatal development. The U.S. Toxic Substances Control Act (TSCA), for instance, identifies four progressive levels of investigational effort depending upon extent, frequency, and nature of chemical use [Ellison, 1977]. However, information on reproduction is only required at level IV, when the product is already on the market. At level II, required as persons are likely to get exposed during production or industrial use of the chemical, the most comprehensive study is the subchronic study. It cannot be ruled out that an acceptable daily exposure set on the basis of such a study may possibly be too high when fertility and reproduction parameters are not included. To obtain infor-

Reproduction Toxicology Section, Department of Biological Toxicology, Institute CIVO – Toxicology and Nutrition TNO, Zeist, The Netherlands.
Address reprint requests to Herman B.W.M. Koëter, Head, Reproduction Toxicology Section, Department of Biological Toxicology, Institute CIVO – Toxicology and Nutrition TNO, P.O. Box 360, 3700 AJ Zeist, The Netherlands.

0271-3586/83/0401-02-0081$02.00 © 1983 Alan R. Liss, Inc.

mation on the significance of fertility and reproduction parameters with regard to the toxicity profile of a compound, data from 51 compounds adequately tested in rats in both a subchronic study and in one or more studies focused on reproduction, were evaluated. The main objective was to find out for how many of the substances had effects on reproduction and/or fertility which were decisive for the no-effect level (NEL).

MATERIALS AND METHODS

The toxicity data used in the present study were obtained from rat studies reported in the relevant literature, and from studies carried out in our institute. In this evaluation the term "subchronic" was interpreted broadly and covered testing periods from 4 to 26 weeks. Studies focused on reproduction could include multigeneration studies, studies of fertility and general reproductive performance (one generation study) and embryotoxicity/teratogenicity studies.

A review of Food and Cosmetics Toxicology (1967–1980), Toxicology and Applied Pharmacology (1967–1980), and of Toxicology (1973–1980) revealed a total of 27 compounds, and the archives of our institute yielded another 17 compounds. An additional computer literature survey of the Medline (Medlars) system from 1978 to 1980 and of Exerpta Medica from 1975 to 1979 revealed only another 7 compounds. Studies were considered only when the routes of administration of the compounds were similar for both the subchronic and the reproduction study. On the other hand, studies on the same compound, carried out in different laboratories possibly at different times, were still included when they were otherwise comparable.

RESULTS

Compounds

For 6 of the 51 initially collected compounds, the dose levels used in the subchronic study could not be compared with the levels used in the study that was focused on reproduction. Besides, for another 8 compounds effect levels could not be assessed at all in any of the studies. Therefore, from the 51 compounds, only 37 were found usable for this evaluation (See Appendix I). As the studies focused on reproduction and development were rather diverse and since for a number of compounds more than one "reproduction" study was reported, the selected compounds were classified into five different categories according to the nature of the experiments (Table I).

Minimal-Effect Level (MEL)

Upon comparison of the MELs in the different studies, it appeared that for a total of 13 compounds (35%) the reproduction study was more sensitive than the subchronic study. For another 13 compounds (35%) the MEL as assessed in the reproduction study corresponded to the MEL found in the subchronic study, whereas the subchronic study revealed the lowest MEL for 11 compounds (30%).

Parameters

Parameters or a combination of parameters affected at the MEL in the most sensitive experiment are listed in Table II. In this table categories A and D as well as B and E (for explanation, see caption of Table I) are brought together because in these categories the teratogenicity study did not appear to be the most sensitive test system. Table II shows further that when the subchronic study was the most sensitive experiment, no

TABLE I. Classification of Compounds According to the Nature of the Experiments, and Comparison of the Minimal-Effect Levels (MEL)

	Number of compounds						
			Type of study yielding the lowest MEL				
Category[a]	Initially selected	Usable	SC	MG	RF	T	Equal MELs in SC and MG, RF, and/or T
A	15	11	1	6	–[b]	–	4
B	8	7	2	–	1	–	4
C	8	6	1	–	–	1	4
D	12	9	4	4	–	0	1
E	8	4	3	–	1	0	0
Total	51	37	11 (30%)		13 (35%)		13 (35%)

[a]Definitions of categories. A, compound tested in a subchronic study (SC) and in a multigeneration study (MG); B, compound tested in SC and in a study of general reproduction and fertility (RF); C, compound tested in SC and in an embryotoxicity/teratogenicity study (T); D, compound tested in SC, MG, and T; E, compound tested in SC, RF, and T.
[b]–) Not relevant.

particular parameter was more frequently affected than others. However, when the reproduction study revealed the lowest MEL, parameters related to fertility, reproduction, or development often determined the MEL. When the different categories were not considered separately, it appeared that, for 8 of the 13 compounds (62%) for which the reproduction study revealed the lowest MEL, parameters related to reproduction were responsible for that MEL (Table III). In addition, for 9 of the 13 compounds (69%) where the subchronic and the reproduction study were equally sensitive, reproduction parameters were at least codecisive for the MEL.

DISCUSSION

The present study has shown that for 35% of the evaluated compounds, the test system aimed at reproduction and development was more sensitive than the subchronic study. In addition, for another 35% of the compounds, the reproduction and the subchronic studies were equally sensitive. Further, for 17 of these 26 compounds (65%) parameters related to reproduction and development were at least codecisive in the assessment of the MEL. For 8 of the total of 37 compounds, that is more than 20%, fertility or reproduction parameters were the most sensitive and thus solely determined the MEL. In these cases should a study on fertility or reproduction not have been carried out, the no-effect level (NEL), based consequently on the MEL in the subchronic study, would have been too high. Apart from missing an effect on fertility or reproduction, which in itself can be considered a serious deficiency, a NEL that is too high can be hazardous to persons exposed since acceptable daily exposure levels are based upon this level.

Because it appears from this study that reproduction in its broadest sense is a very sensitive tool in toxicity testing, it seems appropriate and important to find a mode of introducing fertility and reproduction parameters at an earlier stage into safety testing. One possible means for achieving this would be to incorporate such parameters in relevant situations in the current subchronic study by including satellite groups of small

TABLE II. Parameters or Combination of Parameters Affected at the Minimal-Effect Level (MEL) in the Most Sensitive Experiment

	Number of compounds											
	Categories A + D[a]				Categories B + E				Category C			
	Lowest MEL in:		Equal MELs in:		Lowest MEL in:		Equal MELs in:		Lowest MEL in:		Equal MELs in:	
Parameters	SC[b]	MG[b]	SC	MG	SC	RF[b]	SC	RF	SC	T[b]	SC	T
Body weight and/or food intake	0	2	0	0	2	1	0	0	0	0	1	1
Organ weights	1	1	0	0	0	0	2	0	1	0	0	0
Histopathology	0	0	1	0	2	0	0	1	0	—	2	—
Clinical chemistry and/or hematology	1	1	2	1	1	0	0	0	0	—	0	—
Fertility, reproduction, and/or development	—[c]	6	—	2	—	1	—	1	—	1	—	3
Combination of any of the above *including* fertility, reproduction, and/or development	—	0	—	1	—	0	—	2	—	0	—	0
Combination of any of the above *except* fertility, reproduction and/or development	3	0	2	1	0	0	2	0	0	0	1	0
Total number of compounds	5	10	5		5	2	4		1	1	4	

[a]For explanation of categories see caption of Table I.
[b]SC, subchronic study; MG, multigeneration study; RF, study of general reproduction and fertility; T, embryotoxicity/teratogenicity study.
[c]—) Not determined.

TABLE III. Classification Based on the Minimal-Effect Levels (MEL) and Incidence of Reproduction Parameters Affected at the Lowest MEL

	Number of compounds	Parameters related to reproduction	
		Affected at the lowest MEL	Not affected at the lowest MEL
Collected data usuable for this study	37	17 (46%)	20 (54%)
Subchronic experiment revealed the lowest MEL	11	—[a]	11 (100%)
Study focused on reproduction and development revealed the lowest MEL	13	8 (62%)	5 (38%)
Subchronic and reproduction studies yielded the same MEL	13	9 (69%)	4 (31%)

[a]—) Not relevant.

numbers of female rats. As a matter of course, when such a modification is considered, the reliability of the usual parameters should be warranted.

Another and probably more feasible possibility would be to introduce a simple and relatively cheap screen of fertility and general reproductive performance. Such a screen, to be used as a pilot study at an early stage of toxicity testing, should be generally indicative of possible effects on reproduction and development. In the case of positive results from such a pilot screen, one or more of the currently standard studies focused at special aspects of reproduction could then be carried out. On the other hand, in the case of negative results, one could decide on no further testing of fertility and reproduction. Apart from early warning of possible effects on reproduction at an early stage of testing, a major advantage would be that reproduction testing would be more purposive and target aimed than is the case now.

The sensitivity of the suggested pilot test using small numbers of animals and a limited number of parameters will be studied experimentally in our laboratory with a selected number of compounds in the near future.

ACKNOWLEDGMENTS

This study was carried out at the request of, and was sponsored by, the Ministry of Health and Environmental Protection, Leidschendam, The Netherlands (Project number: M 182).

I am indebted to Dr. Victor J. Feron for helpful discussions and critical review of the manuscript. The help of Margreet van Marwijk and Maria Krakowczyk in collecting the data is greatly appreciated.

REFERENCES

Council on Environmental Quality (1981): "Chemical Hazards to Human Reproduction." Washington, DC: Government Printing Office.

Ellison T (1977): Toxicological effects testing. In Dominguez GS (ed): "Guidebook: Toxic Substances Control Act." Cleveland: CRC Press, pp 8.1–8.22.

Appendix: List of Evaluated Compounds

Compound	Reference
R4254	CIVO: R4254 (1973)[3]; CIVO: R4646 (1975)[4]; CIVO: R4070 (1973)[5]
R4267	CIVO: R4267 (1973)[3]; CIVO: R4756 (1975)[4]; CIVO: R4248 (1973)[5]
R4509	CIVO: R4509 (1974)[3]; CIVO: R5156 (1976)[4]; CIVO: R4854 (1975)[5]
R4264	CIVO: R4264 (1973)[3]; CIVO: R6281 (1979)[4]
R5167	CIVO: R5167 (1976)[3]; CIVO: R6088 (1979)[4]; CIVO: R6111 (1979)[5]
R4253	CIVO: R4253 (1976)[3]; CIVO: R5240 (1977)[4]; CIVO: R4784 (1975)[5]
R5905	CIVO: R5905 (1978)[3]; CIVO: R5948 (1979)[4]
R1894	CIVO: R1894 (1965)[3]; CIVO: R2476 (1967)[4]
n-Octyl gallate	WHO Food add. series #4 (1972)[3]; WHO Food add. series #4 (1972)[4]
Triamiphos	Toxicology 2 (1974) 327–38[3]; Toxicology 2 (1974) 327–38[4]
TCDD	Tox Appl Pharmac 35 (1976) 553–74[3]; Tox Appl Pharmac 50 (1979) 241–252[4]; Fd Cosm Tox 9 (1970) 405–412[5]
Trisodium nitrilotriacetate (Na_3NTA)	Tox Appl Pharmac 18 (1971) 398[3]; Fd Cosm Tox 9 (1971) 509–518[4]
Methyl mercuri chloride	Toxicology 6 (1976) 85–96[3]; Toxicology 6 (1976) 97–106[4]
Metaldehyde	FT/10/60 RIV[2,3]; Toxicology 4 (1975) 97–115[4]
R2170	CIVO: R2170 (1966)[3]; CIVO: R2191 (1966)[6]
R2253	CIVO: R2253 (1966)[3]; CIVO: R2370 (1967)[6]
R6417	Fd Cosmet Tox 5 (1967) 171–177[3]; CIVO: R6417 (1980)[4]; CIVO: R6106 (1979)[5]
R5388	CIVO: R5388 (1977)[3]; CIVO: R5186 (1976)[6]
Loperamide	Arzneimit Forsch 24 (1974) 1645–49[3]; Arzneimit Forsch 24 (1974) 1645–49[5]
Vinylidene chloride	Tox Appl Pharmac 49 (1979) 189–202[3]; Tox Appl Pharmac 49 (1979) 189–202[5]
Propiram-fumarate	Arzneimit Forsch 24 (1974) 624–631[3]; Arzneimit Forsch 24 (1974) 624–631[5]
Hexachlorophene	Tox Appl Pharmac 24 (1973) 239[3]; Fd Cosmet Tox 14 (1976) 421[6]
Mirex	Tox Appl Pharmac 49 (1979) 271–277[3]; Fd Cosmet Tox 14 (1976) 25–29[5]; Tox Appl Pharmac 60 (1981) 235[6]
Floctafenine	Tox Appl Pharmac 36 (1976) 173–185[3]; Tox Appl Pharmac 36 (1976) 173–185[5]
Hexachloro benzene	Tox Appl Pharmac 40 (1977) 529–49[3]; Fd Cosmet Tox 12 (1974) 471–477[5]
Azotrexe	Tox Appl Pharmac 16 (1970) 453–458[3]; Tox Appl Pharmac 16 (1970) 453–458[6]
Cyclohexyl aminehydro chloride	Fd Cosmet Tox 12 (1974) 609–624[3]; Toxicology 6 (1976) 47–65[4]
Tetrasul	Toxicology 1 (1977) 103–112[3]; Toxicology 1 (1973) 63–78[4]
Calcium carrageenan	Lancet 2 (1970) 414–415[3]; Fd Cosmet Tox 15 (1977) 533–538[4]; Fd Cosmet Tox 15 (1977) 539–545[5]
Di-(3-ethylhexyl) phthalate (DEHP)	Tox Appl Pharmac 26 (1974) 253[3]; Tox Appl Pharmac 26 (1973) 253[6]
Butylated hydroxytoluene (BHT)	Austr J Exp Biol Med Sci 39 (1961) 353[3]; Fd Cosmet Tox 3 (1965) 377[4]; Fd Cosmet Tox 3 (1965) 381[5]
Hexachloro-(1,3)-butadiene (B-chloroprene)	Tox Appl Pharmac 47 (1979) 1–14[3]; CIVO: R6225 (1979)[6]; CIVO: R6387 (1980)[5]
Sisomicine	Infection 4 (1976) Suppl 4 349–354[3]; Infection 4 (1976) Suppl 4 349–354[6]; Infection 4 (1976) Suppl 4 349–354[5]
Cefamandole	J Infection Diseases 137 (1978) Suppl 551–550[3]; J Infection Diseases 137 (1978) Suppl 551–559[6]; J Infection Diseases 137 (1978) Suppl 551–559[5]
Chlorfenvinphos (= GC-4072)	Tox Appl Pharmac 17 (1970) 323–36[3]; Tox Appl Pharmac 17 (1970) 323–36[4]
ETMT = 5-ethoxy-3-trichloromethyl-1,2,4 thiadazole	Tox Appl Pharmac 56 (1980) 164[3]; Tox Appl Pharmac 56b (1980) 164[6]
TRIS = 2,3-dibromo propylphosphate	Tox Appl Pharmac 45 (1978) 245[3]; Fd Cosmet Tox 19 (1980) 67[5]

[1]Compounds listed as codenumbers are confidential. Studies with these compounds were carried out under contract at the Institute CVIO-Toxicology and Nutrition TNO.
[2]Internal report of the Dutch National Institute of Public Health (RIV), Bilthoven, the Netherlands.
[3]Subchronic study.
[4]Multigeneration study.
[5]Embryotoxicity teratogenicity.
[6]Study of general reproduction and fertility.

American Journal of Industrial Medicine 4:87–103 (1983)

Biological Monitoring in Surveillance of Exposure to Genotoxicants

Harri Vainio, MD, Marja Sorsa, PHD, and Kari Hemminki, MD

Biological monitoring comprises analysis of samples taken of individuals with presumed or known exposures in order to estimate either the body uptake of a toxicant or to detect reversible effects of the exposure. In medical surveillance, biological monitoring is a preventive tool whose function is either to identify the hazardous exposures, or the early biological effects of such exposures. Biological monitoring is of special importance in identifying exposures to genotoxicants, for which no safe thresholds are known and the effects of which are prone to large interindividual variations. Both the uptake of the genotoxicant(s) and their biological effect should be considered determinants of the hazard potential of the exposure. Thus the occurrence of either one of these phenomena is a reason for preventive health measures.

Key words: genotoxicant, carcinogenicity, mutagenicity, DNA adducts, chromosome damage, sister chromatid exchange

INTRODUCTION: GENOTOXICANTS IN OUR ENVIRONMENT

Man may become exposed to genotoxicants because of his particular life-style (eg, cigarette smoking, drugs, cosmetics) or because these chemicals are present in the life-supporting environment (eg, food, water, and air) we share with other people. Some of the numerous synthetic compounds that are now manufactured enter the environment as a necessary outcome of their use (eg, pesticides). Many of them escape into the workplace environment, and may affect those who work in it. Others are disseminated into the environment as wastes, either intentionally or by accident, from industrial plants. In addition, various industrial combustion processes inadvertently produce organic compounds, some of which are genotoxic, that are disseminated into the environment. From these considerations it is apparent that human populations may become exposed to organic genotoxic compounds by many routes. The occupational genotoxicants are important for a number of reasons. Their hazard to humans may be

Institute of Occupational Health, Helsinki, Finland.
Address reprint requests to Dr. Harri Vainio, Institute of Occupational Health, Haartmaninkatu 1, SF-00290 Helsinki 29, Finland.

0271-3586/83/0401-02-0087$05.00 © 1983 Alan R. Liss, Inc.

identified by the effects on sentinel individuals (workers who are a high-risk group because of heavy exposure or because of sensitive genetic background conducive to carcinogenesis).

The background against which we study genotoxic problems is particularly complicated and frustrating because of our continually changing chemical environment and the consequent alterations in the relative amount and relative importance of the various types of cancer and other genotoxic manifestations.

It is obvious that epidemiological methods are inadequate from the preventive point of view for identifying the genotoxic risks posed by chemicals in the environment. Besides the fundamental fact that the harm has already been done when epidemiological evidence of such damage becomes available, the task of obtaining statistically meaningful material is often formidable. This is why biological monitoring of highly exposed groups of people should be emphasized as a tool in the surveillance of genotoxic manifestations.

Biological monitoring in health surveillance consists of testing individuals with presumed or known exposures in order to determine their body uptake of a toxicant or to detect early effects of a toxicant. As the uptake and the biological effects are the determinants of the hazard potential of an exposure, the occurrence of either one of these phenomena is reason for preventive health measures.

MONITORING OF EXPOSURE

Uptake of Genotoxicants and Their Reactions

Chemicals enter the body via inhalation, ingestion, or skin absorption, and are distributed in the tissues in a manner specific to each chemical. Some chemicals may exert toxicity directly while some are converted into active intermediates through metabolism. The current knowledge on pharmacokinetics of genotoxic agents is meager and limited mainly to experimental animals. The lack of pharmacokinetic information frequently impedes the application of analytical techniques in the measurement of genotoxicants or their reaction products in the body.

Genotoxic agents are capable of causing lesions in the genetic material either directly or indirectly. Most chemical mutagens are electrophiles or metabolized into electrophilic intermediates capable of forming covalent bonds with cellular nucleophiles [Miller and Miller, 1977]. Some mutagens exert their effects through other mechanisms: metals and intercalating agents form tight noncovalent associations with nucleic acids; antimetabolites may interfere with the genetic material indirectly; and spindle poisons are able to disturb the normal segregation of chromosomes during cell division. Recent results on the mutagenicity of natural nucleosides have re-emphasized the role of disturbances in nucleic acid synthesis as an additional mechanism of mutagenesis [Anderson et al, 1981a]. Oxygen radicals triggered by other agents have also been implicated as DNA-damaging agents [Fridovich, 1978].

The lifetime of the covalent modifications in DNA depends on the site modified and the type of the ligand. Some carcinogens dissociate from DNA nonenzymatically as a carcinogen–base complex, leaving an apurinic or apyrimidinic site. Some modifications are removed by DNA repair enzymes. The liberated carcinogen–base complexes, such as 7-methylguanine and aflatoxin B_1-guanine, are excreted in urine.

Information is available on the types of DNA adducts formed by a number of carcinogens and mutagens [Grover, 1979]. According to present evidence, there seems to be a relation between the sites modified by methylating and ethylating agents and their carcinogenic and/or mutagenic potency [see Lawley, 1979]. The data point to the importance of alkylation at O-6 of guanine and at other exocyclic oxygen atoms for mutagenesis and carcinogenesis. With larger mutagens and carcinogens, modifications at other sites appear important. Aflatoxin B_1 and sterigmatocystin, some of the most potent hepatocarcinogens known [Meselson and Russell, 1977; Lutz, 1979], appear to react predominantly at N-7 of guanine. Polycyclic aromatic amines prefer N-2 of guanine, and many aromatic amines react at C-8 of guanine. γ-Carboline derivatives of glutamic acid and tryptophan, which are both potent mutagens, also react at C-8 of guanine [Sugimura et al, 1980]. With cross-linking agents, such as formaldehyde, mustard gas, and haloethylnitrosoureas, linking through exocyclic amino groups and N-7 of guanine appears to be of importance. Thus, the present evidence may be interpreted to suggest, that any covalent modification of DNA in vivo can be indicative of a carcinogenic and mutagenic potential of an exposure.

Determination of a Chemical, Its Metabolite, or Reaction Product

In occupational health, determination of a *chemical or its metabolite* in biological samples is carried out with a limited number of genotoxicants only. The principles and interpretation of such determinations have been discussed in more detail elsewhere [Vainio et al, 1981].

Many carcinogens and mutagens are conjugated metabolically to glutathione and excreted in urine as cysteine derivatives, thioethers [Jones, 1973; Chasseaud, 1976]. Thioether determinations have been applied to a limited extent in occupational health [van Doorn et al, 1981; Vainio et al, 1981]. As it is not possible to determine which part of urinary alkylcysteines is derived from proteins and which from glutathione, the measurement of thioethers provides only indirect information about genotoxic exposure. Since large volumes of urine can easily be collected, the analysis of thioethers in the urine may be of increasing importance as a means of evaluating the absorption of electrophilic reactants. The analytical techniques can be further refined to characterize the alkyl derivatives bound to cysteine.

Direct information on the alkylation products of mutagens and carcinogens can be obtained with determination of *covalent modifications* in proteins. This field has been pioneered by L. Ehrenberg and his co-workers [Ehrenberg et al, 1974]. They have measured the occurrence of alkyl groups in hemoglobin. As there is a certain relationship between the nucleophilic substitution of protein and nucleic acid residues by electrophiles (carcinogens and mutagens), the degree of protein alkylation may be used as an indirect measure of nucleic acid alkylation. The method is sensitive enough to detect, eg, ethylene oxide residues in hemoglobin of workers exposed to 5–10 ppm of ethylene oxide [Calleman et al, 1978]. As yet, the measurement of hemoglobin alkylation has not, however, been adapted for routine use.

In the past few years immunological techniques have been developed whose sensitivity approach the levels required for the *detection of modified DNA* in exposed human populations [Poirier, 1981; Müller and Rajewsky, 1981]. The techniques are based on antibodies (usually produced in rabbits) against modified DNA, polynucleotides, nucleotides, or nucleosides. Polyclonal and monoclonal antibodies have been raised. The assay methods are competitive immunoassays where two identical haptens (eg, modified DNAs) compete for the antibody-binding sites. In radioimmunoassay, the concentration of the radioactive hapten is constant, and various concentrations of the unknown sample (inhibitor, eg, a DNA sample from blood cells) are used to generate an inhibition curve for quantitation of the level of modification in the sample. In solid phase immunoassays a constant quantity of the hapten is bound onto the surface of the wells of a microtiter plate. The antibody is then incubated together with various concentrations of the unknown sample (inhibitor), which competes with the solid phase hapten for the antibody-binding sites. The degree of binding of the antibody to the solid phase is measured after the binding of an enzyme-linked second antibody. The enzymes used generate colored (enzyme-linked immunosorbent assay, ELISA) or radioactive (ultrasensitive enzymatic radioassay, USERIA) reaction products [Harris et al, 1979; Hsu et al, 1980]. Because the levels of the antigen–antibody reaction are enzymatically multiplied in ELISA and USERIA, they are usually, but not always [Müller and Rajewsky, 1980], some orders of magnitude more sensitive than the conventional radioimmunoassays [Harris et al, 1979]. Examples of antigens reported in the literature are listed in Table I. They include alkyl derivatives of guanosine and thymidine, and adducts of acetylaminofluorene and benzo(a)pyrene. Other antibodies raised include the ones against the adducts of aflatoxin B_1 [Haugen et al, 1981], vinyl chloride, and cis-platinum [Working Group on the Development and Possible Use of Immunological Techniques to Detect Individual Exposure to Carcinogens, convened by the International Agency for Research on Cancer, Essen, 1981]. The affinities of the antibodies described vary between 10^6 and 10^{11} liters/mol. With the best available antibodies, detection of one modification per 10^6–10^7 bases is possible [Poirier, 1981]. DNA adducts produced by carcinogens in cultured cells or experimental animals have been successfully measured [Poirier et al, 1979; Poirier et al, 1980]. So far, no positive results have been reported with samples derived from exposed humans. Although the technique is not yet applicable for the monitoring of individuals exposed to carcinogens, it certainly has promising prospectives for such applications in the future.

Mutagenicity in the Urine

The introduction of relatively simple assays for mutagenicity has opened new possibilities for semiquantitative measurement of genotoxic chemicals [Durston and Ames, 1974]. Urine analyses are the most widely used because many mutagenic agents or their metabolites are excreted, usually as conjugates, in the urine. As the conjugates are, with a few exceptions, biologically inactive, they have to be broken down enzymat-

TABLE I. Antibodies to Carcinogen-DNA Adducts*

Carcinogen-DNA modification used as antigen	Affinity constant (1/mol)	Reference
O^6-methylguanosine	3×10^8 -7×10^9	Müller and Rajewsky, (unpublished)
O^6-ethylguanosine	1.8×10^{10}	Müller and Rajewsky, [1980]
O^6-butylguanosine	1×10^{10}	Rajewsky et al [1980]
N^7-methyl GMP	1×10^8	Meredith and Erlanger, [1979]
O^4-ethylthymidine	8.5×10^8	Rajewsky et al [1980]
AAF- and AF-DNA, GMP or deoxyguanosine	2×10^6 -5×10^9	Leng et al [1979], Sage et al [1979]; Poirier et al [1977, 1979]
Benzo(a)pyrene-DNA	1.1×10^8	Poirier et al [1980]

*GMP = guanosine monophosphoric acid, AAF = acetylaminofluorene, AF = aminofluorene

ically. With a liver microsomal activation system (S-9 mix), mutagenicity has been shown in the urine of rodents after administration of benzidine [Bos et al, 1980], 4-nitro-o-phenylenediamine [Ammenheuser and Warren, 1979], acrylonitrile [Lambotte-Vandepaer et al, 1980], and some related dyestuffs [Tanaka et al, 1980]. When benzo(a)pyrene was administered to responsive and nonresponsive strains of mice, higher mutagenic activity was found in the urine of the insensitive mice, a fact which indicated quantitative problems with the test system (Bartsch, unpublished).

In humans, enhanced mutagenic activity has been reported in urine concentrates from smokers [Yamasaki and Ames, 1977], from patients receiving cytostatic therapy [Falck et al, 1979], from anesthesiologists [McCoy et al, 1978], from epichlorohydrin-exposed workers [Kilian et al, 1978], from a group of workers in a rubber factory [Falck et al, 1980], and in a foundry [Schimberg et al, 1981].

The use of mutagenicity tests in analyzing urine concentrates is nonspecific with regard to the exposing agent. It is, however, specific with regard to the genotoxic character of the health hazard. The unspecificity is not necessarily a weakness; in conditions of complex chemical mixtures, such as exposure to rubber vulcanizing fumes [Falck et al, 1980], it may be an advantage. The monitoring of mutagenic activity in the urine can be used in the identification of groups of individuals with exposure to mutagens. In its present form, the method is more suitable for qualitative, rather than quantitative, determinations of exposure to mutagens. Smoking habits should be taken into account in the event of any demonstration of mutagenicity in the urine. Some indication of the cause of mutation can be obtained, eg, by noting bacterial strain specificity [Falck et al, 1980].

SURVEILLANCE OF GENOTOXIC EFFECTS

The growing interest in occupational, biological monitoring through the use of genetic short-term tests is partly based on the inefficiency of routinely used clinical tests in detecting early effects of hazardous exposures before the onset of manifestations of ill health.

The cytogenetic methods have long been used *to detect unstable chromosome damage* in the peripheral blood lymphocytes of persons exposed to ionizing radiation. The linear interdependence of the radiation dose (LET) and the number of structural chromosome aberrations enables the method to be used for biological dosimetry, even at low radiation levels [Evans et al, 1979].

The detection of changes in human germinal mutation frequency is not a practical possibility. Also the cytogenetic monitoring approaches are mainly based on detection of damage in human somatic cells. The application of somatic cell monitoring for detection of early effects of occupational exposure is becoming increasingly popular [Hemminki et al, 1979].

Chromosomal Aberrations

The system most frequently used in monitoring exposure to clastogenic agents is to study mitogen-stimulated lymphocytes in short-term cultures of blood samples. The samples are easily taken from the subjects by venipuncture. Because of the long lifespan of T lymphocytes, the use of phytohemagglutinin (PHA) stimulation offers some possibility of detecting accumulated damage as well, even if the kinetics of damage is unclear. Especially with chemical mutagens, which display wide variation in their mode and specificity of action, the persistency of damage, the specificity of the cellular stage responsive to the mutagenic action, and the possible selectiveness of different subpopulations of lymphocytes responding to the mitogen stimulation are bound to cause uncontrolled variability in the results of cytogenetic studies (see critical review by Dabney [1981]).

Detection of damage in vivo can only be accomplished with direct samples of dividing human cells, eg, from bone marrow or testicular tissue. Because of the difficulties in obtaining these samples, such techniques can only be applied in special cases with strong indications of a genetic health hazard.

Sister Chromatid Exchanges (SCEs)

In addition to the conventional cytogenetic parameters in metaphase chromosomes (mitotic disturbances, incompleteness of chromosomal coiling, chromosome breakage, and rearrangements), recent developments in chromosome-staining methods offer the possibility of observing intrachromosomal exchanges between sister chromatids. Even though the exact molecular mechanism of exchange is not yet certain, it is evident that SCEs are a sensitive indicator of damage to DNA [Perry, 1980].

The efficiency of SCE testing, both in studies in vivo and in human occupational exposures, still needs to be established. In in vitro tests, the induction of SCEs in chromosomes has been found more sensitive than conventional cytogenetics (see recent Gene Tox evaluation, [Latt et al, 1981]). Because the SCE technique is simpler and faster than analysis of structural chromosome aberrations, the method has gained increasing use in biological monitoring of exposure to genotoxic chemicals. However, the results so far obtained are not very encouraging for overall monitoring purposes [Sorsa et al, 1982]. Many clastogenic agents that cause increase of structural chromosome aberrations in occupational exposure, eg, vinyl chloride [Anderson et al, 1981] and styrene [Andersson et al, 1980; Sorsa, unpublished], do not increase the SCE frequency in vivo. Even though concordant findings are the rule in animal experiments in vivo (see Gebhart [1981], Latt et al [1981] for review of data), the reported better sensitivity of the SCE method is largely based on experimentation in vitro. In occupational exposures, an increase of SCEs has hitherto been detected only in the cases of effective alkylating agents, such as ethylene oxide [Garry et al, 1979, Högstedt et al, 1982], or cytostatic drugs (the latter both in therapeutic exposures—Gebhart [1981], Lambert et al [1982]—and in occupational exposures—see next section on *Surveillance of Occupationally Exposed Groups*). The effect of smoking on the SCE frequency has been repeatedly shown in many studies (see next section on *Smoking*), even though consideration of its confounding effect has been neglected in many cytogenetic monitoring studies.

As a result, the SCE techniques should be considered a valuable additional tool in cytogenetic monitoring, and their use is recommended especially if there are experimental data available on the SCE-inducing capacity of the exposing agent in question.

Micronuclei

The microscopically observable micronuclei in the maturing erythrocytes of the bone marrow are nuclear fragments consisting of acentric chromosome fragments or lagging whole chromosomes. The inducing agents can be either clastogens or spindle-inhibiting agents that affect the normal segregation mechanism of chromosomes during karyokinesis.

In human subjects, genetic damage observable as micronuclei can also be observed in bone marrow similar to the micronucleus test in rodents. Because of the difficulty of obtaining the samples, the observations concerning human exposures and referent individuals are few [Knuutila et al, 1978]. Recent findings suggest that in some cases the bone marrow samples may be more sensitive indicators of genetic damage than lymphocyte metaphase chromosomes [Krogh Jensen et al, 1979; Högstedt et al, 1981].

Micronuclei are also observable in rarely dividing lymphocytes, where their spontaneous frequency is about 3–5 per 10^{-3} cells [Countryman and Heddle, 1978]. There is still very little experience about the use of peripheral blood lymphocytes for scoring micronuclei; as a method, it is very much faster than chromosome aberration or SCE analysis. Improvement in the staining methods has given promising results about the applicability of the method, eg, findings of increased frequency of lymphocytic micronuclei in styrene-exposed workers has been observed at much lower exposure level than in induction of structural chromosome aberrations (Högstedt, personal communication).

Point Mutations in Lymphocytes

Recently, several attempts have been made to detect point mutational damage in human somatic cells in vivo. The most promising of these seems to be the measuring of the mutation frequency in a single marker locus (eg, hypoxanthine–guanine–phosphoribosyltransferase, HGPRT).

Strauss and Albertini [1979] have described a method which, through the determination of the frequency of 6-thioguanine-resistant peripheral blood lymphocytes, is supposed to reveal impaired activity of the X-linked HGPRT locus. Even though the method is theoretically appealing, it is quite tedious in its present form with autoradiographical identification of 6-thioguanine-resistant cells. There are, however, possibilities for the automatization of the method with flow cytometry [Amnéus et al, 1982] or computer image analysis systems [Evans and Vijayalaxmi, 1981], which may improve the applicability of the method for monitoring purposes.

There are still uncertainties concerning the primary molecular injury measured in the system as azaguanine-resistant lymphocytes, and there may be both epigenetic [Amnéus et al, 1982] and mutational mechanisms involved.

An analysis of the system by Evans and Vijayalaxmi [1981] in vitro has shown that there is almost exact parallelism in the incidence of HGPRT locus defects (measured as 8-azaguanine resistance) and X-chromosomal SCE events, whereas in cells treated with X rays, the HGPRT defects correlate with gross chromosomal aberrations.

Further research is needed on the basic genetic and selective mechanisms involved before judgement can be passed on the suitability of the human point mutation system for monitoring purposes.

COMPARISON OF RESULTS FROM BIOLOGICAL MONITORING WITH EXPERIMENTAL AND EPIDEMIOLOGICAL DATA

The application of biological monitoring has been limited so far. We have taken as examples some relatively well-characterized exposures, such as cigarette smoke, hair dyes, anesthetic agents, cytostatic drugs, and ethylene oxide. Some other occupational exposures have been discussed elsewhere [Hemminki et al, 1979; Vainio et al, 1981].

The results have been compared with the available epidemiological findings on spontaneous abortions, malformations, and malignancies.

Smoking

Extensive experimental and epidemiological data are available on the harmful effects of smoking. Tobacco smoke contains at least 1200 identified chemical compounds, many of which can be measured in the blood or urine [Falk, 1977]. Tobacco smoke concentrate is clearly mutagenic when tested in bacterial and eukaryotic systems [Kier et al, 1974, Sorsa et al, 1982]. It also causes SCEs and chromosomal aberration in cultured cells [de Raat, 1979; Hopkin and Evans, 1979]. Smoking induces SCEs [Husgafvel-Pursiainen et al, 1980], chromosomal aberrations [Obe and Herha, 1979], and micronuclei (Högstedt, personal communication) in humans. The urine of smokers is mutagenic to *Salmonella typhimurium* [Yamasaki and Ames, 1977; Falck et al, 1979], and it contains an increased amount of thioethers [van Doorn et al, 1979]. Tobacco smoke condensate is carcinogenic to experimental animals, and smoking is thought to cause a large proportion of human cancers.

Smoking is known to induce fetal and placental enzymes that metabolize foreign compounds [Pelkonen et al, 1977], which suggests that smoking might have harmful effects on the developing embryo. The birth weights of smokers' children are smaller than those of nonsmokers' children, and the difference has been reported to persist during the childhood years [Butler and Goldstein, 1973; Hemminki and Starfield, 1977; Pirani, 1978]. Many studies have also associated cigarette smoking and a risk of spontaneous abortions [Himmelberger et al, 1978]. Several studies have also attributed a risk of congenital abnormalities to smoking [Himmelberger et al, 1978], although this has not been confirmed in other studies [Heinonen et al, 1977]. Smoking women appear to have menopause earlier than nonsmoking women [Jick et al, 1977], which is probably caused by the toxicity of the inhaled chemicals to primary oocytes [Mattison and Thorgeirson, 1978]. Smoking has been reported to cause sperm anomalies [Viczian, 1969] and excess perinatal mortality in the children of smoking fathers [Mau and Netter, 1974].

Exposure to Synthetic Hair Dyes

Experimental evidence implies that many constituents of hair dyes, especially the diamino-substituted toluenes, anisoles, and benzenes, exhibit mutagenic and carcinogenic effects [Fahmy and Fahmy, 1977; Kirkland and Venitt, 1976; Perry and Searle, 1977; Van Duuren, 1980]. Systemic absorption of the hair dyes through the skin has been shown both experimentally and in various human studies concerning users of hair dyes [Maibach et al, 1975; Van Duuren, 1980]. The problem is occupational and also concerns the general public, the users of hair dyes. In the United States alone, over 30 million persons, mostly women, have been estimated to dye their hair [see Van Duuren, 1980].

Urinary assay has not indicated increased mutagenicity after use of some specific hair dyes [Burnett et al, 1979]. However, cytogenetic analyses, carried out on professional hair colorists and persons dying their hair frequently, indicated increased structural chromosome damage among women with dyed hair but not among the professional tinters [Kirkland et al, 1978]. However, no significant increase of SCEs was detected after hair dying treatment in a group of volunteer females [Kirkland et al, 1981]. It should be expected that epidemiologic studies, considering the long and extensive use of synthetic hair dyes (20–30 years), and the positive experimental and biological monitoring data, would suggest possible cancer risk in humans.

Several epidemiological studies, both of cohorts of hair dye users as well as of cases of breast cancer with matched referents, have been carried out during recent years. The available epidemiologic studies have been reviewed by Clemmesen et al [1981].

Most of the carefully conducted case referent studies have concluded significant associations to breast cancer to exist in some exposure subgroups of the study [Kinlen et al, 1977; Shore et al, 1979; Nasca et al, 1980] even though further validation of the retrospective data has been requested.

In the case control study carried out by Shore et al [1979], the adjusted relative breast cancer risks of hair dyers were generally increased, but not statistically significantly. The highest risk ratios were obtained for women over 50 years of age and with more than 10 years of hair dying with no other known risk factor associations for breast cancer.

In a case referent study, Nasca et al [1980] found significant excess interaction between breast cancer and hair dye users with previous benign breast disease. Furthermore, a highly significant dose–response relationship was observed among elderly women who had used hair dying to cover gray hair.

The problem of the limited power of epidemiologic studies is evident from the examples of exposure to hair dyes. Considering the systemic absorption through the skin and the experimental data both on carcinogenesis and on mutagenesis of the active hair dye components, the uncontrolled use of hair dyes should be submitted to critical scrutiny.

Anesthetic Agents

Some of the anesthetic agents are known to have mutagenic potential (cf review by Baden and Simmon [1980]). Halothane has been found to induce recessive lethal mutations in *Drosophila* [Clements and Todd, 1981]. Furthermore, halothane has reactive intermediates which can bind covalently to cellular macromolecules [Gandolfi et al, 1980] and which are mutagenic in bacterial tests [Garro and Phillips, 1978].

Another commonly used anesthetic gas, nitrous oxide, has been reported to increase the number of recessive sex-linked lethal mutations in *Drosophila* [Garrett and Fuerst, 1974].

It has been known for many years that female anesthesiologists have a significantly higher risk of certain reproductive failures. The various studies performed have been able to provide reasonably convincing evidence of an increase of the risk of spontaneous abortion in exposed women.

In a recent United States study, male dentists and female chairside assistants who used anesthetic gases (mainly nitrous oxide) were compared with ones who did not [Cohen et al, 1980]. The study showed highly significant association between exposure to anesthetic gases and spontaneous abortions in the chairside assistants (the rate rising from 8.1 per 100 pregnancies in those without exposure to 19.1 per 100 pregnancies in those heavily exposed). Because the concentrations of anesthetic gases during dental surgery are several times higher than those found in general operating rooms, a survey of the reproductive performance of wives of dentists would be expected to be more indicative of the effects of paternal exposure. Interestingly, an association was observed in the dentists' wives (6.7 per 100 for nonexposed against 10.2 per 100 for heavily exposed husbands). No increase in congenital malformations was observed among the children born to exposed dentists.

In summary, it seems reasonably well established that female operation theater staff face an increased risk of abortion (but not in the wives of exposed men) [Vessey and Nunn, 1980]. The recent study done among dentists and chairside assistants and the evidence from experimental animals [Lane et al, 1979] suggest that high concentrations of nitrous oxide may be the causative factor for reproductive failures.

In some studies, eg, those by Cohen et al [1974], Knill-Jones et al [1972], and Göthe et al [1976], a higher risk of malformations has been reported in children born by women exposed to anesthetic gases during pregnancy. No increase in congenital malformations was observed among children born to dentists exposed to inhalation anesthetics [Cohen et al, 1980]—nor among children of female operating room personnel [Ericson and Källen, 1979]. The existing evidence on increased risk of bearing a malformed child is on a much weaker basis than that on spontaneous abortions.

Two retrospective epidemiological studies have been published concerning the deaths among members of the American Society of Anesthesiologists during the years 1947–1966 and 1967–1971 [Bruce et al, 1968; Bruce et al, 1974]. The first suggested that anesthetists were at an increased risk of death from tumors of the lymphatic and reticuloendothelial systems. No such difference was apparent in the latter. A third study done in the United Kingdom concerned over 20,000 male doctors followed for up to 20 years [Doll and Peto, 1977]. Of these, some 1,250 were full- or part-time anesthetists. No excess cancer deaths were observed, though five deaths from cancer of the pancreas were observed among the full-time anesthetists, whereas only 1.7 would have been expected on the basis of the data concerning the other doctors.

In another large American study, morbidity was followed in a questionnaire survey of 73,496 persons. In this study, women, but not men, exposed to anesthetic gases had an above-normal frequency of cancer, especially of leukemia and lymphoma [Cohen et al, 1974].

Ethylene Oxide

Tests in many systems, ranging from bacterial to mammalian, disclose the mutagenicity of ethylene oxide [cf Ehrenberg, 1979; NIOSH, 1981]. Ethylene oxide induces mutations in a linear dose-dependent fashion without any threshold [Ehrenberg and Hussain, 1981]. Ethylene oxide has also been shown to increase the incidence of mononuclear cell leukemia in rats after inhalation exposure [Snellings et al, 1981]. Epidemiological studies of two occupationally exposed populations were suggestive of an association between ethylene oxide and leukemia [Hogstedt et al, 1979a; Hogstedt et al, 1979b].

Surveillance of occupationally exposed groups. Garry et al [1979] have studied people working in a hospital sterilization facility. The measured ambient concentration of ethylene oxide in the sterilizer room was 36 ppm. Four chronically exposed persons who reported upper respiratory and neurologic symptoms also showed significantly increased sister chromatid exchange frequency in their lymphocytes. Similar increases in SCEs have been found by other authors as well [cf NIOSH, 1981; Lambert et al, 1981]. Chromosomal aberrations have also been reported in humans accidentally exposed to high concentrations of ethylene oxide [Ehrenberg and Hällström, 1967] and in employees exposed for more than 20 years to ethylene oxide [Thiess et al, 1981].

Recently, a Swedish study has shown increased rate of micronucleated lymphocytes in peripheral blood of ethylene oxide-exposed workers already at fairly low concentrations [Högstedt et al, 1981]. Out of 28 exposed persons, 15 had never been exposed to ethylene oxide levels exceeding 1 ppm as a TWA of 8 hr. The other 13 persons had, until 2.5 years before the investigation, been exposed to somewhat higher levels, which, however, never had exceeded 5 ppm. The effect observed in the frequency of micronuclei in the peripheral lymphocytes was more pronounced than even that produced by smoking.

Using the monitoring of alkyl groups in hemoglobin as a device, Ehrenberg's group has made risk assessments of occupational ethylene oxide exposure [Calleman et al, 1978]. According to their estimations, the risk of stochastic effects, expressed in units of the corresponding radiation risk is 10 mrad-equ per ppm. Accordingly, a working week of 40 hr in an ethylene oxide concentration of 10 ppm would correspond to a risk of 4 rad-equ, which is about 2 orders of magnitude higher than the risk associated with the maximal permissible dose of radiation (0.1 rad low-LET radiation per week; Ehrenberg and Hussain [1981]).

Careful employment of the biological monitoring in surveillance of ethylene oxide-exposed persons would most probably prevent (some of) the long-term health hazards. The end-effects of exposure to ethylene oxide—cancer in case of somatic mutations and genetic effects in case of germinal mutations—are rare events that appear with time and are difficult to detect with epidemiological studies.

Alkylating Cytostatic Drugs

Promising clinical results have been obtained in chemotherapeutic treatment of many types of cancer, and the use of cytostatic drugs has thus greatly increased during the last decade. Most of the presently used cytostatic drugs have been experimentally proven potent mutagens and carcinogens [Banerjee and Benedict, 1979; Harris, 1976; Lambert, 1982]. Their principal therapeutic mechanism, intended to stop the division of malignant cells, is based on their alkylating capacity, inhibition of cellular metabolism, or functioning of the spindle. Thus direct or indirect damage to the cellular genetic material is characteristic of many chemotherapeutics.

Occupational handling of chemotherapeutic cancer drugs comprises an example of potential work-related exposure to carcinogenic and mutagenic agents. The epidemiologic evidence on cancer risk brought about by exposure to cytostatics is based on increased incidence of secondary malignancies among clinically cured cancer patients, eg, secondary bladder cancer and leukemias after cyclophosphamide treatment [IARC, 1981] and leukemias following chlorambucil treatment [IARC, 1981]. According to patient studies, the etiology of these secondary tumors remains open even to other explanations, such as inherited susceptibility or latent metastases. Until now, however, no studies have been published on possible work-related cancer risks concerning persons who handle cytostatic drugs.

Demonstration of occupational exposure to cytostatics. Urinary mutagenicity analysis has been used to measure possible exposure to mutagens among hospital nurses of oncology units [Falck et al, 1979]. When improvements of occupational hygiene were carried out, eg, protective clothing, masks, and laminar flow chambers, the urinary mutagenic activity in the group of the oncology nurses decreased almost to the level of the unexposed control group working in office occupations or in other duties of the same hospital. The importance of these studies lies in revealing the fact that exposure may occur during the handling of cytostatics if occupational hygiene is not taken care of.

Demonstration of effects of occupational exposure to cytostatics. According to previous discussions cytogenetic response to exogenous mutagenic agents should be considered an early effect, with no known acute ill health associations. Rather, chromosomal lesions represent direct data of damage to human genetic material in somatic cells, the most common object of study being the peripheral blood lymphocytes.

Several studies indicate that hospital personnel in oncology units show increased chromosomal damage measured as SCE frequency [Norppa et al, 1980] or as structural chromosome aberrations [Waksvik et al, 1981]. Such findings should thus be considered early biological responses to exposure to chromosome-damaging agents. Furthermore, observations confirm that the exposure of nurses to cytostatics has proceeded to a stage of inducing a biologic response. Necessary hygienic improvements limiting exposure have been accomplished in many hospitals after these findings.

The results presented give a good example about preventive actions to be taken on the basis of results of biological monitoring, which has demonstrated exposure to potential human carcinogens. The results of after-action monitoring of urinary mutagenicity have shown that hygienic improvements had decreased the urinary mutagenic activity of the nurses nearly to the level of unexposed control persons [Sorsa et al, 1981].

CONCLUSIONS

This paper deals with a relatively specific problem: the possibility of using biological monitoring tests as a means of determining the hazard presented by exposure to genotoxicants. The importance of such a procedure is evident from a number of considerations: the occurrence of genotoxic manifestations is usually delayed by a number of years following exposure to genotoxicants; hence, tests capable of yielding data of their uptake and biological effects are essential for determination of the hazard potential of an exposure and for preventive health measures. Furthermore, there are many examples of synthetic chemicals introduced into the human environment that have not been known to be genotoxicants until long after their introduction into common use. For this reason, and because of the latency time inherent, eg, in the carcinogenic process, we are in a position where immediate evaluation of the health hazards of these exposures are needed. As can be realized for instance from Table II, there is an urgent need for further data on many of the common exposures in our environment. Epidemiologic studies are time consuming, expensive, and difficult to carry out, and the long latency of induced cancer excludes the waiting for epidemiological human data. For preventive actions, it is necessary to identify genotoxic exposures as early as possible.

Several inherent and methodological limitations complicate the revealing of causal connections between exposure and effect. In addition to the long latency often found in genotoxic manifestations, we are hindered by the problem of invalid exposure data. A number of uncontrollable factors may confound the study, or the information may be totally lacking. These factors make final evaluation of the available epidemiological studies, eg, the carcinogenicity of hair dyes, extremely difficult. Under these circumstances, more weight in decision making should be put to other available data, all the way from the chemical structure and reactivity of the exposing agent to the biological monitoring and health surveillance data.

TABLE II. Correlation Between Experimental and Epidemiological Findings[a]

Effect/exposure	Exposing agents: Cigarette smoke	Hair dyes (synthetic)	Anesthetic agents (halothane, nitrous oxide)	Alkylating cytostatic drugs	Ethylene oxide
Genotoxicity of the exposing agent					
Mutagenicity in bacteria	+	+	−	+	+
Mutagenicity in eukaryotes	+	+	+	+	+
SCEs in mammals	+	+	−	+	+
Chromosome aberrations in vitro	+	+	··	+	+
Carcinogenicity in animals	+	+	··	+	+
Studies on exposed populations					
Determination of the chemical(s) in body fluids	+	+	+	+	+
Thioethers in urine	+	··	··	+	··
Mutagenicity of urine	+	··	±	+	··
Covalent adducts					
Hemoglobin alkylation	··	··	··	··	+
DNA adducts	··	··	··	··	··
Cytogenetic parameters					
Structural chromosome aberrations	+	+	··	+	+
SCEs	+	−	−	+	+
Micronuclei	+	··	··	+	+
Epidemiologic outcome					
Spontaneous abortions	+	··	+	··	(+)
Malformations	±	··	±	+	··
Malignancies	+	±	+	+	(+)

[a]+ Positive data, (+) suggestive data, ± contradictory data, − negative data, ·· no data.

REFERENCES

Ammenheuser MM, Warren ME (1979): Detection of mutagens in the urine of rats following topical application of hair dyes. Mutat Res 66:241–245.

Amnéus H, Matsson P, Zetterberg G (1982): Human lymphocytes resistant to 6-thioguanine: Restrictions in the use of a test for somatic mutations arising in vivo studied by flow-cytometric enrichment of resistant cell nuclei. Mutat Res, in press.

Anderson D, Richardson CR, Davies PJ (1981a): The genotoxic potential of bases and nucleosides. Mutat Res 91:265–272.

Anderson D, Richardson CR, Purchase IFH, Evans HJ, O'Riordan ML (1981b): Chromosomal analysis in vinyl chloride exposed workers: Comparison of the standard technique with the sister-chromatid exchange technique. Mutat Res 83:137–144.

Andersson HC, Tranberg EA, Uggla AH, Zetterberg G (1980): Chromosomal aberrations and sister chromatid exchanges in lymphocytes of men occupationally exposed to styrene in a plastic-boat factory. Mutat Res 73:387–401.

Baden JM, Simmon VF (1980): Mutagenic effects of inhalational anesthetics. Mutat Res 75:169–189.

Banerjee A, Benedict WF (1979): Production of sister chromatid exchanges by various cancer chemotherapeutic agents. Cancer Res 39:797–799.

Beek B, Aranda I, Thomson E (1982): Induction of sister chromatid exchanges, cell cycle delay and chromosomal aberrations by human urine concentrates. Mutat Res 92:333–360.

Bos RP, Brouns RME, van Doorn R, Theuws JLG, Henderson PTH (1980): The appearance of mutagens in urine of rats after the administration of benzidine and some other aromatic amines. Toxicology 16:113–122.

Bruce DL, Eide KA, Linde HW, Eckenhoff JE (1968): Causes of death among anesthesiologists—A 20-year survey. Anesthesiology 29:565–569.

Bruce DL, Eide KA, Smith MJ, Seltzer F, Dykes MHM (1974): A prospective survey of anesthesiologists mortality 1967–1971. Anesthesiology 41:71–74.

Burnett CM, Fuchs CM, Corbett JF (1979): Mutagenicity studies on urine concentrates from female users of dark hair color products. Drug Chem Toxicol 2:283–293.

Butler NR, Goldstein H (1973): Smoking in pregnancy and subsequent child development. Br Med J 4: 573–575.

Calleman CJ, Ehrenberg L, Jansson B, Osterman-Golkar S, Segerbäck D, Svensson K, Wachtmeister CA (1978): Monitoring and risk assessment by means of alkyl groups in hemoglobin in persons occupationally exposed to ethylene oxide. J Environ Pathol Toxicol 2:427–442.

Chasseaud LF (1976): Conjugation with glutathione and mercapturic acid excretion. In Arias AM, Jakoby WB (eds): "Glutathione: Metabolism and Function." New York: Raven Press, pp 77–114.

Clements J, Todd NK (1981): Halothane and non-disjunction in Drosophila. Mutat Res 91:225–228.

Clemmesen J (1981): Epidemiological studies into the possible carcinogenicity of hair dyes. Mutat Res 87: 65–79.

Cohen EN, Brown BW, Bruce DL, Cascorbi HF, Corbett TH, Jones TW, Whitcher CH (1974): Occupation disease among operating room personnel: A national study. Anesthesiology 41:321–340.

Cohen EN, Brown BW, Wu ML, Whitcher CE, Brodsky JB, Gift HC, Greenfield W, Jones TW, Driscoll EJ (1980): Occupational disease in dentistry and chronic exposure to trace anesthetic gases. J Am Dent Assoc 101:21–31.

Countryman PJ, Heddle JA (1976): The production of micronuclei from chromosome aberration in irradiated cultures of human lymphocytes. Mutat Res 41:321–332.

Dabney BJ (1981): The role of human genetic monitoring in the workplace. J Occup Med 23:626–631.

Doll R, Peto R (1977): Mortality among doctors in different occupations. Br Med J 1:1433–1436.

van Doorn R, Bos RP, Leijdekkers Ch-M, Wagenaas-Zegers MAP, Thieuws JLG, Henderson PTH (1979): Thioether concentration and mutagenicity of urine from cigarette smokers. Int Arch Occup Environ Health 43:156–166.

van Doorn R, Leijdekkers Ch-M, Bos RP, Brouns RME, Henderson PTH (1981): Detection of human exposure to electrophilic compounds by assay of thioether detoxication products in urine. Ann Occup Hyg 24:77–92.

Durston WD, Ames BN (1974): A simple method for the detection of mutagens in urine of carcinogen-fed rats by means of bacterial mutagenesis. Nature 249:850–852.

Ehrenberg L (1978): Risk assessment of ethylene oxide and other compounds. In McElhen VK, Abrahamson S (eds): "Assessing Chemical Mutagens: The Risk to Humans." Banbury Cold Spring Harbour Laboratory, pp 157–190.

Ehrenberg L, Hiesche KD, Osterman-Golkar S, Wennberg I (1974): Evaluation of genetic risks of alkylating agents: Tissue doses in the mouse from air contaminated with ethylene oxide. Mutat Res 24: 83–103.

Ehrenberg L, Hussain S (1981): Genetic toxicity of some important epoxides. Mutat Res 86:1–113.

Ehrenberg L, Hällström T (1967): Hematologic studies on persons occupationally exposed to ethylene oxide. In "Radiosterilization of Medical Products." Vienna, IAEA Symp. IAEA, pp 327–334.

Ericson A, Källen B (1979): Survey of infants born in 1973 or 1975 to Swedish women working in operating rooms during their pregnancies. Anesth Analog 58:302–305.

Evans HJ, Buckton KE, Hamilton GE, Carothers A (1979): Radiation induced chromosome aberrations in nuclear dockyard workers. Nature 277:531–534.

Evans HJ, Vijayalaxmi (1981): Induction of 8-azaguanine resistance and sister chromatid exchange in human lymphocytes exposed to mitomycin C and X rays in vitro. Nature 292:601–605.

Fahmy MJ, Fahmy OG (1977): Mutagenicity of hair dye components relative to the carcinogen benzidine in Drosophila melanogaster. Mutat Res 56:31–38.

Falck K, Gröhn P, Sorsa M, Vainio H, Heinonen E, Holsti LR (1979): Mutagenicity in urine of nurses handling cytostatic drugs. Lancet i:1250–1251.

Falck K, Sorsa M, Vainio H, Kilpikari I (1980): Mutagenicity in urine of workers in rubber industry. Mutat Res 79:45–52.

Falk HL (1977): Chemical agents in cigarette smoke. In Lee DHT (ed): "Handbook of physiology, section: Reaction to environmental agents." Bethesda, Maryland pp 199–211.

Fridovich I (1978): The biology of oxygen radicals. Science 201:875–880.

Gandolfi AJ, White RD, Sipes G, Pohl LR (1980): Bioactivation and covalent binding of halothane in vitro: Studies with $[^3H]$- and $[^{14}C]$halothane. J. Pharmacol Exp Rev 214:721–725.

Garrett S, Fuerst R (1974): Sex-linked mutations in Drosophila after exposure to various mixtures of gas atmospheres. Environ Res 7:286–293.

Garro AJ, Phillips RA (1978): Mutagenicity of the halogenated olefin, 2-bromo-2-chloro-1,1-difluoroethylene, a presumed metabolite of the inhalation anesthetic halothane. Mutat Res 54:17–22.

Garry VF, Hozier J, Jacobs D, Wade RL, Gray DG (1979): Ethylene oxide: Evidence of human chromosomal effects. Environ Mutagen 1:375–382.

Gebhart E (1981): Sister chromatid exchange (SCE) and structural chromosome aberration in mutagenicity testing. Hum Genet 58:235–254.

Grover PL (ed) (1979): "Chemical Carcinogens and DNA." Boca Raton: CRC Press, Vol 1 and 2, pp 1–236 and 1–211.

Göthe C-J, Dahlgren B-E, Hultén B, Olander L, Övrum P, Westerholm P (1976): Narkosgaser som yrkesrisk. Läkartidningen 73:2553–2563.

Harris CC (1976): The carcinogenicity of anticancer drugs: A hazard in man. Cancer 37:1014–1023.

Harris CC, Yolken RH, Krokan H, Hsu IC (1979): Ultrasensitive enzymatic radioimmunoassay: Application to detection of cholera toxin and rotavirus. Proc Natl Acad Sci USA 76:5336–5339.

Haugen Å, Groopman JD, Hsu IC, Goodrich GR, Wogan JN, Harris CC (1981): Monoclonal antibody to aflatoxin B_1 modified DNA detected by enzyme immunoassay. Proc Natl Acad Sci USA 78:4124–4127.

Heinonen OP, Slone D, Shapiro S (1977): "Birth Defects and Drugs in Pregnancy." Littleton, MA: Publishing Sciences Group Inc.

Hemminki E, Starfield B (1978): Prevention of low birth weight and pre-term birth. Milbank Memorial Fund Q 56:339–361.

Hemminki K, Sorsa M, Vainio H (1979): Genetic risks caused by occupational chemicals. Scand J Work Environ Health 5:307–327.

Himmelberger DU, Brown BW, Cohen EN (1978): Cigarette smoking during pregnancy and the occurrence of spontaneous abortion and congenital abnormality. Am J Epidemiol 108:470–479.

Hogstedt C, Malmqvist N, Wadman B (1979b): Leukemia in workers exposed to ethylene oxide. J Am Med Assoc 241:1132–1133.

Hogstedt C, Rohlen O, Berndtsson BS, Axelson O, Ehrenberg L (1979a): A Cohort Study of Mortality and Cancer Incidence in Ethylene Oxide Production Workers. Br J Ind Med 36:276–280.

Hopkin JM, Evans HJ (1979): Cigarette smoke condensates damage DNA in human lymphocytes. Nature 279:241–242.

Hsu IC, Poirier MC, Yuspa SH, Yolken RH, Harris CC (1980): Ultrasensitive enzymatic radioimmunoassay (USERIA) defects femtomoles of acetylaminofluorene-DNA adducts. Carcinogenesis 1: 455–458.

Husgafvel-Pursiainen K, Mäki-Paakkanen J, Norppa H, Sorsa M (1980): Smoking and sister chromatid exchange. Hereditas 92:247–250.

Högstedt B, Gullberg B, Mark-Vendel E, Mitelman F, Skerfving S (1981): Micronuclei and chromosome aberrations in bone marrow cells and lymphocytes of humans exposed to petroleum vapors. Hereditas 94:179–187.

Högstedt B, Hedner K, Kolnig A-M, Mitelman F, Skerfving S, Gullberg B (1982): Chromosomal damage in bone marrow cells and peripheral blood lymphocytes in humans exposed to ethylene oxide. Mutat Res, in press.

International Agency for Research on Cancer (1981) IARC Monographs on the evaluation of the carcinogenic risk of chemicals to humans. Some antineoplastic and immunosuppressive agents. IARC Lyon, France, Vol. 26, pp 411.

Jick H, Porter J, Morrison AS (1977): Relation between smoking and age of natural menopause. Lancet I:1354–1355.

Jones SR (1973): The metabolism of biological alkylating agents. Drug Metab Rev 2:71–100.

Kier LD, Yamasaki E, Ames BN (1974): Detection of mutagenic activity in cigarette smoke condensates. Proc Natl Acad Sci USA 71:4159–4163.

Kilian DJ, Pullin TG, Conner TH, Legator MS, Edwards HN (1978): Mutagenicity of epichlorohydrin in the bacterial assay system: evaluation by direct in vitro activity of urine from exposed humans and mice. Mutat Res 53:72.

Kinlen LJ, Harris R, Garrod A, Rodriguz K (1977): Use of hair dyes by patients with breast cancer: A case control study. Br Med J 6083:366–368.

Kirkland DJ, Honeycombe JR, Lawler SD, Venitt S, Crofton-Sleigh C (1981): Sister chromatid exchanges before and after hair dyeing. Mutat Res 90:279–286.

Kirkland DJ, Lawler SD, Venitt S (1978): Chromosomal damage and hair dyes. Lancet ii:124–128.

Kirkland DJ, Venitt S (1976): Cytotoxicity of hair colorant constituents: Chromosome damage induced by two nitrophenylenediamines in cultured Chinese hamster cells. Mutat Res 40:47–56.

Knill-Jones RP, Rodrigues LV, Moir DD, Spence AA (1972): Anaesthetic practice and pregnancy: Controlled survey of women anaesthetists in the United Kingdom. Lancet 1:1326–1328.

Knuutila S, Helminen E, Vuopio P, de la Chapelle A (1978): Sister chromatid exchanges in human bone marrow cells. I. Control subjects and patients with leukemia. Hereditas 88:189–196.

Krogh Jensen M, Nyfors A (1979): Cytogenetic effect of methotrexate on human cells in vivo. Comparison between results obtained by chromosome studies on bone marrow cells and blood lymphocytes and the micronucleus test. Mutat Res 64:339–343.

Lambert B, Lindblad A, Holmberg K, Francesconi D (1982): Use of sister chromatid exchange to monitor human populations for exposure to toxicologically harmful agents. In Wolff (ed): "Sister Chromatid Exchange." New York: J. Wiley & Sons, pp 149–182.

Lambotte-Vandepaer M, Duvanger van Bogaert M, de Meester C, Poncelet F, Mercier M (1980): Mutagenicity of urine from rats and mice treated with acrylonitrile. Toxicology 16:67–71.

Lane GA, Nahrwold ML, Tait AR, Taylor BS, Beaudoin AR, Cohen PJ (1979): Nitrous oxide is teratogenic: Xenon is not! Anesthesiology 51:260.

Latt SA, Allen J, Bloom SE, Carrano A, Falck E, Kram D, Schneider E, Schreck R, Tice R, Whitfield B, Wolff S (1981): Sister-chromatid exchanges: A report of the Gene-Tox Program. Mutat Res 87:17–62.

Lawley PD (1979): Approaches to chemical dosimetry in mutagenesis and carcinogenesis: The relevance of reactions of chemical mutagens and carcinogens with DNA. In Grover PL (ed): "Chemical Carcinogens and DNA." Boca Raton: CRC Press, pp 1–36.

Leng M, Sage E, Fuchs RPP, Daune MP (1978): Antibodies to DNA modified by the carcinogen N-Acetoxy-N-2-acetylaminofluorene. FEBS Lett 92:207–210.

Lutz WK (1979): In vivo covalent binding of organic chemicals to DNA as a quantitative indicator in the process of chemical carcinogenesis. Mutat Res 65:289–356.

Maibach HI, Leaffer MA, Skinner WA (1975): Percutaneous penetration following use of hair dyes. Arch Dermatol 11:1444–1445.

Mattison DR, Thorgeirson SS (1978): Smoking and industrial pollution and their effects on menopause and ovarian cancer. Lancet I:187.

Mau C, Netter P (1974): Die Auswirkungen des vaterlichen Zigarettenkomsums auf die perinatale Sterblichkeit und die Missbildungshäufigkeit. Dtsch Med Wochenschr 99:1113–1118.

McCoy EC, Hankel R, Robins K, Rosenkrantz HS, Ginfrida JG, Bizzari DV (1978): Presence of mutagenic substances in the urines of anesthesiologists. Mutat Res 53:71.

Meredith RD, Erlanger BF (1979): Isolation and characterization of rabbit anti-m^7G-5′P antibodies of high apparent affinity. Nucl Acids Res 6:2179–2191.

Meselson M, Russell K (1977): Comparison of carcinogenic and mutagenic potency. In Hiatt HH, Watson JD, Winsten JA (eds): "Origins of Human Cancer." Cold Spring Harbor Laboratory, pp 1473–1481.

Miller JA, Miller EC (1977): Ultimate chemical carcinogens as reactive mutagenic electrophiles. In Hiatt HH, Watson JD, Winsten JA (eds): "Origins of Human Cancer." Cold Spring Harbor Laboratory, pp 605–627.

Müller R, Rajewsky MF (1980): Immunological quantification by high affinity antibodies of O^6-ethyldeoxyguanosine in DNA exposed to N-ethyl-N-nitrosourea. Cancer Res 40:887–896.

Müller R, Rajewsky MF (1981): Antibodies Specific for DNA Components Structurally Modified by Chemical Carcinogens. J Cancer Res Clin Oncol 102:99–113.

Nasca PC, Lawrence CE, Greenwald P, Chorost S, Arbuckle JT, Paulson A (1980): Relationship of hair dye use, benign breast disease, and breast cancer. J Natl Cancer Inst 64:23–28.

National Institute for Occupational Safety and Health (NIOSH) (1981): Ethylene oxide. Curr Intell Bull 35:1–22.

Norppa H, Sorsa M, Vainio H, Gröhn P, Heinonen E, Holsti L, Nordman E (1980): Increased sister chromatid exchange frequencies in lymphocytes of nurses handling cytostatic drugs. Scand J Work Environ Health 6:299–301.

Obe G, Herha J (1978): Chromosomal aberrations in heavy smokers. Hum Genet 41:259–263.

Pelkonen O, Jouppila P, Kärki NT (1977): Effect of cigarette smoking on 3,4-benzpyrene and N-methylaniline metabolism in human fetal liver and placenta. Toxicol Appl Pharmacol 23:399–407.

Perry P (1980): Chemical mutagens and sister-chromatid exchange. In de Serres FJ and Hollaender A (eds): "Chemical Mutagens." New York: Plenum Press, Vol 6, pp 1–39.

Perry P, Searle CE (1977): Induction of sister-chromatid exchanges in Chinese hamster cells by the hair dye constituents 2-nitro-p-phenylene-diamine and 4-nitro-o-phenylene-diamine. Mutat Res 56:207–210.

Pirani BBK (1978): Smoking during pregnancy. Obstet Gynecol Survey 33:1–13.
Poirier MC (1981): Antibodies to carcinogens—DNA adducts. J Natl Cancer Inst 67:515–519.
Poirier MC, Dubin MA, Yuspa SH (1979): Formation and removal of specific acetylaminofluorene-DNA adducts in mouse and human cells measured by radioimmunoassay. Cancer Res 39:1377–1381.
Poirier MC, Santella R, Weinstein IB, Grunberger D, Yuspa SH (1980): Quantitation of benzo(a)pyrene-deoxyguanosine adducts by radioimmunoassay. Cancer Res 40:412–416.
Poirier MC, Yuspa SH, Weinstein IB, Blobstein S (1977): Detection of carcinogen-DNA adducts by radioimmunoassay. Nature 270:186–188.
de Raat WK (1979): Comparison of the induction by cigarette smoke condensates of sister-chromatid exchanges in Chinese hamster cells and of mutations in Salmonella typhimurium. Mutat Res 66:253–259.
Rajewsky MF, Müller R, Adamkiewicz J, Drosdziok W (1980): Immunological detection and quantification of DNA components structurally modified by alkylating carcinogens (ethylnitrosourea). In Pullman B et al (eds): "Carcinogenesis: Fundamental Mechanisms and Environmental Effects." Dordrecht: Reidel, pp 207–218.
Sage E, Fuchs RPP, Leng M (1979): Reactivity of the antibodies to DNA modified by the carcinogen N-acetoxy-N-aminofluorene. Biochemistry 18:1328–1332.
Schimberg RW, Skyttä E, Falck K (1981): Belastung von Eisengiessereiarbeitern durch mutagene polycyclische aromatische Kohlenwasserstoffe. Staub-Reinh Luft 41:421–424.
Shore RE, Pasternack BS, Thiessen EV, Sadow M, Forbes R, Albert RE (1979): A case-control study of hair dye use and breast cancer. J Natl Cancer Inst 62:277–283.
Snellings WM, Weil CS, Maronpot RR (1981): Final Report on Ethylene Oxide Two-year Inhalation Study on Rats, Project Report 44-20, Bushy Run Research Center (formerly Carnegie-Mellon Institute of Research), January 28, 1981. Submitted by Union Carbide Corporation to the U.S. Environmental Protection Agency under Section 8(e) of the Toxic Substances Control Act, on behalf of cosponsors of the study (February 1981).
Sorsa M, Falck K, Norppa H, Vainio H (1981): Monitoring genotoxicity in the occupational environment. Scand J Work Environ & Health 7: suppl 4, 61–65.
Sorsa M, Hemminki K, Vainio H (1982): Biological monitoring of exposure to chemical mutagens in the occupational environment. Teratogenesis, Carcinogenesis and Mutagenesis 2:137–150.
Sorsa M, Norppa H, Leppänen A, Rimpelä M (1982): Induction of sister-chromatid exchange in human lymphocytes by smoke condensates from different brands of cigarette. Mutat Res 103:149–153.
Strauss GH, Albertini RJ (1979): Enumeration of 6-thioguanine-resistant peripheral blood lymphocytes in man as a potential test for somatic cell mutations arising in vivo. Mutat Res 61:353–379.
Sugimura T, Wakabayashi K, Yamada M, Nagao M, Fujino T (1980): Activation of chemicals to proximate carcinogens. In Holmstedt B, Lauwerys R, Mercier M, Roberfroid M (eds): "Mechanisms of Toxicity and Hazards Evaluation." Amsterdam: Elsevier-North Holland, pp 205–217.
Tanaka K-T, Marni S, Mii T (1980): Mutagenicity of extracts of urine from rats treated with aromatic amines. Mutat Res 79:173–176.
Thiess AM, Schwegler H, Fleig I, Stocker WG (1981): Mutagenicity study of workers exposed to alkylene oxides (ethylene oxide/propylene oxide and derivatives). J Occup Med 23:323–347.
Vainio H, Sorsa M, Rantanen J, Hemminki K, Aitio A (1981): Biological monitoring in the identification of the cancer risk of individuals exposed to chemical carcinogens. Scand J Work Environ Health 7:241–251.
Van Duuren BL (1980): Carcinogenicity of hair dye components. J Environ Pathol Toxicol 3:237–251.
Vessey MP, Nunn JF (1980): Occupational hazards of anaesthesia. Br Med J 281:696–698.
Viczian M (1969): Ergebnisse von Spermauntersuchungen bei Zigarettenrauchern. Z Hautkr 44:183–187.
Waksvik H, Brøgger A, Klepp O (1981): Chromosome analyses of nurse handling cytostatic drugs. Cancer Treat Rep 65:607–610.
Yamasaki E, Ames BN (1977): Concentration of mutagens from urine by absorption with the nonpolar resin XAD-2; cigarette smokers have mutagenic urine. Proc Natl Acad Sci USA 74:3555–3559.

MALE AND FEMALE REPRODUCTIVE TOXICOLOGY

American Journal of Industrial Medicine 4:107–112 (1983)

Reproductive Toxicity: Hypothalamic-Pituitary Mechanisms

Carol Grace Smith, PhD

The hypothalamic-pituitary axis serves as the primary control center for reproductive hormones. Several agents can inhibit this control center and thereby disrupt reproductive function. Of major interest among these agents are the neuropharmacologic drugs that alter the central nervous system input to the hypothalamus. Such drug classes as the barbiturates, narcotics, certain tranquilizers, and marijuana can inhibit gonadotropin secretion by this mechanism. Changes in gonadotropin and prolactin levels can result in disruptive effects on the reproductive system including infertility, changes in libido, and problems with sexual function.

Key words: hypothalamic-pituitary axis, gonadotropins, marijuana, alcohol, narcotics, drug abuse, primate neuroendocrinology

INTRODUCTION

Recent studies have reported disruptive effects of a number of environmental agents on the reproductive system. These observations have caused toxicologists to comment that the reproductive system seems particularly vulnerable to the effects of xenobiotics. This observation is not surprising to those who study reproductive physiology. The reproductive system is unusual among the bodily systems in the complexity of the mechanisms that control it and that must operate in order for it to function. In particular, many classes of drugs (both therapeutic and recreational) can disrupt aspects of controlling mechanisms and alter reproductive function. For example, drugs that disrupt nervous function (either voluntary or autonomic) can inhibit reproduction. The hypothalamic nervous pathways that control the secretion of gonadotropins are inhibited by such CNS-active drugs as marijuana, narcotics, barbiturates, and tranquilizers, making this a major mechanism for the effects of drugs on reproductive hormones. Other classes of drugs that can alter reproductive function include agents that inhibit protein or nucleic acid synthesis or cell division; agents that inhibit endocrine or exocrine secretion mechanisms; and agents that alter the biotransformation of hormones by the liver. Few other bodily systems require the intricate interaction of so

Department of Pharmacology, Uniformed Services University of the Health Sciences, Washington, DC.
Address reprint requests to Carol Grace Smith, Department of Pharmacology, Uniformed Services University of the Health Sciences, 4301 Jones Bridge Road, Washington, DC 20814.

0271-3586/83/0401-02-0107$02.00 © 1983 Alan R. Liss, Inc.

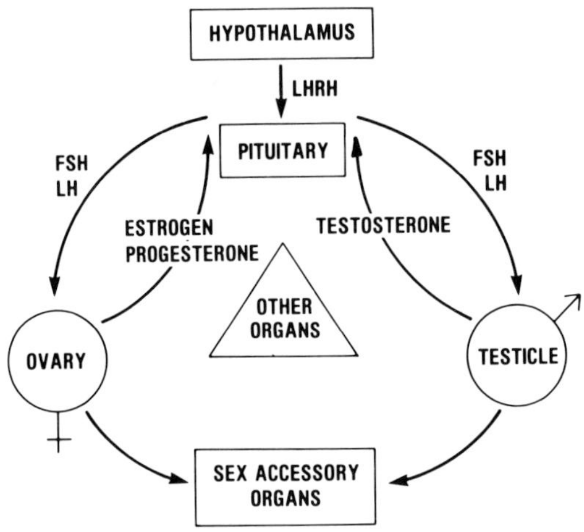

Fig. 1. The normal hypothalamic-pituitary-gonadal relationships that control reproductive functions in males and females. The hypothalamus secretes a hormone referred to as LHRH (luteinizing hormone releasing hormone) that causes the release of the gonadotropic hormones LH and FSH (luteinizing hormone and follicle-stimulating hormone) by the pituitary gland. This complex unit is called the hypothalamic-pituitary axis. LH and FSH are carried by the circulation to the ovaries or testicles. These gonadotropic hormones stimulate the secretion of the sex steroid hormones (estrogen and progesterone in the female and testosterone in the male). The sex steroids are carried by the circulation to the sex accessory organs (eg, prostate and seminal vesicles in the male and uterus and vagina in the female) where they stimulate growth and development and maintain the function of these organs. These sex steroid hormones also complete a feedback loop on hypothalamic-pituitary function, assuring a normal menstrual cycle in females and maintaining spermatogenesis in males.

many components for efficient operation. In addition to the basic physiological control mechanisms, the psychological or emotional inputs to the reproductive system are also apparently vulnerable to the effects of the CNS drugs. We have, over the last several years, attempted to define the effects of certain drugs on the reproductive system using the nonhuman primate as a model. Figure 1 shows a schematic diagram of the most basic controlling mechanisms of the reproductive system. We have attempted to define the effects of certain drugs on the basis of at least these three levels of controls of the reproductive system: the hypothalamic-pituitary axis, the endocrine and ovulatory function of the ovary, and the tissues of the reproductive tract. The hypothalamic-pituitary axis controls the gonadotropins that stimulate the gametogenic and endocrine functions of the testicles or ovaries; and the sex steroids from the gonads complete the feedback loops and stimulate the function of the sex accessory organs. Other organs such as the liver and kidneys are also important to the proper function of the reproductive system. It is important to distinguish between drug effects at these three levels because the level at which the drug effect is produced may well determine the reversibility of the effect. Drugs that produce a reversible disruption of nervous pathways, for example in the hypothalamus, apparently produce a reversible disruption of fertility. Drugs that alter gonadal function directly may produce genetic defects or permanent infertility.

The emphasis of this review of drug effects on reproduction is the drugs that act on hypothalamic-pituitary function. It is now well established that certain of the CNS drugs exert an inhibitory action on hypothalamic/pituitary function. In laboratory animals these effects can be seen as suppression of estrus, ovulation, and fertility in females and inhibition of testosterone and suppression of spermatogenesis in male animals. The primary endocrine effects on the gonadotropins are seen as secondary effects on sex steroids and subsequent effects on the growth, development and function of the sex accessory organs.

The evidence now strongly indicates that the major pathways that are involved in the hypothalamic control of gonadotropins in primates are adrenergic and dopaminergic. There are a multitude of pharmacologic agents that can modify catecholamine levels by altering synthesis, release, receptor activation, reuptake, etc. The best known drugs that produce actions of this type are the neuropharmacologic agents that either inhibit CNS activities—the anesthetics, analgesics, sedatives, and tranquilizers—or that stimulate CNS activities—the antidepressants, stimulants, and hallucinogens. These drugs modify the hypothalamic-pituitary control of the gonadotropins and prolactin because of their action on CNS pathways. The changes in luteinizing hormone (LH), follicle-stimulating hormone (FSH), and prolactin result in the adverse side effects on the reproductive system including changes in libido, impotence, inability to ejaculate, testicular swelling, and gynecomastia. In some cases it is difficult to separate the CNS effects of these drugs from their effects on the peripheral autonomic nervous system.

Drugs of abuse are frequently omitted from the list of CNS drugs that may have effects on reproduction. There is increasing evidence that chronic drug abuse may have serious disruptive effects on the reproductive system. This observation is important because most illicit drug use occurs during the reproductive years.

MARIJUANA

Marijuana consists of a complex group of drugs from the plant Cannabis sativa. The major psychoactive component of marijuana is delta-9-tetrahydrocannabinol (delta-9-THC). Recent studies in man and laboratory animals indicate that chronic, intensive use of marijuana is associated with disruptive effects on the reproductive system. Current evidence indicates that marijuana inhibits the secretion of the gonadotropins LH and FSH by a hypothalamic mechanism [Smith et al, 1979]. A single injection of THC (also crude marijuana extract) produces a marked and long-lasting depression in gonadotropin levels (50–80% decreases for 12–24 hours). The duration of this depression is dose-related, but the effect on gonadotropins outlasts the maximum blood level of THC by 12–24 hours. The effect of THC can be reversed by the administration of the hypothalamic releasing factor, LHRH. This strongly indicates that the effect of THC is produced in the hypothalamus. The inhibitory effect of THC on hypothalamus-pituitary function produces secondary effects on the gonads as well. THC can alter normal ovulatory function in primates and has been shown in studies in rhesus monkeys [Smith et al, 1980]. The rhesus monkeys used in these studies had established normal menstrual cycles. The first day of menses was designated as day 1 of the cycle, and ovulation occurred on about day 15 (average day 15 ± 1 day SD). In order to determine if THC administration would effect ovulation in these monkeys, THC (2.5 mg/kg) or vehicle (3% Tween 80 in saline) was injected daily from day 1 to day 18 of the cycle in three monkeys. Blood was drawn daily from day 8 to day 18 of the cycle and then every other day until the occurrence of menses. The hormones that were measured included total estrogens, LH, and progesterone. Serial laparoscopies were done twice weekly to observe follicular maturation, ovulation, and corpus luteum formation. None of the THC-treated monkeys exhibited estrogen rises or LH surges, and none of the treated monkeys ovulated before the next menses. The first episodes of vaginal bleeding following THC administration were about 21 days, 76 days, and 141 days in the three animals, respectively. Subsequently, two of the three monkeys remained anovulatory for 95 and 51 days. The third monkey had a normal ovulatory cycle the first cycle following THC administration.

Similar results were obtained in studies in which THC was administered to rhesus monkeys during the luteal phase of normal ovulatory cycle [Asch et al, 1979]. Daily administration of THC (2.5 mg/kg) from the day after ovulation until the end of the cycle had no effect on serum progesterone levels or on the length of the luteal phase. The posttreatment period was marked by an absence of normal estrogens, gonadotropins, and progesterone. The prolactin levels recorded during the posttreatment period were four to five times greater than prolactin levels in normal ovulatory cycles. The implications of these results are not fully understood, but it is clear that THC can inhibit normal patterns of gonadotropin and sex steroid secretion and can inhibit ovulation in rhesus monkeys.

The last several years have seen a major controversy in the literature relative to the effects of marijuana use on testosterone levels. The first study by Kolodny showed a significant decrease in testosterone levels in men who were chronic intensive marijuana users compared to age-matched men who had never used marijuana [Kolodny et al, 1974]. In a series of controlled, research-ward studies Mendelson found no evidence of a marijuana-induced depression in testosterone levels [Mendelson et al, 1978]. A further complication to the overall picture comes from another study that showed significant changes in sperm count and morphology without measurable changes in pituitary hormones [Hembree et al, 1976]. These clinical studies have been difficult to compare because of differences in dose levels, time, and duration of administration, experience of drug users and frequency of blood sampling for hormone analysis. Other studies have examined the drug effects on male reproductive hormones using a nonhuman-primate model [Smith et al, 1980]. In these studies, sexually mature male rhesus monkeys were administered a complete dose range of delta-9-THC (as well as other marijuana derivatives). Blood levels of THC were analyzed at the same time intervals that testosterone levels were measured. The THC dose levels that were used produced a blood level of the drug that were within the range of human use of marijuana. In these studies, THC (and marijuana extract containing THC) produced a significant decrease in testosterone and LH. The dose-related effect lasted 6–24 hours and could be reversed by either HCG or GnRH injection. These studies strongly indicated that the primary drug effect is at the level of the hypothalamus with subsequent effects on gonadotropins and testosterone.

NARCOTICS

The effects of narcotic drugs on reproductive function have been reasonably well established in human users [Weiland and Yunger, 1970]. Clinical manifestations such as decreased sexual desire and performance, menstrual irregularities and infertility, and increase fetal loss among narcotic users are attributed in part to altered gonadal functions. Apparently, narcotics exert their primary effect on the hypothalamic-pituitary axis and secondary effects on the gonads and sex accessory organs in both man and animals. Several of the narcotics significantly depress serum LH levels, which in turn produce a marked decrease in serum testosterone levels. In laboratory animals this results in disruption of the functional and structural integrity of the secondary sex organs, primarily the seminal vesicles and prostate gland [Cicero et al, 1979].

In narcotic addiction, testosterone and LH levels are markedly depressed [Mendelson and Mello, 1975; Kley et al, 1977]. According to clinical reports, heroin addicts experience both diminished sexual drive and impairment of sexual function. Sperm count and sperm motility are also decreased. The exact mechanism by which narcotics suppress sexual function and desire is not known. The endorphins may have a role in influencing sexual behavior, but it is also likely that decreased testosterone levels may be responsible for decreased libido. It is clear that the narcotic drugs can decrease gonadotropin secretion and stimulate the secretion of prolactin. Both of these effects are likely to be inhibitory to male sexual function.

If comparisons are made of the effects of roughly equivalent doses of THC and morphine in laboratory primates, it is clear that THC produces a more long-lasting and pronounced depression in testosterone and LH levels than morphine [Smith et al, 1980]. This observation alone might indicate that chronic marijuana use might be associated with even greater disruption of reproduction than narcotic use. However, additional studies need to be done before all of the conflicting information can be resolved. Studies in monkeys strongly indicate that primate animals are in fact vulnerable to the effects of marijuana. Short-term studies, however, show that tolerance develops very quickly to the effects of marijuana on testosterone levels. Questions that remain to be answered include the role of this tolerance, the extent of disruption of hormone levels that must be seen before biologically relevant changes are observed, and importantly, the direct effects of THC and other components of marijuana on spermatogenesis.

ALCOHOL

A third example of a commonly abused CNS drug that affects the reproductive system is alcohol. The effects of alcohol on reproduction have been studied primarily in male subjects. In a study of 16 normal men, LH and testosterone levels were determined during a period of acute alcohol intoxication [Mendelson et al, 1977]. Careful monitoring of both testosterone and LH levels demonstrates clearly that the primary effect of alcohol is on the testicular synthesis and secretion of testosterone. LH levels rise transiently only after the acute fall in testosterone levels. Both ethanol and its major metabolite, acetaldehyde, inhibit the testicular enzymes involved in testosterone synthesis [Johnston et al, 1981]. This testicular effect of alcohol is worsened by the increased metabolic clearance rate of testosterone by the liver also caused by acute alcohol consumption [Gordon et al, 1976]. With chronic alcohol consumption the picture is more complicated. Multiple endocrine abnormalities can be seen in alcoholic cirrhosis [Gordon et al, 1978]. The hypogonadism and gynecomastia that are commonly associated with alcoholic cirrhosis in men are indications of the disruption of reproductive hormones. These observed responses emphasize how important the normal functioning of the liver is to maintaining reproductive function. With destructive liver disease, the pattern of circulating sex steroids changes. Increased levels of estrogenic steroids are observed in these men. These estrogenic steroids secondarily inhibit gonadotropins and testosterone production and inhibit the functioning of sex accessory organs.

CONCLUSION

It is clear that a number of CNS agents including therapeutic drugs and drugs of abuse can inhibit reproductive function by disrupting hypothalamic-pituitary hormones. It is frequently difficult to distinguish between direct drug actions on the hypothalamic-pituitary axis and subsequent effects on gonadal hormones and sex accessory gland function. The distinction is, however, an important one. Most neuroactive drugs produce only transient effects on the central nervous pathways necessary for normal gonadotropin secretion. The disruptive effects of these drugs are likely to be transient and completely reversible. Under these circumstances, normal adults may experience only subtle changes in sexual function. However, individuals with compromised reproductive function may exhibit major problems. It is also likely that adolescents may be at substantial risk of reproductive damage, since the endocrine events associated with puberty are dependent upon the developing hypothalamic-pituitary axis.

REFERENCES

Asch RH, Smith CG, Silver-Khodr TM, Pauerstein CJ (1979): Effects of Δ^9-tetrahydrocannabinol administration on gonadal steroidogenesis activity in vivo. Fertil Steril 32:576–582.

Cicero TJ, Schainker BA, Meyer ER (1979): Endogenous opiods participate in regulation of the hypothalamic-pituitary-luteinizing hormone axis and testosterone's negative feedback control of luteinizing hormone. Endocrinology 104:1286–1291.

Gordon GG, Altman K, Southern AL, Rubin E, Lieber CS (1976): Effect of alcohol (ethanol) administration on sex-hormone metabolism in normal men. N Engl J Med 295:793–797.

Gordon GG, Southern AL, Lieber CS (1978): The effects of alcoholic liver disease and alcohol ingestion on sex hormone levels. Alcoholism: Clinical and Experimental Research 2:259–263.

Hembree WC, Zeidenberg P, Nahas G (1976): Marihuana effects on human gonadal function. In Nahas G, Paton WMD, Idanpaan-Heittila J (eds): "Marihuana: Chemistry, Biochemistry and Cellular Effects." New York: Springer-Verlag, pp 521–532.

Johnston DE, Chiao Y, Gavaler JS, Van Wiel DH (1981): Inhibition of testosterone synthesis by ethanol and acetaldehyde. Biochem Pharmacol 30:1827–1831.

Kley HK, Oellerich M, Wiegelmann W, Herrmann J, Rudorff, KH, Nieschlag E, Kruskemper HL (1977): The effect of methadone on hypophyseal and peripheral glandular hormones during withdrawal. Horm Metab Res 9:484–488.

Kolodny RC, Masters WH, Kolodner RM, Toro G (1974): Depression of plasma testosterone levels after chronic intensive marihuana use. N Engl J Med 290:872–874.

Mendelson JH, Ellingboe J, Kuehnle JC, Mello NK (1978): Effects of chronic marihuana use on integrated plasma testosterone and luteinizing hormone levels. J Pharmacol Expt Ther 207:611–617.

Mendelson JH, Mello NK (1975): Plasma testosterone levels during chronic heroin use and protracted abstinence. A study of Hong Kong addicts. Clin Pharmacol Ther 17:529–533.

Mendelson JH, Mello NK, Ellingboe J (1977): Effects of acute alcohol intake on pituitary-gonadal hormones in normal human males. J Pharmacol Exp Ther 202:676–682.

Smith CG, Besch NF, Smith RG, Besch PK (1979): Effect of tetrahydrocannabinol on the hypothalamic-pituitary axis in the ovariectomized rhesus monkey. Fertil Steril 31:335–339.

Smith CG, Besch NF, Asch RH (1980): Effects of marihuana on the reproductive system. In Thomas JA, Singhal R (eds): "Advances in Sex Hormone Research." Baltimore-Munich: Urban and Schwarzenberg, vol 7, pp 273–294.

Wieland WE, Yunger M (1970): Sexual effects and side effects of heroin and methadone. In "Proceedings Third National Conference on Methadone Treatment." Washington, DC: US Government Printing Office.

American Journal of Industrial Medicine 4:113–125 (1983)

Exercise and Reproductive Function in Women

David C. Cumming, MD, and Robert W. Rebar, MD

Recent studies have documented that women engaged in strenuous exercise programs may develop reproductive defects, including delayed menarche, shortened luteal phase, and amenorrhea. The etiological bases for these abnormalities have not been determined, but physical, hormonal, nutritional, psychological, and environmental factors may be involved and may differ in importance in affected individuals. The consequences of the altered reproductive function associated with exertion will not be identified until prospective long-term investigations are conducted. In the absence of data to the contrary, however, we currently believe that exercise-associated defects in reproduction are not serious and may be reversed by changes in life-style.

Key words: exercise, reproduction, women, athletes, amenorrhea

INTRODUCTION

An increasing number of women are becoming involved in strenuous exercise programs. Although numerous cursory studies of the endocrine effects of exertion have been conducted (see review by Terjung [1979]), in general little is known about the short- and long-term consequences of exercise on reproductive function in women. Older studies found little effect of exertion on reproduction. Zaharieva [1965] investigated 66 athletes competing in the Tokyo Olympic games in 1964 and reported that only one was amenorrheic. Only transient disruption of menstrual cycles was observed in top-grade swimmers [Astrand et al, 1963] and rowers [Erdelyi, 1962] during intense training. In contrast, more recent investigations have identified a number of reproductive abnormalities in women engaging in rigorous physical activities. For example, Anderson [1979] reported that after six months at the US Military Academy at West

Departments of Obstetrics and Gynaecology and Medicine (Division of Endocrinology), University of Alberta, Edmonton, Canada (D.C.C.).
Department of Reproductive Medicine T-002, University of California at San Diego, La Jolla (R.W.R.).
Portions of the studies described were conducted while D.C.C. was a Research Fellow funded by the Medical Research Council of Canada.
Address reprint requests to David C. Cumming, MD, Department of Obstetrics and Gynaecology, University of Alberta, Edmonton, Alberta, Canada T6G 2G3.

0271-3586/83/0401-02-0113$04.00 © 1983 Alan R. Liss, Inc.

Point almost half of the first class of women recruits were amenorrheic. Similarly, Frisch et al [1980] noted a high incidence of delayed menarche, secondary amenorrhea, and irregular menstrual cycles among a population of ballet dancers at three professional schools. It is in fact becoming apparent that delayed menarche, shortened luteal phase, and amenorrhea are observed frequently in exercising women, although the pathophysiologic bases for these abnormalities remain obscure. We shall attempt to review what is known about these defects in women athletes in this survey.

DELAYED MENARCHE, SHORTENED LUTEAL PHASE, AND AMENORRHEA

Malina's group reported that menarche was variably delayed in athletes competing in several sports at different competitive levels [Malina et al, 1973; Malina et al, 1978] (Fig. 1). They questioned whether the physical build associated with the delay of menarche (ie, increased height and longer limbs) might permit better athletic performance, such that these individuals would be selected independent of any effect of exercise on menstrual function. However, it has since been shown that heavy training in the premenarcheal years results in delay of onset of menarche in contrast to women entering athletics after their menses had begun [Frisch et al, 1981]. Although such data suggest that exercise itself or some effect of the increased physical activity is responsible for the delayed menarche, women continuing in athletics are a self-selected group. Those women who undergo early menarche may be unable to continue because of major changes in physique and body composition. Furthermore, early biological maturation may influence socialization away from physical activity and sport within a given chronological age group. Thus, prospective investigation of the effect of exertion on menarche remains necessary to confirm fully the role of premenarcheal exercise on the delay of menarche.

Luteal phase defects have now been observed in a number of athletes. In a single case study of a 30-year-old woman runner, shorter luteal phase and reduced mid-luteal progesterone levels were observed during more intensive training [Shangold et al, 1979]. Daily blood sampling in four elite teenage swimmers revealed abbreviated luteal phases of the menstrual cycle compared with nonexercising girls of similar age [Bonen et al, 1981]. Progesterone levels in the swimmers failed to rise despite an LH surge, suggesting nonluteinization of the follicle. The FSH/LH ratio was much reduced in the swimmers, perhaps accounting for the luteal phase inadequacy. These data suggest that a central mechanism may be responsible for the relative decrease in FSH, which in turn produced inadequate follicular maturation.

Feicht et al [1978] first focused attention on the runner as being particularly likely to develop amenorrhea. Their questionaire study showed that between 6% and 43% of collegiate runners might be amenorrheic depending upon the weekly mileage (Fig. 2). In reply to another questionaire, 46 of 872 respondents (5.3%) indicated that they had become amenorrheic since beginning running [Speroff and Redwine, 1980]. Weekly mileage in this latter group of women was relatively low, and the occurrence of amenorrhea could not be related to the amount of exercise. Other studies have also found a high incidence of amenorrhea in runners [Dale et al, 1979; Schwartz et al, 1981].

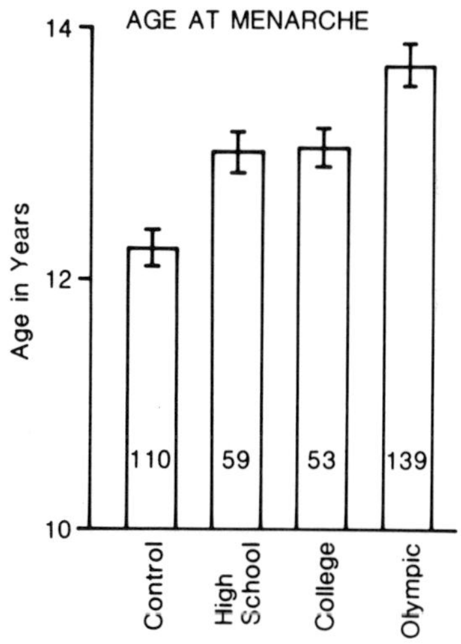

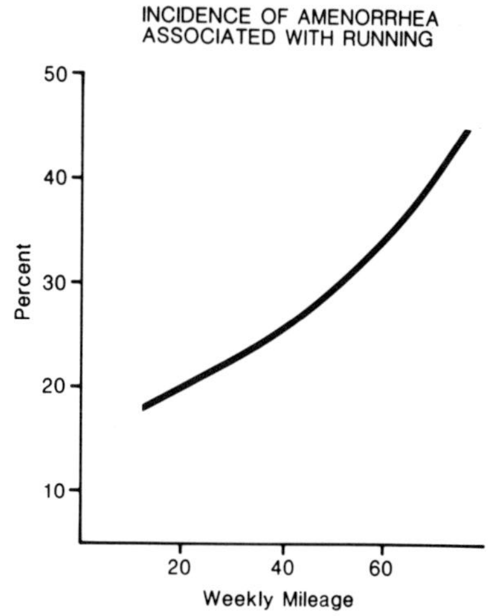

Fig. 1. Recalled age of menarche (± SE) in 110 nonathletes, 59 high school athletes, 53 college athletes, and 139 Olympic athletes (Montreal games, 1976). Adapted from Malina et al [1973, 1978, 1979].

Fig. 2. Relationship between incidence of amenorrhea (defined as less than three cycles in the preceding 12 months) and weekly training mileage of from 20 to 80 miles. Adapted from Feicht et al [1978].

Factors Implicated in Reproductive Defects

While the questions raised by investigations to date related more clearly to athletic activity rather than to occupational exertion, there is a need to identify predisposing factors in women athletes who become amenorrheic. Women are increasingly found in occupations traditionally reserved for men. However, other than those studies involving women at military academies, no studies of the effects of marked changes in the traditional role of women on reproductive function have been made. Although the extrapolation of data from exercise to occupational exertion cannot be made without considerable reservation, questions must arise as to whether comparable industrial exertion may produce similar reproductive problems. Therefore, further investigation of the pathogenesis of amenorrhea of exercise is warranted.

We postulate that several factors may be involved (Table I) and will review current knowledge of the role of each in the pathophysiology of exercise-induced changes in reproductive function.

The Physical and Emotional Stresses of Exercise

Exercise is physically stressful, particularly until adequate fitness is attained. Minor injuries are much more common in women recruits to the military academy than in the males, and over the same time period, the incidence of amenorrhea is highest [Anderson, 1979]. The role of repeated minor trauma and other purely physical stresses in the genesis of amenorrhea remains to be determined. We have investigated the role of psychological stress and found no differences among normal controls, amenorrheic

TABLE I. Factors Implicated in Reproductive Defects in Women Athletes

1. The physical stress of training and/or competing
2. The emotional stress of training and/or competing
3. Changes in the lean/fat ratio
4. Energy drain
5. Loss of weight
6. Dietary changes
7. Acute effects of exertion
8. Chronic effects of repeated exertion

runners and normal cycling women in many objective scales used to measure depression, hypochondriasis, psychastenia, obsessive/compulsive tendencies, or recent stressful events [Schwartz et al, 1981]. Tests used included the Faschingbauer Abbreviated Multiphasic Personality Inventory, the Beck Depression Inventory, the Hopkins Symptom Checklist, and the Schedule of Recent Events. However, when runners were asked to evaluate subjectively the stress associated with running on a scale of 0 to 10 with 10 being maximal, the amenorrheic runners reported significantly more stress than did other runners. Addressing this same question, Warren observed that the timing of menarche in young musicians seeking a professional career in classical music was similar to normal, while in ballet dancers, the onset of menstruation was delayed an average of approximately three years [Warren, 1980]. Assuming that the goal-oriented life-style provided a similar degree of stress for each group, the author suggested that the delay in menarche was "not entirely related to stress."

Body Composition

Body composition has long been considered important in the maintenance of regular menstrual cycles. Frisch and McArthur [1974] suggested that a minimum weight for height was necessary for menarche and the continuation of menstrual cycles. Others have disputed both the repeatability of the original data and the statistical validity of the conclusions that have been drawn from it [Johnston et al, 1975]. Nevertheless, low body weight and weight loss have been associated with secondary amenorrhea [Fries et al, 1974]. We observed that amenorrheic runners had lower percent body fat than normal cycling runners who were, in turn, leaner than unfit women [Schwartz et al, 1981] (Fig. 3). These data were calculated from skinfold thickness at three sites. Ballet dancers with secondary amenorrhea were significantly leaner than those with regular cycles based upon a nomogram calculating body fat from height and weight [Zaharieva, 1965]. Amenorrheic athletes also lose more weight following the onset of running [Speroff and Redwine, 1980; Schwartz et al, 1981] (Fig. 3). However the military recruits at West Point did not have reduced body fat content [Terjung, 1979]. Furthermore, the accuracy of the methods of assessing body fat may be questioned [Reeves, 1979; Jackson et al, 1980], and there clearly is a difference between the amenorrheic sedentary malnourished woman and the amenorrheic athlete with normal lean body weight and low levels of subcutaneous fat [Katch et al, 1980]. Yet, since fat cells and/or stromal cells in the fatty tissues are involved in peripheral steroid metabolism [Siiteri and MacDonald, 1973; Ackerman et al, 1981], alteration in the lean/fat ratio might produce significant changes in steroid metabolism and thus in steroid feedback to the hypothalmic-pituitary unit.

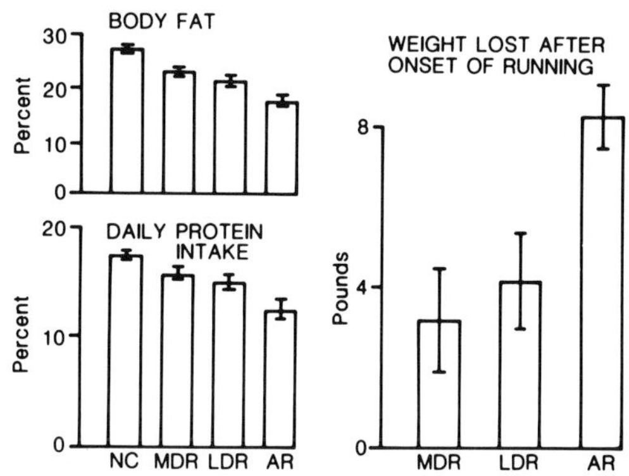

Fig. 3. Percent body fat (± SE) and percent dietary protein intake (± SE) in normal control women (NC), in women running 5 to 30 miles weekly (mid-distance runners, MCR), women running greater than 30 miles weekly (long-distance runners, LDR), and amenorrheic runners (AR) together with weight loss (± SE) following the initiation of running in MDR, LDR, and AR. Adapted from Schwartz et al [1981].

Energy Drain and Weight Loss

Conversely, Warren [1980] has shown that injuries preventing exercise in amenorrheic young ballet dancers precipitated menarche or were followed by resumption of menses in the absence of any change in weight. She postulated that a significant "energy drain" of exercise alone may delay menarche or result in amenorrhea. However, Frisch et al [1981] have provided anecdotal evidence that a relatively small change in weight around the critical level may regulate menstruation since an amenorrheic swimmer was able to control whether or not she had menses through the loss or gain of weight over a range of 5 lb. The authors have also seen runners with this capacity to regulate menstrual cycles by means of a relatively small change in body weight. The mechanism through which such a small weight change can produce such a profound change in menstrual function is obscure, but it seems unlikely that such a change can result from small alterations in the lean/fat ratio.

Diet

Dietary changes also frequently accompany the initiation of exercise. Altered diets were observed in runners beginning training premenarcheally [Frisch et al, 1981]. We have noted significant reductions in the percent of protein content in the diets of amenorrheic runners compared with those who have normal menstrual cycles [Schwartz et al, 1981] (Fig. 3). The role of the various factors related to diet, critical body weight or composition, and loss of weight remains unresolved. Until larger studies accurately define percent body fat using densitometry through underwater weighing of athletes with and without regular cycles, the existing derived data will continue to be a basis for argument.

Hormonal Alterations in Athletes

Basal hormone levels. The amenorrhea associated with exercise has generally been considered to be "hypothalamic" in origin [Speroff, 1981]. Yet others have suggested that exercise-induced changes in circulating hormones may be responsible for the amenorrhea [Shangold et al, 1981]. However, until recently there was relatively little available data examining either circulating hormone levels in women runners or the acute endocrine responses to exercise in women. Baker et al [1981] observed lower levels of plasma estradiol, SHBG, and LH in six amenorrheic runners compared with levels in 14 control runners and 12 normal cycling nonrunners. Levels of dehydroepiandrosterone, dehydroepiandrosterone sulfate (DS), androstenedione, testosterone, cortisol, FSH, and prolactin were also measured, and no differences were found among the groups. We observed higher LH levels in 12 amenorrheic runners compared with cycling runners during the early follicular phase of the menstrual cycle and found a correlation between levels of LH and the distance run per week by amenorrheic runners [Schwartz et al, 1981] (Fig. 4). Although basal steroid levels did not differ among the groups, the ratio of estrone to estradiol was significantly increased in runners compared with nonrunning controls. In comparison to individuals with hypothalamic amenorrhea, amenorrheic runners had significantly higher levels of LH and DS and significantly lower levels of TSH. These basal hormonal values suggest that exercise-associated amenorrhea may be a distinct entity or that the pathophysiological processes leading to amenorrhea may be heterogeneous. In this regard, it is of note that differences in gonadotropin levels may be found in patients with anorexia nervosa who control their weight partially through exercise compared with anorectics who do not exercise [Schwabe et al, 1981].

Alterations in neurotransmitter and neuropeptide levels during exercise in normal subjects have been demonstrated together with other changes in CNS function [Appenzeller and Schade, 1979; Appenzeller et al, 1980; Brenner and Appenzeller, submitted]. It has been suggested that peripheral levels of beta-endorphin progressively increase with exercise and training [Carr et al, 1981]. Alterations in hypothalamic beta-endorphin and dopaminergic regulation have been implicated in the genesis of hypothalamic amenorrhea [Quigley et al, 1980]. However, in a pilot study, none of four amenorrheic runners showed responses to the same endogenous opiate inhibitor (naloxone) previously reported to induce an increment in LH in patients with hypothalamic amenorrhea [Rebar et al, unpublished data] (Fig. 5).

Acute changes with exercise. We have recently examined hormonal responses in various groups of women undertaking exercise. Two prior studies of fit but untrained women had suggested that alterations in ovarian steroid hormones may occur in response to exercise but found little response of gonadotropins [Bonen et al, 1979; Jurkowski et al, 1978]. Normal training activity was reported to increase both prolactin and testosterone [Shangold et al, 1981], and it was postulated that the increased levels of circulating hormones might be, in part, responsible for the genesis of amenorrhea. No study compared the respones in untrained women, exercising women, and amenorrheic athletes.

We established a protocol for examining hormonal changes in exercise and evaluated anterior pituitary, gonadal, and adrenocortical response to exercise in five normal cycling women not involved in any regular exercise program, five normal cycling runners, and six amenorrheic runners [Cumming et al, 1981a,b]. In each case, the runners were running at least 20 miles per week. In the case of normal cycling women, the investigation was carried out in the early follicular phase of the menstrual cycle, and in all cases the investigation began at 1500 h when adrenal steroid concentrations were constant and relatively low. Exercise consisted of cycling at 60 revolutions per minute on an electrically braked bicycle ergometer, and the load was incremental to a

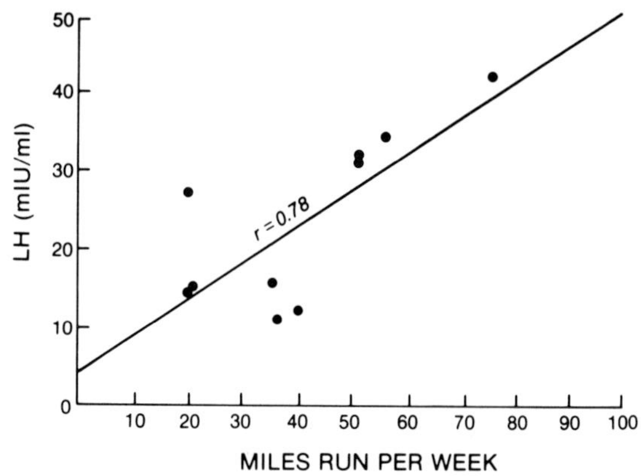

Fig. 4. The correlation of basal LH concentrations with weekly mileage in amenorrheic runners. Constructed from data from Schwartz et al [1981].

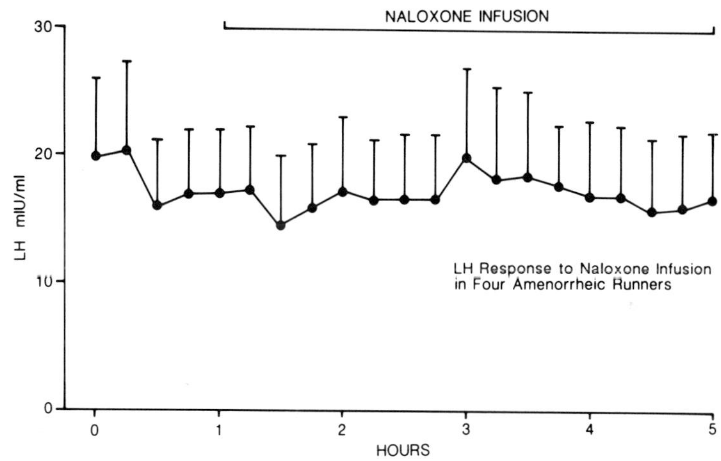

Fig. 5. Effects of intravenous naloxone (1.6 mg/h for 4 h) on serum LH levels (± SD) in four amenorrheic runners. (Unpublished data from Rebar RW, Schwartz B, and Cumming, DC).

symptom-limited maximum. The peak heart rates did not differ between groups, suggesting that the degree of stress was similar, but significantly higher values for maximal oxygen consumption in both groups of runners indicated their anticipated greater fitness.

Basal LH and FSH concentrations were similar in all three groups (Fig. 6). Surprisingly, significant increases in LH were observed in anticipation of exercise. Levels continued to rise to maxima 15 to 20 minutes after the onset of exercise in the two groups of normal cycling women. A significant increment in FSH was seen in response to exercise in all three groups.

Basal levels of testosterone were similar in each of the groups (with marked individual variation), and each group had a significant increment in anticipation of exer-

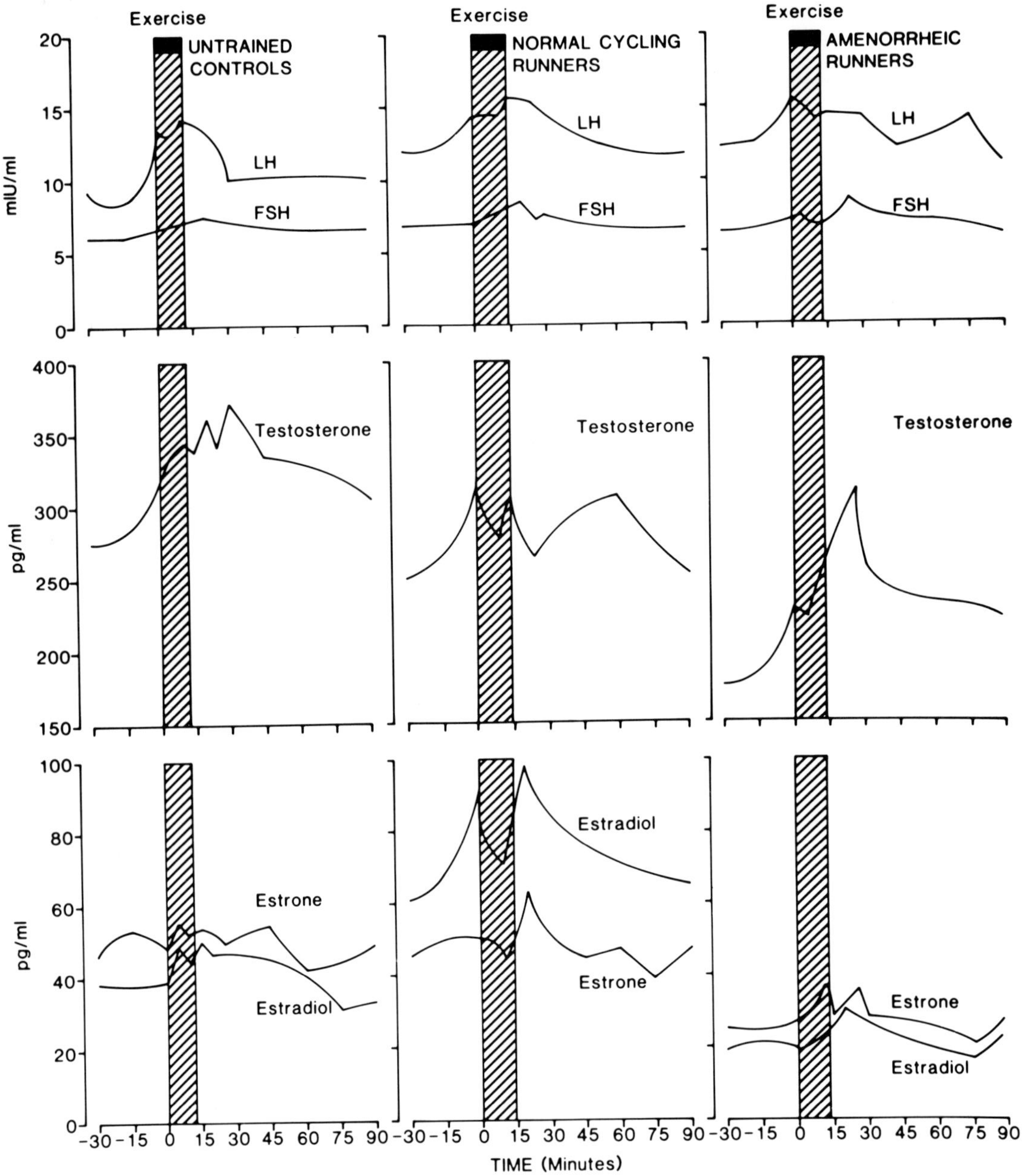

Fig. 6. Levels of LH, FSH, testosterone, estrone, and estradiol before, during, and after symptom-limited incremental exercise on a bicycle ergometer showing responses of normally cycling untrained women, normally cycling women runners, and amenorrheic runners. (Smoothed curves from data accumulated by Cumming et al [1981a,b]).

cise (Fig. 6). A relatively greater overall testosterone release was observed in the amenorrheic compared with the cycling runners in response to exercise.

Baseline estradiol levels were significantly higher in normal cycling runners than in nonrunners who in turn had levels significantly higher than amenorrheic runners (Fig. 6). Estradiol was significantly increased within five minutes of beginning exercise in nonrunners and returned to basal levels after exercise was concluded. Estradiol levels increased significantly in normal cycling runners in anticipation of exercise but not in the amenorrheic groups.

In both normal cycling groups the estrone levels were significantly higher than in the amenorrheic group (Fig. 6). Both groups of runners showed a significant increment in estrone levels as a consequence of exercise, but nonrunners showed no such alteration in levels.

Basal cortisol levels were significantly elevated in both groups of runners (Fig. 7). Serum cortisol initially declined in nonrunners — a constant finding in each subject — and levels then increased to a maximum following exercise. Runners lacked the initial decline in cortisol immediately following the initiation of exercise. However, none of the subsequent changes could be differentiated statistically from those in nonrunners. The adrenal androgens DHEA and androstenedione followed patterns similar to cortisol in each group, although not showing any initial decline with exercise (Fig. 7). Mean peak levels of androstenedione were significantly lower in amenorrheic runners than in the nonexercising women, but otherwise there were no differences among the groups.

In nonrunners, prolactin and growth hormone also increased in the latter part of the exercise period, with the time course similar to that of cortisol (Fig. 8). No significant changes in the levels of TSH were observed throughout the exercise period. A prolactin increment was also seen in normal cycling runners but not in amenorrheic runners. The increment in GH levels was similar in both groups of runners but was significantly smaller in nonexercisers than in the normal cycling runners. A significant increment in TSH was observed only in normal cycling runners.

These data provide for the first time a comprehensive picture of the exercise-induced alterations in anterior pituitary, gonadal, and adrenal cortical hormones in trained and untrained women. The mechanisms involved in such changes are complex and previous findings need further reassessment in view of the detailed picture that has emerged from these studies.

The increase in LH in anticipation of exercise seems likely to be a neuroendocrine event and the increment in testosterone may be related to this increment. Changes in testosterone and other steroids may result from increased production, altered binding or hemoconcentration or from decreased clearance. The absence of significant anticipatory changes in other steroid hormone levels suggests that the increase in testosterone is a specific gonadal event, but the stimulus for this occurrence is obscure. Although primarily suggesting that the exercise-induced changes may be the result of hemoconcentration, Wilkerson et al [1980] found that there was a significant correlation of the level of testosterone preceding exercise with the anticipated load during that exercise in males. It is difficult to envision that hemoconcentration plays a significant role in the increment in testosterone preceding exercise, but postural changes and exercise can result in altered patterns of hepatic perfusion and steroid metabolism [Giusti et al, 1976; Rowell et al, 1964]. However, the runners were sitting before beginning exercise, and the other hormones did not change, again suggesting that the mechanism is

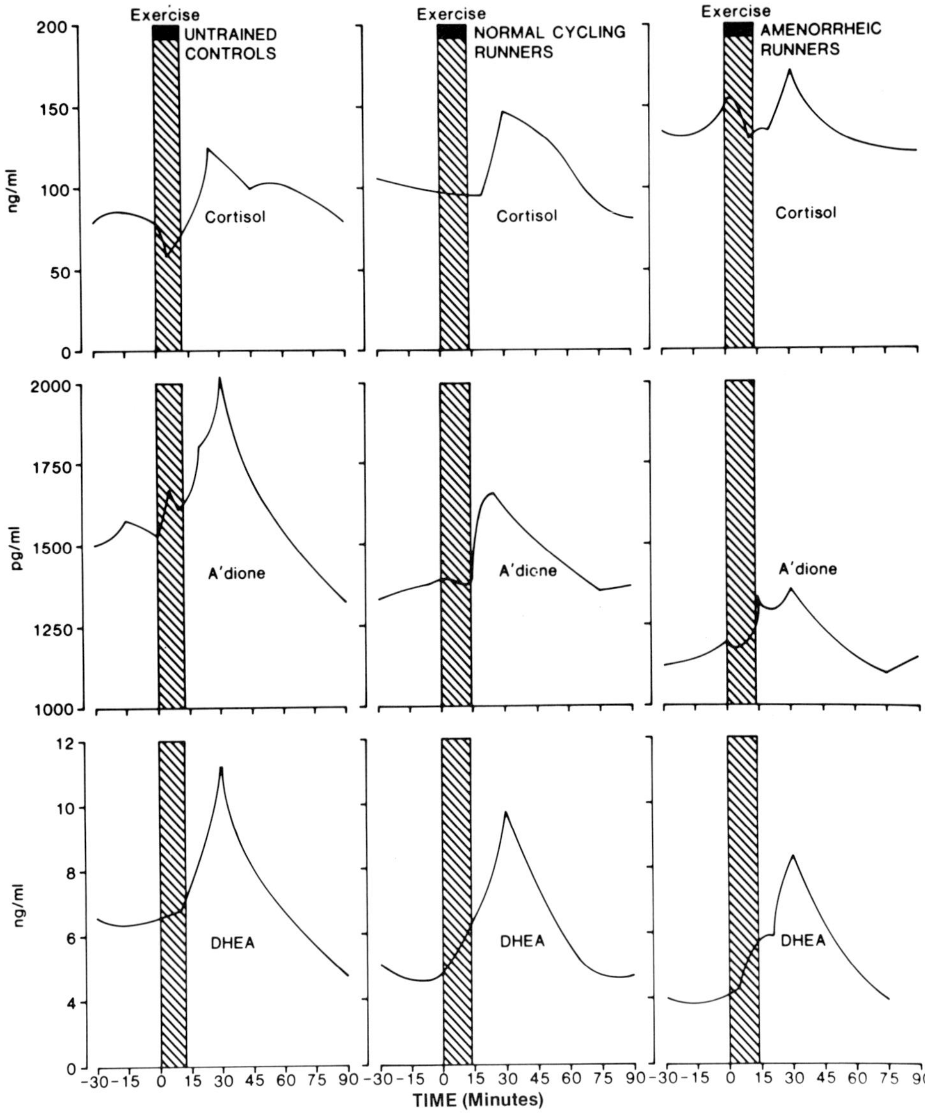

Fig. 7. Levels of cortisol, androstenedione (A'dione), and DHEA before, during, and after symptom-limited incremental exercise on a bicycle ergometer showing responses of normally cycling untrained women, normally cycling women runners, and amenorrheic runners. (Smoothed curves from data accumulated by Cumming et al [1981a,b]).

specific. Later changes are more difficult to explain. A decrease in clearance of estradiol has been demonstrated during brief exercise in women [Keizer et al, 1980]. Presumably, the clearance of other steroids might be altered similarly by exercise. Some of the changes occurring following the initiation of exercise may therefore be due to altered clearance, but the simultaneous increase of cortisol with other stress hormones suggests that activation of the hypothalamic-pituitary axis may be responsible for a good part of the increment of the steroid hormones late in the exercise period.

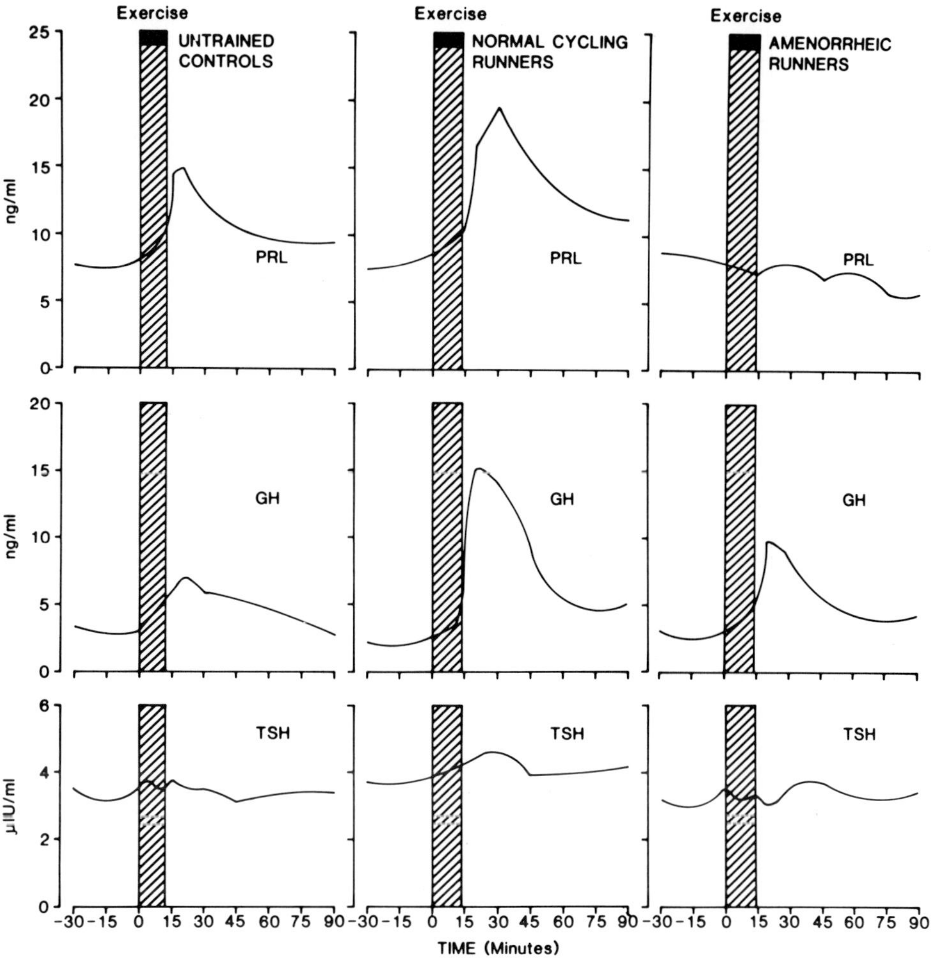

Fig. 8. Levels of prolactin (PRL), growth hormone (GH), and thyroid-stimulating hormone (TSH) before, during, and after symptom-limited incremental exercise on a bicycle ergometer showing responses of normally cycling untrained women, normally cycling women runners, and amenorrheic runners. (Smoothed curves from data accumulated by Cumming et al [1981a,b]).

Regardless of the mechanism(s) of these exercise-induced increases in peptide and steroid hormones, they may exert profound influences on the normal cyclic hormonal changes in women. The chronic effects of repeated exercise must be examined in more detail.

CONCLUSIONS

We have observed that many factors may be involved in exercise-associated amenorrhea. The relative importance of the various factors remains obscure and is probably different in individual patients. We have hypothesized that the exercise-induced changes in reproductive function in women may result from a complex interplay of physical, hormonal, nutritional, psychological, and environmental factors [Rebar and Cumming, 1981]. We have also postulated that exercise-induced amenorrhea results from some alterations originating in the periphery that only secondarily affect the CNS-hypothalamic-pituitary function as well as from central abnormalities. We also surmise that the amenorrhea is reversible, but clearly further investigation of the short- and long-term effects of exercise is necessary.

ACKNOWLEDGMENTS

The authors thank Dr. Samuel S. C. Yen for his continued support and encouragement, all of the women who have participated in our investigations, and Ms. D. Crowe for aid in preparation of the manuscript.

Supported in part by NIH grants HD-12303 and HD-15162.

REFERENCES

Ackerman GE, Smith ME, Mendelson CR, MacDonald PC, Simpson ER (1981): Aromatization of androstenedione by human adipose tissue stromal cells in monolayer culture. J Clin Endocrinol Metab 53:412–417.

Anderson JL (1979): Women's sports and fitness programs at the US military academy. Phys Sports Med 7:72–78.

Appenzeller O, Schade DR (1979): Neurology of endurance training. III. Sympathetic activity during an endurance race. Neurology 29:542.

Appenzeller O, Standefer J, Appenzeller J, Atkinson R (1980): Neurology of endurance training. V. Endorphins. Neurology 30:418–419.

Astrand P-O, Engstrom L, Eriksson BO, Karlberg P, Nylander I, Saltin B, Thoren C (1963): Girl swimmers: Gynaecological aspects. Acta Paediatr Scand [Suppl] 147:33–38.

Baker EM, Mathur RS, Kirk RF, Williamson HO (1981): Female runners and secondary amenorrhea: Correlation with age, parity, mileage, and plasma hormonal and sex hormone binding globulin concentrations. Fertil Steril 36:183–187.

Bonen A, Belcastro AN, Ling WY, Simpson AA (1981): Profiles of selected hormones during menstrual cycles of teenage athletes. J Appl Physiol 50:545–551.

Bonen A, Ling WY, MacIntyre KP, Neil R, McGrail JC, Belcastro AN (1979): Effects of excessive exercise on the serum concentrations of FSH, LH, progesterone and estradiol. Eur J Appl Physiol 42:15–23.

Brenner RP, Appenzeller O: Neurology of endurance training. IV. Brainstem auditory evoked respones (BAERS). Submitted.

Carr DB, Bullen BA, Skrinar GS, Arnold MA, Rosenblatt M, Beitins IZ, Martin JB, McArthur JB (1981): Physical conditioning facilitates the exercise-induced secretion of beta-endorphin and beta-lipotropin in women. New Engl J Med 305:560–563.

Cumming DC, Brunsting L, Greenberg L, Strich G, Ries AL, Yen SSC, Rebar RW (1981a): Patterns of endocrine response to exercise in normal women. 62nd Annual Meeting, Endocrine Society, Abstract #465.

Cumming DC, Strich G, Brunsting L, Greenberg L, Ries AL, Yen SSC, Rebar RW (1981b): Acute exercise-related endocrine changes in women runners and non-runners. Fertil Steril 36:421–422.

Dale E, Gerlach DH, Wilhite AL (1979): Menstrual dysfunction in distance runners. Obstet Gynecol 54: 47–53.

Erdelyi GJ (1962): Gynecological survey of female athletes. J Sports Med Phys Fit 2:174–179.

Feicht CB, Johnson TS, Martin BJ, Sparkes BE, Wagner WW Jr (1978): Secondary amenorrhea in athletes. Lancet 2:1145–1146.

Fries HS, Nillius J, Petterson F (1974): Epidemiology of secondary amenorrhea. II. A retrospective evaluation with special regard to psychogenic factors and weight loss. Am J Obstet Gynecol 118:473–479.

Frisch RE, Gotz-Welbergen A v, McArthur JW, Albright T, Witischi J, Bullen B, Birnholz J, Reed RB, Hermann H (1981): Delayed menarche, irregular cycles and amenorrhea of college athletes in relation to the age of onset of training. JAMA 246:1559–1563.

Frisch RE, McArthur JW (1974): Menstrual cycles: Fatness as a determinant of the minimum weight for height necessary for their maintenance or onset. Science 185:949–951.

Frisch RE, Wyshak G, Vincent LE (1980): Delayed menarche and amenorrhea in ballet dancers. New Engl J Med 303:17–19.

Guisti G, Roncoli E, Forti G, Cattaneo S, Fiorelli G, Pazzagli M, Serio M (1976): Androgenic hormones in hirsutism. In James VHT, Serio M, Giusti G (eds): "The Endocrine Function of the Human Ovary." New York: Academic Press, pp 437–442.

Jackson AS, Pollock ML, Ward A (1980): Generalized equations for predicting body density of women. Med Sci Sports Exerc 12:175–182.

Johnston FE, Roche AF, Schell LM, Wettenall NB (1975): Critical weight at menarche: Critique of a hypothesis. Am J Dis Child 129:19–23.

Jurkowski JE, Jones JL, Walker WC, Younglai EV, Sutton JR (1978): Ovarian hormonal responses to exercise. J Appl Physiol 44:109–114.

Katch FI, Katch VL, Behnke AR (1980): The underweight female. Phys Sports Med 8:55–60.

Keizer HA, Poortman J, Bunnik GSJ (1980): Influence of physical exercise on sex hormone metabolism. J Appl Physiol 48:765–769.

Malina RM, Bouchard C, Shoup RF, Demirjian A, Lariviere G (1979): Age at menarche, family size, and birth order in athletes at the Montreal Olympic Games 1976. Med Sci Sports 11:354–358.

Malina RM, Harper AB, Avent HH, Campbell DE (1973): Age at menarche in athletes and non-athletes. Med Sci Sports 5:11–13.

Malina RM, Spirduso WW, Tate C, Baylor AM (1978): Age at menarche and selected menstrual characteristics in athletes at different competitive levels and in different sports. Med Sci Sports 10:218–222.

Quigley ME, Sheehan KL, Casper RF, Yen SSC (1980): Evidence for increased dopaminergic and opioid activity in patients with hypothalamic hypogonadotropic amenorrhea. J Clin Endocrinol Metab 50:949.

Rebar RW, Cumming DC (1981): Reproductive function in women athletes. JAMA 246:1590.

Reeves J (1979): Estimating fatness. Science 204–881.

Rowell LB, Blackmon JR, Bruce RA (1964): Iodocyanine green clearance and estimated hepatic blood flow during mild to maximal exercise in the upright position. J Clin Invest 43:1677–1690.

Schwabe AD, Lippe BM, Chang RJ, Pops MA, Yager J (1981): Anorexia nervosa. Ann Int Med 94:371–381.

Schwartz B, Cumming DC, Riordan E, Selye M, Yen SSC, Rebar RW (1981): Exercise-associated amenorrhea: A distinct entity. Am J Obstet Gynecol 141:662–670.

Shangold M, Freeman R, Thysen B, Gatz M (1979): The relationship between long distance running, plasma progesterone and luteal phase length. Fertil Steril 31:130–133.

Shangold MM, Gatz ML, Thysen B (1981): Acute effects of exercise on plasma concentrations of prolactin and testosterone in recreational women runners. Fertil Steril 35:699–702.

Siiteri PK, MacDonald PC (1973): Role of extraglandular estrogen in human endocrinology. In Greep RO, Astwood EB (eds): "Handbook of Physiology." Washington, DC: American Physiological Society, Sect. 7, Endocrinology, p 615.

Speroff L (1981): Getting high on running. Fertil Steril 36:149–151.

Speroff L, Redwine DB (1980): Exercise and menstrual function. Phys Sports Med 8:42–52.

Terjung R (1979): Endocrine response to exercise. Exerc Sport Sci Rev 7:153–180.

Warren MP (1980): The effects of exercise on pubertal progression and reproductive function in girls. J Clin Endocrinol Metab 51:1150–1157.

Wilkerson JE, Horvath SM, Gutin B (1980): Plasma testosterone during treadmill exercise. J Appl Physiol 49:249–253.

Zaharieva E (1965): Survey of sportswomen at the Tokyo Olympics. J Sports Med 5:215–222.

American Journal of Industrial Medicine 4:127–134 (1983)

Some Actions of Lead on the Sperm and on the Male Reproductive System

J. A. Thomas, PhD, and W. C. Brogan III, PhD

There is evidence that lead can exert a toxic action upon the male reproductive system. Likewise, there is evidence that some of the effects that lead exerts upon other biological systems might also affect male reproductive capacity. Differences in response to the toxic actions of lead can, in part, be attributed to differences in dose, duration, and method of exposure, species, and lack of comparable experimental designs. Nevertheless, a spermatotoxic and gonadotoxic effect of lead has been seen in a variety of species including man. Furthermore, lead exposure has been linked to chromosomal aberrations in workers.

Key words: male reproductive system, lead, toxicity, spermatotoxicity, gonadotoxicity, chromosomal aberrations, steroidogenesis

INTRODUCTION

Lead poisoning has been associated with reduced fertility, miscarriages, and stillbirths since antiquity [Gilfillan, 1965; Bell and Thomas, 1980]. Lead salts are among the oldest known spermicidal agents and lead compounds have long been known to be abortifacients. Evidence of the deleterious effects of lead on human reproduction dates back to the ancient Roman and Greek civilizations [Gilfillan, 1965]. The presence of lead in storage vessels for water or wine, in pipes that carried water, and in cooking and eating utensils provided increased potential for lead ingestion and poisoning [Henderson, 1824]. Even modern-day civilizations have not entirely escaped lead ingestion since paints, gasoline, storage batteries, and certain industrial processes still represent high risks for poisoning.

Most of the literature dealing with the effects of lead on the reproductive system has dealt with the female since this heavy metal is a known teratogen and abortifacient [Hamilton and Hardy, 1974]. Nevertheless, there is evidence to indicate that lead can also exert deleterious actions upon male reproductive organs and upon male reproduc-

Department of Pharmacology and Toxicology, West Virginia University Medical Center, Morgantown, West Virginia.
Address reprint requests to J. A. Thomas, Travenol Labs, Inc., Deerfield, IL 60015.

0271-3586/83/0401-02-0127$02.50 © 1983 Alan R. Liss, Inc.

tive capacity [Bell and Thomas, 1980]. A number of recent findings about the toxic effects of lead on the male reproductive system warrant a further consideration of this timely topic.

SPERMATOTOXICITY

It is virtually impossible for any species, including man, to avoid the intake of lead. Lead is quite prevalent in the environment [Nriagu, 1980]. The body lead burden includes lead deposited in soft tissue, bone, and blood. In normal rats, lead has been found in various loci in the testes, including the seminiferous tubules, along the edge of the lumen, and at the tails of spermatozoa. Lead-poisoned rats also had lead deposits in the lumina of the seminiferous tubules [Timm and Schulz, 1966]. Lead is not generally believed to have any physiological function in the organism but seems to have only toxic effects (Table I). As early as 1850, de Quatrefages [Mann, 1964] described in great detail the marked toxicity of lead upon mammalian spermatozoa. More recently, Hilderbrand et al [1973] have reported that lead acetate can cause mammalian sperm to become less motile. Recently, Kendall et al [1981] reported that lead interfered with the male reproductive system of adult male ringed turtle doves. They reported an inhibition of spermatogenesis in the lead-treated birds with the seminiferous tubules oftentimes being completely devoid of spermatozoa.

Lead exposure has been associated with a period of temporary infertility in the male rat [Puhac et al, 1963] and reduced fertility in Swiss mice [Varma et al, 1974]. Varma et al [1974] suggested that the reduced fertility in the lead-treated mice was the result of a genetically-related mutagenic change in the spermatozoa. Eyden et al [1978] administered lead acetate to male mice, and although a dietary lead level of 1% had no effect during the first four weeks of exposure, there was a significant increase in the number of abnormal sperm by eight weeks. Unfortunately, this level of exposure to lead caused a severe weight loss in the animals, raising the possibility that the increased incidence of abnormal spermatozoa was secondary to systemic metabolic disturbances rather than to a direct effect on the testes.

Lead-induced gonadal toxicity has been reported by Krasovskii et al [1979] following either the short-term (20–30 days) or the long-term (6–12 months) oral administration of lead acetate to rats. These toxic effects of lead were observed at doses of 0.05 mg/kg. Following short-term exposure, spermatozoa were found to have decreased motility, lower osmotic stability, and reduced time of motility. Small disruptions in the permeability of vessels and dystrophic changes in the Leydig cells were also observed. Long-term exposure (6–12 months) to lead acetate reportedly caused reduced spermatogenesis and some morphological alterations. Whether this represents a direct effect of lead upon the testes or is the result of systemic metabolic disturbances cannot be assessed since neither blood lead levels nor body weight data were reported [Krasovskii et al, 1979].

Little information is available on the effects of low levels of lead on male reproduction. Recent studies, however, were carried out to test the effects of chronic low-level lead toxicity in the rat. Delays in age of vaginal opening were noted in female offspring exposed to lead [Grant et al, 1980]. However, the incidence of abnormal sperm in the male littermates of these female offspring was not significantly different from unexposed controls [Fowler et al, 1980]. Previously, Lancranjan et al [1975] reported

TABLE I. Effects of Lead on the Male Reproductive System*

Species	Effect	References
Rat	Infertility	Puhac et al [1963]
Rat	Germinal epithelium damage	Timm and Schulz [1966]
Rat	Oligospermia and testicular degeneration	Golubovich et al [1968]
Rat	Decreased sperm motility and prostatic hyperplasia	Hilderbrand et al [1973]
Rat	Decreased motility and osmotic stability, dystrophic changes in Leydig cells	Krasovskii et al [1979]
Mouse	Infertility	Varma et al [1974]
Mouse	Increase in abnormal sperm	Eyden et al [1978]
Mouse	Interference with androgen-receptor binding	Donovan et al [1980]
Turtle dove	Inhibition of spermatogenesis	Kendall et al [1981]
Human	Teratospermia, hypospermia, and asthenospermia	Lancranjan et al [1975]

*Adapted from Bell and Thomas [1980].

deleterious effects of lead on human reproduction at subtoxic (moderate) levels. Men occupationally exposed to lead were divided into groups according to blood lead levels. The fertility of the lead-poisoned (74 μg lead/100 mL blood) and moderate absorption (53 μg lead/100 mL blood) groups was decreased; and this reduction correlated with an increased frequency of asthenospermia, hypospermia, or teratospermia. Whether the "moderate" absorption group really represented a subtoxic level of lead depended upon blood lead levels. Blood lead levels may not always be a true indication of susceptibility to lead poisoning. Recently, Raghaven and co-workers [1977, 1980, 1981] have identified a low-molecular weight lead-binding protein in the blood of industrially-exposed workers which appeared to act as a nontoxic compartmentalization of lead. It has been suggested that workers with a diminished capacity for synthesizing this lead-binding protein may be at increased risk for developing lead toxicity at relatively low blood lead levels.

TESTES AND SEX ACCESSORY ORGANS

Aspects of the male reproductive organs have been studied following lead exposure and often have resulted in conflicting findings in the literature. Golubovich et al [1968] reported testicular degeneration and significantly lower ribonucleic acid content in testes of lead acetate-treated rats. They suggested that reduced ribosomal activity and impaired protein synthesis was associated with lead toxicity. In contrast, studies by Hilderbrand et al [1973] and by Der et al [1976] found that lead acetate had no effect upon rat testicular weights. Morgan et al [1975], working with coturnix quail, observed reduced testicular weights in birds receiving 1,000 μg/g lead in the diet; lower levels of lead did not produce this effect. Kendall et al [1981] also observed that lead produced testicular damage in turtle doves.

Der et al [1976] studying other constituents of the male reproductive system failed to observe any significant reduction in the weights of the epididymus or the seminal vesicle from lead-treated rats. Hilderbrand et al [1973] reported similar findings for the rat seminal vesicle following a 30-day treatment period with lead acetate.

Conflicting reports also appear in the literature concerning the effect of lead on the prostate. Hilderbrand et al [1973] reported that lead acetate exposure resulted in significant decreases in the weight of the rat prostate gland and distinct histological hyperplasia. Studies by Der et al [1976], using a different strain of rat, failed to reveal any significant changes in prostatic weights of lead-treated animals. However, when lead acetate was injected directly into the adult male rat prostate, there was a significant reduction in prostatic weight after 60 days [Khare et al, 1978]. In addition, Khare et al [1978] observed edema of prostatic stroma and excessive periprostatic fibrosis, but no change in the weights of the testes, epididymus or seminal vesicle. Donovan et al [1980] reported that lead interfered with the binding of androgens by mouse prostate gland cytosolic receptors.

BIOCHEMICAL AND CHROMOSOMAL ABERRATIONS

In addition to the evidence of lead-induced alterations in spermatogenesis and male reproductive organs, there is evidence of lead-induced changes that indirectly affect the male endocrine system. One such change is the effect of lead on the cytochrome P-450-containing mixed function oxidases (MFO) in the liver [Aitio et al, 1978; Goldberg et al, 1977; Meredith et al, 1977].

The liver can metabolize a variety of lipophilic compounds, making them more water soluble and, hence, facilitate their excretion. The enzyme system responsible for this hepatic bioconversion is the cytochrome P-450-containing mixed function oxidase system. Structurally similar enzyme systems have been found in many organs including steroidogenic cells. In the testes, as well as in the adrenal cortex and ovaries, cytochrome P-450-containing enzymes are responsible for many of the biochemical reactions necessary for the synthesis of steroids. MFO enzymes in steroidogenic organs react with other lipophilic compounds, including certain classes of drugs. Due to the structural similarity of these hepatic and extrahepatic enzyme systems and the avidity of substrates that each is capable of reacting with, a compound found to alter the activity of MFO systems in one tissue may have comparable actions on the MFO enzymes in still other tissues.

The intraperitoneal or intravenous administration of lead has been reported to depress heme synthesis, reduce cytochrome P-450 content, and alter the activities of the cytochrome P-450-containing hepatic enzymes [Alvares et al, 1972; Scoppa et al, 1973; Moore and Meredith, 1979]. The potential for lead to exert similar effects on the cytochrome P-450-containing enzymes of the testes could affect the male reproductive capacity. Krasovskii et al [1979] have observed alterations in testicular oxidizing enzymes following lead exposure of rats. Bruch et al [1972] found that rats inhaling particulate lead oxide exhibited considerable loss in the activity of the benzy(a)pyrene hydroxylating enzyme, a cytochrome P-450-containing mixed function oxidase, from alveolar macrophages and pneumocytes. Further studies are needed to determine the mechanism of this loss of activity, especially since benzo(a)pyrene is also a substrate for

TABLE II. Summary of Studies on Occupational Exposure to Lead and its Relationship to Chromosomal Aberrations*

Exposed subjects	Type of aberration	References
Positive findings[a]		
Lead oxide factory workers	Chromatid and chromosome breaks	Schwanitz et al [1970]
Chemical factory workers	Chromatid gaps, breaks	Gath and Thiess [1972]
Zinc plant workers	Gaps, fragments, rings, exchanges, dicentrics	DeKnudt et al [1973]
Blast-furnace workers, metal grinders, scrap metal workers	Gaps, breaks, hyperploidy, structural abnormalities	Schwanitz et al [1975]
Battery plant workers and lead foundry workers	Gaps, breaks, fragments	Forni et al [1976]
Lead oxide factory workers	Chromatid and chromosome aberrations	Garza-Chapa et al [1977]
Battery melters, tin workers	Dicentrics, rings, fragments	DeKnudt et al [1977]
Ceramic, lead, and battery workers	Breaks, fragments	Sarto et al [1978]
Smelter workers	Gaps, chromatid and chromosome aberrations	Nordenson et al [1978]
Battery plant workers	Chromatid and chromosome aberrations	Forni et al [1980]
Negative findings[b]		
Policemen		Bauchinger et al [1972]
Lead workers		Schmid et al [1972]
Shipyard workers		O'Riordan and Evans [1974]
Smelter workers		Bauchinger et al [1976]
Volunteers (ingested lead)		Bulsma and DeFrance [1976]
Children (near a smelter)	Bauchinger et al [1977]	

*Adapted from IARC Monograph [1980].
[a]Chromosomal aberrations positively linked to lead exposure.
[b]No significant correlation between lead exposure and chromosomal aberrations.

testicular mixed function oxidases. Wetzel and Colby [1982] have recently reported that certain heavy metals (eg, cadmium) affecting testicular metabolism of benzo(a)pyrene had similar effects on testicular steroid synthesis.

Excess lead absorption may cause cytogenetic changes in the male reproductive tract. Table II summarizes some of the recent research that has been carried out to determine if human exposure to lead results in chromosomal aberrations. Epidemiologic studies on workers exposed to lead (and other chemicals/metals) have provided some insight into this potential toxicologic problem. Unfortunately, many of these workers were also exposed to other potentially toxic substances including zinc, cadmium, and arsenic [Evans, 1976]. To date, at least ten studies have reported increased incidence of chromosomal aberrations in workers exposed to lead; at least six studies have not observed such toxic actions. In addition to several such studies where workers were exposed to more than one toxic agent, the statistical analyses of many of these studies have not been satisfactory [DeKnudt, 1978], appropriate controls have not always been available [Gerber et al, 1980], and sample size has often been small [Damstra, 1977]. The major criticism seemed to be that negative results (lead exposure resulting in no significant increase in chromosomal aberrations) were observed in studies using cells cultured for two days, while positive results (lead exposure linked to chromosomal aberrations) were usually reported from cells cultured for three days. Forni et al [1980] studied both the effects of lead exposure and length of time cells were cultured, and reported that lead-induced chromosomal aberrations were not a culture-induced artifact. Lymphocytes from workers occupationally exposed to lead showed significantly increased rates of metaphases, excluding gaps, at both culture times. Further research will determine if this deleterious effect of lead on genetic material also occurs in the male reproductive system.

OTHER ENDOCRINE EFFECTS

Petrusz et al [1979] observed that treatment of newborn male rats with lead by gastric gavage increased pituitary follicle-stimulating hormone (FSH) content but otherwise had no effect on other pituitary or serum gonadotropins. These results suggested that "subclinical" neonatal lead poisoning of the male rat does not have an effect on gonadal development and subsequent function that is mediated by central neuroendocrine mechanisms. Studies by Lancranjan et al [1975] failed to reveal any lead-induced changes in androgens in lead-exposed workers. These investigators reported that total urinary neutral 17-ketosteroid levels were normal in workmen exposed to lead. The finding that total urinary gonadotropin levels were unaffected by lead excluded hypothalamic-adenohypophysial factors despite the presence of altered spermatogenesis. Therefore, germinal epithelial insufficiency was due to a direct toxic action of lead upon the gonads [Lancranjan et al, 1975]. The studies showing lead-induced testicular degeneration [Golubovich et al, 1968] or inhibition of cytochrome P-450-containing enzymes [Alvares et al, 1972] tend to support a direct effect of lead upon the male reproductive system. Thomas et al [1981] have reported that lead can directly inhibit the binding of dihydrotestosterone to the androgen receptor in male sex accessory organs. This inhibition was found to be of intermediate potency with Ca^{2+}, Mg^{2+}, Mn^{2+}, and K^{+} being weaker inhibitors and Cu^{2+}, Hg^{2+}, Zn^{2+}, and Cd^{2+} more effective inhibitors of steroid-hormone binding.

REFERENCES

Aitio A, Ahotupa M, Parkki MG (1978): Inhibition of drug metabolizing enzymes by heavy metals in vitro. Biochem Biophys Res Commun 83:850–856.

Alvares AP, Leigh S, Cohn J, Kappas A (1972): Lead and methyl mercury: Effects of acute exposure on cytochrome P-450 and the mixed function oxidase system in the liver. J Exp Med 135:1406–1409.

Bauchinger M, Schmid E, Schmidt D (1972): Chromosomenanalyse bei Verkehrspolizisten mit erhohter Bleilast. Mutat Res 16:407–412.

Bauchinger M, Schmid E, Einbrodt HJ, Dresp J (1976): Chromosome aberrations in lymphocytes after occupational exposure to lead and cadmium. Mutat Res 40:57–62.

Bauchinger M, Dresp J, Schmid E, Englert N, Krausse C (1977): Chromosome analyses of children after ecological lead exposure. Mutat Res 56:75–80.

Bell JU, Thomas JA (1980): Effects of lead on mammalian reproduction. In Singhal RL, Thomas JA (eds): "Lead Toxicity." Baltimore: Urban and Schwarzenberg, pp 169–186.
Bruch T, Brockhaus A, Dehnen W (1972): In: "Proceedings International Symposium Environmental Health Aspects of Lead." Amsterdam, p 221.
Bulsma JB, DeFrance HF (1976): Cytogenetic investigations in volunteers ingesting inorganic lead. Int Arch Occup Environ Health 40:33–36.
Damstra T (1977): Toxicological properties of lead. Environ Health Perspec 19:297–307.
DeKnudt Gh (1978): PhD Thesis. Cited in: Gerber et al (1980).
DeKnudt Gh, Leonard A, Ivanov B (1973): Chromosome afferations observed in male workers occupationally exposed to lead. Environ Physiol Biochem 3:132–138.
DeKnudt Gh, Manuel Y, Gerber GB (1977): Chromosomal aberrations in workers professionally exposed to lead. J Toxicol Environ Health 3:885–890.
Der R, Fahim Z, Yousef M, Fahim M (1976): Environmental interaction of lead and cadmium on reproduction and metabolism of male rats. Res Commun Chem Pathol Pharmacol 14:689–713.
Donovan MP, Schein LG, Thomas JA (1980): Inhibition of androgen-receptor interaction in mouse prostate gland cytosol by divalent metal ions. Mol Pharmacol 17:156–162.
Evans HJ (1976): Cytological methods for detecting chemical mutagens. In Hollaender A (ed): "Chemical Mutagens," Vol 4. New York: Plenum, pp 1–29.
Eyden BP, Maisin JR, Mattelin G (1978): Long-term effects of dietary lead acetate on survival, body weight and seminal cytology in mice. Bull Environ Contam Toxicol 19:266–273.
Forni A, Cambiaghi G, Secchi GC (1976): Initial occupational exposure to lead. Arch Environ Health 31: 73–78.
Forni A, Sciame A, Bertazzi PA, Alessio L (1980): Chromosome and biochemical studies in women occupationally exposed to lead. Arch Environ Health 35:139–146.
Fowler BA, Kimmel CA, Woods JS, McConnell EE, Grant LD (1980): Chronic low-level lead toxicity in the rat. III. An integrated assessment of long-term toxicity with special reference to the kidney. Toxicol Appl Pharmacol 56:59–77.
Garza-Chapa R, Leal-Garza CH, Molina-Ballesteros G (1977): Chromosome analysis in subjects exposed professionally to lead contamination. Arch Invest Med 8:11–20.
Gath J, Thiess AM (1972): Chromosome studies in chemical workers. Zentrabl Arbeitsmed Arbeitsschutz Prophyl Ergonomie 22:357–362.
Gerber GB, Leonard A, Jacquet P (1980): Toxicity, mutagenicity and teratogenicity of lead. Mutat Res 76: 115–141.
Gilfillan SC (1965): Lead poisoning and the fall of Rome. J Occup Med 7:53–60.
Goldberg A, Meredith PA, Miller S, Moore MR, Thompson GG (1977): Haem biosynthesis and hepatic drug metabolism in lead poisoned rats. Br J Pharmacol 61:134P–135P.
Golubovich E, Aukimenko MM, Chirkovia EM (1968): Taksik Khim Veshchestv 10:64.
Grant LD, Kimmel CA, West GL, Martinez-Vargas CM, Howard JL (1980): Chronic low-level lead toxicity in the rat. II. Effects on postnatal physical and behavioral development. Toxicol Appl Pharmacol 56:42 58.
Hamilton A, Hardy HL (1974): Hereditary lead poisoning. In: "Industrial Toxicology." Acton, Ma: Publishing Sciences Group, pp 119–121.
Henderson A (1824): "A History of Ancient and Modern Wines," p 339.
Hilderbrand BC, Der R, Griffin WT, Fahim MS (1973): Effect of lead acetate on reproduction. Am J Obstet Gynecol 115:1058–1065.
IARC Monograph (1980): Lead and lead compounds. 23:325–415.
Kendall RJ, VEit HP, Scanlon PF (1981): Histological effects and lead concentrations in tissues of adult male ringed turtle doves that ingested lead shot. J Toxicol Environ Health 8:649–658.
Khare N, Der R, Ross G, Fahim M (1978): Prostatic cellular changes after injection of cadmium and lead into rat prostate. Res Commun Chem Pathol Pharmacol 20:351–365.
Krasovskii GN, Vasukovich LY, Chariev OG (1979): Experimental study of biological effects of lead and aluminum following oral administration. Environ Health Perspect 30:47–51.
Lancranjan I, Popescu HI, Gavanescu O, Klepsch I, Serbanescu M (1975): Reproductive ability of workmen occupationally exposed to lead. Arch Environ Health 30:396–401.
Mann T (1964): "The Biochemistry of Semen and of the Male Reproductive Tract." New York: John Wiley and Sons.
Meredith PA, Campbell BC, Moore MR, Goldberg A (1977): The effects of industrial lead poisoning on cytochrome P-450 mediated phenazone (antipyrene) hydroxylation. Eur J Clin Pharmacol 12:235–239.

Moore MR, Meredith PA (1979): The carcinogenicity of lead. Arch Toxicol 42:87–94.
Morgan GW, Edens FW, Thaxtox P, Parkhurst CR)1975): Toxicity of dietary lead in Japanese quail. Poult Sci 54:1636–1642.
Nordenson I, Beckman G, Backman L, Nordstrom S (1978): Occupational and environmental risk in and around a smelter in northern Sweden. IV. Chromosomal aberrations in workers exposed to lead. Hereditas 88:263–268.
Nriagu JO (1980): Lead in the atmosphere and its effect on lead in humans. In Singhal RL, Thomas JA (eds): "Lead Toxicity." Baltimore: Urban and Schwarzenberg, pp 483–503.
O'Riordan ML, Evans HJ (1974): Absence of significant chromosome damage in males occupationally exposed to lead. Nature 247:50–53.
Petrusz P, Weaver CM, Grant LD, Mushak P, Krigman MR (1979): Lead poisoning and reproduction: Effects on pituitary and serum gonadotropins in neonatal rats. Environ Res 19:383–391.
Puhac I, Hrgovic N, Stankovic M, Popvic S (1963): Laboratory investigation on the possibility of employing lead compounds as raticides by decreasing the reproductive capacity of rats. Acta Verterin 13: 3–9.
Raghaven SRV, Gonick HC (1977): Isolation of low molecular weight lead-binding protein from human erythrocytes. Proc Soc Exp Biol Med 155:164–167.
Raghaven SRV, Culver BD, Gonick HC (1980): Erythrocyte lead-binding protein after occupational exposure. I. Relationship to lead toxicity. Environ Res 22:264–270.
Raghaven SRV, Culver BD, Gonick HC (1981): Erythrocyte lead binding protein after occupational exposure. II. Influence on lead inhibition of membrane Na^+, K^+-adenosinetriphosphatase. J Toxicol Environ Health 7:561–568.
Sarto F, Stella M, Acqua A (1978): Cytogenetic studies in 20 workers occupationally exposed to lead. Med Lav 69:172–180.
Schmid E, Bauchinger M, Pietruck S, Hall G (1972): Die cytogenetische Wirkung von Blei in menschlichen peripheren Lymphocyten in vitro und in vivo. Mutat Res 16:401–406.
Schwanitz G, Lehnert G, Gebhart E (1970): Chromosomenschaden bei beruflicher Bleibelastung. Dtsch Med Wochenschr 95:1636–1641.
Schwanitz G, Gebhart E, Rott HD, Schaller KH, Essing HG, Lauer O, prestele H (1975): Chromosomenuntersuchungen bei Personen mit beruflicher Bleiexposition. Dtsch Med Wochenschr 100:1007–1011.
Scoppa P, Roumengous M, Penning W (1973): Hepatic drug metabolizing activity in lead-poisoned rats. Experientia 29:970–972.
Thomas JA, Donovan MP, Waalkes MP, Curto KA (1981): Interactions between protein and heavy metals in the rodent prostatic cell. In: "The Prostatic Cell: Structure and Function," Part A. New York: Alan R. Liss, pp 459–474.
Timm F, Schulz G (1966): Hoden und Schwermetalle. Histochemistry 7:15–21.
Varma MM, Joshi SR, Adeyemi AO (1974): Mutagenicity and infertility following administration of lead sub-acetate to Swiss male mice. Experientia 30:486–487.
Wetzel LT, Colby HD (1982): Comparative effects of cadmium administration on testicular and hepatic microsomal metabolism in rats (Abstract). Fed Proc 41:1569.

American Journal of Industrial Medicine 4:135–147 (1983)

Adaptive Biochemical Repair Response Toward Germ Cell DNA Damage

I. P. Lee, PhD

The final expression of chemical toxicity is determined by pharmacokinetic, pharmacodynamic, homeostatic, and adaptive factors. In order to understand differential toxicity at the cellular and organ levels, one must consider the main factors which may modulate the toxic effects of chemicals. In the male gonads, such modifying factors are the blood-testis barrier which restricts the testicular penetration of many foreign chemicals, differentially distributed mixed-function oxidase(s) between the germ cell and interstitial cell compartments, and the presence of differing DNA repair capacities in the various stages of spermatogenic cells. DNA repair systems are present in spermatogonia and spermatocytes, while more mature spermiogenic cells (spermatids and testicular sperms) appear deficient in DNA repair capabilities. Furthermore, strain and species differences in the germ cell's ability to repair DNA suggest a diverse response of germ cells to DNA-damaging agents. The capacity of prespermiogenic cells to respond to damage induced by certain chemical mutagens accounts for stage-specific spermatogenic cell toxicity. Monofunctional alkylating agents such as methylmethane sulfonate (MMS) cause single strand DNA breaks in prespermiogenic cells which are repaired expeditiously, while the same germ cell types are unable to repair DNA damage induced by bifunctional or polyfunctional alkylating agents. DNA damage induced by polyfunctional alkylating agents involves inactivation of DNA template as a result of inter- and/or intrastrand DNA, which is more slowly repaired with greater error frequency in the newly-synthesized DNA. Thus, both bifunctional and polyfunctional alkylating agents are more toxic to male germ cells than are the monofunctional alkylating agents. DNA damage measured by the alkaline elution analysis appears to be a relatively simple, sensitive indicator of germ cell DNA damage. The DNA repair system appears to be another protective mechanism against monofunctional alkylating agents such as MMS, while the same adaptive system is ineffective in overcoming the effects of bifunctional or polyfunctional alkylating agents. DNA repair rates need to be quantified and factored into the pharmacokinetic model being developed for extrapolation from laboratory animals to man.

Key words: germ-cell damage, DNA repair

Laboratory of Reproductive and Developmental Toxicology, National Institute of Environmental Health Sciences, Research Triangle Park, North Carolina.

Address reprint requests to I.P. Lee, Laboratory of Reproductive and Developmental Toxicology, National Institute of Environmental Health Sciences, Research Triangle Park, NC 27709.

0271-3586/83/0401-02-0135$04.00 **© 1983 Alan R. Liss, Inc.**

INTRODUCTION

In recent years, there has been increasing concern that drugs and other chemicals may damage germ cells which, as a consequence, may produce infertility or genetically alter the offspring. Indeed, there is a great deal of information available concerning the effects of chemicals on male reproduction in experimental animals and in man [Fox and Fox, 1967; Gomes, 1977; Jackson, 1973; Lee, 1981; Lucier et al, 1977; Neuman et al 1976; Sherins and DeVita, 1973; Sieber and Adamson, 1975]. The underlying mechanisms of toxin action, however, are not clearly understood. Spermatogenesis is a unique system in which germ cells synchronously differentiate into only one type of cell—the spermatozoa. The kinetics of spermatogenesis are precise and the cellular differentiation is hormone dependent. Thus, spermatogenic cells can be directly or indirectly affected by chemicals. In addition, by affecting Sertoli cells either directly or indirectly, spermeation may be premature or delayed. The ultimate result of damage to spermatogenic cell differentiation is interference with the production of adequate numbers of normal spermatozoa or more importantly, mutated spermatozoa perhaps associated with pregnancy wastage, congenital malformations, and/or in heritable genetic diseases.

The expression of chemical toxicity is determined by fundamental pharmacokinetic, pharmacodynamic, homeostatic, and adaptive factors in the target cells. Pharmacokinetic factors such as absorption, distribution, binding or localization in cells, metabolic activation and inactivation, and excretion determine the concentrations of chemicals at its receptor. The pharmacodynamic response agreement involves molecular events triggered by an adequate amount of active chemicals at its receptor and may alter normal cellular, biochemical, and/or physiological processes. Cellular processes respond to any induced alterations in an effort to maintain homeostasis. There may be adaptive factors such as repair which can compensate for the altered biological state. Circulating levels of chemicals are determined by the balance of systemic and pharmacokinetic factors (see Fig. 1). In most cases the parent chemical and various metabolites in blood readily gain access to the testicular lymph. In contrast, germ cells in the seminiferous tubules are protected by a biological barrier similar to the blood-brain barrier. Enzymatic toxication and detoxication processes are present in both the seminiferous epithelial cells and the interstitial cell compartment. These enzymes and others respond to chemical and other biological factors. Unscheduled DNA synthesis has been reported in spermatogonia and spermatocytes. Spermatids and spermatozoa apparently lack an effective DNA repair capacity. Chemically-exposed germ cells entering the posttesticular compartment may be unaffected or repaired with or without errors, or lethal mutation may arise, resulting in cell death or early implantation loss. Like the interstitial space, chemicals gain easy entry to the posttesticular compartment. DNA repair, testicular enzymatic toxication and detoxication of chemicals, and the blood-testis barrier (BTB) are all important factors which can modify the final expression of toxic effects. This paper, however, will focus mainly on DNA repair systems in male germ cells.

DNA REPAIR IN SPERMATOGENIC CELLS

There is considerable experimental evidence that eukaryotic cells can recognize and remove DNA damaged by exposure to a variety of chemical mutagens. Moreover, there are apparently two distinct mechanisms by which this repair of eukaryotic cells

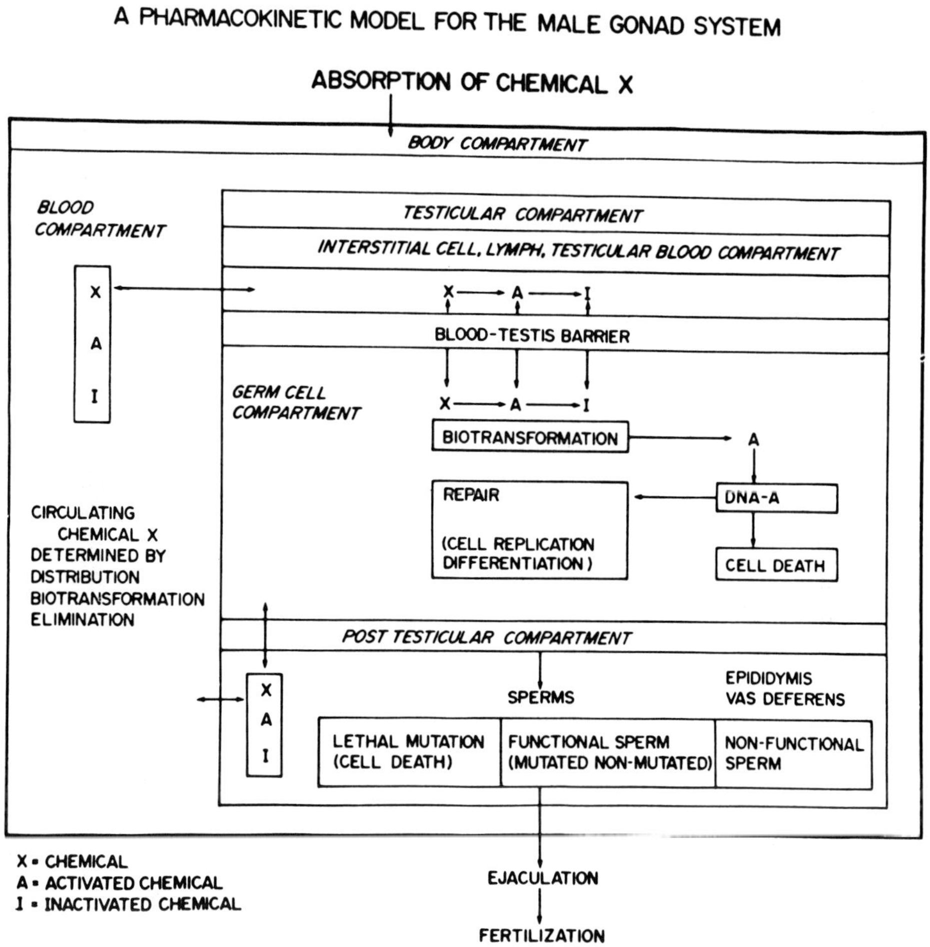

Fig. 1. A model for pharmacokinetic and adaptive factors for male gonad system (from Lee and Dixon, 1978).

is achieved. The first is referred to as excision repair; the alkylated region of DNA is recognized by a complex enzyme system that excises the modified base from the DNA strands, degrades it, resynthesizes the damaged section and then inserts this newly synthesized piece of DNA into preexisting DNA to reconstitute the complimentary DNA strands. Evidence for this form of repair has come from measurements of the loss of mutagen-induced substituents on DNA [Cranthorn and Roberts, 1966] and from the detection of nonsemiconservative DNA synthesis in alkylated cells [Ayad et al, 1969]. The uptake of ^{3}H-thymidine into cells that are not in the S-phase of the cell cycle is generally indicative of unscheduled DNA synthesis [Lieberman et al, 1971]. The second type of repair in eukaryotic cells is postreplication repair [Lehman, 1972]. It was found using thymidine labeling that DNA strands synthesized in irradiated eukaryotic cells are initially of lower than normal molecular weight. This suggests synthesis only along the undamaged portions of the DNA template. With further incubation, these low molecular weight nascent chains join into strands of original molecular weight following cell division. The pyrimidine dimers remain in the original DNA templates. Therefore, this type of DNA repair mechanism, which bypasses damage in the DNA templates, is thought to be another important mechanism for eukaryotic cell DNA repair.

It has been shown that both physical (UV and X irradiation) and chemical agents can damage DNA molecules. Damage inflicted on the DNA templates, unless repaired, may interfere with transcription and cell replication or result in lethal mutations (cell death). Mutations may also result in transformed cells, or genetic changes that can be accumulated, in the case of germ cells, by generations of stem cell replication. DNA repair deficiency in eukaryotic somatic cells has been shown to be related to genetic disease and cancer. Xeroderma pigmentosum, ataxia telangiectasia, Fanconi's anemia, and several other genetic and tumorigenic diseases are associated with a deficiency in DNA repair [Strong, 1977]. We have determined that male mouse strains also differ in the ability of male gonads to repair chemically induced DNA damage [Lee and Suzuki, 1981]. Radiation-induced DNA repair synthesis in spermatogonia and spermatozoa of mice [Ono and Okada, 1977] and chemically induced unscheduled DNA synthesis in prespermiogenic cells have been reported by several investigators [Sega, 1974; Lee, 1978; Schmid et al, 1978].

In an attempt to better understand the mechanism of action and to detect germ cell damage induced by chemicals, DNA repair activity in various spermatogenic cell types has been investigated. A combination of several biochemical techniques including velocity sedimentation cell separation [Lee and Dixon, 1972], alkaline sucrose gradient centrifugation [Elkind, 1971], and alkaline elution analysis [Kohn and Grimek-Ewig, 1973] was applied to spermatogenic cells. Unscheduled DNA synthetic activities and alkaline elution analysis were used to determine the dose-response relationship for methylmethane sulfonate (MMS)-induced DNA single strand breaks and the time response for the reversal of effect [Lee and Dixon, 1978].

ISOLATION AND HISTOCHEMISTRY OF PREPUBERTAL PRESPERMIOGENIC CELLS

Prepubertal prespermiogenic cells isolated by a combination of brief enzymatic digestion of the minced seminiferous tubules followed by albumin-gradient centrifugation resulted in an enriched prespermiogenic cell population relatively free from cell debris and interstitial cells. It has been demonstrated that carefully-teased seminiferous tubules exposed to 0.1% trypsin at room temperature for 10 min damaged interstitial cells extensively without visible damage to prespermiogenic cells [Dufau, 1979]. Furthermore, Sertoli cells are extremely resistant to collagenase trypsin-chicken serum (CTC)-induced release from tubular myoid cell membranes under the condition described [Tung et al, 1975]. To assess the purity of prespermiogenic cell population, a specific marker enzyme for interstitial cells, Δ^5-3β-hydroxysteroid dehydrogenase, was used [Ballie and Griffith, 1964]. The suspension of isolated prespermiogenic cells (10^6 cells/ml) was incubated with 3β-hydroxypregnen-5-en-20-one (3×10^{-5} M) in propylene glycol, 1.2×10^{-3} M nitro blue tetrazolium at 32°C for 60 min. The sites of enzyme activity were marked by the deposition of characteristic dark blue, densely aggregated granules of diformazan in the interstitial cells. The histochemical results showed only 2% of the total cell population to be the interstitial cells; the remaining cells appeared to be prespermiogenic cells which included spermatogonia, leptotene, zygotene, pachytene, and diplotene spermatocytes. These isolated cells are designated as prespermio-

TABLE I. Effects of Hydroxyurea Concentrations on the Semiconservative DNA Synthesis of Prespermiogenic Cells of CD-1 Mouse Strain

Hydroxyurea (mM)	Incubation time (hr) (dpm/5 × 10^5 cells)[a] 1	2	4
0	2,310 ± 76 (10)	4,332 ± 126 (10)	7,499 ± 118 (10)
1	635 ± 66 (3)[b]	794 ± 80 (3)[b]	2,934 ± 480 (3)[b]
2.5	433 ± 68 (3)[b]	664 ± 34 (3)[b]	1,688 ± 290 (3)[b]
5	394 ± 31 (3)[b]	535 ± 47 (3)[b]	1,045 ± 21 (3)[b]
10	332 ± 39 (3)[b]	536 ± 47 (3)[b]	865 ± 57 (3)[b]
20	228 ± 22 (11)[b]	470 ± 35 (11)[b]	781 ± 84 (8)[b]

[a]The values represent mean ± SE; numbers in parentheses indicate the number of experiments. Varying concentrations of HU were added and incubated for 30 min prior to the addition of ^{3}H-thymidine.
[b]Significantly different from the controls ($p < 0.05$). Experimental values were analyzed by means of the Student's t-test [Lee and Suzuki, 1981].

genic cells. Cell viability of the isolated prespermiogenic cells was 95% based on trypan blue cell exclusion.

HYDROXYUREA INHIBITION OF S-PHASE DNA SYNTHESIS IN CD-1 MOUSE PRESPERMIOGENIC CELLS

The effects of hydroxyurea (HU) on semiconservative DNA synthesis in isolated prespermiogenic cells from CD-1 is shown in Table I. Semiconservative DNA synthesis in the isolated prespermiogenic cells was reasonably linear up to 4 hr of incubation at 32°C. In vitro exposure of 10^6 cells to varying concentrations showed concentration-dependent inhibition of semiconservative DNA synthesis in isolated prespermiogenic cells. At the highest HU concentration, the extent of inhibition of semiconservative DNA synthesis in vitro was approximately 90%, irrespective of incubation time with ^{3}H-thymidine. Although an HU concentration as high as 20 mM failed to completely inhibit semiconservative DNA synthesis, 20 mM HU concentration has been useful in studying nonsemiconservative DNA synthesis.

IN VITRO DETERMINATION OF METHYLMETHANE SULFONATE-INDUCED NONSEMICONSERVATIVE DNA SYNTHESIS IN PRESPERMIOGENIC CELLS

In vitro exposure of 10^6 cells to varying doses of MMS induced a significant degree of nonsemiconservative DNA synthesis during the incubation period (Table II). The in vitro MMS-induced nonsemiconservative DNA synthesis was relatively linear during the first 4 hr of incubation, irrespective of MMS dose. However, at MMS concentrations greater than 0.4 mM, the nonsemiconservative DNA synthesis appears to approach V_{max}. In contrast, both testicular and epididymal spermatozoa failed to demonstrate any nonsemiconservative DNA synthetic activity following MMS exposure. Although no discernible nonsemiconservative DNA synthesis was detected in late stages of spermatids, other researchers have observed nonsemiconservative DNA synthesis by early spermatids [Schmid et al, 1978; Sega et al, 1976]. It thus appears that as

TABLE II. The Unscheduled DNA Synthesis in the Prespermiogenic Cells of CD-1 Mouse Strain Following Varying MMS Concentrations

		Incubation time (hr) (dpm/5 × 10^5 cells)[a]		
MMS (mM)	(mg/l)	1	2	4
0[b]	0	227 ± 22 (11)	470 ± 35 (11)	752 ± 84 (8)
0.05	5.6	360 ± 18 (3)[c]	756 ± 90 (3)[c]	1,172 ± 222 (4)[c]
0.10	11.3	458 ± 26 (3)[c]	1,008 ± 90 (3)[c]	2,253 ± 109 (4)[c]
0.20	22.5	508 ± 29 (4)[c]	1,437 ± 99 (4)[c]	3,199 ± 84 (4)[c]
0.40	45.0	536 ± 21 (5)[c]	1,688 ± 124 (3)[c]	3,265 ± 309 (3)[c]
0.80	90.0	589 ± 46 (4)[c]	1,948 ± 163 (4)[c]	3,078 ± 117 (4)[c]

[a]Values represent mean ± SE, and the number in the parentheses indicates the number of experiments.
[b]Hydroxyurea (20 mM) was incubated for 30 min prior to addition of ^{3}H-thymidine. HU suppressed 90% of semiconservative DNA synthesis of spermatogenic cells. To obtain the net unscheduled DNA synthetic activity, the total dpm of HU control was subtracted from the total activity obtained with MmS plus HU.
[c]Significantly different from the controls ($p < 0.05$). Experimental values were analyzed by means of the Student's t-test [Lee and Suzuki, 1981].

spermiogenic cell DNA condenses as these cells differentiate, late spermiogenic cells lose their ability to repair DNA damage.

ALKALINE ELUTION TECHNIQUE TO DETERMINE DNA SINGLE STRAND BREAKS

The alkaline elution technique for the measurement of DNA strand breaks is based on the behavior of large DNA strands released from mammalian cells by lysis on membrane filters [Kohn and Grimek-Ewig, 1973]. This technique measures the rate at which DNA single strands, released in alkali, are able to pass through filters; under certain conditions this depends on strand length and on the presence of interstrand or DNA-protein crosslinks. Two basic phenomena appear to control the ability of DNA single strands to penetrate filters under alkaline conditions: 1) The filter mesh retards the penetration of long DNA strands in a size-dependent manner; 2) linkage of DNA to protein retards penetration of the DNA, apparently because of adsorption of protein to the filter. The kinetics of elution of DNA in an alkaline solvent pumped slowly through the filter can provide sensitive and convenient evidence of DNA strand breaks and crosslinks.

In the various forms of alkaline elution technique, cells are first loaded on a membrane filter, then washed with a physiological salt solution and lysed with detergent. DNA is released in native double-stranded form and is retained on the filter, while most of the cell protein and RNA are washed through by the detergent-lysis solution. A proteinase digestion can be inserted at this step to reduce or eliminate the effects of DNA-protein crosslinks. The filter is then eluted with an alkaline solution which is pumped slowly through as fractions are collected to determine the kinetics of DNA elution from the filter. For the determination of crosslinks, the cells are either X irradiated or treated with methylmethane sulfonate (MMS) prior to lysis in order to introduce single strand breaks and to measure the extent of crosslinks. The greater the crosslinks between interstrand(s) and/or intrastrands of DNA, the slower the DNA single strand elution from the filter.

In a typical procedure, prepubertal animals are injected IP with 0.5 μCi ^{3}H-thymidine (Spec. Act: 41–45 Ci/mmol) per gm body weight four times daily for five days. Eighteen hours after the last injection of ^{3}H-thymidine, test chemical mutagens are administered IP. Animals are sacrificed by cervical dislocation, and testes are removed for prespermiogenic cell isolation. The isolated cells (10^5 cells) are diluted in cold phosphate-buffered saline and are filtered through a stack of 2 polyvinyl chloride millipore filters (2 μm pore size and 25 mm diameter). The cells are washed on the filters with cold phosphate buffered saline containing cold thymidine (10^{-6} M) at 20 ml/min. Cell lysis is accomplished by adding 5 ml of a solution consisting of 2 M NaCl, 0.02 M Na_3EDTA, 0.08% deoxycholate, and 0.2% Sarkosyl (pH 10). The lysing solution is allowed to flow through the filter by gravity and is collected in a scintillation vial. When the lysing solution has almost ceased flowing, the filters are washed with 5 ml of cold 0.02 M EDTA. The outflow tube from the filtering funnel is then connected to tubing that passes through a peristaltic pump to a fraction collector. The filters are then eluted with 0.18 M tetrapropyl ammonium hydroxide, 0.02 M H_4EDTA, pH 12.2, at a flow rate of 0.04 ml per min. Since DNA is vulnerable to the effects of light both in alkali and in intact cells [Elkind, 1971; Friedman et al, 1975; Gantt et al, 1978], the entire procedure is conducted in dim light and the elution procedure is carried out in a black lucite box.

Radioactivity is determined by scintillation counting of fraction samples with 10 ml of Aquasol containing 0.5% acetic acid. To determine the radioactivity remaining on the filter, the filter is digested with 0.5 ml of 1 N HCl at 70°C. After cooling to room temperature, 2.0 ml of 0.05 N NaOH is added and the filter is allowed to stand in this alkaline solution for 30 min. After vigorously shaking the filter, 10 ml of Aquasol is added and the radioactivity is determined.

Figure 2 presents a typical alkaline elution profile following MMS treatment. The elution rates of single-strand DNA fragments after 0, 22.5, 45, or 90 mg/kg body weight of MMS administered IP decreased with time. Single-strand DNA fragments were measured at 2, 4, 8, 12, 24, 36, 48, and 72 hr following treatment, and the effects observed appear dose dependent. The elution rates 24 hr after treatment with any of the test doses of MMS indicate partial DNA repair. By 48 hr, the elution rates of single-strand DNA fragments at all doses approached those of controls. After 72 hr, the elution rates at all MMS doses approached control levels, suggesting complete DNA repair. Further studies with DNA repair time have demonstrated that the durations for complete DNA repair in prespermiogenic cells following a single IP administration of MMS at 22.5, 45, and 90 mg/kg were found to be 28, 46, and 70 hr, respectively (Fig. 3). Figure 4 presents a dose-response relationship between the total amount of ^{3}H-thymidine labeled DNA single strands eluted versus the MMS doses at 11.25, 22.5, 45, and 90 mg/kg (2 hr following the IP administration). The fractions of total ^{3}H-DNA eluted were 23, 42, and 90% of the total radioactivity of ^{3}H-DNA, respectively, and thus, the effects were clearly dose related.

In order to compare the effects of other chemical mutagen treatment on the alklaline elution of DNA prelabeled with ^{3}H-thymidine, prepubertal animals were treated once intraperitoneally with a monofunctional alkylating agent (MMS, 90 mg/kg), polyfunctional alkylating agents (mitomycin C, 5 mg/kg; triethylene melamine, 1 mg/kg, busulfan, 43.5 mg/kg), DNA intercalators (adriamycin, 10 mg/kg; proflavin, 10 mg/kg; acriflavin, 10 mg/kg), antimetabolites (methotrexate, 200 mg/kg; cytosine arabinoside, 500 mg/kg), and a mitotic inhibitor (vincristine, 5 mg/kg). Unlike the elu-

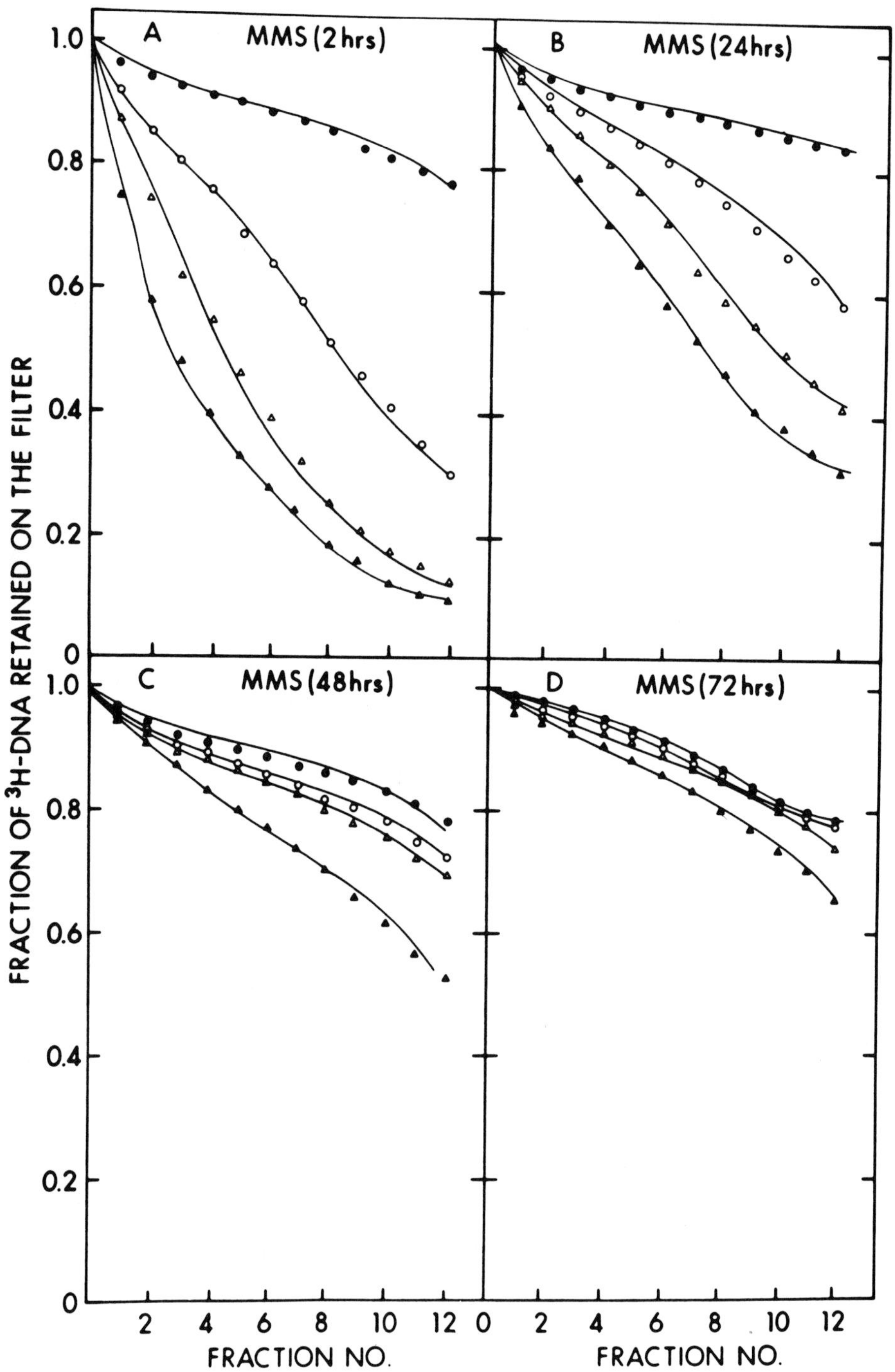

Fig. 2. Alkaline elution rates of a single-strand DNA following MMS treatment. Ordinate is fraction of DNA retained on the filter. Animals were treated intraperitoneally with a single dose of MMS at 0 (●), 22.5 (○), 45 (△), or 90 (▲) mg/kg body weight. Panels A–D represent elution rates at 2, 24, 48, and 72 hr after MMS treatment, respectively.

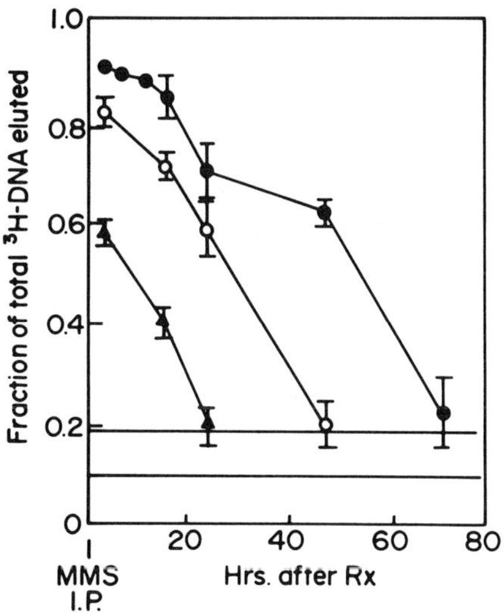

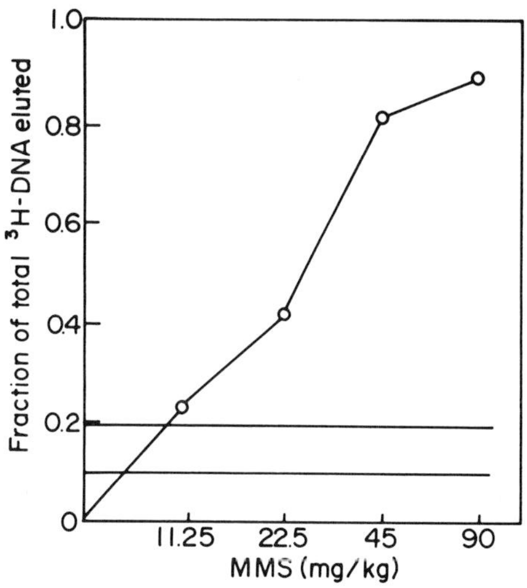

Fig. 3. Fraction of total DNA eluted at various times after a single intraperitoneal treatment of MMS at 22.5 (▲), 45 (○), and 90 (●) mg/kg; (=) represents the range of controls. Each point represents the mean of 3–5 experiments [Lee and Dixon, 1978].

Fig. 4. Fractions of total DNA eluted as function of dose 2 hr after varying doses of MMS treatment. (=) represents the range of controls. Each point represents the mean of three experiments [Lee and Dixon, 1978].

tion rate of germ cell DNA following monofunctional alkylating agents, the elution rate of germ cell DNA following polyfunctional alkylating agents was significantly slower than those of controls and the MMS-treated group (Fig. 5). These results suggest that germ cell DNA following treatment with polyfunctional alkylating agents is very resistant to shearing caused by experimental handling of the germ cells as indicated in controls (percent elution due to DNA shear is approximately 23% of total radioactivity). Furthermore, following polyfunctional alkylating agents, germ cell DNA also becomes resistant to both X irradiation and MMS-induced single strand breaks (Table III). The resistance to alkaline elution following these chemical treatments may be attributed to crosslinking of inter-, intrastrand DNA, and/or DNA-protein. Similar findings with somatic cells have been reported [Ewig and Kohn, 1977]. Moreover, recent work with somatic cells has suggested that polyfunctional alkylating agents produce two kinds of crosslinks: Proteinase resistant (presumably DNA interstrand links) and proteinase sensitive (interpreted as DNA-protein links). Bifunctional alkylating agents (busulfan) were found to be more effective DNA crosslinkers against MMS-induced DNA single strand breaks than TEM or mitomycin C. DNA intercalating agents appeared to retard alkaline elution rates following MMS treatment, but the degree of resistance against MMS was not significantly different from that of controls. In contrast, non-DNA interacting agents failed to demonstrate any significant effects on alkaline elution kinetics. The results, therefore, clearly demonstrate that chemical interactions between prespermiogenic cell DNA differ greatly with various chemical mutagens.

Similarly, a homologous series of alkylsulfonates affect different stages of spermatogenic cells based on dominant lethal studies. For example, MMS affects only

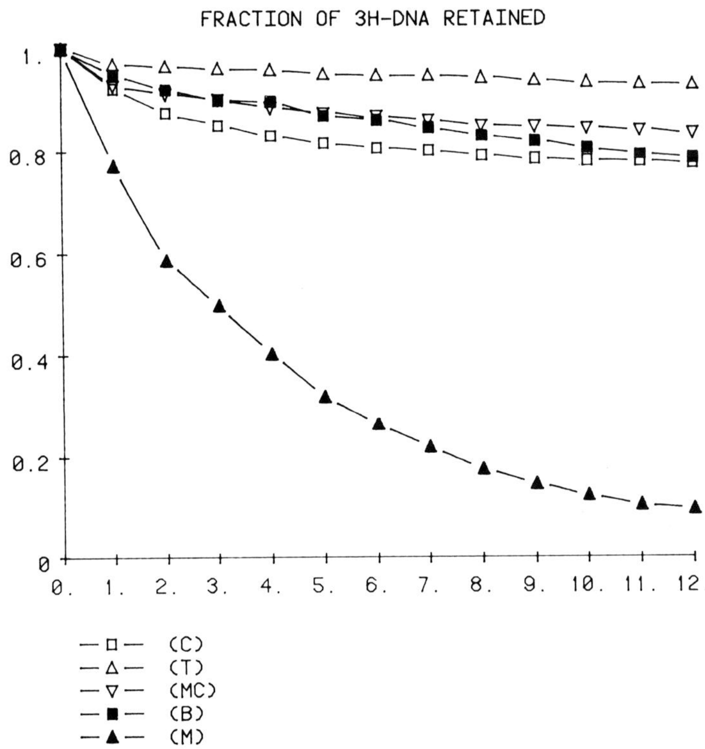

Fig. 5. Alkaline elution rates of single strand DNA 2 hr following 0 control (□), 90 mg/kg MMS (▲), 43.5 mg/kg busulfan (■), 1 mg/kg triethylenemelamine (△), or 5 mg/kg mitomycin C (▽) treatment (a single intraperitoneal administration). Ordinate is fraction of DNA retained on the filter; abscissa is the fraction number. Each point represents the mean of 3 experiments.

spermatozoa and spermatids, while busulfan affects prespermiogenic cells only (Figs. 6, 7). These differential stage-specific effects on spermatogenic cells by two alkylating agents have not been satisfactorily elucidated. Autoradiographic studies offer no evidence for selective localization of alkylating agents in any specific cell stages of spermatogenesis. Since MMS should be capable of alkylating nucleophilic sites irrespective of cell type, the selective dominant lethal effects associated with spermatids and spermatozoa might be attributed to differences in the expression of toxicity by alkylated cells. For example, DNA repair deficiency in spermatids and spermatozoa could result in stage-specific dominant lethality. Alkaline elution studies with prespermiogenic cells suggest that the possible mechanism for the lack of dominant lethal effects of spermatogonia and spermatocytes is due to their ability to repair DNA damage, while spermatids and spermatozoa are unable to repair DNA damage [Lee and Dixon, 1978; Schmid et al, 1978; Sega et al, 1976] and thus express lethal mutations. The lack of DNA repair activity in spermatids and spermatozoa may be attributed to an absence of any DNA repair enzymes and also to the physical state of highly condensed sperm DNA. In Contrast, busulfan-induced dominant lethal mutation is associated with prespermiogenic cells which failed to show any single DNA strand breaks, and DNA repair processes are

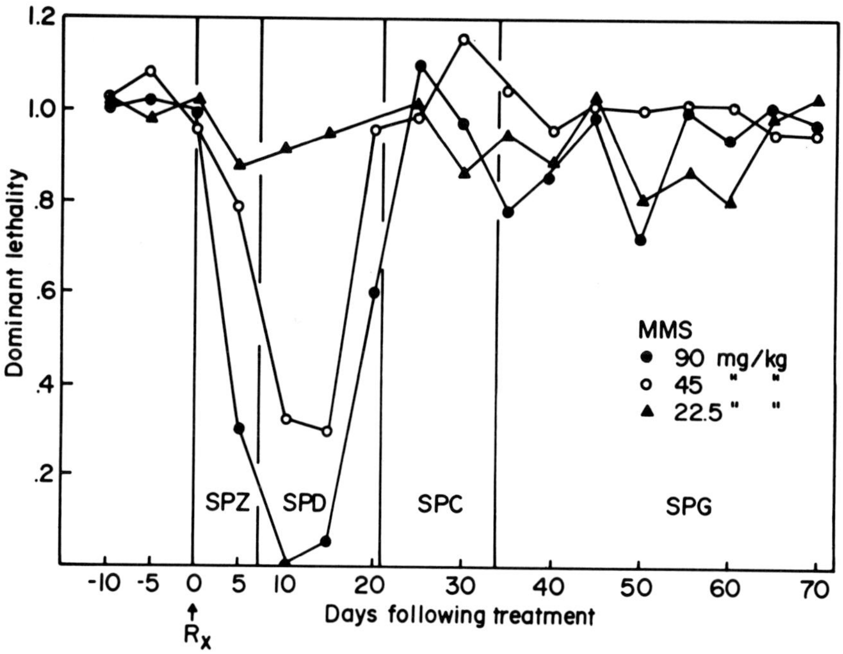

Fig. 6. Dominant lethality data following a single intraperitoneal administration of MMS at 22.5 (▲), 45 (○), or 90 (●) mg/kg body weight. Dominant lethality is expressed as total live embryos of experimental animals/total live embryos of the controls. Each point represents 10 animals. SPZ, spermatozoa; SPD, spermatids; SPC, spermatocytes; SPG, spermatogonia including stem cells [Lee and Dixon, 1978].

TABLE III. Effects of Various Classes of Chemicals on the Prespermiogenic Cell DNA

Chemicals	Dose (mg/kg)	None		MMS (90 mg/kg)	
		% E	% F	% E	% F
Control					
Saline		23.8	76.2	92.2	7.8
$DMSO_4$	20%	21.0	79.0	77.7	22.8
Polyfunctional alkylating agents					
Mitomycin C	5	6.0	93.9	63.5	36.5
TEM	1	7.2	92.8	43.4	56.8
Busulfan	43.5	5.6	94.4	40.2	58.6
DNA-Intercalating agents					
Adriamycin	10.0	9.1	90.9	74.4	25.6
Acriflavin	10.0	20.1	79.9	83.7	28.5
Proflavin	10.0	15.2	84.8	73.8	25.2
Non-DNA interacting agents					
Methotrexate	200	22.4	77.6	82.4	17.6
Ara C	500	16.5	83.5	81.1	18.9
Vincristine	5	22.3	77.7	79.8	20.2

Prepubertal male rats were treated with various classes of chemicals. Four hr after the chemical treatment, 90 mg/kg of methylmethane sulfonate was administrered for 2 hr before sacrifice. Each value presents the mean of two experiments. E, % of total DNA eluted; F, % of total DNA remaining on the filter.

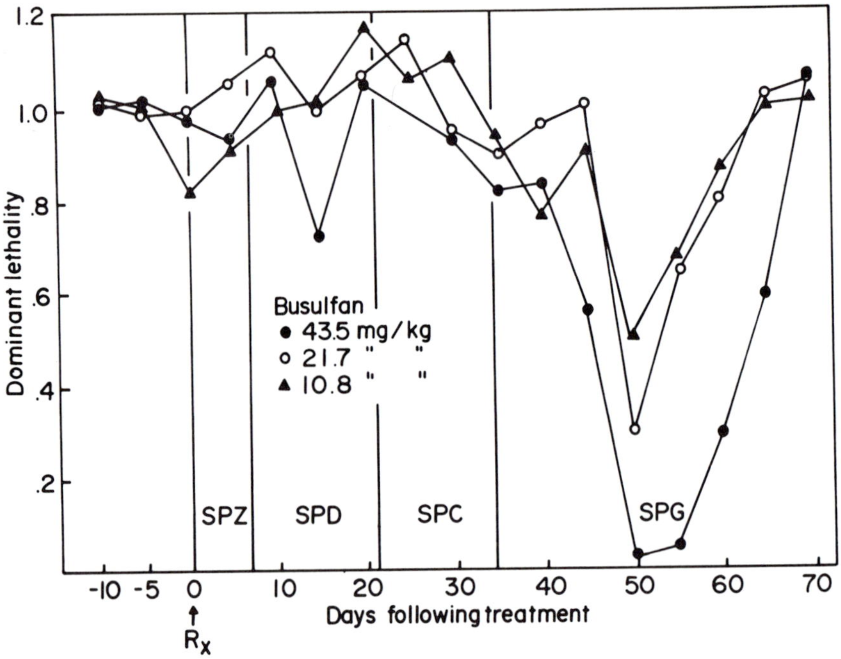

Fig. 7. Dominant lethality data following a single intraperitoneal administration of busulfan at 10.8 (▲), 21.7 (○), or 43.5 (●) mg/kg body weight. Dominant lethality is expressed as total live embryos of experimental animals/total live embryos of the controls. Each point represents 10 animals. SPZ, spermatozoa; SPD, spermatids; SPC, spermatocytes; SPG, spermatogonia including stem cells.

much slower. Dominant lethal effects thus associated with spermatogonia are attributable to difficulties in conserving the fidelity of both semi- and nonsemiconservative DNA synthesis. Roberts et al [1971] suggested that the lethal event following treatment of cells with bifunctional alkylating agents involves inactivation of the DNA template as a result of crosslinking. This suggests that error probability introduced into a complimentary DNA strand should be much greater than that of monofunctional alkylating agents during both semi- and nonsemiconservative DNA synthesis. In addition, the lack of dominant lethal effects associated with spermatids and spermatozoa may be due to the inability of bifunctional alkylating agents to penetrate to DNA nucleophilic sites.

It thus appears that the DNA repair system is another protective mechanism against monofunctional alkylating agents such as MMS, while the same adaptive system may be ineffective in overcoming the bifunctional alkylating agents such as busulfan. DNA repair rates need to be quantified and factored into the pharmacokinetic model described. The DNA damage measured by the alkaline elution analysis appears to be a relatively simple, sensitive indicator of chemical injury in germ cells.

REFERENCES

Ayad SR, Fox M, Fox BW (1969): Nonsemiconservative incorporation of labelled 5-bromo-2′-deoxyuridine in lymphoma cells treated with low doses of methyl methanesulfonate. Mutation Res 8:639–645.

Ballie AH, Griffith K (1964): 3-Beta-hydroxysteroid dehydrogenase activity in the mouse Leydig cells. J Endocrinol 29:9–17.
Cranthorn AR, Roberts JA (1966): Mechanism of the cytotoxic action of alkylating agents on mammalian cells and evidence for the removal of alkylated groups from deoxyribonucleic acid. Nature 211:150–153.
Dufau M (1979): Personal communications.
Elkind MM (1971): Sedimentation of DNA released from Chinese hamster cells. Biophys J 11:502–510.
Ewig RAG, Kohn KW (1977): DNA damage and repair in mouse leukemia L-1210 cells treated with nitrogen mustard, 1,3-bis(2-chloroethyl)-1-nitrosourea, and other nitrosoureas. Cancer Res 37:2114–2122.
Friedman CA, Kohn KW, Erickson LC (1975): DNA chain growth during replication of asynchronous L1210 cells with alkaline sedimentation studies. Biochemistry 14:4018–4023.
Fox BW, Fox M (1967): Biochemical aspects of the actions of drugs on spermatogenesis. Pharmacol Rev 19:21–57.
Gantt R, Parshad R, Ewig RAG, Sanford KK, Jones GM, Tarone R, Kohn KW (1978): Fluorescent light-induced DNA crosslinkage and chromatid breaks in mouse cells in culture. Proc Natl Acad Sci USA 75:3809–3812.
Gomes WR (1977): Pharmacological agents and male fertility. In Johnson AD, Gomes WR (eds): "The Testis." New York: Academic Press, pp 605–628.
Jackson H (1973): Comparative effects of some antispermatogenic chemicals. In Segal SJ (ed): "The Regulation of Mammalian Reproduction." Springfield: Thomas, p 257.
Kohn KW, Grimek-Ewig KA (1973): Alkaline elution analysis, a new approach to the study of DNA single stranded interruptions in cells. Cancer Res 33:1849–1853.
Lee IP, Dixon RL (1972): Antineoplastic drug effects of spermatogenesis studied by velocity sedimentation cell separation. Toxicol Appl Pharmacol 23:20–41.
Lee IP, Dixon RL (1978): Factors influencing reproduction and genetic toxic effects on male gonads. Environ Health Pers 24:117–127.
Lee IP, Suzuki K (1981): Differences in methylmethane sulfonate induced unscheduled DNA synthesis in prespermiogenic cells of various mouse strains. Mutat Res 80:201–211.
Lee IP (1981): Effects of drugs and chemicals on male reproduction. In Spira A, Jouannet P (eds): "Human Fertility Factors." INSERM Monograph 103:311–334.
Lehman AR (1972): Post replication repair of DNA in ultraviolet irradiated mammalian cells. J Mol Biol 66:319–338.
Lieberman MW, Baney RN, Lee RE, Sell S, Farber E (1971): Studies on DNA repair in human lymphocytes treated with proximate carcinogens and alkylating agents. Cancer Res 31:1307–1312.
Lucier GW, Lee IP, Dixon RL (1977): Effects of environmental agents on male reproduction. In Johnson AD, Gomes WR (eds): "The Testis." New York: Academic Press, pp 578–604.
Newman F, Diallo FA, Hasan SH, Schenk B, Traore L (1976): The influence of pharmaceutical compounds on male fertility. Andrologia 8:203–235.
Ono T, Okada S (1977): Radiation induced DNA single strand scission and its rejoining in testicular cells of mouse. Mutat Res 36:213–222.
Roberts JJ, Brent TP, Crathorn AR (1971): Evidence for the inactivation and repair of the mammalian DNA template after alkylation by mustard gas and half mustard gas. Eur Cancer Res 7:515–524.
Schmid B, Lee IP, Zbinden G (1978): DNA repair processes in germ cells demonstrated in ejaculated sperms of rabbits treated with methyl methane sulfonate. Arch Toxicol 40:37–43.
Sega GA (1974): Unscheduled DNA synthesis in the germ cells of male mice exposed in vivo to the chemical mutagen ethylmethane sulfonate. Proc Natl Acad Sci USA 71:4955–4959.
Sega GA, Owns JG, Cummings R (1976): Studies on DNA repair in early spermatid stage of male mice after in vivo treatment with methyl, ethyl, propylm and isopropyl methane sulfonate. Mutat Res 36: 193–212.
Sherins RJ, DeVita VT (1973): Effect of drug treatment for lymphoma on male reproductive capacity. Ann Intern Med 79:216–220.
Sieber SM, Adamson RH (1975): Toxicity of antineoplastic agents in man: Chromosomal aberration, antifertility effects, congenital malformations, and carcinogenic potential. Adv Cancer Res 22:57–155.
Strong LC (1977): Genetic etiology of cancer. Cancer 40:438–444.
Tung PS, Dorrington JS, Fritiz IB (1975): Structural changes induced by follicle-stimulating hormone or dibutyrl cyclic AMP on presumptive Sertoli cells in culture. Proc Natl Acad Sci USA 72:1838–1842.

American Journal of Industrial Medicine 4:149–161 (1983)

Gonadal Injury Resulting From Chemotherapy

Ramona M. Chapman, MD

Radiation, alkylating agents and the vinca alkaloids produce variable degrees of gonadal dysfunction in humans; additional drugs have been linked to gonadal damage in males only. Age at time of exposure and total dose of drug administered are associated with the degree of gonadal injury. Irreversible sterility develops at lower doses in men than in women. The infertile men usually retain sexual function, whereas the infertile women experience premature menopause and sexual dysfunction. Subfertility with normal sexual function is the sequela of lesser degrees of gonadal injury in men and women. For a given drug dose, children apparently sustain less gonadal damage than their adult counterparts. Rigorous reporting of basic essential data will enhance our understanding of drug-induced gonadal damage; incomplete reports only cloud the issue.

Key words: gonadal damage, chemotherapy, radiation, drugs, sexual dysfunction

INTRODUCTION

Now that cancer patients are living longer, the toxic effects of chemotherapeutic drugs on the health and quality of life of these survivors have become important issues in the field of clinical oncology. Patients with non-malignant immune disorders are also being treated with cytotoxic agents, and they too must live with the long-term side effects. Some chemotherapeutic drugs are known to produce gonadal dysfunction (Table I). However, investigators have not yet evaluated all these drugs for this sort of damage.

Many reports of drug-induced gonadal dysfunction describe studies that are focused on one small window of time during which patients of varying ages receive varying total doses of drugs, at times on different schedules. The majority of studies are retrospective; in some series follow-up continued after the closing date of the study. Questions abound concerning the immediate and delayed observable toxicity of cytotoxic agents on gonadal function in prepubertal, pubertal, and adult patients. What is the critical dose at which gonadal damage occurs? At what dose does damage become irreversible? How do the drugs differ in their effects on ovaries and testes? Is it possible that testicular dysfunction is more often reversible than ovarian dysfunction? Since the germ cells of the testes are continually being regenerated, partial destruction of the spermatogonia need not result in permanent azoospermia. The surviving spermatogonia, given time, may repopulate the testes and restore fertility. On the other hand, all the ova that a woman will ever have are presesent at birth. Destruction of ova is an irre-

St. John's Regional Oncology Center, Joplin, Missouri.
Address reprint requests to Ramona M. Chapman, MD., Director, St. John's Regional Oncology Center, 2727 McClelland Blvd., Joplin MO 64801.

0271-3586/83/0401-02-0149$04.00 **© 1983 Alan R. Liss, Inc.**

TABLE I. Drugs That Are Gonadotoxic in Humans

Males	Females
Radiation	Radiation
Busulfan	Busulfan
Chlorambucil	Chlorambucil
Cyclophosphamide	Cyclophosphamide
Nitrogen mustard	Nitrogen mustard
Vinblastine	Vinblastine
Adriamycin	
Corticosteroids	
Cytosine-arabinoside	
Methotrexate	
Procarbazine	
Vincristine	

versible phenomenon. We know the importance of age at time of drug exposure and its relation to gonadal damage. For example, treated women over 25 years of age may become menopausal because they have a smaller reserve of ova than younger women. No study of any of these drugs has been reported in which serial observations of patients have been made over enough years to determine if ovarian damage for a given drug is equal at all ages. Does the greater reserve of ova in the younger women merely give the *appearance* of less ovarian damage than that which is evident in their older sisters? Do the majority of women, regardless of age at time of treatment, have observable histological changes in the ovaries? Will most drug-exposed young women have physiologic ovarian dysfunction by the time they reach their midthirties? Do prepubertal ovaries have a greater resistance to destruction of ova than postpubertal ovaries?

Male Patients

In men and boys germ-cell destruction and abnormalities have been associated with certain classes of cytotoxic agents: radiation, the alkylators (nitrogen mustard, cyclophosphamide, chlorambucil, busulfan), the vinca alkaloids (vincristine, vinblastine), procarbazine, two of the antimetabolites, methotrexate and cytosine-arabinoside, adriamycin, and the corticosteroids. Bleomycin, hydroxyurea, thio-TEPA, CCNU, and 5-fluorouracil have demonstrated toxicity in laboratory models [Lu and Meistrich, 1979; Lambert and Eriksson, 1979], but there have been no reports of similar toxic effects in the germ cells of men. Of course, these and other chemotherapeutic agents have not been studied specifically for their effect on fertility and gonadal function. However, many of these drugs have been used to treat acute leukemia. The general undocumented opinion is that successfully treated men with leukemia do not have a reduced fertility rate or other evidence of gonadal dysfunction.

The toxic drugs mentioned above have been shown to have variable effects on gonadal function. The most common findings have been azoospermia, oligospermia, reduced sperm motility, abnormal serum gonadotropin levels, and germ cell aplasia. Practically speaking, permanent sterility is a common side effect, but sexual potency and libido are usually retained. Prepubertal boys sustain less gonadal injury from cytotoxic agents than men. They retain partial spermatogenesis and normal gondadotropin levels after comparable doses.

Oral cyclophosphamide in total doses ranging from 0.7 to 52 g caused gonadal toxicity in only 10 of 63 prepubertal boys (16%); one boy had an elevated serum FSH and the other nine had abnormal testicular histology [Pennisi et al, 1975; Lentz et al, 1977; Arneil, 1972; Kirkland et al, 1976; DeGroot et al, 1974; Hyman and Gilbert, 1972; Penso et al, 1974].

Pubertal boys do not fare as well as their younger brothers. Of 15 boys treated with 2.4 to 51.9 g of cyclophosphamide, 10 (66%) had some evidence of gonadal dysfunction: elevated serum gonadotropin levels, abnormal semen analysis, and abnormal testicular histology [Kirkland et al, 1976; Pennisi et al, 1975; DeGroot et al, 1974; Penso et al, 1974; Rapola et al, 1973].

Guesry et al [1978] provided a careful report on the gonadal effects of chlorambucil in prepubertal and pubertal boys. The total dose of chlorambucil was 2.4 to 79 mg/kg and the mean time from the end of therapy to time of gonadal evaluation was just over six years. The gonadal toxicity was equivalent in both groups: azoospermia was always present in boys who had received more than 25 mg/kg. There was no evidence of Leydig-cell compromise except in one boy who had received a total of 75 mg/kg.

The toxic effects of leukemia versus those of multiple drugs on the testes are not clear. Shalet et al [1977] and Haghbin et al [1980] described 17 prepubertal boys treated for leukemia. These children received up to nine different drugs and some had cranial radiation. Three of these 17 boys (18%) had low serum testosterone levels and one of them had elevated serum FSH.

Lendon et al [1978] compared testicular histology of boys who survived multiple drug therapy for leukemia (group A) with that of boys who died of leukemia early on, before they received much therapy (group B). Testicular tissue from normal boys who died accidentally served as the control. All these children were prepubertal or pubertal. The investigators found significant destruction of spermatogonia in the testes of all the leukemic children, worse in group A. The extent of gonadal damage in group A was significantly related to the total dose of cyclophosphamide, the total dose of cytosine-arabinoside, and the length of time between the end of therapy and the time of biopsy. No information was given about the possible relation between pubertal status at the end of therapy and germ cell damage.

Standard combination MOPP therapy (see Table II for acronyms of drug regimens) for Hodgkin's disease appears to be less toxic in the prepubertal gonads. Six prepubertal boys so treated had no evidence of endocrine or gonadal dysfunction: none had gynecomastia or abnormal serum hormone levels. In contrast, 9 of 13 pubertal boys (16%) in the same study developed gynecomastia, germinal aplasia, and abnormal serum gonadotropin and testosterone levels [Sherins et al, 1978].

More studies on cytotoxic-induced gonadal dysfunction have been reported on adult men than on any other group of humans. Despite this flurry of publications, large gaps of knowledge on this subject persist.

Cyclophosphamide has been used often as a single agent for men with nephrotic syndrome as well as in men with malignancies. In reviewing 89 of such cases reported by several authors, one cannot draw firm conclusions about incidence of infertility, recovery rate, or the dose of drug that produces irreversible azoospermia; the individual studies are too small to permit statistical significance. However, it is safe to say that all men develop azoospermia after taking 18 g or more of cyclophosphamide orally on an extended basis [Fairley et al, 1972; Kumar et al, 1972; Qureshi et al, 1972; Buchanan et al, 1975]. Cyclophosphamide is also administered intravenously either as a single agent or in combination with other drugs. In these circumstances the incidence of azoospermia or reversibility thereof remains uncertain.

Other alkylators used as single agents have been implicated with gonadal toxicity. Autopsy reports on men who died during the first trials of nitrogen mustard describe testicular atrophy [Spitz, 1948]. Richter et al [1970] prospectively studied eight men during chronic chlorambucil therapy. Two men received a total dose of 400 mg or 210 mg; the former man developed temporary oligospermia and the latter man had a nor-

TABLE II. Acronyms of Drug Regimens

CVP	COP	MOMP	MVPP	MOPP
Cyclophosphamide	Vincristine (Oncovin) for vinblastine in CVP	Methotrexate for procarbazine in MOPP	Nitrogen mustard	Nitrogen mustard
Vinblastine			Vinblastine	Vincristine
Prednisone			Procarbazine	Procarbazine
			Prednisolone	Prednisone

mal semen analysis. However, three months later he died and at autopsy his testes were shown to contain few spermatogonia, spermatocytes, and sperm cells. The other six men received more than a total of 400 mg of chlorambucil. All six became azoospermic. Three men underwent testicular biopsy; peritubular fibrosis was present in all three cases, and no germ cells were observed. This study shows that reversible gonadal injury occurs at lower doses of chlorambucil, but irreversible changes occur at doses greater than 400 mg or 5 mg/kg for a 70-kg man.

The gonadal toxicity of busulfan in men has received scant attention [Richter et al, 1970]. However, it is associated with gynecomastia and germinal aplasia [Kenis et al, 1956; Shalet, 1980].

Since the vinca alkaloids are never used alone today, there is no direct evidence of their effect on the testes or on mature sperm. Fertility dysfunction has occurred when vincristine or vinblastine were used in combination with alkylating agents (see below). However, the incidence of azoospermia or subfertility in men treated with vincristine in combination with adriamycin, bleomycin, and DTIC (ABVD) is less than that reported when vincristine is combined with alkylating agents. This indicates that the primary source of the trouble is the alkylating agents.

The vinca alkaloids produce mitotic arrest; thus, they would be expected to have this effect on dividing spermatogonia. These drugs also bind to microtubular proteins such as those present in the flagella of mature sperm, so they may interfere with sperm motility [Greenwald, 1973]. None of these potential toxic effects of the vinca alkaloids has been satisfactorily investigated.

Procarbazine has been studied in laboratory animals, but little is known about its effect on reproduction in humans. However, beginning at the age of two days, male monkeys were given subcutaneous or intraperitoneal procarbazine at total doses of 7.0 to 101 g over a period ranging from 35 to 68 months. At the end of therapy all five monkeys (100%) had atrophic testes; microscopic examination of the testicular tissue revealed complete loss of the germinal epithelium [Sieber et al, 1978].

Methotrexate has an immediate effect on maturing sperm. Semen samples taken from patients during active methotrexate therapy have often been abnormal. In 12 of the reported cases [Grunnet et al, 1977; El-Beheiry et al, 1979; Sussman and Leonard, 1980], ten men (83%) had abnormalities of sperm count, motility, or morphology. These changes are reversible by 70 to 90 days after the discontinuation of therapy.

Adriamycin is one of the most gonadotoxic drugs in male mice. [Lu and Meistrich, 1979; da Cunha, 1982]. It is less toxic in human testes, but to date no study has established the incidence of sterility. The toxic doses and duration of toxicity are not known [da Cunha et al, 1979].

Adrenocorticosteroids are included in many multidrug chemotherapeutic protocols, and they are widely considered not to be gonadotoxic. Indeed, prednisone has been prescribed in the treatment of infertility in men. Good evidence, however, suggests that corticosteroids have adverse effects on spermatogenesis and can even damage germinal epithelium. Ten of ten men (100%) treated chronically with topical cortisone for psoriasis had abnormal semen samples because of absent or depressed motility, and four of ten had oligospermia with less than 50 million sperm per ejaculate [Grunnet et al, 1977]. Mancini et al [1966] demonstrated depressed sperm counts and motility in normal men after only 15 days of dosing with 30 mg/d of prednisolone. All patients had severe oligospermia and motility of sperm had dropped from 80–90% before the trial to 20% by 30 days. Recovery had not occurred at six months after stopping the drug. Testicular tissue taken after prednisolone was stopped compared with tissue taken before the drug was administered showed spermatogenic arrest and some sloughing of germinal cells. Significant decreases occurred in spermatogonia types A, B, and N and in spermatozoa.

It should come as no surprise that the combined use of multiple chemotherapeutic drugs produces irreversible gonadotoxic effects more often than the use of the same drugs singly [Sherins and DeVita, 1973]. There is confusion in the literature over the incidence of male infertility after MVPP or MOPP therapy. A review published in 1980 [Schilsky et al, 1980] states that 80% of men develop azoospermia after MOPP therapy. This figure is based on a retrospective report [Sherins and DeVita, 1973] which dealt with only 11 men, two of whom had evidence of spermatogenesis at seven and four years after cessation of therapy. These men received either MOPP or the less toxic MOMP; total sperm counts and sperm motility are not provided.

Chapman et al [1979a] published a retrospective study on 74 men who had received MVPP therapy. They reported that 27/27 men (100%) had azoospermia in the *first 12 months after therapy*. Whitehead et al [1982] reported that 5/5 men (100%) were azoospermic in the first year after MVPP therapy and 11/12 men (92%) were azoospermic during the second year after therapy. The exceptional man produced a speci men containing less than 1 million dead sperm. Therefore, all men are infertile for at least the first year after therapy.

Chapman et al [1981] gave further proof in a prospective study that 100% of men treated with MVPP develop azoospermia when they showed that 14/14 men were azoospermic after only *two* cycles of MVPP. Other observers have confirmed this finding [da Cunha, 1982].

The azoospermia is usually permanent. Chapman and Sutcliffe reported in early 1979 that only 4/64 men (6.25%) had reversibility of azoospermia after a median of 27 months after therapy. Follow-up of these men (median more than four years) has not revealed any subsequent spermatogenic recovery [Waxman, 1981]. DeVita [1981] reported spermatogenesis in only 3/45 men (6.6%) who had been treated with MOPP therapy. More recently, Whitehead et al [1982] reported recovery of less than 10 million sperm/ml in 7/49 men (14%) after MVPP therapy. Based on 158 cases, we conclude that irreversible sterility is produced in over 90% of men treated with MOPP or MVPP.

There is confusion about the indicators of spermatogenic recovery. Roeser et al [1978] reported that falling follicle-stimulating hormone (FSH) levels indicate recovery of spermatogenesis. He based this statement on the cases of ten men who had received CVP or COP on various dose schedules. Seven of these men had FSH levels which were normal. Only four of these seven men had confirmatory evidence of spermatogenesis by semen analysis, one of whom was oligospermic. In the other three men no specimens were analyzed. Elsewhere in the paper we are told that one of six men with Hodgkin's disease had oligospermia after MOPP or MVPP therapy. This man had an elevated FSH level. Based on these sparse observations the authors concluded that falling FSH levels are "evidence of recovery". I have observed normal FSH levels in 4 of 64 men posttherapy, yet only 1 of the 4 men who had recovery of viable sperm had a normal FSH level; and none of more than 20 men who had falling FSH levels have had recovery of fertility (Personal observation). At best one can state that a falling FSH level *may* be an indicator of spermatogenic recovery. It could also be evidence of pituitary failure.

Compensated Leydig-cell failure occurs in men treated with MVPP. However, serum testosterone levels have not been low enough to indicate androgen therapy for men with depressed libido. The exceptions are men treated with radiation below the diaphragm in addition to aggressive chemotherapy [Chapman et al, 1979a; Whitehead et al, 1982].

Female Patients

Ovarian damage is known to be caused by therapy with radiation, alkylating agents (cyclophosphamide, busulfan, chlorambucil, nitrogen mustard), and vinblastine. Other drugs have not been reported to cause ovarian dysfunction, many of them not having been studied for this side effect. Women surviving acute leukemia treated by adriamycin, thioguanine, methotrexate, and 6-mercaptopurine are thought not to have significant impairment of fertility or sexuality. However, no study has substantiated this clinical impression.

The ovaries of prepubertal girls are fairly resistant to the toxic effects of chemotherapeutic drugs. There are reports involving 23 girls chronically treated with oral cyclophosphamide in total doses ranging from 1.0 to 36 g [Arneil, 1972; Pennisi et al, 1975; Lentz et al, 1977]. None sustained ovarian damage, and all developed regular menses at the normal age of puberty. Normal levels of serum gonadotropins were found in 11 girls in which they were measured. Twelve of the 23 girls died of their primary diseases and autopsy studies showed "normal ovaries" [Arneil, 1972].

Few data exist on the use of single agents in pubertal girls. Only six such girls have been assessed and reported after cyclophosphamide therapy. They received total doses of 3.0 to 27 g. All went on to develop regular menstrual cycles; all had normal levels of serum gonadotropins [Pennisi et al, 1975; Lentz et al, 1977].

Himelstein-Braw et al [1978] examined the ovaries of 31 prepubertal girls who had died of acute leukemia after cytotoxic therapy and of 28 normal children who had died accidentally. Inhibition of follicular development was noted in all ovaries of the leukemic children. This is in contrast to the destruction of ova seen at times in adults after chemotherapy. This inhibitory effect was thought to have been produced by the cytotoxic drugs used: prednisone and vincristine in most cases, and other agents, which variably included 6-mercaptopurine, methotrexate, cytosine arabinoside, L-asparaginase.

The majority of prepubertal girls seem to escape long-term ovarian toxicity from the use of multiple chemotherapeutic agents for acute leukemia. Of 17 girls so treated, only 1 (6%) had depressed serum levels of FSH and estradiol, and this result seems to indicate a hypothalamic-pituitary malfunction rather than ovarian damage [Siris et al, 1976]. When an alkylating agent was included in the leukemia therapy, 3/12 girls (25%) had evidence of ovarian damage [Shalet, 1980].

Eleven pubertal girls who received multiple drugs did not fare as well [Siris et al, 1976]. Seven went on to develop normal menses, but four failed to undergo menarche.

However, three of these girls had depressed levels of serum gonadotropins, again indicating a hypothalamic-pituitary abnormality. The fourth girl had been treated with busulfan and other drugs for chronic granulocytic leukemia; her serum gonadotropins were elevated. However, at autopsy her ovaries revealed normal numbers of ova with inhibition of follicular maturation.

In women, cytotoxic therapies have variable effects on fertility and gonadal function. Teenagers may sustain impaired fertility while menstruation and serum hormone levels remain normal [Chapman and Sutcliffe, 1981b]. Some women in their early twenties develop clinical evidence of ovarian dysfunction which may be reversible, depending on the drug, the schedule, and total dose. In some clinical circumstances women previously treated continue to show signs of progressive ovarian dysfunction and eventually develop complete ovarian failure. Abnormal ovarian function may be manifested by irregular menstrual cycles, anovulatory cycles, abnormalities in levels of serum gonadotropins, and estradiol both in and out of the normal ranges. On occasion, symptoms of menopause occur (see below).

From their late twenties on women exposed to cytotoxic drugs have an increased risk of sustaining complete ovarian failure. The clinical picture includes amenorrhea, postmenopausal levels of gonadotropins and estradiol as well as associated symptoms: hot flushes, insomnia, irritability, depressed libido. The occurrence and intensity of these symptoms vary according to the age at which menopause occurs. They are worse in young women, but they may be troublesome even for some women in their late forties.

The critical age at which the majority of women are likely to develop ovarian failure after treatment with individual drugs is not known. However, the use of drug combinations produces a high risk of ovarian failure. For example, 85% of women over the age of 25 years develop complete irreversible ovarian failure after six cycles of MOPP or MVPP therapy [Chapman et al, 1979b; Horning et al, 1981; Schilsky et al, 1981].

Cyclophosphamide alone may produce ovarian dysfunction or complete ovarian failure; both of these conditions may be reversible. Amenorrhea, abnormal hormone levels, or destruction of ova may occur [Uldall et al, 1972; Kumar et al, 1972; Warne et al, 1973; Rose and Davis, 1977]. The incidence and reversibility of this ovarian toxicity are related to age [Samaan et al, 1978; Schulz et al, 1979] and probably to dose as well. From available data it is difficult to determine the critical dose levels for various age groups. However, some generalizations may provide a rough guide until more refined studies are reported.

A median total dose of approximately 27g (range 12g to 63g) of cyclophosphamide produced permanent amenorrhea in 11 of 14 women younger than 29 years of age [Uldall et al, 1972; Warne et al, 1973]. For 21 women of 30 years or older the median total dose given before the onset of amenorrhea ranged from 12g [Uldall et al, 1972] to 24g [Warne et al, 1973]. A third of these older women developed amenorrhea after only 3.0g or less.

Ovarian dysfunction may occur with a lower total dose when cylcophosphamide is used in combination with other drugs. Rose et al [1977] evaluated premenopausal women (ages 32 to 51) after they had received an approximate total dose of 16g of cyclophosphamide plus 5-fluorouracil, methotrexate and vincristine: all developed ovarian dysfunction or failure. Schulz et al [1977] reported ovarian dysfunction in all 15 premenopausal women (100%) by the time they had received an approximate total dose of 10g of cyclophosphamide together with adriamycin or methotrexate and 5-fluorouracil; a third of these women became amenorrheic after they had received approximately 14g of cyclophoshamide. The effect of age is reflected by the fact that the same authors found complete ovarian failure after only about 4g in all seven women who were perimenopausal (ca 50 years of age) at the beginning of combination therapy.

Reports of permanent amenorrhea due to busulfan therapy have appeared since the 1950s [Louis et al, 1956; Galton et al, 1958; Belohorsky et al, 1960; Smalley and Wall, 1966; Wilkinson and Turner, 1959]. This side effect is age-related [Samaan et al, 1978; Schulz et al, 1979; Fisher et al, 1979]; but it is not possible from available data to estimate the influence of total dose. Ovarian atrophy has been documented at autopsy in at least three women previously treated with busulfan [Belohorsky et al, 1960; Smalley and Wall, 1966].

Scant data link chlorambucil with chemical castration in women. Ezdinli and Stutzman [1965] reported amenorrhea in two women treated with chlorambucil, but the number of premenopausal women at risk was not provided. Freckman et al [1964] stated that "castration can be achieved with chlorambucil, especially in women over 40". Two months of therapy at 14–20 mg/d were usually necessary to achieve amenorrhea. The women's ages were not reported.

The vinca alkaloids are not widely associated with ovarian dysfunction. However, Sobrinho et al [1971] described three cases in which vinblastine was the only agent in use when amenorrhea occurred. Amenorrhea developed in a 22-year-old woman after six months of weekly injections; she resumed menstruation two months after stopping vinblastine therapy. Two other women, ages 32 and 40 years at time of therapy, developed permanent amenorrhea.

The gonadal effect of nitrogen mustard alone in women has not been reported. However, this alkylating agent appears to affect ovarian function significantly. Sobrinho et al [1971] reported four women—ages 25, 26, 30, and 31 years—who had received about 26 mg of nitrogen mustard followed by 6–8 mg/d of chlorambucil. The two younger women immediately developed irregular menses followed by amenorrhea within five months or less. The 31-year-old woman had no menses after the nitrogen mustard injection, and the 30-year-old woman was amenorrheic after only one month during which she also received about 150 mg of chlorambucil. At autopsy no ova were seen in the ovaries of the 26- and 31-year-old women.

The use of two or more gonadotoxic drugs sequentially [Sobrinho et al, 1971] or in combination significantly increases the incidence of ovarian failure. The risk is related significantly to age and possibly to the total dose [Chapman et al, 1979b; Horning et al, 1981; Schilsky et al, 1981]. With multiple-drug therapy the full extent of gonadal toxicity is not immediately apparent; progressive ovarian dysfunction and failure continues even after cessation of therapy [Chapman et al, 1979b; Schilsky et al, 1981]. Fertility is significantly reduced, even in groups of women with regular menstrual cycles [Chapman and Sutcliffe 1981b; Horning et al, 1981]. An erratic menstrual pattern does not necessarily preclude pregnancy, though conception in this circumstance is less likely. Pregnancy does not establish lack of gonadal damage; young women have developed erratic menses or become amenorrheic postpartum [Chapman et al, 1979b; Smalley and Wall, 1966].

Impact of Gonadal Failure

Iatrogenic ovarian failure exacts a price in terms of quality of life by sexual, emotional, and psychological instability. Sympathetic counseling will ameliorate the psychologic and emotional difficulties encountered by these women [Chapman et al,

1979c; Chapman, 1982]. Early menopause places young women at a greater risk of developing osteoporosis. Sequential estrogen and progesterone replacement therapy largely eliminate the symptoms of menopause [Chapman et al, 1979c] and may preserve normal bone homeostasis [Lindsay et al, 1976; Nordin et al, 1980; Paganini-Hill et al, 1981]. Also, sequential therapy with these hormones mimics the normal physiologic ovarian function and probably reduces the risk of uterine cancer associated with unopposed estrogen therapy [Gambrell, 1978; Whitehead et al, 1981]. The premature onset of estrogen deprivation may also place successfully treated women at a greater risk for cardiovascular disease.

Men also encounter unwelcome sequelae of the gonadal effects of cytotoxic drugs. Most men experience a decrease in libido during combination chemotherapy and a third of them experience impaired sexual performance and pleasure after conclusion of therapy [Chapman et al, 1979a; Whitehead et al, 1982; Chapman, 1982]. Because of the high incidence of irreversible infertility, successfully treated young men often face a childless future. Pretherapy sperm banking should be offered to those men who have a good prognosis and a normal sperm count. Ejaculates containing less than 50 million sperm or sperm with less than 40% motility should not be stored.

Protection of adult gonads from toxic drug effects may be possible [Chapman et al, 1979a; Chapman et al, 1979c; Chapman and Sutcliffe, 1981b]. The hypothesis is based on the apparently greater resistance of the prepubertal gonads as compared to those of adults. Hormonal suppression of adult gonads may turn back the physiologic clock to simulate the prepubertal gonadal state. Animal and human studies on this problem are now in progress.

Every good physician should alert his patients to the possible consequences of cytotoxic therapy, and he should continue to observe them for clinical evidence of gonadal dysfunction. Anticipation and intervention on the part of the physician can improve the quality of life, even for patients who fail cancer therapy [Chapman, 1982].

ENVOI

Large number of adults in the reproductive span of life as well as children are treated with chemotherapeutic drugs for benign and malignant disease. Yet our present knowledge of the toxic effects on sexuality and fertility remains inadequate. It is urged that the medical community report more precise data on this subject. Without this information clinicians cannot provide patients with an informed opinion about the gonadal effect of many therapies.

A few examples of the problems in the literature will suffice. Some articles report the lack of drug toxicity when the number of cases in each pubertal stage studied are too small for general conclusions, and the drugs used are not those known to be associated with long-term gonadal dysfunction [Blatt et al, 1981]. Some authors provide accurate data showing gonadal toxicity but tone the article with inappropriate optimism about the effects of these drugs on fertility [Horning et al, 1981; vs Chapman et al, 1981a]. Some authors further the confusion by denigrating good work that demonstrates the risks of chemotherapy [Blatt et al, 1981; vs Shalet et al, 1981].

The quality of data can be improved. When an author reports observations about gonadal function after cytotoxic therapy, he must include certain essential facts. Otherwise the observation is useless or even confusing. For example, Wilimas et al [1980] reported that 5/10 girls (50%) with Hodgkin's disease developed amenorrhea after therapy with one of two schedules of cyclophosphamide and vincristine together with radiation. The radiation ports and the drug schedules used on the two groups of girls were not provided. These data conflict with other reports which indicate much less toxicity of cyclophosphamide on prepubertal ovaries [Arneil, 1971; Pennisi et al, 1975; Lentz et al, 1977; Shalet, 1980]. In fact, all of the girls with Hodgkin's disease [Wilimas et al, 1980] who developed amenorrhea had radiation through ports below the diaphragm, while none of the girls who retained their menses had such radiation [Wilimas, 1982]. This suggests that in this instance chemotherapy was not the prime gonadotoxin.

Some reports lump patients of varying stages of sexual maturity and then report general conclusions about the gonadal effect of the therapy. For example, the otherwise excellent study by Lendon et al [1978] is marred by the lack of precise information regarding pubertal status through the time of therapy. The gonads of a prepubertal child whose treatment concludes before puberty may not sustain as much damage as those of a prepubertal child whose treatment continues into puberty; and the gonads of children beginning and continuing therapy all through puberty and even beyond may show yet another grade of toxicity.

We will never complete this puzzle until all the pieces are provided—observations about gonadal function; detailed data about age and pubertal stage at beginning, during, and at the end of therapy, and information about the drugs, dose schedules, and total doses used. And when numbers of cases are too small to achieve statistical significance, the conclusions should be interpreted with caution. Even when individual cases are reported with adequate information, multiple reports can be combined to achieve statistical significance.

I have concentrated on the problem of gonadal toxicity in children, but the same strictures can be leveled against many of the reports of therapy in women and men. My criticisms are intended as a challenge to investigators in this field.

REFERENCES

Arneil CC (1972): Cyclophosphamide and the prepubertal testes. Lancet 2:1259–1260.

Belohorsky B, Siracky J, Sandor L, Klauber E (1960): Comments on the development of amenorrhea caused by myleran in cases of chronic myelosis. Neoplasma 4:397–402.

Blatt J, Poplack DG, Sherins RJ (1981): Testicular function in boys after chemotherapy for acute lymphoblastic leukemia. N Engl J Med 304:1121–1124.

Buchanan JD, Fairley KF, Barrie JU (1975): Return of spermatogenesis after stopping cyclophosphamide therapy. Lancet 2:156–157.

Chapman RM (1982): Effect of cytotoxic therapy on sexuality and gonadal function. Semin Oncol 9: 84–94.

Chapman RM, Sutcliffe SB, Rees LH, Edwards CRW, Malpas JS (1979a): Cyclical combination chemotherapy and gonadal function. Retrospective study in males. Lancet 1:285–289.

Chapman RM, Sutcliffe SB, Malpas JS (1979b): Cytotoxic-induced ovarian failure in women with Hodgkin's disease: I. Hormone function. JAMA 245:1877–1881.

Chapman RM, Sutcliffe SB, Malpas JS (1979c): Cytotoxic-induced ovarian failure in Hodgkin's disease: II. Effects on sexual function. JAMA 242:1882–1884.

Chapman RM, Sutcliffe SB, Malpas JS (1981): Male gonadal dysfunction in Hodgkin's disease: A prospective study. JAMA 245:1323–1328.

Chapman RM, Sutcliffe SB, Crosby WH (1981a): Reproductive potential after treatment for Hodgkin's disease. N Engl J Med 305:890–892, 1359.

Chapman RM, Sutcliffe SB (1981b): Protection of ovarian function by oral contraceptives in women receiving chemotherapy for Hodgkin's disease. Blood 58:849–851.

da Cunha MF (1982): Personal communication. M.D. Anderson Hospital, Houston, Texas.

da Cunha MF, Meistrich ML, Reid HL, Powell ML (1979): Effect of chemotherapy on human sperm production. Proc Am Assoc Cancer Res 20:100.

DeGroot GW, Faiman C, Winter JSD (1974): Cyclophosphamide and the prepubertal gonad: A negative report. J Pediatr 84:123–125.

DeVita VT (1981): The consequences of the chemotherapy of Hodgkin's disease: The 10th David A. Karnofsky Memorial Lecture. Cancer 47:1–13.

El-Beheiry A, El-Mansy E, Kamel N, Salama N (1979): Methotrexate and fertility in men. Arch Androl 3: 177–179.

Ezdinli EZ, Stutzman L (1965): Chlorambucil therapy for lymphocytic leukemia. JAMA 191:100–106.

Fairley KF, Barrie JU, Johnson W (1972): Sterility and testicular atrophy related to cyclophosphamide therapy. Lancet 1:568–569.

Fisher B, Sherman B, Rockette H, Redmond C, Margolese R, Fisher ER (1979): 1-Phenylalanine mustard (L-PAM) in the management of premenopausal patients with primary breast cancer. Cancer 44: 847–857.

Freckman HA, Fry HL, Mendez FL, Maurer ER (1964): Chlorambucil-prednisolone therapy for disseminated breast carcinoma. JAMA 189:23–26.

Galton DAG, Till M, Wiltshaw E (1958): Busulfan (1,4-Dimethanesulfonyloxybutane, myleran): Summary of clinical results. Ann NY Acad Sci 68:967–973.

Gambrell RD (1978): The prevention of endometrial cancer in post-menopausal women with progestogens. Maturitas 1:107–112.

Greenwald ES (1973): Pharmacology of cancer chemotherapy. In "Cancer Chemotherapy." New York: Medical Examination Publishing Co., 28–79.

Grunnet E, Nyfors A, Hansen KB (1977): Studies on human semen in topical corticosteroid-treated and in methotrexate-treated psoriatics. Dermatologica 154:78–84.

Guesry P, Lenoir G, Broyer M (1978): Gonadal effects of chlorambucil given to prepubertal and pubertal boys for nephrotic syndrome. J Pediatr 92:299–303.

Haghbin M, Murphy ML, Tan CC, Clarkson BD, Thaler HT, Passe S, Burchenal J (1980): A long-term clinical follow-up of children with acute lymphoblastic leukemia treated with intensive chemotherapy regimens. Cancer 46:241–252.

Himelstein-Braw R, Peters H, Faber M (1978): Morphological study of the ovaries of leukemic children. Br J Cancer 38:82–87.

Horning SJ, Hoppe RT, Kaplan HS, Rosenberg SA (1981): Female reproductive potential after treatment for Hodgkin's disease. N Engl J Med 304:1377–1382.

Hyman LR, Gilbert EF (1972): Testicular atrophy in a prepubescent male after cyclophosphamide therapy. Lancet 2:426–427.

Kenis Y, Dustin P, Henry J A, Tagnon H J (1956): Action du myleran dans 22 cas de leucémie myéloïde chronique. Review Francaise d'Etudes Cliniques et Biologiques 1:435–442.

Kirkland RT, Bongiovanni AM, Cornfeld D, McCormick JB, Parks JS, Tenore A (1976): Gonadotropin responses to luteinizing releasing factor in boys treated with cyclophosphamide for nephrotic syndrome. J Pediatr 89:941–944.

Kumar R, Biggart JD, McEvoy J (1972): Cyclophosphamide and reproductive function. Lancet 1: 1212–1214.

Lambert B, Eriksson G (1979): Effects of cancer chemotherapeutic agents on testicular DNA synthesis in the rat: Evaluation of a short-term test for studies of the genetic toxicity of chemicals and drugs *in vivo*. Mutat Res 68:275–289.

Lendon M, Palmer MK, Hann IM, Shalet SM, Morris Jones PH (1978): Testicular histology after combination chemotherapy in childhood for acute lymphoblastic leukemia. Lancet 2:439–441.

Lentz RD, Bergstein J, Steffes MW, Brown DR, Prem K, Michael AF, Vernier RL (1977): Postpubertal evaluation of gonadal function following cyclophosphamide therapy before and during puberty. J Pediatr 91:385–394.

Lindsay R, Hart DM, Aitken JM, MacDonald EB, Anderson JB, Clarke AC (1976): Long-term prevention of post-menopausal osteoporosis by oestrogen: Evidence for an increased bone mass after delayed onset of oestrogen treatment. Lancet 1:1038–1041.

Louis J, Limarzi LR, Best WR (1956): Treatment of chronic granulocytic leukemia with myleran. AMA Arch Intern Med 97:299–308.

Lu CC, Meistrich ML (1979): Cytotoxic effects of chemotherapeutic drugs on mouse testis cells. Cancer Res 39: 3575–3582.

Mancini RE, Lavieri JC, Muller F, Andrada JA, Saraceni DJ (1966): Effect of prednisolone upon normal and pathologic human spermatogenesis. Fertil Steril 17:500–513.

Nordin BEC, Horsman A, Crilly RG, Marshall DH, Simpson M (1980): Treatment of spinal osteoporosis in postmenopausal women. Br Med J 280:451–454.

Paganini-Hill A, Ross RK, Gerkins VR, Henderson BE, Arthur M, Mack TM (1981): Menopausal estrogen therapy and hip fractures. Ann Intern Med 95:28–31.

Pennisi AJ, Grushkin CM, Lieberman E (1975): Gonadal function in children with nephrosis treated with cyclophosphamide. Am J Dis Child 129:315–318.

Penso J, Lippe B, Ehrlich R, Smith FG (1974): Testicular function in prepubertal and pubertal male patients treated with cyclophosphamide for nephrotic syndrome. J Pediatr 84:831–836.

Qureshi MSA, Goldsmith HJ, Pennington JH, Cox PE (1972): Cyclophosphamide therapy and sterility. Lancet 2:1290–1291.

Rapola J, Koskimies O, Huttunen NP, Floman P, Vilska J, Hallman N (1973): Cyclophosphamide and the pubertal testis. Lancet 1:98–99.

Richter P, Calamera JC, Morgenfeld MC, Kierszenbaum AL, Lavieri JC, Mancini RE (1970): Effect of chlorambucil on spermatogenesis in the human with malignant lymphoma. Cancer 23:1026–1030.

Roeser HP, Stocks AE, Smith AJ (1978): Testicular damage due to cytoxic drugs and recovery after cessation of therapy. Aust NZ J Med 8:250–254.

Rose DP, Davis TE (1977): Ovarian function in patients receiving adjuvant chemotherapy for breast cancer. Lancet 1:1174–1176.

Samaan NA, DeAsis DN, Buzdar AU, Blumenschein GR (1978): Pituitary-ovarian function in breast cancer patients on adjuvant chemoimmunotherapy. Cancer 41:2084–2087.

Schilsky RL, Lewis BJ, Sherins RJ, Young RL (1980): Gonadal dysfunction in patients receiving chemotherapy for cancer. Ann Intern Med 93:109–114.

Schilsky RL, Sherins RJ, Hubbard SM, Wesley MN, Young RC, DeVita VT (1981): Long-term follow-up of ovarian function in women treated with MOPP chemotherapy for Hodgkin's disease. Am J Med 71:552–556.

Schulz KD, Schmidt-Rhode P, Weymar P, Künzig H-J, Geiger W (1979): The effect of combination chemotherapy on ovarian, hypothalamic, and pituitary function in patients with breast cancer. Arch Gynecol 227:293–301.

Shalet SM (1980): Effects of cancer chemotherapy on gonadal function of patients. Cancer Treat Rev 7:141–152.

Shalet SM, Beardwell CG, Twomey JA, Morris PH (1977): Endocrine function following the treatment of acute leukemia in childhood. J Pediatr 90:920–923.

Shalet SM, Lendon M, Morris-Jones PH (1981): Testicular function after chemotherapy for acute lymphoblastic leukemia. N Engl J Med 305:520.

Sherins RJ, DeVita VT (1973): Effect of drug treatment for lymphoma on male reproductive capacity. Ann Inter Med 79:216–220.

Sherins RJ, Olweny CLM, Ziegler JL (1978): Gynecomastia and gonadal dysfunction in adolescent boys treated with combination chemotherapy for Hodgkin's disease. N Engl J Med 299:12–16.

Sieber SM, Correa P, Dalgard DW, Adamson RH (1978): Carcinogenic and other adverse effects of procarbazine in nonhuman primates. Cancer Res 38:2125–2134.

Siris ES, Leventhal BG, Vaitukaitis JL (1976): Effects of childhood leukemia and chemotherapy on puberty and reproductive function in girls. N Engl J Med 294:1143–1146.

Smalley RV, Wall RL (1966): Two cases of busulfan toxicity. Ann Intern Med 64:154–164.

Sobrinho LG, Levine RA, DeConti RC (1971): Amenorrhea in patients with Hodgkin's disease treated with anti-neoplastic agents. Am J Obstet Gynecol 109:135–139.

Spitz S (1948): The histological effects of nitrogen mustards on human tumors and tissues. Cancer 1: 383–398.

Sussman A, Leonard JM (1980): Psoriasis, methotrexate, and oligospermia. Arch Dermatol 116:215–217.

Uldall PR, Kerr DNS, Tacchi D (1972): Sterility and cyclophosphamide. Lancet 1:693–694.

Warne GL, Fairley KF, Hobbs JB, Martin FIR (1973): Cyclophosphamide-induced ovarian failure. N Engl J Med 289:1159–1162.

Waxman J (1981): Personal observations. St Bartholomew's Hospital, London, England.

Whitehead MT, Townsend PT, Pryse-Davies J, Ryder TA, King RJB (1981): Effects of estrogens and progestins on the biochemistry and morphology of the postmenopausal endometrium. N Engl J Med 305:1599–1605.

Whitehead E, Shalet SM, Blackledge G, Todd I, Crowther D, Beardwell CG (1982): The effects of Hodgkin's disease and combination chemotherapy on gonadal function in the adult male. Cancer 49:418–422.

Wilimas J (1982): Personal communication. St Jude's Hospital for Children, Memphis, Tennessee.

Wilimas J, Thompson E, Smith KL (1980): Long-term results of treatment of children and adolescents with Hodgkin's disease. Cancer 46:2123–2125.

Wilkinson JS, Turner RL (1959): Chemotherapy of chronic myeloid leukemia, with special reference to myleran. Prog Hematol 2:225–238.

American Journal of Industrial Medicine 4:163–173 (1983)

Estrogenic Action of DDT Analogs

William H. Bulger, PhD, and David Kupfer, PhD

The estrogenic action of DDT analogs, among these methoxychlor, is discussed. Emphasis is placed on investigations of estrogenic properties of DDT derivatives, their long-term effects on reproductive processes, and their mechanism of action. Kepone is discussed briefly as an example of another structural class of insecticides with estrogenic activity.

Key words: estrogen , DDT, methoxychlor, Kepone, uterus, testis, neonate, estrogen receptor, ornithine decarboxylase

INTRODUCTION

The information concerning the estrogenic properties of DDT analogs covers a time span from 1950 to the present. In an effort to convey a historic perspective, an attempt was made to present the subject matter in a chronological fashion. This review is divided into sections that describe the development of our knowledge about the estrogenic properties of DDT derivatives, the long-term consequences of exposure to DDT estrogens, and more recent investigations concerning the mechanism of action of these compounds. The section on Kepone was included as an example of another structural class of estrogenic chlorinated hydrocarbon insecticides and is not reviewed extensively. Also, various related topics have been excluded from this review. The most noteworthy omission concerns the effect of chlorinated hydrocarbon insecticides on the hepatic monooxgenase system and the subsequent influence on the metabolism of endogenous and exogenous steroid hormones. This subject is discussed in a review by Kupfer [1975]. Other reviews that may interest the reader are Ware [1975], Kupfer and Bulger [1976a], Huff and Gerstner [1977], Nelson el al [1978], Kupfer and Bulger [1980], and Kupfer and Bulger [1982].

ESTROGENIC PROPERTIES

One of the first observations in support of an estrogenic action for DDT was made in 1950 by Burlington and Lindeman when they noted that Leghorn cockerels,

Worcester Foundation for Experimental Biology, Shrewsbury, Massachusetts.
Address reprint requests to David Kupfer, Worcester Foundation for Experimental Biology, 222 Maple Avenue, Shrewsbury, MA 01545.

0271-3586/83/0401-02-0163$03.50 © 1983 Alan R. Liss, Inc.

treated with what was probably technical-grade DDT demonstrated a decrease in testis growth and a diminished development of secondary sexual characteristics. Reasoning that these results might easily have been duplicated by administration of an estrogen, these workers suggested an estrogen like action for DDT. However, Fisher et al [1952] failed to demonstrate estrogenic activity with DDT in ovariectomized rats.

These observations attracted little attention until an attempt to use technical-grade DDT as an inducer of monooxygenase to diminish the estrogenic activity of administered estradiol led to the chance discovery that DDT was uterotropic (increased uterine weight) in the rat [Levin et al, 1968; Welch et al, 1969]. Subsequently, it was established that the o,p′-isomer of DDT, which is present as an 11%–29% contaminant of technical-grade DDT, [Metcalf, 1955; Okey and Page, 1974] was responsible for the uterotropic activity [Welch et al, 1969]. About the same time, Bitman et al [1968] determined that o,p′ DDT was estrogenic in the rat, chicken, and quail. They observed increases in weight, water content, glycogen, and RNA in uteri and oviducts from rats and birds treated with o,p′ DDT. However, p,p′ DDT (the insecticidal component in technical-grade DDT) was found to be inactive [Bitman et al, 1968]. Subsequently, other investigators confirmed these initial observations. Singhal et al [1970] demonstrated that o,p′ DDT administration to ovariectomized rats mimicked estrogens in elevating uterine weight, glycogen content, and the activities of various enzymes involved in glycolysis and the hexose monophosphate shunt pathway. Similar treatment with the p,p′-isomer enhanced, though to a lesser extent, most of the uterine enzyme activities. However, uterine weight and glycogen content were not affected. Duby et al [1971] confirmed that o,p′ DDT was more active than p,p′ DDT as a uterotropic agent in the rat and that the uterotropic activity of technical-grade DDT was related to its o,p′ DDT content. They also observed that o,p′ DDT, but not p,p′ DDT, was uterotropic in intact and ovariectomized mink. Clement and Okey [1972] reported that o,p′ DDT administered orally to immature rats stimulated uterine growth as assessed by increases in organ weight (wet and dry) and in glycogen and protein content. Gellert et al [1972] observed that o,p′ DDT treatment caused precocious vaginal opening, cyclicity, and increases in uterine and ovarian weight in young rats and caused cornification of vaginal epithelium as well as induction of a typical estrogenic response pattern in the endometrium of uteri from ovariectomized rats. They also reported that serum LH, but not FSH, levels were reduced in ovariectomized rats after o,p′ DDT treatment and that p,p′ DDT was estrogenic (although less potent than o,p′ DDT) in the ovariectomized rat.

EFFECT ON FERTILITY

Feeding experiments with DDT have yielded equivocal results with respect to effects on reproduction. Ottoboni [1969] found no effects on fertility or fecundity in rats fed a diet of DDT. Similarly, Wren et al [1971] observed that o,p′ DDT feeding did not

Chemical names of chlorinated hydrocarbons: p,p'DDT, 1,1-bis(p-chlorophenyl)-2,2,2-trichloroethane; o,p'DDT, 1-(o-chlorophenyl)-1-(p-chlorophenyl)-2,2,2-trichloroethane; p,p'DDE, 2,2-bis(p-chlorophenyl)-1,1-dichloroethylene; p,p'DDD, 2,2-bis(p-chlorophenyl)-1,1-dichloroethane; o,p'DDD, 1-(o-chlorophenyl)-2,2-dichloroethane; p,p'DDA, 1,1-bis(p-chlorophenyl)acetic acid; methoxychlor, 2,2-bis(meth oxyphenol)-1,1,1-trichloroethane; HPTE, 2,2-bis(p-hydroxyphenyl)-1,1,1-trichloroethane; Kepone (chlordecone), decachlorooctahydro-1,3,4-metheno-2H-cyclobuta (cd) pentalen-2-one; mirex, dodecachlorooctahydro-1,3,4-metheno-2H-cyclobuta (cd) pentalene.

affect the onset of sexual maturity in female rats and found no effect on fertility and fecundity. This finding was in contrast with an earlier study by the same investigators, Wren et al [1970], in which force-feeding of rats at an earlier age with lower doses of o,p′DDT was found to advance vaginal opening. In an earlier study, Ware and Good [1967] demonstrated that the feeding of DDT to mice produced alterations in fertility as judged by an increase in the time required for first litter. Subsequently, Lundberg [1973] demonstrated a prolongation of the estrous cycle and a decreased frequency of ova implantation after mating in mice treated with pure p,p′DDT (o,p′DDT and p,p′DDD were in amounts of less than 1%). These discrepancies probably reflect differences in the dose of compound administered and the duration of exposure. Also, differences in the level of estrogenic component in different batches of technical-grade DDT may have influenced the results of such studies.

LONG-RANGE EFFECTS

The recognition of the estrogenic properties of DDT and reports, [Harris, 1964] that exposure of newborn female rats to estrogens and androgens markedly perturbed reproductive processes in later life prompted Heinrichs and Gellert [1971] to investigate the chronic effects of DDT in the rat. The administration of o,p′DDT to neonatal females resulted in permanent sterility associated with advanced puberty, persistent vaginal estrus (PVE) after a period of normal estrous cycles, follicular cysts, corpora lutea reduction, a diminished uterotropic response to administered estradiol, and a slight disturbance in mating behavior. In a subsequent study, the same investigators, Gellert et al [1974], observed an inverse relationship between the age at first vaginal opening (onset of puberty) and an increase in the dose of o,p′DDT given neonatally. These investigators also established that PVE appeared earlier in life as the dose of o,p′DDT administered to the neonate increased and that the feedback rise in serum gonadotropin in response to ovariectomy was diminished after treatment of neonatal rats with o,p′DDT. However, treatment of male rats in neonatal life with o,p′DDT or other DDT analogs had no effect on reproductive processes in later life.

In a recent study, Fry and Toone [1981] observed that the injection of o,p′DDT into gull eggs, at concentrations similar to those found in contaminated seabird eggs in 1970, induced in *male* embryos abnormal changes similar to those observed with estradiol. The abnormalities (feminization) observed were oviduct induction and the development of an ovarian-like tissue in the left testis. The o,p′-isomer of DDT was found to be the most potent DDT analog with respect to feminization of male embryos. The authors speculated that contamination of eggs with DDT in the early 1970s was responsible for the skewed sex ratio (3.85 females/male) and poor breeding success observed among gulls on Santa Barbara Island in California.

MECHANISTIC STUDIES

Other investigations have addressed the question of whether estrogenic DDT derivatives and the steroidal estrogens share a common mechanism of action. Estrogens affect numerous biochemical parameters in the uterus that can be measured [Segal and Scher, 1967]. Some of the biochemical markers have proven useful in mechanistic studies of estrogenic DDT derivatives. They include, estrogen and progesterone receptors, DNA-biosynthesis, and marker protein induction including enzyme activity.

Interaction With Estrogen Receptor

Indirect evidence suggesting that DDT derivatives bind to estrogen receptor was obtained by Welch et al [1969] when they observed that o,p′ DDT treatment of immature rats diminished the uterine uptake of subsequently administered [^{3}H]-estradiol. Additional evidence to support this notion was obtained by Cecil et al [1971] when they observed that administration of MER-25 (an antiestrogen that acts by competing for estrogen receptor binding sites) to immature rats reduced the uterine response to o,p′ DDT. The first evidence that DDT derivatives could act in a manner similar to estradiol by binding to uterine estrogen receptor was obtained by Nelson [1974], who demonstrated that o,p′ DDT inhibits the binding of [^{3}H]-estradiol to rat uterine cytosolic estrogen receptor in vitro. Analysis of this data by double reciprocal plot (reciprocal-bound estradiol vs reciprocal-free estradiol in the presence and absence of o,p′ DDT) led Nelson to the conclusion that the binding inhibition was competitive in nature. Subsequent studies by Forster et al [1975] and in our laboratory [Kupfer, 1975; Kupfer and Bulger, 1976a, 1976b] utilized sucrose density-gradient sedimentation analysis to demonstrate that the 8-9S cytosolic estrogen receptor was the site of the o,p′ DDT-mediated inhibition of [^{3}H]-estradiol binding. Scatchard plot analysis of the binding of [^{3}H]-estradiol to rat uterine cytosol in the presence of o,p′ DDT revealed the same number of binding sites as in the absence of o,p′ DDT, indicating that the binding inhibition was not caused by an o,p′ DDT-mediated destruction of receptor [Kupfer and Bulger, 1976b]. The Scatchard plot analysis also confirmed that the inhibition of [^{3}H]-estradiol binding was of the competitive type. In subsequent experiments utilizing cytosols obtained from human breast and uterine tumors, Kupfer and Bulger [1977] established that o,p′ DDT inhibited [^{3}H]-estradiol binding to the 8-9S estrogen receptor in a competitive manner without altering the number of binding sites for estradiol. On the basis of that study, we speculated that the human body burden of o,p′ DDT had the potential to elicit o,p′ DDT binding to estrogen receptors in humans [Kupfer and Bulger, 1977].

Forster et al [1975] demonstrated an inhibition of [^{3}H]-estradiol binding to nuclei after cell-free incubation of rat uterine homogenates with o,p′ DDT and [^{3}H]-estradiol. Similar results were obtained by Turner and Eliel [1978] with o,p′ DDT and quail oviduct homogenate. Evidence that o,p′ DDT could actually translocate estrogen receptor (movement of receptor from cytosolic to nuclear compartment, presumably the initialing step in estrogen action) in a manner similar to estradiol was obtained when we observed an increase in uterine nuclear estrogen receptor levels three to eight hours after administration of o,p′ DDT to immature rats [Kupfer and Bulger, 1976b]. In a subsequent study by Kupfer and Bulger [1980], the levels of rat uterine cytosolic and nuclear estrogen receptors were monitored simultaneously. Within two hours after treatment with o,p′ DDT, cytosolic receptor levels decreased and remained diminished for 8 hours, returning to control values within 16 hours. Concurrently, nuclear receptor levels were elevated 1 hour after treatment, reached maximal concentration at 4 hours, and gradually returned to control levels by 72 hours. That o,p′ DDT could translocate estrogen receptor in vitro was demonstrated when the incubation of o,p′ DDT with isolated rat uteri resulted in the movement of estrogen receptor from the cytosolic to the nuclear compartment [Kupfer and Bulger, 1979]. However, p,p′ DDT and p,p′ DDE were inactive in this respect [Kupfer and Bulger, 1980].

The demonstration of high-affinity estrogen binding protein (EBP), presumably an estrogen receptor, in rat *testicular* cytosol by Brinkmann et al [1972] and the observation of Burlington and Lindeman [1950] that DDT interfered with *male* sex accessory organ development and reproduction prompted us to investigate whether o,p′DDT would interfere with [^{3}H]-estradiol binding to testicular EBP [Bulger et al, 1978a]. Results revealed that o,p′DDT, but not p,p′DDT, inhibited [^{3}H]-estradiol binding to rat testicular 8S EBP in a competitive manner and indicated that, as in the case of uterine estrogen receptor, only estrogenic DDT derivatives bind to testicular estrogen "receptor." It has also been reported that o,p′DDT inhibits the binding of 5α-dihydrotestosterone to specific receptor proteins in prostate cytosol from the rat [Wakeling and Visek, 1973].

DNA Synthesis

Ireland et al [1980] observed an increase in uterine DNA synthesis (incorporation of [^{3}H]-thymidine) in immature and ovariectomized/adrenalectomized rats treated with o,p′DDT. Evidence that estradiol and o,p′DDT interact with a common "receptor" was obtained by the same workers when they observed that the stimulation of uterine DNA synthesis and wet-weight increase in response to maximal doses of the two estrogens were not additive. They also confirmed the earlier observation by Bitman et al [1968] that o,p′DDT increases uterine DNA content.

Marker Proteins

The biological significance of uterine induced protein (IP) as an early marker of estrogenic action has been the subject of debate since its discovery [Baulieu, 1974]. In an investigation by Stancel et al [1980], it was observed that uterine IP production was stimulated one hour after o,p′DDT administration to immature rats. Also the incubation of o,p′DDT with isolated uteri for one hour resulted in enhanced IP formation, which could be blocked by actinomycin D. IP was not induced by p,p′DDT. The induction by o,p′DDT of uterine cytosolic progesterone receptor as a marker for estrogenic activity was studied by Mason and Schulte [1980]. The treatment of ovariectomized adult rats with o,p′DDT increased a progesterone binding component in uterine cytosol that had [^{3}H]-progesterone binding characteristics similar to those observed in cytosols from control and estradiol-treated rats.

Singhal et al [1970] were among the first investigators to utilize enzyme activities as indicators of o,p′DDT-mediated estrogen action. The administration of o,p′DDT to mature ovariectomized rats elevated activities of uterine hexokinase, phosphofructokinase, aldolase, pyruvatekinase, glucose-6-phosphate dehydrogenase, and 6-phosphogluconate dehydrogenase twofold to threefold within 16 hours of treatment.

The observations by Cohen et al [1970] and Kaye et al [1971] that estradiol administration markedly elevated rat uterine ornithine decarboxylase (ODC) prompted us to consider the effects of o,p′DDT on this activity in the uterus [Bulger and Kupfer, 1978]. Ovariectomized rats injected with o,p′DDT demonstrated peak ODC activity (130-fold increase) within 6 hours of administration. Similar results were obtained with immature rats. In this study, the order of decreasing potency, with respect to uterine ODC induction in immature rats, of DDT analogs was o,p′DDT, o,p′DDD, and p,p′DDT. The p,p′DDD and p,p′DDE derivatives were inactive. Similar results were obtained in ovariectomized rats with the exception that p,p′DDD was only slightly active. Prior treatment of ovariectomized rats with actinomycin D or cycloheximide inhibited the o,p′DDT-mediated rise in uterine ODC. The results with actinomycin D and cycloheximide are similar to those obtained by Singhal et al [1970] with respect to inhibition of o,p′DDT-mediated stimulation of uterine carbohydrate metabolizing enzymes and those of Kaye et al [1971] with respect to blockade of estrogen-mediated elevation of rat uterine ODC activity. In a subsequent study, Bulger et al [1978b], the kinetic characteristics of uterine ODC were determined in preparations from control and from estradiol-, o,p′ DDT-, and diethylstilbestrol-treated immature rats. Based on similar K_M and K_I values (inhibition of ODC by in vitro addition of putrescine) obtained with ODC preparations from the various treatment groups, it was concluded that the induced ODC probably represents an increase in "constitutive" enzyme.

METHOXYCHLOR

When contrasted with DDT, methoxychlor (bis-p-methoxy DDT) has certain properties that make it attractive as an alternate pesticide. It is less persistant in the biosphere [Kapoor et al, 1970], has a short half-life in mammals [Kunze et al, 1950], and is not as toxic as DDT [Hayes, 1971; Metcalf et al, 1971; Metcalf, 1976]. The diminished toxicity of methoxychlor in mammals appears to be related to the ability of he insecticide to undergo rapid metabolism to more polar derivatives [Weikel, 1957].

Kapoor et al [1970] conducted the first study of methoxychlor metabolism in the mouse and found that O-demethylation of methoxychlor was a major metabolic pathway. The bis-phenolic derivative of methoxychlor (HPTE) was recovered as a major metabolite both in vivo and in vitro (incubation with liver microsomes). The monophenolic methoxychlor derivative was also recovered as an in vitro metabolite. Subsequent investigations were restricted to in vitro studies with rat liver microsomes and established the mono- and bis-phenolic derivatives of methoxychlor as metabolites in the rat [Nelson et al, 1976, 1978; Bulger et al, 1978c; Ousterhout et al, 1979, 1981].

Methoxychlor, like DDT, is estrogenic. Welch et al [1969] obtained a uterotropic response after methoxychlor administration to immature rats and noted that methoxychlor was less potent than o,p′ DDT or technical-grade DDT. They also observed that methoxychlor treatment inhibited the uptake of [^{3}H]-estradiol by rat uteri. Employing the 18-hour glycogen response in immature rat uteri, Bitman and Cecil, [1970] confirmed the estrogenic activity of methoxychlor and reported that it was about 16-fold less active than o,p′ DDT. They also observed that technical-grade methoxychlor was fourfold more active than a purified preparation. Subsequently, Nelson [1974] reported that a technical preparation of methoxychlor was more potent than purified methoxychlor as an inhibitor of [^{3}H]-estradiol binding to rat uterine cytosolic estrogen receptor. In an initial experiment employing sucrose density-gradient sedimentation analysis, we observed that laboratory-grade methoxychlor (99% + pure) suppressed [^{3}H]-estradiol binding to the 8S rat uterine estrogen receptor [Bulger et al, 1978d]. However, when the sucrose gradient experiment was repeated with purified laboratory-grade methoxychlor (washed with aqueous base and recrystallized) no suppression of [^{3}H]-estradiol binding to the estrogen receptor was obtained. Moreover, the base washings, presumably phenolic contaminants of methoxychlor, were found to inhibit [^{3}H]-estradiol binding to the estrogen receptor, [Bulger et al, 1978d]. Subsequently, we observed that purified methoxychlor incubated in vitro with rat uteri failed to translocate estrogen receptor into the nuclear compartment [Kupfer and Bulger, 1979]. Recently, Lamoureux and Feil [1980] have characterized numerous impurities from technical-grade methoxychlor. Several of these compounds were phenolic. However, whether they represent the acidic contaminants observed by us in laboratory grade methoxychlor remains to be established.

The elevation of uterine ornithine decarboxylase (ODC) and wet weight in ovariectomized rats was employed to assess the in vivo estrogenic activity of purified and unpurified laboratory-grade methoxychlor. Results revealed that the purified preparation was only slightly less estrogenic than the unpurified one [Bulger et al, 1978c]. These results and the observation by Kapoor et al [1970] that methoxychlor is O-demethylated in vivo suggested that methoxychlor is a proestrogen undergoing metabolic activation in vivo. Additional support for this notion was obtained when formaldehyde formation was observed after incubation of methoxychlor with rat liver microsomes fortified with NADPH, indicating that O-demethylation had occurred [Bulger et al, 1978c]. The metabolic products were subsequently resolved into three components that were more polar than methoxychlor. One of these was chromatographically identical to bis-phenol (HPTE). Analysis revealed that each component suppressed [^{3}H]-estradiol binding to the 8S estrogen receptor and that the product chromatographically identical to HPTE was equal in potency to authentic HPTE. At about the same time, Nelson et al [1978] and Ousterhout et al [1979, 1981] obtained similar results with respect to in vitro inhibition of [^{3}H]-estradiol binding to rat uterine cytosol by microsomal metabolities of methoxychlor. However, they found that the monophenolic derivative was the primary in virtro metabolite of methoxychlor. In a subsequent study, the incubation of purified methoxychlor in the presence of isolated rat uteri failed to translocate the uterine estrogen receptor into the nuclear compartment; however, when liver microsomes fortified with NADPH were added to a similar preparation, polar metabolites of methoxychlor were formed, and translocation of estrogen receptor occurred Kupfer and Bulger [1979].

The estrogenic properties of HPTE were first reported by Fisher et al [1952] who found the compound to maintain estrus in ovariectomized rats. Additionally, the administration of HPTE to ovariectomized rats increased uterine weight and caused induction of uterine ODC [Bulger et al, 1978c]. Based on ODC induction, HPTE was about 100-fold more active than purified laboratory-grade methoxychlor. Recently Ousterhout et al [1981] determined the in vitro relative affinities of HPTE for rat uterine estrogen receptor and reported that HPTE was about 100 times less active than diethylstilbestrol, about 10 times more active than o, p′DDT, and slightly more active than the monophenolic derivative of methoxychlor. Also, the in vitro incubation of HPTE with rat uteri caused estrogen receptor translocation to take place [Kupfer and Bulger 1979]. However, purified methoxychlor was inactive. Evidence that methoxychlor may exert an estrogenic action in the male via its metabolite(s) was obtained when HPTE (but not methoxychlor) was found to suppress, by competitive inhibition, the in vitro binding of [^{3}H]-estradiol to rat testicular 8-9S estrogen receptor [Bulger et al 1978a].

KEPONE

The DDT group of insecticides does not appear to be the only structural class of pesticide with estrogenic activity. Kepone (chlordecone), a chlorinated polycyclic ketone, has demonstrated an estrogen-like action in the uterus of the mouse [Huber, 1965; Eroschenko and Mousa, 1979], the quail oviduct [McFarland and Lacy, 1969; Eroschenko and Wilson, 1975], the chicken oviduct, [Palmiter and Mulvihill, 1978], and the rat uterus [Bulger et al, 1979; Hammond et al, 1979]. Moreover the Kepone analog mirex was not uterotropic in the rat [Hammond et al, 1979]. However, mirex has estrogenic potential because it undergoes degradation to Kepone in nature [Carlson et al 1976].

Evidence suggesting that Kepone has an "estrogen-like" effect on quail testes was reported by McFarland and Lacy [1969] and Eroschenko and co-workers [Escherenko and Wilson, 1975; Escherenko, 1978]. Also, abnormal spermatogenesis has been observed among the victims of Kepone toxicity at a production facility in Hopewell, Virginia [Guzelian, 1976; Huff and Gerstner, 1977].

CONCLUSION

Over the past two decades a vast amount of data has accumulated, establishing that certain DDT derivatives or their metabolites are estrogenic. In mammals, these compounds appear to mimic in all essential aspects the action of estradiol, with the exception that the DDT analogs exhibit lower potency. The observations that Kepone is estrogenic indicates that (among insecticides) estrogenic action is not restricted to one structural class of compounds. This notion, recently illustrated by Katzenellenbogen et al [1980], can be extended beyond the realm of pesticides to include a wider range of estrogenic compounds with structural diversity. Furthermore, though these compounds are generally weak estrogens, a situation might arise where the level of exposure of animals and man to these compounds is high enough to elicit an estrogenic response. The detrimental effect on the reproductive processes of industrial workers inadvertently exposed to Kepone serves to illustrate this point.

ACKNOWLEDGMENTS

The studies mentioned in this review, that were conducted in our laboratory were supported by a USPHS grant from the National Institute of Environmental Health Sciences (ES-00834).

REFERENCES

Baulieu E-E (1974): The mode of action of steroid hormones: Some recent findings. In Coutinho EM, Fuchs E (eds): "Physiology and Genetics of Reproduction, Part A." New York: Plenum, pp 113–141.

Bitman J, Cecil HC, Harris SJ, Fries GF (1968): Estrogenic activity of o,p'DDT in mammalian uterus and avian oviduct. Science 162:371–372.

Bitman J, Cecil HC (1970): Estrogenic activity of DDT analogs and polychlorinated biphenyls. J Agric Food Chem 18:1108–1112.

Brinkmann AO, Mulder E, Lamers-Stahlhofen GJM, Mechielsen MJ, Van Der Molen HJ (1972): An estradiol receptor in rat testis interstitial tissue. FEBS Lett 26:301–305.

Bulger WH, Kupfer D (1978): Studies on the induction of rat uterine ornithine decarboxylase by DDT analogs. I. Comparison with estradiol-17β activity. Pest Biochem Physiol 8:253–262.

Bulger WH, Muccitelli RM, Kupfer D (1978a): Interactions of chlorinated hydrocarbon pesticides with the 8S estrogen-binding protein in rat testes. Steroids 32:165–177.

Bulger WH, Muccitelli RM, Kupfer D (1978b): Studies on the induction of rat uterine ornithine decarboxylase by DDT analogs. II. Kinetic characteristics of ornithine decarboxylase induced by DDT analogs and estradiol. Pest Biochem Physiol 8:263–270.

Bulger WH, Muccitelli RM, Kupfer D (1978c): Studies on the in vivo and in vitro estrogenic activities of methoxychlor and its metabolites: Role of hepatic mono-oxygenase in methoxychlor activation. Biochem Pharmacol 27:2417–2423

Bulger WH, Muccitelli RM, Kupfer D (1978d): Interactions of methoxychlor, methoxychlor base-soluble contaminant, and 2,2-bis(p-hydroxyphenyl)-1,1,1-trichloroethane with rat uterine estrogen receptor. J Toxicol Environ Health 4:881–893.

Bulger WH, Muccitelli RM, Kupfer D (1979): Studies on the estrogenic activity of chlordecone (Kepone) in the rat: Effects on uterine estrogen receptor. Mol Pharmacol 15:515–524.

Burlington H, Lindeman VF (1950): Effect of DDT on testes and secondary sex characters of white leghorn cockerels. Proc Soc Exp Biol Med 74:48–51.

Carlson DA, Konyha KD, Wheeler WB, Marshall GP, Zaylskie RG (1976): Mirex in the environment: Its degradation to Kepone and related compounds. Science 194:939–941.

Cecil HC, Bitman J, Harris SJ (1971): Estrogenicity of o,p'DDT in rats. J Agric Food Chem 19:61–65.

Clement JG, Okey AB (1972): Estrogenic and anti-estrogenic effects of DDT administered in the diet of immature female rats. Can J Physiol Pharmacol 50:971–975.

Cohen S, O'Malley BW, Stastny M (1970): Estrogenic induction of ornithine decarboxylase in vivo and in vitro. Science 170:336–338.

Duby RT, Travis HF, Terrill CE (1971): Uterotropic activity of DDT in rats and mink and its influence on reproduction in the rat. Toxicol Appl Pharmacol 18:348–355.

Eroschenko VP, Wilson WO (1975): Cellular changes in the gonads, liver, and adrenal glands of Japanese quail as affected by the insecticide Kepone. Toxicol Appl Pharmacol 31:491–504.

Eroschenko VP (1978): Alterations in the testes of Japanese quail during and after the ingestion of the insecticide Kepone. Toxicol Appl Pharmacol 43:535–545.

Eroschenko VP, Mousa MA (1979): Neonatal administration of insecticide chlordecone and its effects on the development of the reproductive tract in the female mouse. Toxicol Appl Pharmacol 49: 151–159.

Fisher AL, Keasling HH, Schueler FW (1952): Estrogenic action of some DDT analogues. Proc Soc Exp Biol Med 81:439–441.

Forster MS, Wilder EL, Heinrichs WL (1975): Estrogenic behavior of 2(o-chlorophenyl)-2-(p-chlorophenyl)-1,1,1-trichloroethane and its homologues. Biochem Pharmacol 24:1777–1780.

Fry DM, Toone CK (1981): DDT-induced feminization of gull embryos. Science 213:922–924.

Gellert RJ, Heinrichs WL, Swerdloff RS (1972): DDT hamologues: Estrogen-like effects on the vagina, uterus, and pituitary of the rat. Endocrinology 91:1095–1100.

Gellert RJ, Heinrichs WL, Swerdloff RS (1974): Effects of neonatally administered DDT homologs on reproductive function in male and female rats. Neuroendocrinology 16:84–94.

Guzelian PS (1976): Fourteen workers exposed to pesticide Kepone are probably sterile, researchers report. Occup Health Saf Lett 6:2.

Hammond B, Katzenellenbogen BS, Krauthammer N, McConnell J (1979): Estrogenic activity of the insecticide chlordecone (Kepone) and interaction with uterine estrogen receptors. Proc Natl Acad Sci USA 76:6641–6645.

Harris GW (1964): Sex hormones, brain development and brain function. The Upjohn lecture of the endocrine society, Endocrinology 75:627–648.

Hayes, WJ (1971): "Clinical Handbood on Economic Poisons." Chamblee, Ga: Environmental Protection Agency Pesticide Program.

Heinrichs WL, Gellert RJ (1971)): DDT administered to neonatal rats induces persistent estrus syndrome. Science 173:642–643.

Huber JJ (1965): Some physiological effects of the insecticide Kepone in the laboratory mouse. Toxicol Appl Pharmacol 7:516–524.

Huff JE, Gerstner HB (1977): Kepone. I. A literature summary, Black SA Kepone. II. An abstracted literature collection 1952–1977. Bethesda, Maryland, National Library of Medicine, Toxicology Information Program. Toxicology Information Response Center, ORNL/TIRC-76/3, September.

Ireland JS, Mukku VR, Robison AK, Stancel GM (1980): Stimulation of uterine deoxyribonucleic acid synthesis by 1,1,1-trichloro-2-(p-chlorophenyl)-2-(o-chlorophenyl)ethane (o,p'DDT). Biochem Pharmacol 29:1469–1474.

Kapoor IP, Metcalf RL, Nystrom RF, Sangha GK (1970): Comparative metabolism of methoxychlor, methiochlor, and DDT in mouse, insects, and in a model ecosystem. J Agric Food Chem 18:1145–1152.

Katzenellenbogen JA, Katzenellenbogen BS, Tatee T, Robertson DW, Landvatter SW (1980): The chemistry of estrogens and antiestrogens: Relationships between structure, receptor binding, and biological activity. In McLachlan JA (ed): "Estrogens in the Environment." New York: Elsevier/North Holland, pp 33–51.

Kaye AM, Icekson I, Lindner HR (1971): Stimulation by estrogens of ornithine and S-adenosylmethionine decarboxylases in the immature rat uterus. Biochim Biophys Acta 252:150–159.

Kupfer D (1975): Effects of pesticides and related compounds on steroid metabolism and function. Crit Rev Toxicol 4:83–124.

Kupfer D, Bulger WH (1976a): Interactions of chlorinated hydrocarbons with steroid hormones. Fed Proc 35:2603–2608.

Kupfer D, Bulger WH (1976b): Studies on the mechanism of estrogenic actions of o,p'DDT: Interactions with the estrogen receptor. Pest Biochem Physiol 6:561–570.

Kupfer D, Bulger WH (1977): Interactions of o,p'DDT with the estrogen-binding protein (EBP) in human mammary and uterine tumors. Res Commun Chem Pathol Pharmacol 16:451–462.

Kupfer D, Bulger WH (1979): A novel in vitro method for demonstrating proestrogens: Metabolism of methoxychlor and o,p'DDT by liver microsomes in the presence of uteri and effects of intracellular distribution of estrogen receptors. Life Sci 25:975–984.

Kupfer D, Bulger WH (1980): Estrogenic properties of DDT and its analogs. In McLachlan JA (ed): "Estrogens in the Environment." New York: Elsevier/North Holland, pp 239–263.

Kupfer D, Bulger WH (1982): Estrogenic actions of chlorinated hydrocarbons. In Chambers JE, Yarbrough J (eds): "Effects of Chronic Exposure to Pesticides on Animal Systems." New York: Raven Press, in press.

Kunze FM, Laug EP, Prickett CS (1950): The storage of methoxychlor in the fat of the rat. Proc Soc Exp Biol Med 75:415–416.

Lamoureux CH, Feil VJ (1980): Gas chromatographic and mass spectrometric characterization of impurities in technical methoxychlor. J Assoc Off Anal Chem 63:1007–1037.

Levin W, Welch RM, Conney AH (1968): Estrogenic action of DDT and its analogs. Fed Proc 27:649 (abst 2440).

Lundberg C (1973): Effects of long-term exposure to DDT on the oestrus cycle and the frequency of implanted ova in the mouse. Environ Physiol Biochem 3:127–131.

Mason RR, Schulte GJ (1980): Estrogen-like effects of o,p'DDT on the progesterone receptor of rat uterine cytosol. Res Commun Chem Pathol Pharmacol 29:281–290.

McFarland LZ, Lacy PB (1969): Physiologic and endocronologic effects of the insecticide Kepone in the Japanese quail. Toxicol Appl Pharmacol 15:441–450.

Metcalf RL (1955): "Organic Insecticides: Their Chemistry and Mode of Action." New York: Interscience.

Metcalf RL, Kapoor IP, Hirwe AS (1971): Biodegradable analogues. Bull WHO 44:363–377.

Metcalf RL (1976): Organochlorine insecticides, survey, and prospects. In Metcalf RL, McKelvey JJ (eds): "The Future for Insecticides." New York: John Wiley and Sons, pp 223–285.

Nelson JA (1974): Effects of dichlorodiphenyltrichloroethane (DDT) analogs and polychlorinated biphenyl (PBC) mixtures on 17β-[^{3}H] estradiol binding to rat uterine receptor. Biochem Pharmacol 23: 447–451.

Nelson JA, Struck RF, James R (1976): Estrogenically active forms of o,p'DDT and methoxychlor. Pharmacologist 18:247 (abst 730).

Nelson JA, Struck RF, James R (1978): Estrogenic activities of chlorinated hydrocarbons. J Toxicol Environ Health 4:325–340.

Okey BA, Page DJ (1974): Acute toxicity of o,p'DDT to mice. Bull Environ Contam Toxicol 11:359–363.

Ottoboni A (1969): Effect of DDT on reproduction in the rat. Toxicol Appl Pharmacol 14:74–81.

Ousterhout JM, Struck RF, Nelson JA (1979): Estrogenic properties of methoxychlor metabolites. Fed Proc 38:537 (abst 1625).

Ousterhout JM, Struck RF, Nelson JA (1981): Estrogenic activities of methoxychlor metabolites. Biochem Pharmacol 30:2869–2871.

Palmiter RD, Mulvihill ET (1978): Estrogenic activity of the insecticide kepone on the chicken oviduct. Science 201:256–258.

Segal SJ, Scher W (1967): Estrogens, nuclei acids, and protein synthesis in uterine metabolism. In Wynn RM (ed): "Cellular Biology of the Uterus." New York: Appleton-Century-Crofts, pp 114–150.

Singhal RL, Valadares JRE, Schwark WS (1970): Metabolic control mechanism in mammalian systems. IX. Estrogen-like stimulation of uterine enzymes by o,p'-1,1,1-trichloro-2-2-bis(p-chlorophenyl) ethane. Biochem Pharmacol 19:2145–2155.

Stancel GM, Ireland JS, Mukku VR, Robison AK (1980): The estrogenic activity of DDT: in vivo and in vitro induction of a specific estrogen inducible uterine protein by o,p'DDT. LIfe Sci 27:1111–1117.

Turner RT, Eliel LP (1978): o,p'DDT as an estrogen: An evaluation of its ability to compete with ^{3}H-estradiol for nuclear estrogen receptor sites in the quail oviduct. Bull Environ Contam Toxicol 19: 139–142.

Wakeling AE, Visek WJ (1973): Insecticide inhibition of 5α-dihydrotestosterone binding in the rat ventral postate. Science 181:659–661.

Ware GW, Good EJ (1967): Effects of insecticides on reproduction in the laboratory mouse. Toxicol Appl Pharmacol 10:54–61.

Ware GW (1975): Effects of DDT on reproduction in higher animals. Residue Rev 59:119–140.

Weikel JH (1957): The metabolism of methoxychlor (1,1,1-trichloro 2,2-bis(p-methoxyphenyl) ethane. I. The role of the liver and bilary excretion in the rat. Arch Int Pharmacodyn Ther 110:423–432.

Welch RM, Levin W, Conney AH (1969): Estrogenic action of DDT and its analogs. Toxicol Appl Pharmacol 14:358–367.

Wren TR, Wood JR, Fries GF, Bitman J (1970): Tests of estrogenicity in rats fed low levels of o,p'DDT. Bull Environ Contam Toxicol 5:61–66.

Wren TR, Weyant JR, Fries GF, Bitman J (1971): Effect of several dietary levels of o,p'DDT on reproduction and lactation in the rat. Bull Environ Contam Toxicol 6:471–480.

American Journal of Industrial Medicine 4:175–190 (1983)

Female Germ Cell Loss From Radiation and Chemical Exposures

R. Lowry Dobson, MD, PhD, and James S. Felton, PhD

Female germ cells in some mammals are extremely sensitive to killing by ionizing radiation, especially during development. Primordial oocytes in juvenile mice have an LD_{50} of only 6–7 rad, and the germ cell pool in squirrel monkeys is destroyed by prenatal exposure of 0.7 rad/day. Sensitivity varies greatly with species and germ cell stage. Unusually high sensitivity has not been found in macaques and may not occur in man, but this has not been established for all developmental stages. The exquisite oocyte radiosensitivity in mice apparently reflects vulnerability of the plasma membrane, not DNA, which may have implications for estimating human genetic risks. Germ cells can be killed also by chemicals. Such oocyte loss, with similarities to radiation effects, is under increasing study, including chemotherapy observations in women. More than 75 compounds have been tested in mice, with in vivo toxicity quantified by oocyte loss; certain chemicals apparently act on the membrane.

Key words: oocyte destruction, germ cell loss, primate oocytes, radiation effects, transplacental exposure

INTRODUCTION

The susceptibility of oocytes to destruction by X-rays was observed in rabbits 70 years ago [Lacassagne, 1913]. Since then, experimental results have been accumulating, including clinical observations in humans, and are providing a clearer, but still incomplete, picture of female germ-cell radiosensitivity in mammals (for review see Baker and Neal [1977]). Sensitivity varies with species and with germ cell maturation stage [Oakberg and Clark, 1964]. It can also depend on the animal's age (eg, see Dobson et al [1978]). In the juvenile mouse, oocytes have a striking radiosensitivity that is among the highest known for any mammalian cell.

Ovarian germ cell loss can also be caused by exposure to chemical agents. This too was described many years ago [Marchant, 1957]. While data for chemicals are not as extensive as for radiation, they are increasing rapidly and include clinical findings from human chemotherapy (eg, see Chapman et al [1979]). Certain parallels between responses to radiation and chemicals have been noted [Dobson et al, 1977, 1978].

Biomedical Sciences Division, Lawrence Livermore National Laboratory, University of California, Livermore.
Address reprint requests to Dr. R. Lowry Dobson, Biomedical Sciences Division, L-452, Lawrence Livermore National Laboratory, University of California, P.O. Box 5507, Livermore, CA 94550.

0271-3586/83/0401-02-0175$04.50 © 1983 Alan R. Liss, Inc.

Considered here are selected relevant observations on radiation and chemical exposures that may help to integrate findings and interpretations. Genetic and other sublethal damage to germ cells is not specifically dealt with. Emphasis is on cell killing and loss, which in certain instances is such a prominent feature as to be of special interest in itself. Indeed, as we shall see, its magnitude and character in the mouse may bear on the evaluation of genetic risk in the human female.

IONIZING RADIATION

Female Germ Cell Killing by Radiation

If a young female mouse is exposed to the modest gamma ray dose of 40 rad and its ovaries are removed a few days later for microscopic study, the remarkable finding is that essentially no immature, primordial oocytes are present; only growing and maturing ones remain, those that normally constitute about 10% of the germ cells. The primordial oocytes that usually make up 90% of the young animal's supply and constitute the germ cell pool will be missing; they will have been destroyed and removed from the ovary [Oakberg, 1962; Dobson and Kwan, 1977]. Effectively all (99.9%) of these oocytes in a 14-day-old juvenile mouse are quickly killed by 40 rad of gamma rays. This is seen in Figure 1. Their LD_{50} is only 6 rad.

If a developing squirrel monkey is continuously exposed in utero to a modest concentration (3 μCi/ml) of tritium in body water (via ^{3}HOH in the pregnant mother's drinking water), giving a radiation dose of only 0.7 rad/day, its ovaries at birth, grossly shrunken, will be almost devoid of germ cells; 99% of the gametic cell pool of this primate will have been lost [Dobson et al, 1978]. This is even more remarkable because discernible effects are not noted by histopathological examination of other organs or tissues [Jones et al, 1980].

Of particular interest is the extraordinarily high radiosensitivity in the female germ lines of certain animals. Among the questions arising are: In what species are female germ cells especially vulnerable? Which germ cell stages are most radiosensitive? What mechanisms are involved in high sensitivity when it does occur? What is the situation in man, and to what extent are animal data relevant to the human female?

Radiosensitivity

Variation with species, germ cell stage, and age. Ovarian germ cells of different mammals show widely differing radiation responses. Very high radiosensitivity has up to now been clearly demonstrated in relatively few species, notably in the mouse [Russell et al, 1959; Oakberg, 1962; Peters, 1966; Dobson and Cooper, 1974], the prenatal pig [Erickson, 1978], and prenatal squirrel monkey [Dobson et al, 1978]. In mice, it is the immature, small oocytes in primordial follicles that are so very sensitive; maturing oocytes in growing follicles are not easily killed by radiation. Although a transient elevation of germ cell sensitivity occurs prenatally at about day 11 of gestation (corresponding to peak oogonial mitotic activity), the high sensitivity in murine oocytes is a postnatal phenomenon. In Swiss-Webster mice it appears shortly after birth and develops rapidly during the first few postnatal days to a plateau lasting from about day 5 to day 17, when it decreases slightly toward its (still remarkably high) level in the adult. This changing pattern is illustrated in Figures 1 and 2. Details vary somewhat with strain. The rat displays a similar picture [Beaumont, 1962; Mandl and Beaumont, 1964; Matsumoto, 1971] but is considerably less radiosensitive.

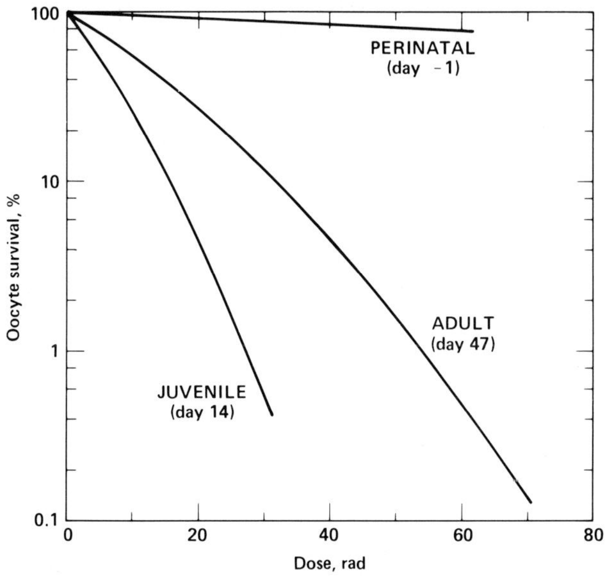

Fig. 1. Survival of primordial oocytes (percentage of control numbers) as functions of ^{60}Co gamma ray dose (given at 1 rad/min) in Swiss-Webster mice of selected ages: perinatal, 1 day before birth (exposed in utero); juvenile, 14 days old; and adult, 47 days old. Germ cells were microscopically enumerated in serial sections of ovaries removed in each case 5–7 days after irradiation. (From Dobson et al, unpublished data).

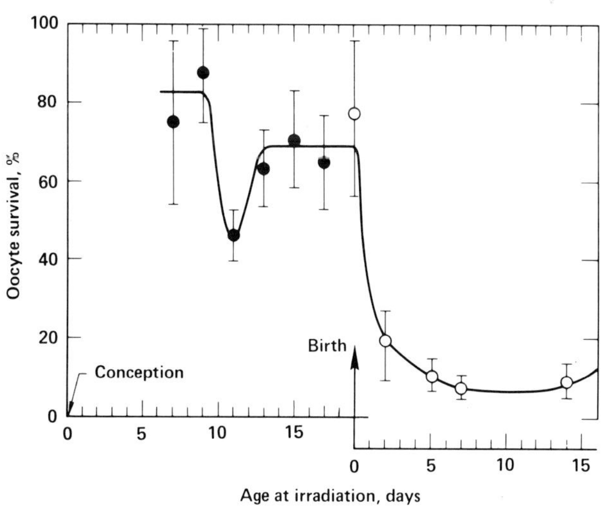

Fig. 2. Surviving primordial oocytes (percentage of control numbers) enumerated in 49-day-old Swiss-Webster mice that had been irradiated at various pre- and postnatal ages. Each animal received a single 18-rad dose of ^{60}Co gamma rays (at 1 rad/min). Confidence limits, ± 2 SEM. (From Wuebbles [1978] and Dobson et al, unpublished data).

In contrast to both mouse and rat, the guinea pig does not show high sensitivity in primordial oocytes, and its maturing ones are more radiosensitive than its primordials, the reverse of what is seen in mice and rats [Oakberg and Clark, 1964]. While the prenatal period deserves more study, very high sensitivity has not been reported in the guinea pig. Thus, general rules concerning female germ cell radiosensitivity are not readily found. Sensitivity is high for certain germ cell stages in certain rodents but not in others.

Such contrasts are found also among primates. In squirrel monkeys, as noted, female germ cells are demolished prenatally by exposure to 0.7 rad/day; however, the same exposure to adults causes much smaller germ cell loss [Dobson et al, unpublished]. The dramatic effect of in utero ^{3}HOH exposure in squirrel monkeys is illustrated in Figure 3, and comparison is made with effects seen in mice.

The magnitude of this prenatal germ cell loss in a monkey generates concern regarding possibilities of similar in utero sensitivity in other primates, including man. A study undertaken to test this in rhesus and bonnet monkeys has, however, failed to detect any substantial effects in either species of macaque from continuous ^{3}HOH exposure during the last two trimesters of gestation, where the radiation dose rate was 0.88 rad/day [Dobson et al, unpublished]. While the possibility that important germ cell killing may occur during the first trimester needs further examination, the negative re-

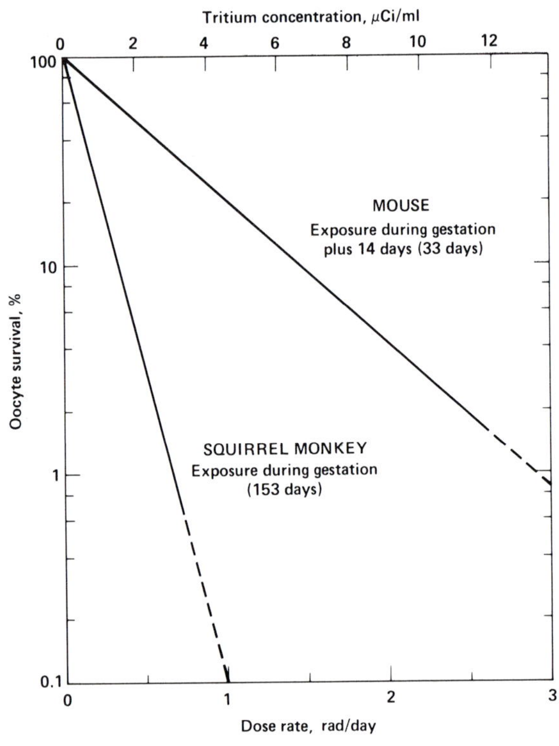

Fig. 3. Oocyte survival (for total oocytes in monkey, primordial oocytes in mouse) as functions of body water ^{3}HOH concentration and corresponding radiation dose rate, for continuous exposure over periods indicated. (Modified from Dobson [1979a]).

sults found so far match well those from rhesus monkeys irradiated acutely at various gestational times [Baker and Beaumont, 1967] as well as results from adult animals [Baker, 1966]. Other macaque experiments have shown significant germ cell loss only after single or accumulated prenatal doses of the order of 200 rad [Andersen et al, 1972, 1977].

Quantitative data for man on numbers of oocytes remaining in the ovary after irradiation are scarce, but sterility induction in the human female requires doses in the neighborhood of 400 rad [Baker, 1971], with generally larger doses needed in younger women (see Baker and Neal [1977]). Not surprisingly, ovarian failure sometimes does follow abdominal radiotherapy, including that for childhood cancer [Stillman et al, 1981; Shalet et al, 1976], but pregnancies have occurred despite very large previous exposures (eg, see Gans et al [1963]; Vuksanovic [1966]). Early menopause or sterility, which might be expected with high enough doses, has not been clearly linked to prenatal radiation exposure, either in clinical experience (but see Meyer et al [1969]) or in the Japanese A-bomb survivors; however, this has not been fully examined yet. Human fetal ovaries studied in organ culture have not shown high radiosensitivity [Baker and Neal, 1969; Baker, 1969]. Thus, while the data are still incomplete, high radiosensitivity of female germ cells in man has not been found, and available evidence suggests it is unlikely (as in the rhesus monkey). The possibility, however, of its occurrence at some time during prenatal development has not been ruled out. Interestingly, in the pig, where adult germ cell sensitivity is not remarkable, continuous gamma irradiation at 1 rad/day throughout gestation results in a 95% oocyte deficiency in the newborn [Erickson, 1978], a dramatic effect that rivals that seen in the squirrel monkey. Somewhat similar studies have been carried out also on dogs [Andersen et al, 1961] and cows [Erickson, 1966; Erickson and Reynolds, 1978] without revealing unusually striking germ cell effects.

Variation with type of radiation. Most investigations have involved X or gamma rays. But other radiations have been used, including: (1) alpha rays, from ^{210}Po injected into mice [Samuels, 1966]; (2) beta rays, from ^{3}H administered as tritiated water to mice and monkeys (eg, see Dobson [1979b]), or as tritiated thymidine ($^{3}HTdR$) administered to rats [Haas and Fliedner, 1971] or to mice (eg, see Baker and McLaren [1973]); (3) beta and gamma radiations from various radionuclides (for review see Baker and Neal [1977]); (4) mesons and neutrons administered to mice [Coggle et al, 1976]; and (5) neutrons of various energies administered to mice [Straume, 1982]. Studies specifically comparing the effectiveness of different ionizing radiations in killing female germ cells are few in number and essentially confined to mice. Indeed, available data (from whatever source) usable for quantitative comparisons are largely limited to the mouse. So attention here is on murine germ cells, and particularly on comparing oocyte killing by gamma, tritium, and neutron radiations.

Owing to their extreme sensitivity, mouse oocytes have allowed measurements of relative biological effectiveness (RBE) of tritium and gamma radiation at low, chronically administered doses, and of RBE dependence on dose and dose rate (eg, see Dobson [1979b]). Results, illustrated in Figure 4, demonstrate that the low-energy tritium beta rays are approximately threefold more effective at low doses in killing these cells than are gamma rays. Comparisons were then extended to neutrons [Straume, 1982], using particles of energies 0.4 and 15 meV, as well as ^{252}Cf fission neutrons (average energy 1.8 meV). Most interestingly, 0.4-meV neutrons, which are

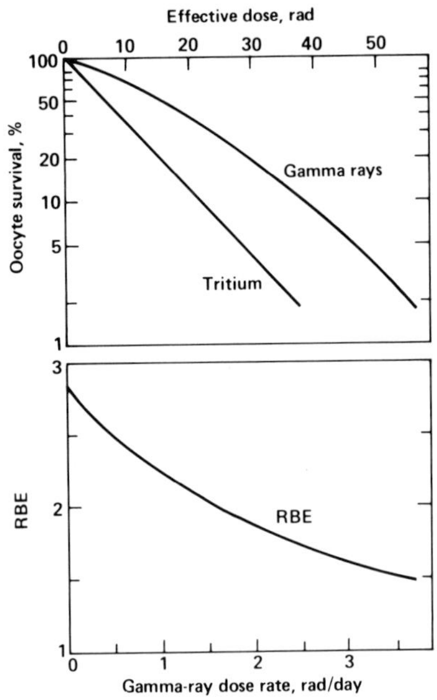

Fig. 4. Survival curves for primordial oocytes in mice chronically exposed under comparable conditions to ^{3}HOH and to ^{60}Co gamma rays, and the RBE of tritium compared with gamma radiation, as functions of dose and dose rate. (From Dobson [1979]).

much more damaging than gamma rays for many biological endpoints, are only slightly more effective in mouse oocyte killing than are gamma rays; the RBE at low doses is less than 2 [Straume, 1982], contrasting greatly to values of 100 or more reached for certain other end points (eg, see Bateman et al [1972]; Schellabarger et al [1974]). The underlying purpose of the neutron studies on oocytes was to obtain information on "target" size, considered next.

The Radiosensitive Target in Mouse Oocytes

To understand the extremely high radiosensitivity of mouse primordial oocytes, it is important to know, for example, whether ionization "hits" in an oocyte's adjacent follicle cells contribute to the observed germ cell lethality, ie whether the target has a cross-sectional area larger than that of the oocyte itself; or whether it is about the size of the oocyte, or perhaps of subcellular dimensions. From the neutron results and microdosimetric considerations, the presenting area of the target can be calculated. Assuming it to be circular, its diameter turns out to be about 10 μm [Straume, 1982], approximating that of the oocyte; and the target volume, derived from gamma ray data, proves to be about 4 μm^3. These two values match the plasma membrane geometry so closely as to suggest that it (or something of similar form) is the vulnerable target in these particular cells.

Other evidence suggesting the plasma membrane as the target comes from an entirely independent source—tritium data. If DNA were the target (for this view see Baker and Neal [1977]; for another view see Davies and Evans [1966]), the short-ranged beta rays of ^{3}H (average < 1 μm) would be expected to be more effective in oocyte killing when ^3H is concentrated in the nucleus (administered as ^3HTdR) than when generally distributed (chronically administered as ^3HOH). But it is not. The opposite is true. Routinely found in radioautographs of mouse oocytes labeled during prenatal development with ^3HTdR (eg, see Borum [1967]; Peters [1968]; Baker and McLaren [1973]) are numerous beta ray tracks over germ cell nuclei; and these are observed in mice old enough for the cells to have spent significant periods of time as highly radiosensitive immature oocytes, being irradiated all the while by the ^3H in their DNA. Consider a specific example: Baker and McLaren [1973] have provided both radioautographs and oocyte survival data and show an illustration (their Fig. 3) in which at least seven to nine grains can be identified over the nucleus of a unilayer-follicle oocyte in a 5-month-old mouse. Now even if one considers only 5 grains (and the 14-day radioautographic exposure that was used), the calculated dose for this nucleus (see Feinendengen [1967]) would be some 300 to 400 rad or more! Baker and McLaren's data show 60% survival of immature oocytes in these mice, whereas ^3HOH data (see Fig. 4) indicate that for such a dose well over 99.9% of immature oocytes would be killed. So there is an incompatibility between ^3HTdR and ^3HOH data. This seeming paradox is resolved if the plasma membrane, which is traversed by only a small fraction of the beta ray tracks originating in the nucleus, is in fact the radiosensitive target.

Possible implications for genetic risk estimation. If indeed the lethality target of these cells is thus distinctly different from that for genetic effects, then certain consequences may follow. The essential points needing consideration are: (1) estimates of human genetic risk from radiation are presently based mainly on data from mice [UNSCEAR, 1977; BEIR, 1980]; (2) because of its long duration (months in mouse, decades in man), the immature, resting oocyte is the principal female germ cell stage at risk; (3) these cells in mice are extremely easily killed by radiation, apparently through a non-DNA mechanism (and any genetic changes induced in such doomed cells are not transmitted); (4) whereas, in women these oocytes are not easily killed, so mutation transmission is presumably not interfered with as in the mouse.

Taken together, these points (which are aside from considerations of DNA repair) warn of the possibility that genetic risk in the human female from radiation—when it involves exposure of primordial oocytes—may be underestimated by mouse data unless the differing targets and cell killing are accounted for. This is under further study. It is possible that membrane vulnerability may also play a role in certain chemical effects on germ cells, at least in mice, and perhaps in other species.

CHEMICAL AGENTS

Oocyte Killing by Chemicals

Certain chemicals can, like radiation, cause primordial oocyte depletion in mice. Interpretation of these chemical effects, however, requires special considerations. While radiation dose to the ovary can often be clearly defined, effective chemical dose may be influenced by factors such as mode of exposure (oral, intraperitoneal, inhalation, etc), tissue-, species-, and strain-specific metabolism, physiological transport of molecules, and stability of compounds. Such factors can affect the toxicity observed in germ cells.

Until recently, the study of chemical effects on oocytes has mainly involved polycyclic aromatic hydrocarbons (PAHs). One such compound, 7,12-dimethylbenz(a)anthracene (DMBA), was early linked to both ovarian tumor induction and oocyte destruction in mice [Marchant et al, 1954; Howell et al, 1954; Marchant, 1957; Mody, 1960; Jull et al, 1966; Krarup, 1970]. Certain other PAHs, eg, 3-methylcholanthrene (MC) and benzo(a)pyrene (BP), were also shown to destroy oocytes (and induce ovarian tumors); but they are somewhat less potent than DMBA (eg, see Felton and Dobson [1982]; Mattison and Thorgeirsson [1978a]). Also, certain similarities were noted between MC- and radiation-induced murine response patterns: Oocyte loss from the two agents is additive [Dobson et al, unpublished]; effects on the ovary at the cytological level are strikingly similar—primordial oocytes and not maturing ones are killed, with effects not seen in follicle cells; and postnatal age-related sensitivity relations show parallels [Dobson et al, 1977; Dobson et al, 1978]. Taking advantage of both the high sensitivity of immature oocytes and the elevated response in juvenile animals, we applied a "standard" oocyte-depletion test for assessing the gametic toxicity of various chemical agents [Felton and Dobson, 1982].

Primordial Oocyte Destruction in Juvenile Mice

Our "standard" assay uses: (1) juvenile mice 7–21 days of age; (2) six graded doses of the chemical being tested, including one dose that reaches or exceeds the whole-animal LD_{50}; (3) administration by ip injection; and (4) at least five animals per dose point. The resulting dose–response curves are sigmoidal, and LD_{50}s are easily obtained from them [Felton and Dobson, 1982]. We have examined 77 chemical compounds in this way. Detailed results will appear elsewhere [Felton et al, manuscript in preparation], but a general picture can be obtained from Table I, where the 77 compounds are represented by 11 somewhat arbitrary classes. Of the 77 chemicals tested, 21 were positive in the assay (ie, caused significant oocyte loss), 56 were negative. Positive compounds were found in 7 of the 11 classes, notably among the PAHs, alkylating agents, esters, epoxides and carbamates, fungal toxins and antibiotics, and nitrosamines. Interestingly, all compounds positive for oocyte killing proved to be agents with known mutagenic–carcinogenic properties, and 21 of the 43 mutagen–carcinogens tested were positive

TABLE I. Primordial Oocyte Killing in Juvenile Mice by Chemicals of Various Classes

	Number of compounds		
Class of compound	Tested	Known mutagen–carcinogen[a]	Kill oocytes[b]
Polycyclic aromatic hydrocarbons	14	7	4
Aromatic amines	9	8	0
Alkyl halides	9	4	3
Aryl halides	3	2	0
Esters, epoxides, and carbamates	4	4	4
Nitrosamines	3	3	2
Fungal toxins and antibiotics	9	9	5
Metals	4	0	0
Steroids, alcohols, and aldehydes	5	1	0
Heterocyclics and nitrogen compounds	11	4	2
Miscellaneous compounds	6	1	1
Total	77	43	21

[a]As classified by McCann et al [1975].
[b]Significantly different from control (solvent only) at $p < 0.05$.

(22 were not). Four of the 11 classes were negative—the aromatic amines, aryl halides, metals, and steroid group.

The PAHs were the largest group examined, and in addition to DMBA, MC, and BP, dibenz(a,h)anthracene was positive. Some rather weak carcinogens, eg, benzene and benz(a)anthracene, were negative. The aromatic amines were negative, and the lack of response to such compounds as 2-acetylaminofluorene (AAF) and 3-amino-1-methyl-5H-pyrido(4,3-b)indole (Trp-P-2) is likely due to especially effective hepatic metabolism, sparing the ovary. Supporting this notion are preliminary pharmacokinetic results with radiolabeled molecules [Felton et al, unpublished] that suggest smaller fractions of AAF and its metabolites, than of BP and its metabolites, actually reach the ovary. (Note also that while ovarian BP metabolism is known to be adequate for oocyte toxicity [Mattison et al, this volume], little is known about ovarian AAF metabolism.) Ethidium bromide and acridine, both of which intercalate well with DNA, were negative.

Of the mutagen–carcinogen group of alkyl halides tested, only ICR-191 was negative. The important agricultural pesticide 1,2-dibromo-3-chloropropane (DBCP) was weakly positive. Three aryl halides tested, although among the most generally toxic compounds examined (eg, tetrachlorodibenz(p)dioxin, TCDD), were all negative. The next class in Table I, esters, epoxides, and carbamates, included methyl and ethyl methanesulfonate, busulfan, and urethane. All are mutagen–carcinogens and all were positive; but the doses of urethane needed were very high. Among the nitrosamines, the weakly mutagenic but strong liver carcinogen, N,N-dimethylnitrosamine, was negative even at the whole-animal toxic dose of 340 μmol/kg.

The fungal toxin and antibiotics class contained some of the most generally potent compounds studied, yet four of the nine agents examined were negative for oocyte killing. These included actinomycin D and chromomycin A_3, which interact strongly with DNA. Another notable negative member of this class is aflatoxin B_1, which, like the aromatic amines, is strongly metabolized in the liver. Among the heterocyclics and nitrogen compounds, two were positive, procarbazine HCl and 4-nitroquinoline-1-oxide. Sodium azide, a strong mutagen, was negative. Of the miscellaneous group, which included the vehicle controls, normal saline, DMSO, and corn oil, only trimethyl phosphate was positive.

Representative dose–response curves are shown in Figure 5. They illustrate comparisons of oocyte-killing potency. At the two extremes are bleomycin, which kills

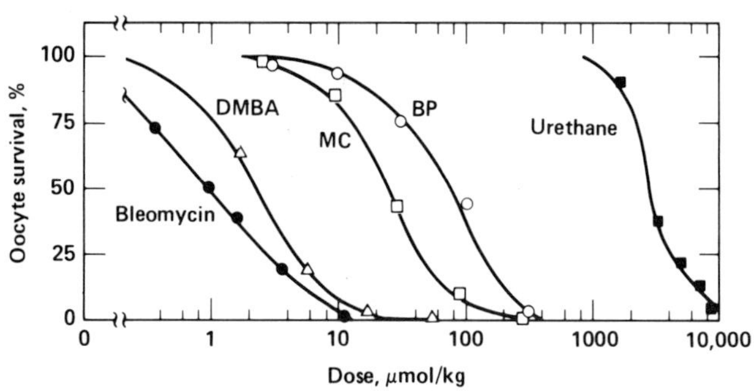

Fig. 5. Dose–response curves for murine primordial oocyte killing by bleomycin, DBMA (7,12-dimethylbenz(a)anthracene, MC (3-methylcholanthrene), BP (benzo(a)pyrene), and urethane.

mouse oocytes at lower doses than any other compound tested, and urethane, with doses four orders of magnitude larger, the least potent of the 21 positive agents. Regarding bleomycin, it is noteworthy that the rather drawn-out curve in Figure 5 corresponds well with decreased effectiveness at high doses reported for other cells [Terasima et al, 1978]. The three PAHs, DMBA, MC, and BP, make an interesting comparison. DMBA kills 50% of a juvenile mouse's oocytes at a 20-fold lower dose than does BP. The relative oocyte-killing effectiveness of these three compounds is related in the same manner as their known mutagenicity–carcinogenicity, DMBA > MC > BP.

Oocyte toxicity index. Quantitative comparisons of effectiveness between compounds can be made in terms of oocyte LD_{50} or of "oocyte toxicity index" (OTI) [Felton and Dobson, 1982]. We define OTI as the ratio of the whole-animal toxic dose (taken for this purpose as the dose that kills at least two but not more than three of five animals) to the oocyte LD_{50}. Shown in Table II are oocyte LD_{50}s and OTIs for representative compounds from each of the positive classes of Table I. The LD_{50}s range more than a thousandfold from that of bleomycin to that of urethane [Felton and Dobson, 1982]. The OTI range is less. Note that compounds with high OTIs are not necessarily the most oocytotoxic. Rather they represent chemicals that cause oocyte depletion at much lower doses than are toxic to the whole animal. Bleomycin and N-ethyl-N-nitrosourea kill murine primordial oocytes at 117- and 54-fold smaller doses, respectively, than they kill mice.

Variation of Effect With Species and Exposure Route

Intraperitoneal exposure. The majority of studies on chemically induced oocyte loss have been in the mouse, and mostly with ip administration. This animal has proven very sensitive to PAHs (as discussed above). The rat and guinea pig are more resistant to chemical oocyte killing. Mattison and Thorgeirsson [1978] reported no detectable oocyte loss in Sprague-Dawley rats given doses of MC that kill 90% of the oocytes in C57BL/6N mice. Furthermore, the lack of killing could not be explained by any deficiency in ovarian PAH metabolism. Rats are also resistant to ovarian tumor development following PAH treatment [Jull, 1973], probably related to lack of oocyte depletion [Krarup, 1967].

TABLE II. Whole-Animal Toxic Dose, Oocyte LD_{50}, and Oocyte Toxicity Index (OTI) for Selected Compounds*

Compound	Whole-animal toxic dose[a]	Oocyte LD_{50}[a]	OTI[b]
7,12-dimethylbenz[a]anthracene	96	2.7	36
Dibromochloropropane	1,300	690	1.9
Urethane	13,500	2,500	5.4
N-ethyl-N-nitrosourea	1,400	26	54
Bleomycin	110	0.9	117
Procarbazine	1,200	810	1.5
4-Nitroquinoline-1-oxide	130	47	2.8

*One compound from each of the positive classes of Table I.
[a]Micromoles per kilogram.
[b]Ratio of whole-animal toxic dose (see text) to the oocye LD_{50}.

Oral and transplacental exposures. Using MC as a test compound, with a standard dose of 160 mg/kg, we compared in juvenile C57BL/6 mice the effectiveness of oral, subcutaneous, and ip injection routes of exposure. The ip was most effective, subcutaneous the least, both in magnitude and speed. With ip injection, 50% of the primordial oocytes were lost within 2 days, 100% by 6 days. With oral administration, 50% were lost by 8 days, and only 80% loss was reached even at 1 month. With subcutaneous injection, 50% were lost only after 21 days, and 5 weeks took this just to 55%.

Of particular interest are experiments in which female offspring are transplacentally exposed to PAHs. We have used the maternal–fetal system to explore effects on oocyte numbers in offspring following oral or ip administration of PAHs to the mother [Felton and Dobson, 1982]. When MC or BP is given to either C57BL/6 or DBA/2 mice during gestation, oocyte loss is found not only in the treated mothers, but in the offspring; and this transplacentally induced oocyte loss is substantially greater than that in the mothers themselves, as seen in Figure 6. It is even greater than that seen after oral administration to the very sensitive juvenile mouse. Similar studies have been done by Erickson [1981] in guinea pigs and rats as well as in mice. In his experiments the mothers were fed DMBA for extended periods including the whole of gestation and

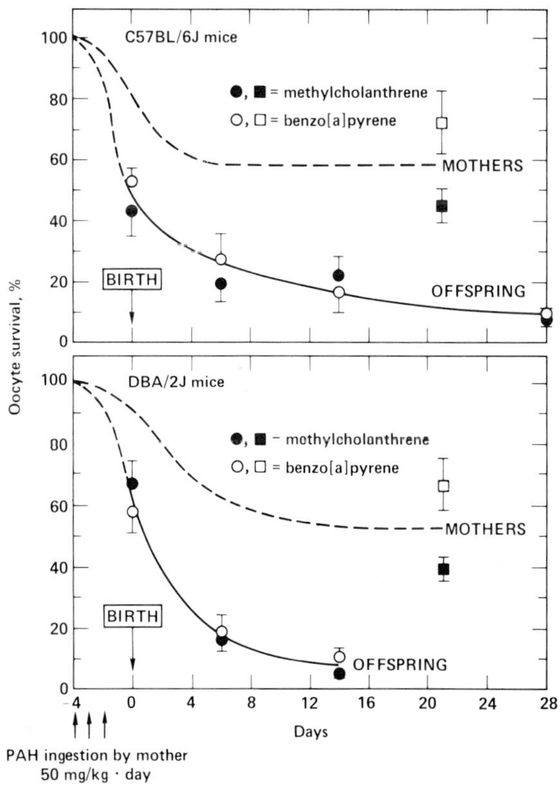

Fig. 6. Survival of primordial oocytes (percentage of control numbers) in treated mothers and their offspring at various times after maternal exposure (by gavage in late pregnancy) to MC or BP in two strains of mice. Shown are mean values ± SEM; interpolated portions of curves (dashed lines) are shaped to conform with available time–response data.

intervals before and after. Interestingly, no effect was seen in the mothers under these chronic-exposure conditions. In the rat, no oocyte loss was observed in either the adult or the transplacentally exposed offspring. Erickson [1982, personal communication] has also used DMBA to examine oocyte sensitivity in adult pigs and sheep, but no detectable effects have been seen. Additional data on transplacental exposure would be useful.

Human germ cell sensitivity to chemicals. Whether or not female germ cells in man are noticeably decreased in number by chemical exposures in the human environment is an important question. It has been suggested, for example, that cigarette smoking may be related to early onset of menopause [Jick et al, 1977], an idea that is further considered by Mattison and Thorgiersson [1978b]. Definitive answers to such questions, with an understanding of the mechanisms involved, are highly desirable.

A main source of direct human data is clinical experience. Ovarian failure has been reported by a number of investigators in young as well as middle-aged women following chemotherapy (eg, see Chapman et al [1979]; Warne et al [1973]; Sobrinho et al [1971]; Rose and Davis [1977]). That sterility can be induced is not surprising in view of the large doses of highly cytotoxic drugs that are required in cancer treatment. It is noteworthy, however, that Chapman et al [1979], in ovarian biopsies from four of six patients after multiple cycles of chemotherapy for Hodgkin's disease, found significant (though substantially reduced) numbers of primordial follicles present. This suggests that primordial oocytes in man are not unusually sensitive to these powerful drugs, just as they are not unusually sensitive to radiation. Of interest also is the observation that young women tolerate significantly larger doses of the alkylating drug cyclophosphamide before manifesting sterility than do older women [Mattison, 1982, personal communication], a relationship again similar to that seen with radiation therapy.

CONCLUSIONS

The stage has been reached now where it appears justified to attempt, at least in broad outline, an integrative interpretation of female germ-cell killing by radiation and chemicals. For this purpose, attention here is called to certain key points and working hypotheses which seem to accommodate a number of otherwise confusing combinations of findings. Though they are incomplete and likely inaccurate in detail, hopefully these ideas may speed progress.

(1) In the mouse, which is the most extensively studied species, female germ cells are not especially radiosensitive before birth. Shortly after birth, however, primordial oocytes (all of which are formed during gestation, as in most mammals) become exquisitely sensitive to radiation killing. This remarkable sensitivity, characteristic of mouse primordial oocytes, relates not to nuclear damage but to great vulnerability of the plasma membrane. (It is realized that the target may not be the plasma membrane itself; but it is something of similar geometry and location and is probably the membrane). This situation contrasts sharply with that in most cell types studied, especially mitotically multiplying cells, where the usual radiosensitive target is DNA.

(2) Differing oocyte radiosensitivities may be compared with each other, and with sensitivities of other cells in the animal, by means of oocyte LD_{50} s and by what may be called the "oocyte radiosensitivity index" (ORI), the ratio of the whole-animal LD_{50} to the oocyte LD_{50}. (The ORI is thus analogous to the OTI.) While murine primordial oocytes just before birth have an unremarkable LD_{50} of the order of 175 rad and an ORI

about 3, at two weeks after birth the oocyte LD_5 has dropped to approximately 6 rad (see Fig. 1) and the ORI is close to 100.

(3) Observed similarities between chemically and radiation-induced cell loss in the postnatal mouse suggest that much of the murine oocyte killing caused by exposure to certain reactive, cytotoxic compounds—the PAHs, MC and DMBA, for example—are due to attack on the plasma membrane (a more vulnerable and "weaker link" in these cells than DNA). This would be expected especially for compounds showing high OTIs, eg, bleomycin (OTI = 117), DMBA (OTI = 36), MC (OTI = 23), and cyclophosphamide (OTI = 10). But it would not be expected for otherwise highly toxic agents that show low OTIs, eg, actinomycin D and chromomycin A_3. These latter two drugs have OTIs less than unity. Though they react strongly with DNA and interrupt cellular processes, they apparently do not attack the oocyte plasma membrane and so do not show high *specific* toxicity for these nondividing, "resting" murine cells.

(4) In many other species, however, including man, evidence is lacking for such a "special" target that leads to unusually high sensitivity of oocytes either to radiation or chemicals. Human oocytes can indeed by killed by radiation and by chemotherapeutic drugs, but they do not appear to be *especially* sensitive, compared, say, with bone marrow cells. In fact, there is reason to believe that some of the germ cell loss in human ovaries from such exposures may (in contrast to that in mice) be by way of damage to multiplying follicle cells, leading to atresia of growing follicles. This could explain the temporary sterility sometimes seen in man, but which does not occur in mice, where the pool of immature germ cells is preferentially destroyed, making recovery impossible.

(5) Germ cell losses from prenatal radiation and chemical exposures, particularly the latter, which in some species and for certain types of exposure can substantially exceed losses induced postnatally, clearly involve additional factors. These need to be understood better.

ACKNOWLEDGMENTS

Appreciation is expressed to the many individuals who contributed to these investigations. We are especially grateful to Chinnie Kwan and Mark Knize for essential experimental work and to Tore Straume for key neutron studies. This work was performed under the auspices of the U.S. Department of Energy by the Lawrence Livermore National Laboratory under Contract W-7405-ENG-48 with support from the U.S. Environmental Protection Agency under Interagency Agreement EPA-IAG-D5-E681-AQ.

REFERENCES

Andersen AC, Schultz FT, Hague TJ (1961): The effect of total body x-irradiation on reproduction of the female beagle to 4 years of age. Radiat Res 15:745–753.

Andersen AC, Nelson VG, Simpson ME (1972): Fractionated x-radiation damage to developing monkey ovaries. J Med Primatol 1:318–325.

Andersen AC, Hendrickx AG, Momeni MH (1977): Fractionated x-radiation damage to developing ovaries in the bonnet monkey (Macaca radiata). Radiat Res 71:398–405.

Baker TG (1966): The sensitivity of oocytes in post-natal rhesus monkeys to x-irradiation. J Reprod Fertil 12:183–192.

Baker TG (1969): The sensitivity of rat, monkey, and human oocytes to x-irradiation in organ culture. In Sikov MR, Mahlum DD (eds): "Radiation Biology of the Fetal and Juvenile Mammal." Washington: US Atomic Energy Commission.

Baker TG (1971): Radiosensitivity of mammalian oocytes with particular reference to the human female. Am J Obstet Gynecol 110:746–761.

Baker TG, Beaumont HM (1967): Radiosensitivity of oogonia and oocytes in the foetal and neonatal monkey. Nature 214:981–983.

Baker TG, McLaren A (1973): The effect of tritiated thymidine on the developing oocytes of mice. J Reprod Fertil 34:121–130.

Baker TG, Neal P (1969): The effects of x-irradiation on mammalian oocytes in organ culture. Biophysik 6:39–45.

Baker TG, Neal P (1977): Actions of ionizing radiations on the mammalian ovary. In Zuckerman PL, Weir BJ (eds): "The Ovary." New York: Academic Press, vol 3, pp 1–58.

Bateman JL, Rossi HH, Kellerer AM, Robinson CV, Bond VP (1972): Dose-dependence of fast neutron RBE for lens opacification in mice. Radiat Res 51:381–390.

Beaumont HM (1962): The radiosensitivity of germ cells at various stages of ovarian development. Int J Radiat Biol 4:581–590.

BEIR, Committee on the Biological Effects of Ionizing Radiations, National Research Council (1980): "The Effects on Populations of Exposure to Low Levels of Ionizing Radiation: 1980." Washington: National Academy Press, pp 4–6.

Borum K (1967): Oogenesis in the mouse: A study on the origin of the mature ova. Exp Cell Res 45:39–47.

Chapman RM, Sutcliffe SB, Malpas JS (1979): Cytotoxic-induced ovarian failure in women with Hodgkin's disease. I. Hormone function. J A M A 242:1877–1881.

Coggle JE, Gordon MY, Lindop PJ, Shewell J, Mill AJ (1976): Some in vivo effects of π^--mesons in mice. Br J Radiol 49:161–165.

Davies DR, Evans HJ (1966): The role of genetic damage in radiation-induced cell lethality. In Augenstein LG, Mason R, Zelle M (eds): "Advances in Radiation Biology." New York: Academic Press, vol 2, pp 243–353.

Dobson RL (1979a): New radiobiological findings bearing on the 1977 ICRP recommendations. In: "Application of the Dose Limitation System for Radiation Protection, Practical Implications." Vienna: International Atomic Energy Agency, pp 593–605.

Dobson RL (1979b): The toxicity of tritium. In: "Biological Implications of Radionuclides Released from Nuclear Industries." Vienna: International Atomic Energy Agency, vol I, pp 203–211.

Dobson RL, Cooper MF (1974): Tritium toxicity: Effect of low-level ^3HOH exposure on developing female germ cells in the mouse. Radiat Res 58:91–100.

Dobson RL, Felton JS, Kwan TC (1977): Destruction of female germ cells in the mouse by 3-methylcholanthrene: Effect of animal's age. Mutat Res 53:100–101.

Dobson RL, Koehler CG, Felton JS, Kwan TC, Wuebbles BJ, Jones DCL (1978): Vulnerability of female germ cells in developing mice and monkeys to tritium, gamma rays, and polycyclic aromatic hydrocarbons. In Mahlum DD, Sikov MR, Hackett PL, Andrew FD (eds): "Developmental Toxicology of Energy-Related Pollutants." Washington: US Department of Energy, pp 1–14.

Dobson RL, Kwan TC (1977): The tritium RBE at low-level exposure—Variation with dose, dose rate, and exposure duration. Curr Top Radiat Res Q 12:44–62.

Erickson BH (1966): Development and radio-response of the prenatal bovine ovary. J Reprod Fertil 11: 97–105.

Erickson BH (1978): Interspecific comparison of the effects of continuous ionizing radiation on the primitive mammalian stem germ cell. In Mahlum DD, Sikov MR, Hackett PL, Andrew FD (eds): "Developmental Toxicology of Energy-Related Pollutants." Washington: US Department of Energy, pp 57–67.

Erickson BH (1981): Comparison of germ-cell response to polycyclic aromatic hydrocarbons in the mouse, rat, and guinea pig. In Mahlum DD, Gray RH, Felix WD (eds): "Coal Conversion and the Environment." Washington: US Department of Energy, pp 410–418.

Erickson BH (1982): Personal communication.

Erickson BH, Reynolds RA (1978): Oogenesis, follicular development and reproductive performance in the prenatally irradiated bovine. In: "Late Biological Effects of Ionizing Radiation." Vienna: International Atomic Energy Agency, vol II, pp 199–205.

Feinendegen LE (1967): "Tritium-Labeled Molecules in Biology and Medicine." New York: Academic Press, pp 179, 315.

Felton JS, Dobson RL (1982): The mouse oocyte toxicity assay. In Waters MD, Sandhu SS, Lewtas JH, Nesnow S, Claxton L, Chernoff N (eds): "Application of Short-Term Bioassays in the Analysis of Complex Environmental Mixtures." New York: Plenum Press (in press).

Gans B, Bahry C, Levie B (1963): Ovarian regeneration and pregnancy following massive radiotherapy for dysgerminoma. Obstet Gynecol 22:596–600.
Haas RJ, Fliedner TM (1971): The effect of tritiated thymidine on the oocytes of foetal rats following maternal infusion in pregnancy. Int J Radiat Biol 19:197–200.
Howell JS, Marchant J, Orr JW (1954): The induction of ovarian tumors in mice with 9,10-dimethyl-1,2-benzanthracene. Br J Cancer 8:635–646.
Jick H, Porter J, Morrison AS (1977): Relation between smoking and age of natural menopause. Lancet 1:1354–1355.
Jones DCL, Krebs JS, Sasmore DP, Mitoma C (1980): Evaluation of neonatal squirrel monkeys receiving tritiated water throughout gestation. Radiat Res 83:591–606.
Jull JW (1973): Ovarian Tumorigenesis. Methods Cancer Res 7:131–186.
Jull JW, Streeter DJ, Sutherland L (1966): The mechanism of induction of ovarian tumors in the mouse by 7,12-dimethylbenz(a)anthracene. I. Effect of steroid hormones and carcinogen concentration in vivo. J Natl Cancer Inst 37:409–420.
Krarup T (1967): 9,10-dimethyl-1,2-benzanthracene induced ovarian tumors in mice. Acta Pathol Microbiol Scand 70:241–248.
Krarup T (1970): Effect of 9,10-dimethyl-1,2-benzanthracene on the mouse ovary. Ovarian tumorigenesis. Br J Cancer 24:306–312.
Lacassagne A (1913): "Etude Histologique et Physiologiquc dcs Effets Produits sur l'Ovaire par les Rayons X." Lyon: Thèse Médecine.
Marchant J (1957): The chemical induction of ovarian tumors in mice. Br J Cancer 11:452–464.
Marchant J, Orr JW, Woodhouse DL (1954): Induction of ovarian tumors with 9,10-dimethyl-1,2-benzanthracene. Nature 173:307.
Matsumoto A (1971): Effects of perinatal x-ray irradiation on subsequent development of ovaries in rats. Annot Zool Jpn 44:99–104.
Mattison DR, Thorgeirsson SS (1978a): Gonadal aryl hydrocarbon hydroxylase in rats and mice. Cancer Res 38:1368–1373.
Mattison DR, Thorgeirsson SS (1978b): Smoking and industrial pollution and their effects on menopause and ovarian cancer. Lancet 1:187–188.
Mattison DR (1982): Personal communication.
Mandl AM, Beaumont HM (1964): The differential radiosensitivity of oogonia and oocytes at different developmental stages. In Carlson WD, Gassner FX (eds): "Effects of Ionizing Radiation on the Reproductive System." New York: Pergamon Press, pp 311–321.
McCann J, Choi E, Yamasaki E, Ames BN (1975): Detection of carcinogens as mutagens in the Salmonella/microsome test: assay of 300 chemicals. Proc Natl Acad Sci USA 72:5135–5139.
Meyer MB, Mertz T, Diamond EL (1969): Investigation of the effects of prenatal x-ray exposure of human oogonia and oocytes as measured by later reproductive performance. Am J Epidemiol 89:619–635.
Mody JK (1960): The action of four carcinogenic hydrocarbons on the ovaries of IF mice and the histogenesis of induced tumors. Br J Cancer 14:256–266.
Oakberg EF (1962): Gamma-ray sensitivity of oocytes in immature mice. Proc Soc Exp Biol Med 109: 763–767.
Oakberg EF, Clark E (1964): Species comparisons of radiation response of the gonads. In Carlson WD, Gassner FX (eds): "Effects of Ionizing Radiation on the Reproductive System." Oxford: Pergamon Press, pp 11–24.
Peters H (1968): Radioautographs. In Rough R: "The Mouse, Its Reproduction and Development." Minneapolis: Burgess, p 293.
Peters H (1966): Age dependence of radiosensitivity of mouse oocytes. In Lindop PJ, Sacher GA (eds): "Radiation and Ageing." London: Taylor and Francis, p 281–292.
Rose DP, Davis TE (1977): Ovarian function in patients receiving adjuvant chemotherapy for breast cancer. Lancet 1:1174–1176.
Russell WL, Russell LB, Steele MH, Phipps EL (1959): Extreme sensitivity of an immature stage of the mouse ovary to sterilization by irradiation. Science 129:1288.
Samuels LD (1966): Effects of polonium-210 on mouse ovaries. Int J Radiat Biol 11:117–129.
Shalet SM, Beardwell CG, Jones PHM, Pearson D (1976): Ovarian failure following abdominal irradiation in childhood. Br J Cancer 33:655–658.
Schellabarger CJ, Kellerer AM, Rossi HH, Goodman LJ, Brown RD, Mills RE, Rao AR, Shanley JP, Bond VP (1974): Rat mammary carcinogenesis following neutron or x-radiation. In "Biological Effects of Neutron Irradiation." Vienna: International Atomic Energy Agency, pp 391–401.

Sobrinho LG, Levine RA, DeConti RC (1971): Amenorrhea in patients with Hodgkin's disease treated with antineoplastic agents. Am J Obstet Gynecol 109:135–139.

Stillman RJ, Schinfeld JS, Schiff I, Gelber RD, Greenberger J, Larson M, Jaffe N, Li FP (1981): Ovarian failure in long-term survivors of childhood malignancy. Am J Obstet Gynecol 139:62–66.

Straume T (1982): "Biological Effectiveness of Neutron Irradiation on Animals and Man." Davis: PhD thesis, University of California.

Terasima T, Watanabe M, Takebe Y (1978): Upward-concave dose-response relationship in bleomycin lethality of mammalian cells. In Hecht SM (ed): "Bleomycin: Chemical, Biochemical, and Biological Aspects." New York: Springer-Verlag, pp 297–308.

UNSCEAR, United Nations Scientific Committee on the Effects of Atomic Radiation (1977): "Sources and Effects of Ionizing Radiation." New York: United Nations, pp 8–9, 537.

Vuksanovic MM (1966): Pregnancy following ovarian irradiation. Am J Roentgenol Radium Ther Nucl Med 97:951–956.

Warne GL, Fairley KF, Hobbs JB, Martin FIR (1973): Cyclophosphamide-induced ovarian failure. N Engl J Med 289:1159–1162.

Wuebbles BJ (1978): "Radiation Damage in the Mouse Female Germ Cell: A Two-Part Study." Hayward: MA Thesis, California State University.

American Journal of Industrial Medicine 4:191–202 (1983)

Oocyte Destruction by Polycyclic Aromatic Hydrocarbons

Donald R. Mattison, MD, Kenji Shiromizu, MD, PhD, and Maria S. Nightingale, MS

Certain polycyclic aromatic hydrocarbons, ubiquitous environmental pollutants, destroy oocytes. Oocyte destruction by polycyclic hydrocarbons requires distribution of the parent hydrocarbon to the ovary where ovarian enzymes metabolize the compound to reactive intermediates responsible for ovotoxicity. Descriptive assays of polycyclic hydrocarbon metabolic activation such as the aryl hydrocarbon (benzo(a)pyrene) hydroxylase assay (AHH) are not good predictors of strain or species differences in sensitivity to polycyclic hydrocarbon ovotoxicity. Using benzo(a)pyrene as a probe of ovarian metabolic processing suggests that the rate of formation of metabolites along the metabolic pathway to the 7,8-dihydrodiol-9,10-epoxide may be the appropriate measure of the role of metabolic activation in strain or species differences in sensitivity to oocyte destruction. The multistep metabolic pathway involved in ovarian metabolic processing of benzo(a)pyrene may represent a useful model for exploring the roles of metabolic activation, detoxification, intrinsic sensitivity, and repair in reproductive toxicity.

Key words: reproductive toxicity, ovarian toxicity, ovotoxicity, oocyte destruction, polycyclic aromatic hydrocarbons, benzo(a)pyrene

INTRODUCTION

Polycyclic aromatic hydrocarbons (PAH) are ubiquitous environmental pollutants, produced by synfuel production, fossil fuel combustion, and contained in cigarette smoke [Santodonato et al, 1981; Mattison, 1982]. Several investigators have demonstrated that selected polycyclic hydrocarbons destroy murine oocytes in a strain-, species-, age-, dose-, and metabolism-dependent fashion [Krarup, 1969; Felton et al, 1978; Mattison and Thorgeirsson, 1978, 1979]. Polycyclic aromatic hydrocarbons have also been demonstrated to produce ovarian tumors [Marchant, 1957; Jull, 1973],

Pregnancy Research Branch, NICHD, NIH, Bethesda, Maryland.

Address reprint requests to Dr. Donald R. Mattison, Bldg. 10, Rm. 8C313, Pregnancy Research Branch, NICHD, NIH, Bethesda, MD 20205.

0271-3586/83/0401-02-0191$03.50 © 1983 Alan R. Liss, Inc.

errors in oocyte meiosis [Bassler and Rohrborn, 1976], and decrease fertility [Krarup, 1970; Mattison et al, 1980b; Mackenzie and Angevine, 1981]. Because polycyclic hydrocarbons produce reproductive toxicity by more than one mechanism [Mattison, this issue, see page 65], and because they represent common environmental pollutants, we are exploring in detail the mechanism by which they produce one form of reproductive toxicity, oocyte destruction, in experimental animal systems.

Our initial experiments verified previous observations demonstrating that PAH destroy murine oocytes [Mattison and Thorgeirsson, 1978, 1979; Gulyas and Mattison, 1979]. These observations also demonstrated a significant strain difference in sensitivity to oocyte destruction by PAH (Fig. 1) and suggested that differences in metabolic activation were responsible for this difference. These observations set the stage for the experiments reviewed here, which were designed to explore the role of metabolic activation in oocyte destruction by PAH. Although these experiments focus on ovarian toxicity, the insights they provide may also represent a useful model for many other forms of toxicity.

THE ROLE OF OVARIAN METABOLISM IN OVOTOXICITY

Although PAH are known to require metabolism to reactive intermediates before they produce adverse effects in other biological systems [Gelboin, 1980], the requirement for metabolism in ovotoxicity had been questioned [Krarup and Loft, 1971]. To explore the requirement for metabolism in ovotoxicity we treated mice with α-naptho-flavone (α-NF) before and after treatment with benzo(a)pyrene, 3-methylcholanthrene, or 7,12-dimethylbenz(a)anthracene [Mattison and Thorgeirsson, 1979]. Since α-NF is a competitive inhibitor of PAH metabolism [Weibel et al, 1971], it could have one of two effects on ovotoxicity: if oocyte destruction is produced by the parent compound, α-NF treatment should increase the rate and extent of ovotoxicity observed. If, on the other hand, ovotoxicity is produced by a metabolite of the parent compound, oocyte destruction should be decreased. As we expected, treatment with α-NF blocked oocyte destruction by each of the three PAH tested (Fig. 2). This suggests that oocyte destruction by PAH requires metabolism of the parent compound to a reactive, ovotoxic intermediate.

The enzymes responsible for metabolic processing of PAH to reactive intermediates—microsomal cytochrome P-450-dependent monooxygenases, and epoxide hydrase—are widely distributed throughout the body including the gonads [Heinrichs and Juchau, 1980], and found in high levels in the liver [Miller and Miller, 1976; Brookes, 1980]. Previous experiments have demonstrated metabolic interaction between the liver and target organs for xenobiotic toxicity [Somogyi et al, 1971; Okey, 1972; Miller and Miller, 1974] and also demonstrated that reactive metabolites formed in isolated hepatocytes could produce mutations in bacterial and mammalian cells [Burke et al, 1977; Jones et al, 1978; Nordenskjold et al, 1981]. These observations suggested that PAH ovotoxicity could result from metabolic processing of the parent compound in the liver, or other extraovarian organ, in the ovary, or some combination requiring metabolic cooperation between the ovary and another organ. To explore these possibilities we treated mice with bilateral or unilateral intraovarian injections of polycyclic hydrocarbons [Shiromizu and Mattison, 1982]. Bilateral intraovarian injection of benzo(a)pyrene destroyed oocytes in a dose-dependent fashion. Unilateral intraovarian injection of benzo(a)pyrene only destroyed oocytes in the treated ovary, pro-

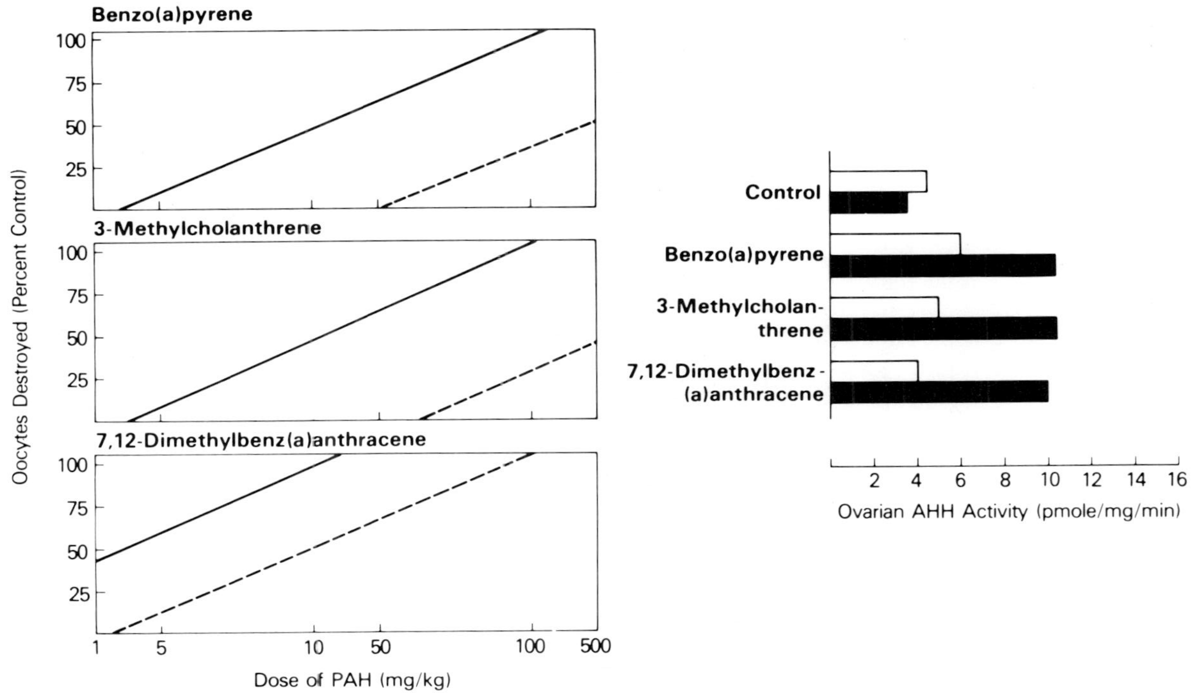

Fig. 1. The effect of benzo(a)pyrene, 3-methylcholanthrene, or 7,12-dimethylbenz(a)anthracene on oocyte destruction (left panel) and ovarian AHH activity (right panel). Dose-response curves for small oocyte destruction (types I to III of Pederson and Peters [1968]) in C57BL/6N (B6, solid line) and DBA/2N (D2, dashed line) inbred mice. All mice were treated with intraperitoneal injections of the indicated polycyclic hydrocarbon in doses ranging from 1 to 500 mg/kg on day 0 and sacrificed on day 5. Ovarian aryl hydrocarbon (benzo(a)pyrene) hydroxylase activity (AHH, EC 1.14.14.1) was measured in ovarian 9,000 × g homogenate in control (corn-oil) and polycyclic-hydrocarbon treated mice. D2 represented by open bars, B6 by closed bars. Data modified from Mattison and Thorgeirsson [1979] and Mattison and Nightingale [1980].

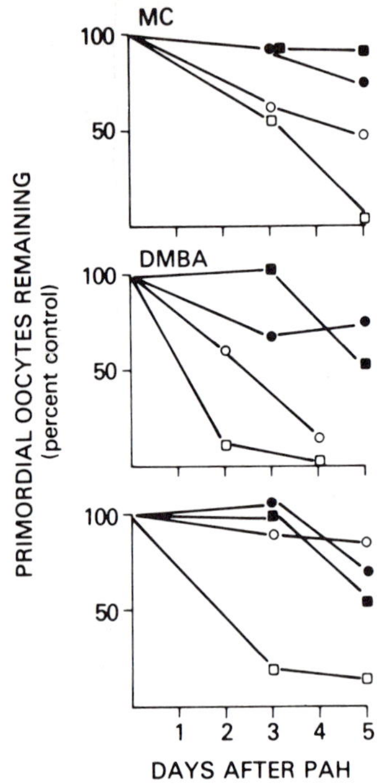

Fig. 2. Effect of α-napthoflavone (α-NF) on oocyte destruction produced by 3-methylcholanthrene (MC), 7,12-dimethylbenz(a)anthracene (DMBA), or benzo(a)pyrene. Open figures represent mice treated with α-NF and the indicated PAH; closed figures represent mice treated with the indicated PAH alone at a dose of 80 mg/kg; squares represent B6 mice; circles represent D2 mice. (From Mattison and Thorgeirsson [1979] with permission).

ducing no effect on oocyte number in the contralateral corn-oil treated ovary. The extent of oocyte destruction in treated ovaries following unilateral or bilateral intra-ovarian injection was similar at the doses tested. These data suggest that ovarian enzymes are fully capable of metabolizing PAH to reactive ovotoxic intermediates.

Taken together these two sets of experiments suggest that oocyte destruction by PAH requires distribution of the parent compound to the ovary, where ovarian enzymes metabolize the parent PAH to one or more reactive intermediates (Fig. 3). These reactive intermediates then produce oocyte destruction either by direct effects on the oocyte, or indirectly through toxicity to granulosa cells that support the oocyte in the follicle [Mattison and Ross, 1982]. If ovarian enzymes are responsible for metabolic activation of PAH to ovotoxic intermediates, then differences in ovarian metabolism may account for strain or species differences in sensitivity to oocyte destruction.

THE ROLE OF METABOLIC ACTIVATION IN OVOTOXICITY

Aryl Hydrocarbon (Benzo(a)pyrene) Hydroxylase Activity

In several murine systems the level, or inducibility, of aryl hydrocarbon hydroxylase (AHH, EC 1.14.14.1) activity correlates with PAH toxicity, mutagenicity, or carcinogenicity [Nebert and Jensen, 1979]. However, in other systems the correlation is not as clear [Schreck and Latt, 1980]. We explored the relationship between ovarian AHH activity and oocyte destruction using several experimental approaches: (1) a

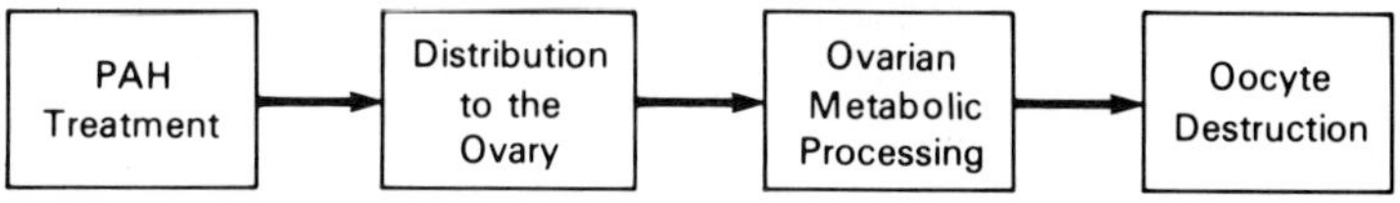

Fig. 3. Pathway to murine oocyte destruction by ovotoxic polycyclic aromatic hydrocarbons. After treatment with PAH, distribution to the ovary and ovarian metabolic processing of the parent compound to reactive intermediates produces oocyte destruction.

survey of ovarian AHH activity and oocyte destruction in seven inbred murine strains [Mattison and Nightingale, 1980]; (2) the effect of 2,3,7,8-tetrachlorodibenzo-p-dioxin (TCDD) a potent inducer of ovarian AHH on oocyte destruction; (3) the genetics of the inducibility of ovarian AHH activity and oocyte destruction using DBA/2N and C57BL/6N mice [Mattison and Nightingale, 1982]; and (4) comparison of ovarian AHH activity and oocyte destruction in rats and mice [Mattison et al, 1980a].

Murine Strain Survey

In this survey, the basal and induced ovarian AHH activity and extent of oocyte destruction after treatment with benzo(a)pyrene (BP) was determined [Mattison and Nightingale, 1980]. Among the seven strains surveyed, four were inducible with two- to threefold increases in ovarian AHH activity after BP treatment. Oocyte destruction among these strains varied from 50 to 100%. Comparison of induced ovarian AHH activity with the extent of oocyte destruction demonstrated a correlation coefficient (r) of 0.38, suggesting that approximately 20% of the variation in oocyte destruction was due to differences in ovarian AHH activity. A similar low degree of correlation was observed in the comparison of the ratio of induced:basal ovarian AHH activity to oocyte destruction.

The Effect of TCDD on Oocyte Destruction

TCDD is a potent inducer of rodent ovarian AHH activity in both DBA/2N and C57BL/6N mice, but is not an ovotoxin following short exposures [Mattison and Nightingale, 1980]. If ovarian AHH activity plays a major role in determining sensitivity to PAH ovotoxicity, then induction of ovarian AHH activity by pretreatment with TCDD should increase both the rate and extent of oocyte destruction produced by an ovotoxic PAH. Treatment of DBA/2N and C57BL/6N mice with TCDD had no effect on either the rate or extent of oocyte destruction following treatment with 3-methylcholanthrene [Mattison and Nightingale, 1980].

Genetics of Ovarian AHH Activity and Oocyte Destruction

Earlier studies have used murine genetics to dissect the relationship between AHH activity and PAH-induced pathology [Nebert and Jensen, 1979; Nebert, 1980]. We utilized a similar approach to explore the relationship between ovarian AHH activity and oocyte destruction in crosses between C57BL/6N (B6) and DBA/2N (D2) mice.

Ovarian AHH activity is inducible by 3-methylcholanthrene in B6 but not D2 mice (Fig. 4A). The F_1 heterozygote offspring of a D2 × B6 mating also have PAH-inducible, ovarian AHH activity. If the F_1 heterozygotes are backcrossed with D2 mice, the offspring fall into two groups, half of which are inducible and half noninducible. These data suggest that inducibility of ovarian AHH activity is inherited as an auto-

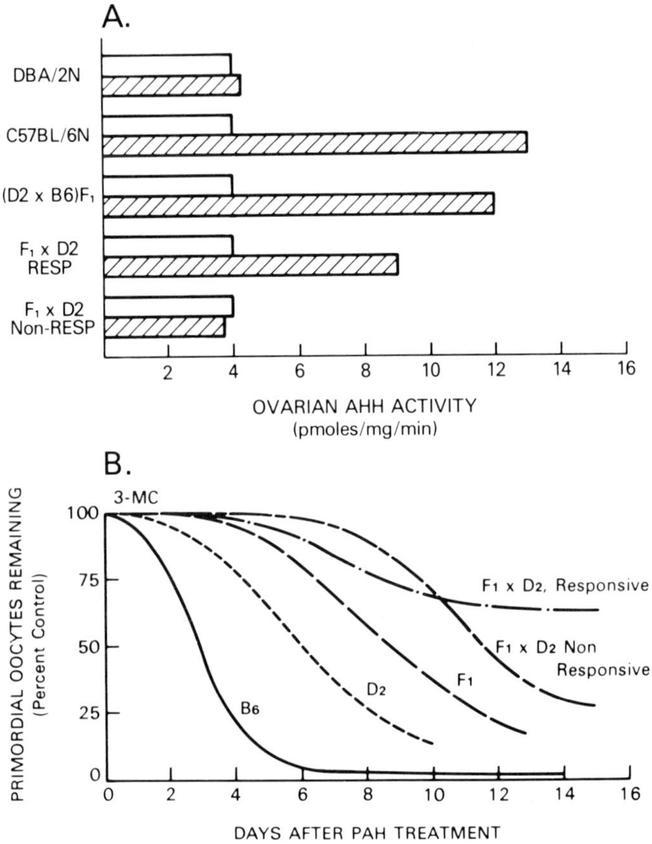

Fig. 4. Genetics of the inducibility of ovarian AHH activity (A) and oocyte destruction (B) by 3-methylcholanthrene (3-MC). A. Ovarian AHH activity in DBA/2N, C57BL/6N, $(D2 \times B6)F_1$, $F_1 \times D2$-responsive, and $F_1 \times D2$-nonresponsive mice. Open bars represent control; hatched bars represent 3-MC (100 mg/kg IP) treated mice. B. Oocyte number following 3-MC treatment. Mice were treated with 100 mg/kg IP on day 0 and sacrificed at the time indicated on the legend for determination of oocyte number (data modified from Mattison and Nightingale [1982]).

somal dominant trait, as previously demonstrated for the inducibility of hepatic AHH in these two strains and their crosses [Nebert and Jensen, 1979].

Although inducibility of ovarian AHH activity was inherited as an autosomal dominant trait, sensitivity to oocyte destruction was inherited in a different pattern. The C57BL/6N mouse was very sensitive to oocyte destruction, having a rapid decrease in oocyte number following treatment with 3-methylcholanthrene (Fig. 4B). The inbred DBA/2N (D2) strain was less sensitive than the inducible C57BL/6N (B6) strain with slower and less extensive oocyte destruction. The inducible $(D2 \times B6)F_1$ heterozygote appeared to be no more sensitive to oocyte destruction than the noninducible D2 strain, and was considerably less sensitive than the inducible B6 strain. Both of the groups of $F_1 \times D2$ backcross mice had similar sensitivity to oocyte destruction, independent of their inducibility. The $F_1 \times D2$ backcross mice in fact appeared much more resistant to oocyte destruction than both parental strains and the F_1 heterozygote [Mattison and Nightingale, 1982]. This suggests that although inducibility of ovarian AHH activity is inherited as an autosomal dominant trait, sensitivity to oocyte destruction is inherited as a polygenic trait and does not correlate with ovarian AHH activity.

Rat and Mouse Ovarian AHH Activity and Oocyte Destruction

All rat strains surveyed have inducible ovarian [Mattison and Thorgeirsson, 1978] and hepatic [Nebert et al, 1982] AHH activity. An additional test of the relationship between ovarian AHH activity and oocyte destruction can be provided by comparison of rat and mouse sensitivity to oocyte destruction by PAH. Sprague-Dawley rats like B6 mice have inducible ovarian AHH activity [Mattison et al, 1980a]. However, their sensitivity to oocyte destruction following treatment with one of three PAH (benzo(a)pyrene, 3-methylcholanthrene or 7,12-dimethylbenz(a)anthracene) is much less than observed in either inducible B6 mice, or noninducible D2 mice [Mattison, 1979].

THE ROLE OF OVARIAN AHH ACTIVITY IN OOCYTE DESTRUCTION

These experiments suggest that although distribution to the ovary, and metabolic processing in the ovary is necessary for oocyte destruction (Fig. 3), factors other than ovarian AHH activity determine sensitivity to oocyte destruction. Similar conclusions have been reached in exploring the relationship between human lymphocyte AHH activity and susceptibility to lung cancer [Nebert, 1980], as well as other solid tumors and leukemias [Levine et al, 1982].

The discrepancy between ovarian AHH activity and oocyte destruction could result from one or more factors: (1) differences in hormonal factors that modulate the access of toxins to the gonad [Janson, 1975; Setchell and Sharpe, 1981]; (2) AHH activity may not measure ovarian metabolism along the path leading to the ultimate ovotoxin; (3) metabolic activation of PAH to ovotoxins may play only a minor role in ovo toxicity, with detoxification mechanisms determining tissue sensitivity [Speilberg et al, 1981]; (4) tissue responses to toxicity, such as repair of PAH-produced damage may also influence strain or species differences in ovotoxicity [Pederson and Mangia, 1978; Generoso et al, 1979]; and (5) other differences, such as availability of binding sites for the reactive species, as suggested by differences in chromosomal organization and sensitivity to oocyte destruction following irradiation, may also influence differences in sensitivity to PAH produced oocyte destruction [Mandl, 1964; Baker, 1976].

To explore these issues we have decided to focus first on the identification of pathways responsible for metabolic activation of polycyclic hydrocarbons within the ovary. This has been approached from two directions: (1) oocyte destruction produced by derivatives of benzo(a)pyrene, and (2) ovarian production of BP metabolites measured using high-performance liquid chromatography (HPLC).

Oocyte Destruction by BP Derivatives

Benzo(a)pyrene (BP) is metabolized by microsomal monooxygenases and epoxide hydrases to more than 20 different metabolites [Gelboin, 1980]. Oocyte destruction by BP is the result of metabolic processing of the parent compound to one or more toxic metabolites that produce the observed oocyte destruction. Treatment of mice with metabolites of BP that are in the metabolic pathway to the ovotoxin should produce greater oocyte destruction than treatment with the parent compound. By contrast, treatment with metabolites not in the metabolic path to an ovotoxin should pro-

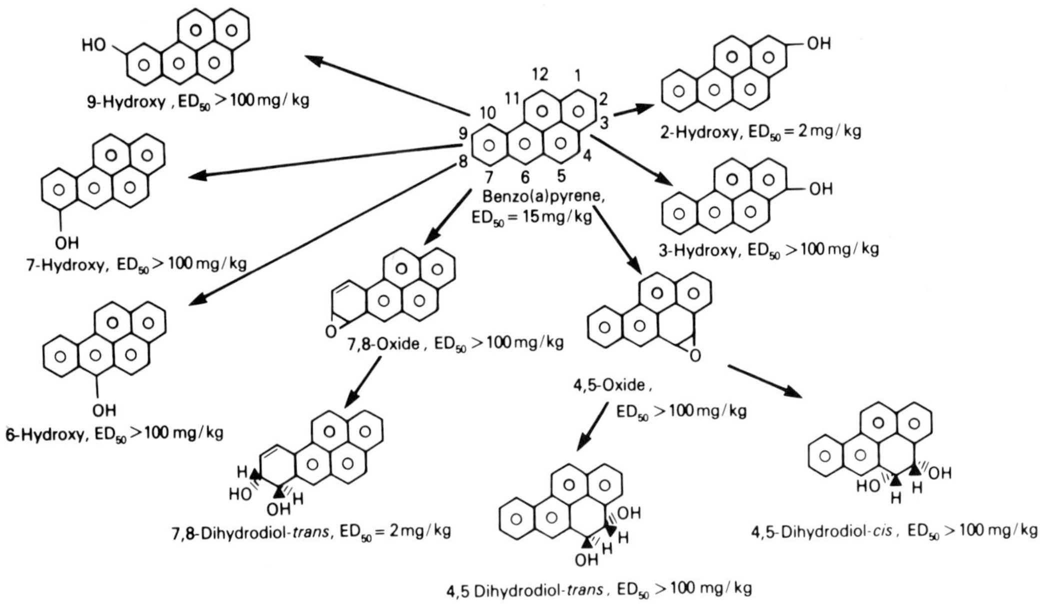

Fig. 5. Oocyte destruction by benzo(a)pyrene and ten derivatives. C57BL/6N mice were treated intraperitoneally with the indicated PAH at doses ranging from 1 to 100 mg/kg, and oocyte destruction determined. ED_{50}s were computed following probit and log dose transformations with a least-squares approach.

duce less ovotoxicity than the parent compound. Using this logical framework, we measured the ED_{50} for oocyte destruction after intraperitoneal treatment of B6 mice with BP or one of ten derivatives at doses ranging from 1 to 100 mg/kg. The ED_{50} for oocyte destruction following BP treatment was 15 mg/kg in B6 mice (Fig. 5). Only two of the tested derivatives had an ED_{50} for oocyte destruction less than the parent compound: the 2-hydroxy derivative, and the *trans*-7,8-dihydrodiol derivative both with ED_{50}s of approximately 2 mg/kg. The eight additional compounds tested had ED_{50}s for oocyte destruction greater than 100 mg/kg, the highest dose tested. Of special interest with respect to our earlier investigations of the relationship between ovarian AHH activity and oocyte destruction are the ED_{50}s of the 3- and 9-hydroxy derivatives, the phenolic products measured in the AHH assay. Both of these metabolites had ED_{50}s greater than BP, supporting our earlier data detailing the lack of correlation between ovarian AHH activity and oocyte destruction.

HPLC Characterization of Ovarian BP Metabolism

The previous data suggests that concentrating on metabolic processing in the 7,8-region and the 2-region of BP might be better measures of the ability of the ovary to activate the parent compound to the ovotoxic product. Because HPLC techniques cannot independently resolve the 2-phenol from several other phenols, and because other investigators have suggested that this product is not formed in vivo, our attention will focus on the metabolic processing of BP in the 7,8,9,10-region [Gelboin, 1980]. This focus is also consistent with a large body of evidence suggesting that the mutagenicity and carcinogenicity of BP is the result of formation of the *trans*-7,8-dihydrodiol-9,10-epoxide metabolite [Gelboin, 1980].

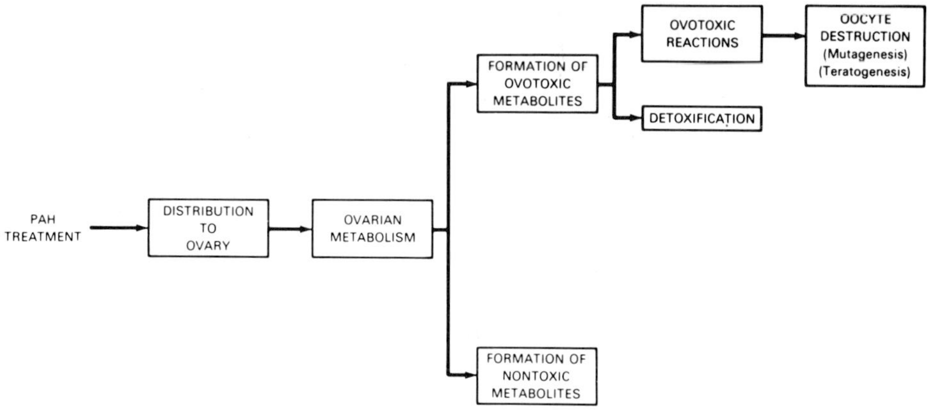

Fig. 6. Schematic of the pathway leading to oocyte destruction following treatment with polycyclic hydrocarbons. Oocyte destruction follows distribution of the parent xenobiotic to the ovary, metabolic activation to a reactive intermediate, and toxic interaction with essential cellular macromolecules in the oocyte itself or surrounding granulosa cells which support the oocyte in the follicle. Strain or species differences in metabolic processing, metabolic activation, or detoxification will account for some differences in ovotoxicity. Other factors including "receptors" for toxic interactions and repair of damage will also modulate ovotoxicity.

Using this approach, initially, the difference in Sprague-Dawley rat and B6 mouse ovarian metabolic processing of BP was explored [Mattison et al, 1979]. Although both inbred rodents assayed (B6 mice and Sprague-Dawley (SD) rats) have PAH-inducible ovarian AHH activity, they have considerably different sensitivity to PAH-induced oocyte destruction. Metabolism of BP in the 4,5-region was greater in SD rat ovarian microsomes than mouse ovarian microsomes. Formation of the 1,6-, 3,6-, and 6,12-diones was similar in rat and mouse ovarian microsomes, as was formation of the 9,10-diol. Formation of the 7,8-diol occurred approximately five times faster in B6 mouse than SD rat ovarian microsomes. Formation of the 3- and 9-phenols occurred approximately three times faster in mouse than rat ovarian microsomes. These data suggest that metabolic processing of BP in the 7,8,9,10-region may represent a better measure of the role of metabolic activation in oocyte destruction.

If the 7,8-dihydrodiol-9,10-epoxide is an ultimate ovotoxin, then three enzymatic steps are responsible for this metabolic route to ovotoxicity: formation of the 7,8-epoxide, hydration to the 7,8-dihydrodiol, and oxidation to form the putative ultimate ovotoxin the 7,8-dihydrodiol-9,10-epoxide. Strain or species differences in one or all three of these steps may be responsible for some of the strain and species differences in ovotoxicity observed.

CONCLUSIONS

Oocyte destruction by polycyclic aromatic hydrocarbons is a complex toxicological problem. Understanding strain and species differences in ovotoxicity requires dissection of multistep pathways for metabolic activation of the parent compound to an ultimate ovotoxic product. Differences in the rate of formation of the ultimate ovotoxic product may not correlate well with observed differences in sensitivity to oocyte destruction because they do not include detoxification, nor do they consider intrinsic differences in sensitivity, or response to injury (Fig. 6). A complete understanding of biological factors underlying differences in sensitivity to ovotoxicity will only come when the biological path from exposure to the parent compound to oocyte destruction has been fully dissected. This complex biological problem is important for several reasons: (1) polycyclic hydrocarbons, the prototype ovotoxins used in these experiments, are ubiquitous environmental pollutants; (2) cigarette smoking, a common source of human exposure to PAH, including BP, has been demonstrated to decrease in a dose-related fashion the age of menopause, a manifestation of oocyte depletion; and (3) several drugs used in treatment of human diseases destroy oocytes. A more complete understanding of the mechanism of oocyte destruction by polycyclic hydrocarbons will allow us to develop a better understanding of human susceptibility to ovotoxins.

ACKNOWLEDGMENTS

We would like to thank Ms. Linda Baldwin for consistent, high-quality secretarial assistance.

REFERENCES

Baker TB (1976): The effects of ionizing radiation on the mammalian ovary with particular reference to oogenesis. In Greep RO, Astwook EB (eds): "Handbook of Physiology, Endocrinology II, Part I." Washington, DC: American Physiological Society, pp 349–361.

Bassler A, Rohrborn G (1976): Chromosome aberrations in oocytes of NMRI mice and bone marrow cells of Chinese hamsters induced with 3,4-benzopyrene. Mutat Res 38:327–332.

Brookes P (1980): Chemical carcinogenesis. Br Med Bull 30:1–104.

Burke MD, Vadi H, Jernström B, Orrenius S (1977): Metabolism of benzo(a)pyrene with isolated hepatocytes and the formation and degradation of DNA binding derivatives. J Biol Chem 252:6424–6431.

Felton JS, Kwan TC, Wuebbles BJ, Dobson RL (1978): Genetic differences in polycyclic aromatic hydrocarbon metabolism and their effects on oocyte killing in developing mice. In Mahlum DD, Skiov MR, Hackett PL, Andrew FS (eds): "Developmental Toxicology of Energy Related Pollutants." DOE Symposium series 47, pp 1526–1532.

Gelboin HV (1980): Benzo(a)pyrene metabolism activation and carcinogenesis: Role and regulation of mixed function oxidases and related enzymes. Physiol Rev 60:1107–1166.

Generoso W, Cain K, Krishner M, Huft SW (1979): Genetic lesions induced by chemicals in spermatozoa and spermatids of mice are repaired in the egg. Proc Natl Acad Sci USA 76:435–437.

Gulyas BJ, Mattison DR (1979): Degeneration of mouse oocytes in response to polycyclic aromatic hydrocarbons. Anat Rec 193:863–882.

Heinrichs WL, Juchau MR (1980): Extrahepatic drug metabolism: The gonads. In Gram TE (ed): "Extrahepatic Metabolism of Drugs and Other Foreign Compounds." New York: SP Medical and Scientific Books, pp 319–322.

Janson PO (1975): Effects of the luteinizing hormone on blood flow in the follicular rabbit ovary, as measured by radioactive microspheres. Acta Endocrinol 79:122–133.

Jones CA, Moore BP, Cohen GM, Fry JR, Bridges JW (1978): Studies on the metabolism and excretion of benzo(a)pyrene in isolated adult rat hepatocytes. Biochem Pharmacol 27:693–702.

Jull JW (1973): Ovarian tumorigenesis. Methods Cancer Res 7:131–186.

Krarup T (1969): Oocyte destruction and ovarian tumorigenesis after direct application of a chemical carcinogen (9:10-dimethyl-1:2-benzanthracene) to the mouse ovary. Int J Cancer 4:61–75.

Krarup T (1970): Effect of 9:10-dimethyl-1:2-benzanthracene on the mouse ovary. Impaired fertility and its relation to ovarian tumorigenesis. Acta Endocrinol 64:489–507.

Krarup T, Loft H (1971): Presence of DMBA-^{3}H in the mouse ovary and its relation to ovarian tumor induction. Acta Pathol Microbiol Scand [A] 79:139–149.

Levine AS, McKinney CE, Echelberger C, Kouri RE, Edwards BK, Nebert DW (1982): Aryl hydrocarbon hydroxylase (AHH) inducibility among primary relatives of children with leukemia or solid tumors. American Association for Cancer Research Abstracts (Abstract #191) 23:49.

Mackenzie KM, Angevine DM (1981): Infertility in mice exposed in utero to benzo(a)pyrene. Biol Reprod 24:183–191.
Mandl AM (1964): The radiosensitivity of germ cells. Biol Rev 39:288–371.
Marchant J (1957): The chemical induction of ovarian tumors in mice. Br J Cancer 11:452–464.
Mattison DR (1979): Differences in sensitivity of rat and mouse primordial oocytes to destruction by polycyclic aromatic hydrocarbons. Chem Biol Interact 28:133–137.
Mattison DR (1982): The effects of smoking on reproduction from gametogenesis to implantation. Environ Res 28:410–433.
Mattison DR (1982): The mechanisms of action of reproductive toxins. Amer J Indust Med 3:65–80.
Mattison DR, Menard RH, Thorgeirsson SS (1980a): Ovarian aryl hydrocarbon hydroxylase activity and oocyte destruction. In Coon MJ, Conney AH, Estabrook RW, Gelboin HV, Gillette JR, O'Brien PJ (eds): "Microsomes and Drug Oxidation." New York: Academic Press, pp 1219–1222.
Mattison DR, Nightingale MS (1980): The biochemical and genetic characteristics of murine ovarian aryl hydrocarbon (benao(a)pyrene) hydroxylase activity and its relationship to primordial oocyte destruction by polycyclic aromatic hydrocarbons. Toxicol Appl Pharmacol 56:399–408.
Mattison DR, Nightingale MS (1982): Oocyte destruction by polycyclic aromatic hydrocarbons is not linked to the inducibility of ovarian aryl hydrocarbon (benzo(a)pyrene) hydroxylase activity in (DBA/2N × C57BL/6N) F_1 × DBA/2N backcross mice. Pediatr Pharmacol 2:11–21.
Mattison DR, Ross GT (1982): Laboratory methods for evaluating and predicting specific reproductive dysfunctions. Oogenesis and ovulation. In Nelson N (ed): "Laboratory Methods for Evaluating and Predicting Specific Reproductive Dysfunctions." (In press.)
Mattison DR, Thorgeirsson SS (1978): Gondal aryl hydrocarbon hydroxylase in rats and mice. Cancer Res 38:1368–1373.
Mattison DR, Thorgeirsson SS (1979): Ovarian aryl hydrocarbon hydroxylase activity and primordial oocyte toxicity of polycyclic aromatic hydrocarbons in mice. Cancer Res 39:3471–3475.
Mattison DR, West DM, Menard RA (1979): Differences in benzo(a)pyrene metabolic profile in rat and mouse ovary. Biochem Pharmacol 28:2101–2104.
Mattison DR, White NB, Nightingale MS (1980b): The effect of benzo(a)pyrene on fertility, primordial oocyte number and ovarian response to pregnant mares serum gonadotropin. Pediatr Pharmacol 1: 143–151.
Miller EC, Miller JA (1974): Biological mechanisms of chemical carcinogenesis. In Busch H (ed): "The Molecular Biology of Cancer." New York: Academic Press, pp 347–402.
Miller ED, Miller JA (1976): The metabolism of chemical carcinogens to reactive electrophiles and their possible mechanisms of action in carcinogenesis. In Searle CE (ed): "Chemical Carcinogens." Washington, DC: American Chemical Society, pp 737–762.
Nebert DW (1980): Pharmacogenetics: An approach to understanding chemical and biological aspects of cancer. J Natl Cancer Inst 64:1279–1290.
Nebert DW, Jensen NM (1979): The *Ah* locus: Genetic regulation of the metabolism of carcinogens, drugs and other environmental chemicals by cytochrome P-450 mediated monooxygenases. CRC Crit Rev Biochem 6:401–403.
Nebert DW, Jensen NM, Shinozuka H, Kunz HW, Gill TJ (1982): The *Ah* phenotype. Survey of 48 rat strains and 20 inbred mouse strains. Genetics 100:79–87.
Nordenskjold M, Svensson SA, Jernstrom B, Moldeus P, Dock L, Soderhall S (1981): Studies on the in vitro transfer of DNA binding benzo(a)pyrene metabolites from rat hepatocytes to human fibroblasts. Carcinogenesis 2:1151–1160.
Okey AB (1972): Dimethylbenzanthracene induced mammary tumors in rats: Inhibition by DDT. Life Sci 11:833–843.
Pederson R, Mangia F (1978): Ultraviolet light induced unscheduled DNA synthesis by resting and growing mouse oocytes. Mutat Res 49:425–429.
Pederson T, Peters H (1968): Proposal for a classification of oocytes and follicles in the mouse ovary. J Reprod Fertil 17:555–557.
Santodonato J, Howard P, Basu D (1981): Health and ecological assessment of polynuclear aromatic hydrocarbons. J Environ Pathol Toxicol 5:1–364.
Schreck RR, Latt SA (1980): Comparison of benzo(a)pyrene metabolism and sister chromatical exchange induction in mice. Nature 288:407–408.
Setchell BP, Sharpe RM (1981): Effect of injected human chorionic gonadotrophin on capillary permeability, extracellular fluid volume and the flow of lymph and blood in the testes of rats. J Endocrinol 91:245–254.

Shiromizu K, Mattison DR (1982): Oocyte destruction by intraovarian injection of benzo(a)pyrene. In Sato R, Kato R (eds): "Microsomes, Drug Oxidations and Drug Toxicity." New York: Wiley-Interscience, pp 559–560.

Somogyi A, Kovacs K, Solymos B, Kuntzman R, Conney AH (1971): Suppression of 7,12-dimethylbenz(a)anthracene produced adrenal necrosis by steroids capable of inducing aryl hydrocarbon hydroxylase. Life Sci 10:1261–1271.

Speilberg SP, Gordon BG, Blake DA, Goldstein DA, Herlong HF (1981): Predisposition to phenytoin hepatotoxicity assessed in vitro. N Engl J Med 305:722–727.

Weibel FJ, Lentz JC, Diamond L, Gelboin HV (1971): Aryl hydrocarbon (benzo(a)pyrene) hydroxylase in microsomes from rat tissues: Differential inhibition and stimulation by benzoflavones and organic solvents. Arch Biochem Biophys 144:78–86.

PRENATAL, PERINATAL, AND POSTNATAL TOXICOLOGY

American Journal of Industrial Medicine 4:205–244 (1983)

Perinatal Toxicology: Its Recognition and Fundamentals

Richard K. Miller, PhD

Perinatal toxicology is the study of aberrant or toxic responses to environmental agents when exposure occurs from conception through the neonatal period. The current increased interest in perinatal toxicology reflects the concern of both society and the individual in the resultant deficits induced by exposure to agents in the workplace, home, environment and by therapeutic intervention during early development. Not only is differentiation during organogenesis (the first eight weeks in human gestation) a highly susceptible period to the induction of malformations, but the fetal/neonatal developmental phases are just as sensitive for certain developmental deficits. Long-term postnatal evaluations are in most instances critical for the documentation of fetal/neonatal functional defects, which include carcinogenesis, behavioral impairment, endocrine and immune dysfunction. Ethylnitrosourea, estrogens (diethylstilbestrol), and cadmium are discussed in relationship to specificity of their fetal effects, long-term follow-up, and possible mechanisms of toxic action.

Key words: behavioral teratology, cadmium, diethylstilbestrol, ethylnitrosourea, methyl mercury, placenta, transplacental carcinogenesis, trophoblast

PERINATAL TOXICOLOGY: A PROCESS OF RECOGNITION

Pregnancy is a time of recognition for the woman, the conceptus, and society. Today a pregnant woman in the workplace is recognized to be at *potentially* greater risk for toxic response than other workers. Such a toxic response may include loss of pregnancy or the birth of an impaired or abnormal offspring. Although abnormalities, especially physical ones, have been viewed with curiosity and concern since pre-Egyptian times, the association and recognition that the environment may have a significant impact on a developing organism is a relatively recent revelation. Joseph Warkany [1978] has reviewed these historical developments in vivid detail.

The recognition of pregnancy by the mother is the result of many exquisitely sensitive physiological interactions between mother and the embryo-placental unit. The trophoblast releases its signals (protein and steroid hormones) to convert and adapt the maternal physiology to the immediate needs of the conceptus for intrauterine survival. In concert with the corpus luteum and other maternal processes, the embryo-placental unit can undertake its own development and cellular recognition.

Departments of Obstetrics-Gynecology and Pharmacology-Toxicology, University of Rochester, School of Medicine and Dentistry, Rochester, NY.

Address reprint requests to Dr. Richard K. Miller, Department of Obstetrics-Gynecology, University of Rochester, School of Medicine and Dentistry, Rochester, NY 14642.

0271-3586/83/0401-02-0205$10.50 © 1983 Alan R. Liss, Inc.

This entire process of cellular differentiation which extends from preimplantation throughout in utero development has engendered substantial research but little thorough understanding of how an organism develops. In normal development, the recognition of signposts, eg, chemical mediation within or between cells, represents one of the most sensitive processes, which when perturbed, leads to abnormal development.

One may appropriately extrapolate from a historical perspective to our current understandings, beliefs, hypotheses, and prejudices concerning the science of abnormal embryonic development. Our forefathers in this scientific quest have concentrated on abnormal structural development as the primary measure for evaluation. However, the developmental biologist/toxicologist/clinician of today must be concerned with a greater breadth of biological responses to developmental perturbations (Fig. 1). In many instances, these responses due to prenatal exposure may not be recognized at birth or in utero, eg, transplacental carcinogenesis, or mental compromise/dysfunction, and other responses may only be recognized by biochemical monitoring, eg, immunological or endocrine dysfunctions. Thus, the parameters for determining whether an abnormality may have been associated with in utero complications necessitates an unending search to develop more sensitive and reliable analytical testing tools. Furthermore, careful examination of the probes by which mechanisms for producing these abnormalities can be definitively established, is another obvious research goal.

This chapter will explore the scientific basis for current perinatal toxicology testing and the demands of society placed upon the medical scientist to insure that the unborn do not risk needless exposure to physical, chemical, or disease processes. The premise for such concern is that the period of in utero development is one of the most sensitive periods in the life of the mammal and that extreme care must be given to the maternal/conceptal unit to reduce any significant risk. The only way to assure a reduction or elimination of such risk is to determine the incidence of such toxic responses as listed in Figure 1, by establishing the time of exposure to a particular agent or group of

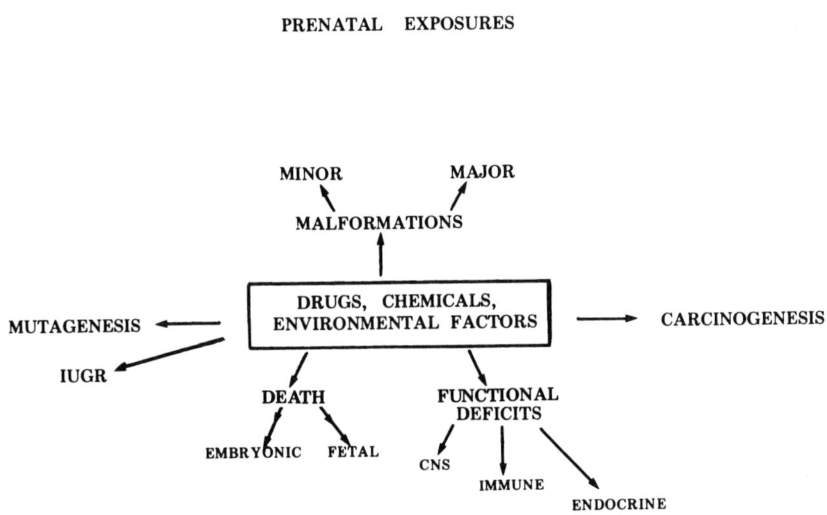

Fig. 1. Effects of prenatal exposures to environmental agents. Many of these effects may result from exposures of a single agent at different times in gestation. IUGR, intrauterine growth retardation; CNS, Central nervous system.

agents, the genetics involved, and the mechanisms by which this agent/agents may be producing or inducing the abnormal response. Especially with therapeutic agents as well as other xenobiotic exposures, intercurrent disease may modify the expression of the perinatal/lifetime abnormalities [Miller, 1981].

The primary goals of the clinical/basic scientists are to establish and prevent the exposure to toxic levels of agents in man or animal during the perinatal period. Perinatal is defined as the time from conception through the neonatal period. In vivo and in vitro testing procedures in animals and human tissues during this period of development have been explored as screening technics to prevent or reduce human and general environmental/animal exposures.

Specifically, this chapter will discuss I) animal and human development and testing procedures; II) physiology of the pregnant female and the conceptus; and III) mechanisms by which selected agents may produce perinatal toxicity: perinatal carcinogenesis, behavioral teratology, reproductive function and perinatal immunology.

This review will examine the neonatal period only in reference to selected developing organ systems for species comparisons and will not discuss the important topics of lactation, nutrition and transmammary exposure to chemicals. The reader is referred to excellent neonatal reviews by Vorherr [1974], Staves [1978], Schwarz and Jaffe [1979], and Clarkson and Nordberg [1983].

ANIMAL AND HUMAN DEVELOPMENT

Ideally, the reason for selecting a species for the study of perinatal toxicity is that it will mimic the responses seen in the human. In reality, the species selected for the study often represents the animal most readily available or easiest or most economical to study rather than the most sensitive for determining a mechanism for the suspected induced abnormalities. Currently, the Food and Drug Administration of the United States requires that agents be tested in both a rodent and non-rodent species. The rat usually is selected because of its relatively short gestational period (21-22 days), cost, size of fetuses, and number of fetuses per litter. The non-rodent species has usually been a lagamorph-rabbit. These specific testing protocols have been reviewed in depth by Palmer [1981], Dixon [1980], Harbison [1980], and the Interagency Regulatory Liason Group [1981]. Current interest is developing to explore the use of the ferret as an alternative to the rabbit, cat, or dog. A recent conference on the ferret as an alternative species in teratological and toxicological studies [Teratology 24(2), 1981] demonstrated the potential usefulness of the ferret for such studies. Other countries have guidelines similar to the United States', although some do require more extensive evaluations, eg, behavioral assessment in the United Kingdom, France, and Japan [Council on Environmental Quality, 1981; Vorhees and Butcher, 1982].

Yet even with behavioral assessments, are all functional systems appropriately evaluated? Animals from multigenerational studies which are used to evaluate reproductive capacity also can be used for immune, endocrine dysfunctions, and tumorigenesis evaluations, if routine schedules are developed [Swenberg, 1979]. This is becoming especially critical for exposures during the fetal period, where few gross structural defects are induced but rather functional alterations may be apparent (see ethylnitrosourea, diethylstilbestrol, methyl mercury). Thus, attention must be given to environmental exposures, whether therapeutic or work-related, at any stage of gestation and not only during the first trimester [Miller, 1981].

Detailed records for each woman before and throughout her pregnancy are essential in evaluating any potential risk of environmental exposures to a woman or her unborn. To assist in the evaluation, the National Institutes of Health has a data base at Research Triangle Park, NC, for teratology literature searches. The American Medical Association is also developing such a program. In addition the Institute for Scientific Information (Current Contents) in Philadelphia has available a Teratology Printout Service (Teratology ASCA Profile) for keeping abreast of recent developments [Brent, 1982].

The landmarks establishing pregnancy can have considerable variation and the lengths of gestation are different (days in rats and months in man). However, the time of implantation between rats and man are rather similar—six days post conception. It is impossible to duplicate all aspects of perinatal development in both human and opos-

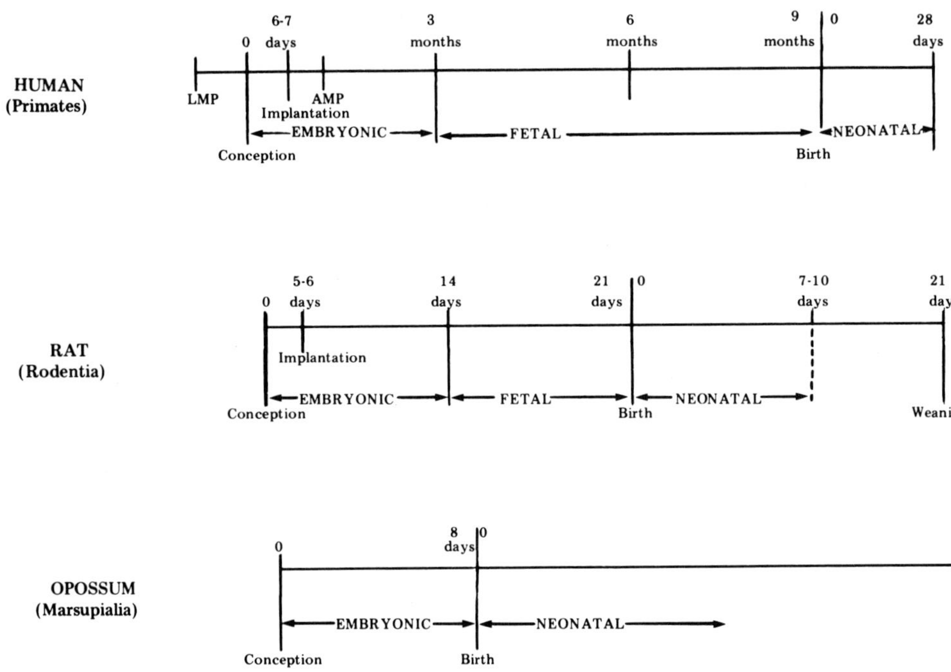

Fig. 2. Comparative perinatal development in the human, rat and opossum. The periods of development are defined for each species. LMP, last menstrual period; AMP, anticipated menstrual period. The menstrual cycle in the average human female is 28 days with ovulation occurring about 14 days. As noted for the rat, since in utero development is abbreviated compared to the human (21–22 days vs 9 months), the maturation of rat newborn is substantially different than in the human. Thus it has been proposed that the 7–10-day-old rat neonate may be more comparable to the human at birth. The opossum presents an entirely different developmental pattern: (1) there is no chorioallantoic placenta, and (2) the neonate is still in an embryonic form. Even with these differences, the effects of environmental agents in all three species can be established; however, the timing of exposures, route of administration, and developmental stage must be considered when comparing effects among species. For specific discussion of the opossum, see Jurgelski et al, 1974; Jurgelski and Porter, 1974; Jurgelski, 1974.

sum by examining only prenatal development in the rat as a test system; however, many similarities can be observed and investigated (Fig. 2). In man, the normal length of gestation is 38–40 weeks. However, the dating system used by the clinician often can be confusing to the basic scientist since the clinician usually documents pregnancy based upon physical signs, eg, the last menstrual period. Sonography is becoming another useful tool to measure the size of the conceptus and head (biparietal diameter). Such documentation forms the record establishing the date of conception plus or minus one week. In the laboratory, documentation of conception can be determined with a variation of plus or minus four hours. With whatever system or species used, the initial factor to be established is the date of conception.

In the human without the estimated date of conception, all epidemiological data are difficult to evaluate. Beside the dating of conception, the use of controls in studies, whether retrospective, prospective, or cohort, is an essential component for any human or animal study.

Many reports are only case studies because of the limited number of exposures without appropriate control information, and these reports concern only major malformations. An example of this reporting problem relates to the chronic exposure to ethanol during pregnancy. It was not until the late 1960's that French and American teams of medical scientists independently reported the Fetal Alcohol Syndrome [Lemaine et al, 1967; Jones et al, 1973]. Since that time, numerous clinical reports and studies have confirmed this syndrome of dysmorphology, growth retardation, mental retardation, and cardiovascular defects. Yet, certainly ethanol is not a newly introduced chemical. Ethanol and the fetal alcohol syndrome exemplify the necessity to evaluate and reevaluate both morphological/functional deficits in mother and progeny with all agents. These investigations must focus on the mechanisms and sites of toxicity: maternal, placental, embryonic, and fetal.

Maternal Factors

Among the factors modifying the distribution of compounds are alterations in maternal physiology (Fig. 3). During the course of pregnancy, maternal blood volume increases by approximately 20%; the plasma proteins are selectively increased, eg, increased thyroid-binding globulins by estrogen production in the trophoblast. Throughout pregnancy, maternal nutritional deprivation, altered hepatic metabolism, uteroplacental ischemia, and uterotropic action, both central (prolactin) and direct action (prostaglandins), can directly alter the survival of the conceptus [Page et al, 1981].

Substances are normally administered on an adult body-weight basis; however, both maternal and fetal toxicity can be substantially altered based upon the ratio of uterine weight/maternal weight and the placental permeability of the agent under study. If a toxic agent rapidly penetrates into the conceptus, then increased fetal toxicity and possibly decreased maternal toxicity may be seen, eg, methyl mercury. However, if the agent does not easily reach the fetus then higher maternal tissue concentrations of the compound may lead to increased maternal toxicity based on distribution data alone, eg, cadmium [Magos and Webb, 1983]. Thus the site of toxic action may be the same as in the nonpregnant female, but the distribution of the agent to and into the pregnant uterus may account for the exposure differences and therefore toxic response.

Metabolism of xenobiotics by the mother can be an important detoxifying mechanism; however, through the induction of enzyme activity, increased metabolism of environmental agents to bioreactive metabolites may lead to increased toxicity

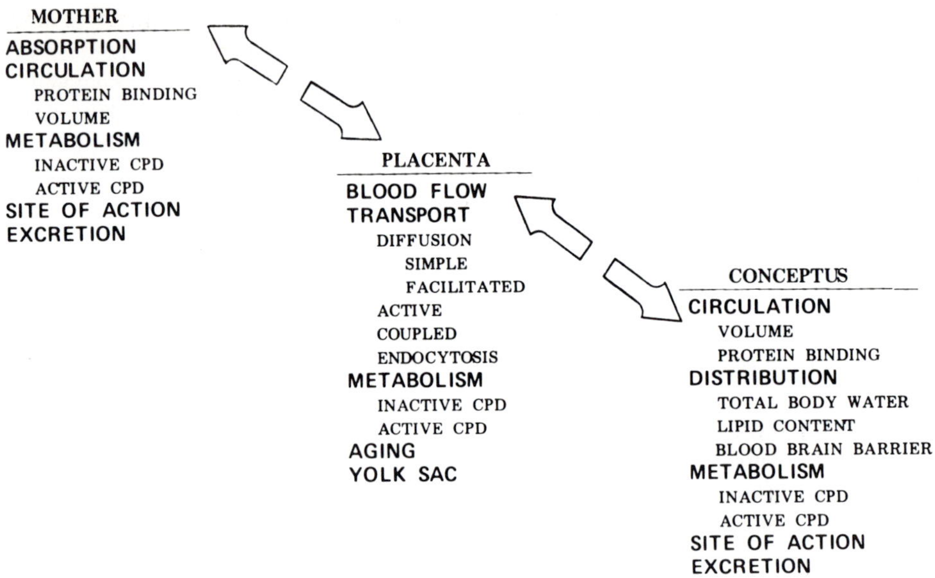

Fig. 3. Biochemical and physiological considerations in prenatal environmental exposures.

[Oesch, 1975; Pelkonen, 1981, 1982; Nebert, 1981]. For example, cigarette smoking has been associated with increased embryonic and fetal lethality, placental pathology and intrauterine growth retardation in man [Yerusholmy, 1971; USPHS, 1973; Meyer and Tonascia, 1977; Papoz et al, 1982; Naeye, 1981; Buchet et al, 1978; Asmussan, 1979]. Such toxicity has been attributed to the constituents of the smoke, eg, polycyclic hydrocarbons (benz(α)pyrene), carbon monoxide, cyanide, nicotine, heavy metals (see Cadmium).

Many of these constituents of cigarette smoke alone have been associated with alterations in the conceptus. However, many of these same constituents can be produced by the metabolism of other compounds besides cigarette smoke.

Polycyclic hydrocarbons can be inducers of maternal, placental, and embryonic enzymes [Nebert, 1981; Filler and Lew, 1981]. Shum et al [1979] have demonstrated that the teratogenicity of benz(α)pyrene is dependent primarily upon induced enzymes and the genetics of the concepto-placental unit. The presence of these xenobiotic metabolizing enzymes and their induction are both stages of gestation- and species-dependent. This is particularly true for the monoxygenases, P_{450} enzymes, and phase II inactivation enzymes [Juchau, 1982; Nebert and Gelboin, 1969; Pelkonen, 1982].

Carbon monoxide can induce biochemical and behavioral deficits which result in altered growth and development [Longo, 1977]. Part of these effects of carbon monoxide have been attributed to its high affinity for fetal hemoglobin [Delivoria-Papadopoulos et al, 1972; Longo and Hill, 1977]. Carbon monoxide, though, may be produced by chemical metabolism other than by incomplete combustion. Anders and Sunram [1982] demonstrated that methylene chloride can be metabolized to carbon monoxide, but more important, that in the pregnant rat, even when the maternal blood levels of methylene chloride were greater than in the near-term fetal blood, the fetal and maternal blood concentrations of carbon monoxide were the same. Thus, the potential for carbon monoxide to induce alterations in the conceptus is present following endogenous metabolism of xenobiotics.

Besides polycyclic hydrocarbons and carbon monoxide, drug metabolism of certain nitriles will produce cyanide. Recent studies by Willhite and associates [1981, 1982] have dramatically demonstrated the teratogenicity of acrylnitriles, which are used in the plastics industry, acetonitrile, and of laetrile, which has been used by some for cancer therapy. This teratogenicity of nitriles in hamsters has been directly related to the metabolism of these nitriles to cyanide as evidenced by antagonism of the teratogenicity by thiosulfate administration to the mother. Thus, even if the parent compound is not toxic, the metabolites, whether reactive intermediates (epoxides, arene oxides, semiquinones) or cyanide, can produce developmental deficits.

The quantity of reactive metabolites is dependent upon the maternal, placental, and embryo/fetal capacity to metabolize these chemicals and the local reactivity of the embryonic/maternal cells. For certain target tissues, the degree of metabolism and therefore deficit may be only correlated with the enzymatic activity in that organ. Therefore, one must evaluate the complete metabolic processes for xenobiotics in both mother and conceptus. Remember that the mother has both phase I and phase II enzyme systems, whereas the conceptus may be partially deficient in phase II enzymes, eg, epoxide hydrase, glutathione synthetase, glucuronyl transferase. Further, many toxicologists consider that the mixed function oxidases are the primary metabolizing systems for xenobiotics; however, these monoxygenases are not the only ones. Xenobiotics, peroxidases, and prostaglandin synthetases also metabolize [McLachlan et al, 1981].

Placental Factors

The trophoblast performs many essential tasks besides the metabolism of xenobiotics. It regulates the transfer of nutrients, waste products, and xenobiotics [Miller et al, 1976, 1983; Waddell and Marlowe, 1981; Young et al, 1981]; the synthesis and release of steroid and protein hormones [Klopper, 1980]; the immunological responsiveness of the concepto-placental unit [Faulk, 1981; Beer and Sio, 1982]; and the production and release of protein carriers [Ng et al, 1981; Miller and Thiede, 1981; Wild, 1981].

Without this interface between mother and conceptus, the gestational period would be no more than eight to ten days. Nature has provided such a mammal which does not have a chorioallantoic placenta: It is the marsupial, eg, kangeroo or opossum. Such animal models have been investigated to determine the relevance for toxicity testing (Fig. 2). There is no fetal period and a portion of the neonatal period corresponds to organogenesis. Only the visceral yolk sac is available for sustaining the opossum embryo in utero. To study the direct effects of chemicals on the developing conceptus in most species, we will have to contend with the placenta and/or visceral yolk sac; however, the newborn opossum has proved most useful in the teratogenic/carcinogenic study of ethylnitrosourea [Jurgelski et al, 1976, 1979].

The visceral yolk sac is an important organ in comparative developmental toxicology. Besides the opossum, the mouse, rat, and rabbit are dependent upon a functioning yolk sac early in gestation and then near term for the movement of large molecules, eg, IgG-immunologic competence [Brambell, 1970; Miller et al, 1976; Wild, 1981]. In the human, a role for the yolk sac beyond the first trimester has not been identified. However, for the rodent, the yolk sac has been implicated in perinatal toxicology. There are substantial differences in the placental permeability of toxicants in early gestation compared with late gestation in the rodent. This has been partially attributed to the fact that the trophoblast-ectoplacental cone does not play a major transport or regulatory role until about day 11 of gestation in the rat. Until that time, nutrition is primarily supplied by the yolk sac. Beck [1976, 1982] has called this histiotrophic nutrition, a period where the embryo is relying primarily upon local nutritional supplies.

The hemotrophic period is predominantly when the maternal and fetal circulations are transferring substances through the chorioallantoic placenta. For the rat, hemotrophic activity begins about day 11 while histiotrophic activity (yolk sac) can persist throughout gestation. Specific toxicity has been related to the histiotrophic function of the yolk sac. Trypan blue, a bisazo dye, and specific antisera induce malformations and resorptions by inhibiting histiotrophic activity, ie, endocytosis and catabolism of macromolecules by the yolk sac [Williams et al, 1975a,b, 1976; New and Brent, 1972; Brent and Jensen, 1977; Beck, 1982; Williams, 1982; Freeman et al, 1981]. This developmental toxicity for trypan blue and antisera are substantially reduced after day 10–11, which is consistent with the proposed mechanism of action and involvement with yolk sac function. Freeman et al [1981] have further demonstrated that like trypan blue, these antisera disrupt endocytosis and degradation of proteins to essential amino acids.

Metabolism of selected chemicals (benzo(α)pyrene dimethylbenzanthracene, N-Z-fluorenylacetamide) by the placenta has resulted in the production of reactive intermediates. This has been particularly true for the cytochrome P_{448} monoxygenases in placentae from mothers who smoke. Arylhydrocarbon hydroxylase in the human placenta from cigarette smokers is induced eight- to tenfold when compared with nonsmokers [Welch et al, 1969]. This is the enzyme that metabolizes benzo(α)pyrene. Thus, the polycyclic hydrocarbons, eg, benzo(α)pyrene, in cigarette smoke may be related to its alterations in fetal development; however, a dose-response relationship needs to be established both for these effects of smoking and for benzo(α)pyrene as a teratogen and perinatal carcinogen (Table I). Human placental enzyme preparations catalyzed the metabolism of benzo(α)pyrene which resulted in metabolites bound covalently to DNA [Berry et al, 1977] and a positive mutagenic response in the Ames test [Jones et al, 1977; Juchau et al, 1978]. These data certainly indicate the potential for the placenta to be a production site for toxic substances, especially when these enzymes are induced by environmental agents [Juchau, 1980; Pelkonen, 1982].

Regulation of placental transport and function can occur at the level of uteroplacental circulation or at the umbilical circulation in the chorioallantoic placenta [reviews by Rankin, 1978; Miller et al, 1976; Goodman et al, 1982; Miller et al, 1983].

A number of vasoreactive agents inhibit uteroplacental blood flow and induce fetal compromise: prostaglandins [Rankin, 1978], serotonin [Robson and Sullivan, 1966; Honey et al, 1967; Craig, 1966] and cigarette smoke [Lehtovista and Forss, 1978]. In addition, uterine vascular clamping also produces malformations, growth retardation, and/or lethality of the conceptus in the rat depending upon the length of clamping, body temperature of the mother, and stage of gestation [Brent and Franklin, 1960; Franklin and Brent, 1964; Wigglesworth, 1964; Bruce, 1972; Leist and Graviler,

TABLE I. Perinatal Carcinogens in Experimental Animals*

Aflatoxin-B_1	Ethylurea + sodium nitrite
o-Aminoazotoluene	N-Ethyl-N-nitrosobiuret
Azoethane	N-Ethyl-N-nitrosourea
Azoxyethane	Furylfuramide
Azoxymethane	4-Hydroxybutyl butylnitrosamine
Benz(α)pyrene	2-Hydroxypropyl propylnitrosamine
bix(2-Hydroxypropyl)nitrosamine	Methoxyazomethanol
N-n-Butyl-N-nitrosourea	1-Methyl-2-benzylhydrazine
Carcinolipin	3-Methylcholanthrene
Cycad meal	Methyl methanesulphonate
Dibutylnitrosamine	N-Methyl-N-nitrosourea
3,3′-Dichlorobenzidine	N-Methyl-N-nitrosourethane
1,2-Diethylhydrazine	Methylpropylnitrosamine
Diethylnitrosamine	4-Nitroquinoline-1-oxide
3,3-Diethyl-1-phenyltriazene	Nitrosohexamethyleneimine
3,3-Diethyl-1-pyridyltriazene	Nitrosopiperidine
Diethylstilbestrol	2-Oxopropyl propylnitrosamine
Diethylsulphate	Procarbazine
4-Dimethylaminoazobenzene	Propane sulphate
7,12-Dimethylbenz(α)anthracene	N-n-Propyl-N-nitrosourea
Dimethylnitrosamine	Safrole
3,3-Dimethyl-1-phenyltriazene	o-Toluidine
Dimethylsulphate	Urethan
Dimethyltrosamine	Vinylchloride
Dipropylnitrosamine	
Ethyl methanesulphonate	

*Experimental animals are rat, mouse, hamster, gerbil, patas monkey, dog, pig, and opossum. This table is compiled from reviews by Rice [1981] and Kleihues [1982] and data presented in this chapter are not necessarily comprehensive listings.

1974]. The increased susceptibility of the rat conceptus to uterine vascular clamping which occurs after day 10 has been attributed to the dependence of the conceptus on the chorioallantoic circulation and its decreased capacity for anaerobic glycolysis [Leist and Graviler, 1979]. Following uterine vascular clamping, substantial embryonic/fetal death (30% or greater) and placental pathology (congestion, hemorrhage) does occur less than three hours after clamping [Franklin and Brent, 1964; Wigglesworth, 1964; Leist and Graviler, 1974; Bruce, 1972].

The time to produce placental pathology is considerably shorter (within minutes) for agents like serotonin [Robson and Sullivan, 1966]. Placental necrosis has also been induced by feeding peroxidized lipid [Kaunitz et al, 1962], vitamin E deficiency [Kaunitz et al, 1962], and by injections of endotoxin [McKay and Wong, 1962] or cadmium [Parizek, 1964, 1965; see Cadmium]. The histological changes in the rodent placentae associated with these conditions are similar and may represent terminal changes that are related to local vascular responses to trophoblastic injury or to direct effects of the agents on the uterine vasculature. A description of the trophoblastic injury which may occur following environmental exposures are presented for cadmium (see Cadmium). Injury to the trophoblast may be the result of lipid peroxidation, anoxia, endothelial damage, or direct cellular injury, eg, mitochondrial damage.

The above examples implicate the trophoblast directly in the production of perinatal toxicity. Perhaps placental toxicology represents a major concern for early miscarriages as well as fetal losses due to environmental exposures.

The placenta is the only organ which has two separate blood supplies and therefore two separate protein solutions having different characteristics (fetal vs maternal red blood cells; α-fetoproteins, fetal>maternal; albumin, maternal>fetal). Such unique characteristics for an organ with a defined and limited life span present many opportunities for evaluation. Unfortunately, few studies have critically evaluated toxicity or concentration of chemicals in the placenta. Heavy metals have been one exception as noted in Table II. Many locations around the world have been randomly sampled and compared with one another for the levels of cadmium, mercury, and lead. The use of the placenta as an index for environmental exposure may be of some use if the appropriate environmental variables are considered, eg, occupation, smoking, water, and food exposures. Cadmium is an example of wide variations in these reports, which now can bedirectly related to degree of smoking by the mother [Lauwerys et al, 1978; Miller and Gardner, 1981; Kuhnert et al, 1982]. Occupational exposures, eg, painters in the pottery industry [Khera, 1980], must be determined with controls from the same area, sufficient numbers of patients, but most important, standardized sensitive analytical technics to assure reliability. Such issues have been recently examined for reproductive and developmental toxicity of metals in general [Clarkson and Nordberg, 1983].

Thus, the placenta and its extraembryonic membranes are the essential interfaces between mother and conceptus. These membranes perform vital nutritive, transport, hormone-controlling functions and in addition may be monitors for environmental exposures, eg, polycyclic hydrocarbons or metals, but more important, sites for toxic action. For additional reviews of placental toxicology see Kelman [1979], Panigel [1981], Miller and Thiede [1981], and Goodman et al [1982].

TABLE II. The Placenta as an Exposure Index: Content of Lead, Cadmium, and Mercury in Human Placentae* From Around the World

Location	Cadmium	Lead	Mercury: Total	Mercury: Inorganic	Mercury: Organic	Zinc	Copper	Reference
Galveston, TX	0.056 ± 0.020	—	—	—	—	15.73 ± 0.91	1.024 ± 0.054	Dawson et al [1968]
Los Angeles, CA	0.055	—	—	—	—	13.01	0.267	Creason et al
Salt Lake City, UT	0.050	—	—	—	—	14.53	0.218	Creason et al
Charlotte, NC Birmingham, AL	0.026	—	—	—	—	11.65	0.500	Creason et al
Charlotte, NC	0.028 ± 0.004	—	—	—	—	11.57 ± 0.64	1.821 ± 0.139	Karp and Robertson [1977]
Birmingham, AL	0.030 ± 0.005	—	—	—	—	9.95 ± 0.67	1.457 ± 0.148	Karp and Robertson [1977]
Augusta, GA	0.053 ± 0.004	—	—	—	—	9.42 ± 1.06	1.572 ± 0.47	Karp and Robertson [1977]

(Continued on next page)

TABLE II. (Continued)

Location	Cadmium	Lead	Mercury Total	Inorganic	Organic	Zinc	Copper	Reference
Riverhead, NY Elizabeth, NJ	0.051	–	–	–	–	13.43	0.231	Creason et al
Columbia, MO	–	0.07 ± 0.001	–	–	–	–	–	Fahim et al [1976]
Rühr Valley, FRG	0.176 ± 0.051	–	–	–	–	–	–	Thürauf et al [1975]
Nürnberg, W Ger	0.129 ± 0.037	–	–	–	–	–	–	Thürauf et al [1975]
Bavarian Forest, W Ger	0.152 ± 0.052	–	–	–	–	–	–	Thürauf et al [1975]
Altotting, W Ger	0.018 ± 0.001	0.100 ± 0.003	–	–	–	–	–	Thieme et al [1974]
München, W Ger	0.021 ± 0.004	0.350 ± 0.030	–	–	–	–	–	Thieme et al [1974]
Nashville, TN	0.017 ± 0.013	0.305 ± 0.427	0.022 ± 0.027	–	–	–	–	Baglan et al [1974]
Libramont, Belgium	0.0114	0.111	0.010	–	–	–	–	Hubermont et al [1978]
Belgium								
Nonsmoker	0.0125 ± 0.0086	0.083 ± 0.051	0.016 ± 0.015	–	–	–	–	Lauwerys et al [1978]
Smoker	0.0157 ± 0.0092	0.085 ± 0.044	0.012 ± 0.010	–	–	–	–	
Rochester, NY								
Nonsmoker	0.0042 ± 0.0028	–	–	–	–	11.76 ± 0.64	1.24 ± 0.08	Miller and Gardner [1981]
Smoker	0.0192 ± 0.0046	–	–	–	–	11.94 ± 0.64	1.36 ± 0.08	
Cleveland, OH								
Nonsmoker	0.0137 ± 0.0064	–	–	–	–	–	–	Kuhnert et al [1982]
Smoker	0.0181 ± 0.0073	–	–	–	–	–	–	
	–	–	0.0067	0.0014 ± 0.0011	0.0053 ± 0.0033	–	–	Kuhnert et al [1981]
Iowa City, IA	–	–	0.002 ± 0.002	–	–	–	–	Pitkin et al [1975]
Tokyo, Japan	–	–	0.071 ± 0.027	–	–	–	–	Suzuki et al [1971]
UK	–	1.03 ± 0.65	–	–	–	–	–	Wibberley et al [1977]
Stokes on Trent, UK								Khera et al [1980]
Pottery Factory								
Painters	–	0.540 ± 0.090	–	–	–	–	–	
Transferers	–	0.340 ± 0.160	–	–	–	–	–	
Lithographers	–	0.340 ± 0.140	–	–	–	–	–	
Miscellaneous	–	0.250 ± 0.070	–	–	–	–	–	

*μg/g placental weight.

Embryonic Factors

The embryonic period has been divided into two stages: preimplantation and organogenesis.

Preimplantation. The preimplantation period has been classically described as "the all or none period" [Wilson and Fraser, 1977; Brent and Harris, 1976], based on the observations that the embryo will either continue to grow normally or perish after an insult. During this time, the embryonic cell mass has not differentiated other than into the trophectoderm and can recover and differentiate at a later time if sufficient cell mass remains viable. At this stage of development, the largest quantity of embryonic tissue is trophoblast. The toxicity related to this all-or-none effect has been established using a number of agents, eg, X-irradiation [Brent and Gorson, 1972; Brent, 1973], hormones [Eibs et al, 1982; McLachlan, 1981], and other agents [Mattison, 1983]. Ethylnitrosourea also increases the incidence of embryo lethality during the preimplantation period of the rat (Fig. 4).

Despite its limited differentiation, the preimplantation embryo is already secreting hormones (hCG).

In addition, besides the presence and accumulation of environmental agents (nicotine, salicylates, caffeine, benz(α)pyrene) in the preimplantation embryo [Lutwak-Mann, 1973; Fabro, 1973], Filler and Lew [1981] have demonstrated that the preimplantation embryo can metabolize xenobiotics. The preimplantation mouse embryo can metabolize benzo(α)pyrene to diols, phenols, and quinones. This metabolism

Fig. 4. The effects of transplacental exposure to ethylnitrosourea (ENU)

$$(ON = N \begin{matrix} CH_2 - CH_3 \\ CO - NH_2 \end{matrix})$$

following single administrations of ENU (20 mg/kg, IV) on individual days of gestation in the rat. The embryo toxic effects of ENU (---) were embryolethality and atrophy of the allantois. The teratogenic effects of ENU (—) were hydrocephaly, exencephaly, and anophthalmia. The carcinogenic effects of ENU (-•••) were mainly neurogenic tumors. With permission from Napalkov [1973].

of benzo(α)pyrene is strain-sensitive and can be induced by 2, 3, 7, 8-tetrachlorodibenzo-p-dioxin (TCDD). Such reactivity of the early preimplantation embryo was not anticipated. However, the ability to produce reactive intermediates by these early embryonic cells as well as be induced by coincident exposure to other mixed function oxidase inducers (TCDD), increases one's sensitivity to the sanctity of the all-or-none period. Further investigations are certainly needed to extend this observation that polycyclic hydrocarbons are metabolized by the early embryonic cells and that perhaps mutagenic, carcinogenic or other effects may be related to this preimplantation exposure period.

Organogenesis. Organogenesis is the embryonic period in which the greatest amount of cellular differentiation is occurring. In the rat, organogenesis begins about the time of implantation [Witschi, 1968; Shepard, 1980] and involves primarily the skeleton and major organ systems (Fig. 2). When administered during organogenesis, numerous agents will induce structural abnormalities [Wilson and Fraser, 1977; Schwarz and Jaffe, 1979; Shepard, 1980; Juchau, 1981; Miller, 1981). For example (Fig. 4), when a single dose of the ethylnitrosourea (ENU) is administered during organogenesis, there is a high incidence of malformation especially in the CNS.

Species differences are important to an understanding of the mechanisms involved in teratogenesis. Some agents, eg, glucocorticoids, will increase the incidence of cleft lip/palate in sensitive strains of mice [Pratt and Solomon, 1981]. Two other articles (Tuchmann-Duplessis and Vainio et al, this issue) and two recent volumes [Juchau, 1981; Snell, 1982] have addressed these issues in detail.

Fetal Factors

The fetal period has been defined as the stage of development when the skeletal form and differentiation has been substantially completed. Many have considered complete formation of most organs, eg, renal, hepatic, cardiac, and gross structure, as the end of the embryonic period and the beginning of the fetal period. However, certain organ systems are not completed during the embryonic period, eg, CNS, urogenital, immune, endocrine.

One of the major difficulties in species comparisons arises because of the unique kinetics associated with in utero vs neonatal development. Besides the differential organ systems development, the interaction of mother, placenta, and fetus can be critical to expressed toxicity (Fig. 4). Fetal exposure to substances should be considered in two classifications: (1) labor-delivery exposures and (2) prelabor-delivery exposures, since the closer the exposure is to the delivery of the conceptus, the greater the potential risk for continued neonatal exposure. The major reason for such classifications is that the placenta is the primary means for excreting waste products, xenobiotics, and other substances from the fetus. Once delivery has occurred, the neonate must rely on its relatively immature hepatic and renal functions to reduce its body burden of these substances. If body burdens for these xenobiotics are excessive during the labor-delivery period, substantial toxicity can be observed during the neonatal period, eg, apnea, hypotonia which results from diazepam and ethanol [Scher et al, 1972; Omen et al, 1972; Cree et al, 1973; Abel, 1980]; hemorrhage from aspirin and warfarin [Raivio et al, 1977]; hypoglycemia from chlorpropamide and propranalol [Berkowitz, 1980]. Furthermore, the metabolism of some agents may not reduce their toxicity but maintain or increase their effects depending upon the types of metabolites formed, eg, diethylstilbestrol [Metzler, 1981; Miller et al, 1982a], methylene chloride [Anders and

Sunram, 1982], acrylonitrile [Willhite et al, 1981], laetrile [Willhite, 1981] and diazepam [Sher et al, 1972].

The development of the nervous system, the reproductive system and the immune system during the fetal period will be dicussed in greater detail. Many of the reported effects of various agents on these organ systems are due to exposures occurring during the fetal period. In addition, perinatal carcinogenesis will be discussed as a separate fetal problem.

NERVOUS SYSTEM – BEHAVIORAL TERATOLOGY

Behavior represents the integration of multiple organ systems functioning at varying levels of activity. In the adult, this integration is dynamic. Yet, if one superimposes the development of the organ systems and their manipulation by environmental agents (Table III), the resultant behavior can have endless possibilities. Thus, behavioral analysis can determine abnormal functions, but the intriguing factors are the subtleties in the timing of the perinatal exposures.

The nervous system is developing throughout the perinatal period with bursts of morphological and biochemical activity at specific times [Rodier, 1980; Black, 1982]. Evaluation of the behavioral parameters can be obvious, eg, cerebral palsy with methyl mercury, or more subtle, eg, dominance and group-activity modification with hormones [Weiss and Doherty, 1976; Reinisch et al, 1978].

TABLE III. Environmental Agents (Nondrugs) Reported To Be Perinatal Behavioral Toxins in Animals and/or Man

Allura red AC
Aspartame
Brominated vegetable oil
Butylated hyroxyanisole
Butylated hydroxytoluene
Cadmium
Caffeine
Carbofuran
Carbon disulphide
Carbon monoxide
Chloroform
Diazinon
Ethanol
Excess iron
Lead
Methylazoxymethanol
Methyl mercury
Microwaves
Monosodium glutamate
Perchloroethylene
Polychlorinated biphenyls
Stress
X-irradiation

This table is compiled from data reviewed by Vorhees and Butcher [1982], Adams and Buelke-Sam [1981] and data presented in this chapter.

Morphological sequellae can be documented when available in the human, but this requires autopsy material. Selected cases for neuropathological evaluation have been reviewed for methyl mercury poisoning [Matsumato et al, 1965; Choi et al, 1978; Choi, 1983] and ethanol use [Jones et al, 1976].

Morphological development can be compromised by numerous agents. Ethanol can reduce brain growth [Jones et al, 1976; Abel, 1980; Chernoff, 1977]; methoxy-azomethanol has been demonstrated to reduce the growth of the cerebellum of both rat and ferret when administered during organogenesis [Haddad et al, 1979; Woodward et al, 1976]. Cerebral ataxia and other behavioral deficits are present [Haddad et al, 1979]. Such interference with differentiation can be expected; however, Rodier has demonstrated that interference with cell proliferation can be just as significant [Rodier and Gramann, 1979; Rodier, 1980].

In rodent studies, both fetal and neonatal exposure to agents can give different behavioral responses depending upon when the agent is administered. Obviously, interference with cell proliferation results in altered righting reflex, permanent deficit, and locomotor coordination and hypoactivity or hyperactivity depending upon day of administration [Rodier, 1980; Rodier et al, 1979, 1980]. Such deficits have been associated with 5-azacytidine, methyl mercury and anesthetics [Rodier et al, 1979; Rodier et al, 1982].

Methyl mercury poisoning whether in Minamata Bay, Japan, or Iraq, Canada, or the US has now accounted for over 55 cases of transplacental exposure [Harada, 1975; Amin-Zaki et al, 1974]. These exposures are the result of microbial biotransformation of mercury in water and its entrance into the food chain. In Iraq, seed grain treated with a methyl mercury fungicide was used to make bread. Thus, multiple sources are available for human exposure. Most of the adverse effects in the progeny relate to neural alterations rather than morphology. A partial explanation for these effects may be found in the physical properties, especially its high lipid solubility and the fetal physiology involved.

During fetal development, the cardiovascular system is developed to minimize lung function and maximize placental exhange of O_2 and CO_2. Further, 40% of the cardiac output is diverted to the developing brain. [Stave, 1978] . Just these two factors alone can partially account for increased CNS toxicity in the fetus of selected agents. However, these characteristics are combined with (1) lipid content being a much greater percentage of total body mass, (2) the CNS composing a much greater percentage of lipid content because there is little body fat present, and (3) a reduced blood-brain barrier [Stave, 1978].

In combination, these factors may partially account for the cerebral palsy caused by MeHg [Amin-Zaki et al, 1974], the behavioral teratogenicity of diazepam [Kellogg, 1980, 1983], and newborn apnea and behavioral effects with anesthetics [Rodier and Koeter, 1982] However, the rapid proliferation of this CNS tissue and the relative immaturity of the fetal-neonatal renal and hepatic systems are also contributing factors [Pelkonen, 1982; Hook and Baillie, 1979].

Besides the high affinity and presentation of methyl mercury to the fetal CNS, the neuropathology from the human cases [Choi, 1983] combined with the neonatal mouse studies of Sager et al [in press] demonstrate that the abnormal neuronal migration, early cell loss, and persistent reduction in cell number appear to be related to the disruption of spindle fiber microtubules in the developing CNS.

Neurochemical alterations due to environmental exposures have also been explanations for observed deficits in sensory, learning, memory, activity, and problem-solving behaviors. However, the developmental studies have been few in number due to the difficulty of measuring neurotransmitters regionally in the brain. Whole brain levels do not truly indicate how certain neural systems and circuits are functioning or are compromised. Changes in catecholamines have been transiently noted for prenatal exposure to d-amphetamine [Middaugh et al, 1974; Zemp and Middaugh, 1975]. Yet, adult behavior was modified even though the dopamine levels were not. Prenatal exposure to methadone resulted in transient decreases in catecholamines in whole brain of 21-day-old pups but not in adults [Rech et al, 1980]. One explanation for the decreased norepinephrine at 21 days is the persistence of methadone for four weeks in the neonatal brain [Levitt et al, 1981].

Regional brain analyses have been performed for the evaluation of the behavioral teratogenicity for diazepam. Both transient and persistent behavioral alterations in the pups are noted following prenatal diazepam exposure [Kellogg et al, 1980, 1983]. These behavioral changes are correlated with the hypothalamic norepinephrine content and turnover in dose-response related manner while dopamine levels did not change [Simmons et al, 1982]. Thus, regional brain analysis can determine specificity in these evaluations. By day 10 postpartum, no diazepam is present in the neonates [Simmons et al, 1983]. Thus, the depression of neuroamines cannot be explained by persisting diazepam.

If the agent will induce major effects at one dose, it is likely that a lower dose may produce behavioral change. This is especially true for psychoactive agents. The plasticity and unique time-responsiveness of the various regions of the developing CNS [see Rodier, 1980] represent a range of behaviors, which may be totally different depending upon day of administration (day 14 in utero vs day 2 postnatal). Thus, the complexity of behavioral testing criteria is evidenced in the inability of many regulatory agencies to establish criteria statements and standard protocols. This does not mean these tests are not needed, but rather, more experience is required before a fixed regimen is chosen. A detailed discussion of these testing problems are presented by Vorhees and Butcher [1982] and Adams and Buelke-Sam [1981].

PERINATAL CARCINOGENESIS

It is becoming increasingly apparent that the perinatal period in the life of any organism may be one of the most sensitive to tumor induction. The predisposition of the unborn or newborn to the induction of childhood- or adult-onset tumors by environmental agents was not widely recognized 25 years ago. However with the arrival of the DES era, such possibilities are generally discussed not only in scientific meetings but also in many social settings around the world. Since the first report of tumor induction (lung) following prenatal exposure to urethan in 1947 [Laren, 1947], at least 50 compounds have been reported to have carcinogenic capabilities when adminstered to various species during the prenatal period (Table I). This number of agents is low compared to the number of known carcinogens.

This section is not devoted to a discussion of whether a carcinogen can induce tumors following perinatal exposure, but rather whether the developing organism may be more sensitive or responsive to tumor induction following perinatal exposure when compared to the adult. Is this enhanced carcinogenic response to agents in the workplace, home, or following therapy due to the rapidily developing and differentiating stage of the tissue, the particular chemical interactions with these cells, or the unique distributions of these agents in the feto-placental unit (Fig. 3)?

Swenberg [1979] has recommended a testing routine for rodents which capitalizes on the rapid proliferation of the developing organism both pre- and postnatally. Unfortunately, no uniform exposure has been established for all species. In fact, implementation of these testing procedures are in question due to the expenditures involved. Among the concerns of any testing program is the separation of the teratogenicity and mutagenicity from carcinogenic responses. This is particularly true for many perinatal carcinogens which are also teratogens. The expression of the carcinogenic action may be masked for at least two reasons: (1) the teratogenicity may inhibit or kill neoplastic cells and the whole organism (see Ethylnitrosourea, Diethylstilbestrol), and (2) the inherent tumorigenicity may not be expressed until a promoter is added.

In many instances, depending upon the dose-exposure of an agent, the agent may be an abortifacent, a teratogen, and a carcinogen. Ethylnitrosourea (Fig. 4) and diethylstilbestrol (Fig. 5) are examples of these characteristics. Yet, is the same mechanism of action responsible for all of these toxic responses?

The mechanisms by which carcinogens act have been divided in two general categories: genotoxic and epigenetic. A discussion of these classes and general in vitro and in vivo carcinogen testing has been recently reviewed by Weisburger and Williams [1981] and Farber [1981].

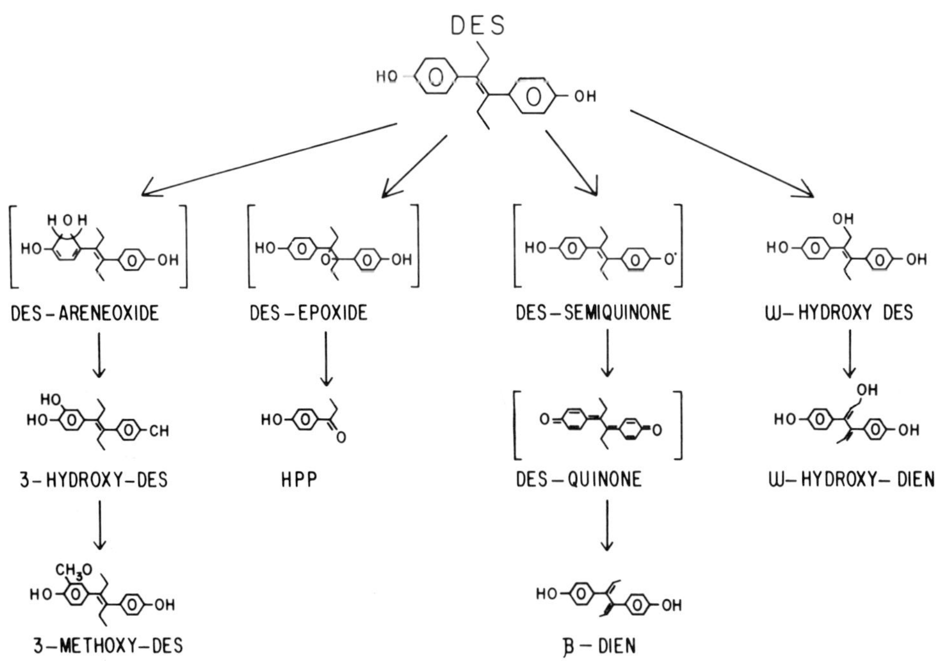

Fig. 5. Proposed scheme for DES metabolism compounds in brackets represent proposed intermediates. DIEN, dienestrol; HPP, 4-OH-propiophenone. With permission from Miller et al [1982].

Genotoxic agents are proposed to interact with DNA or alter the fidelity of DNA replication. This may occur by direct actions of the agent or may require metabolism, ie, procarcinogen. Such genotoxic agents acting in utero are alkylnitrosoureas (see Ethylnitrosourea), ethylurea + sodium nitrate [Rustia and Shubik, 1974; Fox et al, 1977], nitrosamines [Alexandrov, 1968; Althoff et al, 1977; Mohr et al, 1966], cycasin [Spatz and Laquer, 1967] and methylazoxymethanol (aglycone of cycasin [Laquer and Spatz, 1973]).

Epigenetic agents are substances which are not usually genotoxic and have proposed mechanisms of action which do not directly involve DNA. Such mechanisms include, (1) direct physical form interactions with mesenchymal cells; (2) alteration of endocrine function; (3) immunosuppression, which permits neoplastic growth; (4) cocarcinogenesis, which enhances another genotoxic agent possibly through metabolism when both are given at the same time; and finally (5) promotion, which enhances the effect of another genotoxic agent that had been previously administered. Estrogens are suspected epigenetic agents acting perinatally.

Genotoxic and epigenetic chemicals can all be metabolized to form active or inactive compounds (see Fig. 3). Whether maternal, placental, or fetal tissues are the site of this metabolism can be important in determining whether an agent will be a transplacental carcinogen. This specifically relates to procarcinogens and cocarcinogens. Since the fetal tissues often have poor xenobiotic metabolism (see Fetal Period), a procarcinogen may be less effective in inducing tumors in the offspring compared with the adult. This point is critical in that the reactive intermediates/metabolites must reach the fetal target organ to induce the tumor. However, reduced metabolism in the fetus is coupled with a cell system that is rapidly dividing. Increased cellular division is considered the essential factor by which the fetus is more susceptible than the adult to tumor induction by any carcinogen, especially direct-acting agents, eg, ethylnitrosourea.

Procarcinogens may interact with a wide variety of cellular constituents, including DNA. The actual cellular localization of these agents depends upon their affinity for the tissue, metabolism by the tissue, and localization within the tissue.

Many excellent reviews and symposia on perinatal carcinogenesis are available. The three symposia which have established this field are the International Agency for Research on Cancer Meeting in Hannover, FRG, In October 1971 [Tomatis and Mohr, 1973]; National Cancer Institute meeting in Tampa, FL, in January 1976 [Rice, 1979]; and a journal symposium in 1981 [Int. J. Biol. Res. Preg., 1981]. Additional recent reviews by Rice [1981] and Kleihues [1982] are recommended. These references review the large number of agents studied to date and the species and organ toxicity noted for each compound. In this review only ENU and DES will be evaluated further (see ENU; DES).

REPRODUCTIVE FUNCTION

Besides carcinogenesis and behavioral alterations, reproductive function is a critical parameter to consider following perinatal exposures. Even though environmental agents may not dramatically alter fertility in the exposed adult, numerous examples are now appearing that indicate the fetus and/or newborn may be at greater risk for fertility problems [Mattison, 1983; McLachlan et al, 1981; Dixon, 1980].

Among the agents noted to induce decreased fertility, are the sex hormones (see diethylstilbestrol); however, other compounds (methoxychlor; dimethylbensanthracene; methyl methanesulfonate; procarbazine, cyclophosphamide, busulfan, and cyanoketone) reduce fertility or altered cycling following prenatal exposure in rodents. Many chemicals, eg, some pesticides, metals, drugs, and food additives, also impair reproductive structure/function following postnatal exposure [Mattison, 1982; McLachlan et al, 1981].

Reproductive function is dependent upon a complete neural-pituitary-gonadal axis (see articles by Takizawa and Mattison, Smith, and Cumming and Rebar in this issue). Thus, interruption during development of any of these sites can produce infertility. Since the formation of all oocytes occurs during the fetal period, chemicals that interfere with DNA can be expected to affect the germ cell. If the pituitary or reproductive tissues are impaired as noted following DES, then malformations of the reproductive tract may be an important factor for both males and females. (see DES).

Forced breeding studies in conjunction with anatomical and biochemical evaluations have been useful tools for studying reproductive toxicity following perinatal exposure. Yet of primary concern for both the male and female of mammalian species is that reproduction is dependent upon an integrated pituitary-gonadal axis which is highly dependent upon a behavioral component. Thus an integrative evaluation (structure, function, behavior) of reproduction toxicity is essential for establishing any perinatal relationship.

PERINATAL IMMUNOTOXICOLOGY

Perinatal immunotoxicology has been acknowledged only recently as a field of study. Such delayed recognition is partially attributed to (1) the recent and rapidly improving understanding of the immune system from an expanded data base generated by new technics; (2) the delayed observations of many immune deficiencies which often appear only when appropriately challenged during adult life; and (3) the recent confirmation that agents will specifically alter permanently the developing immune system [Park and Good, 1974; Spyker, 1975; Roberts and Chapman, 1981].

The immune system, has two lines of interacting immunocompetent cells: (1) thymus-derived lymphocytes (T cells) which are responsible for cell-mediated immunity and (2) bursa equivalent-derived lymphocytes (B cells) which eventually become immunoglobulin-producing plasma cells. A complete immune system also requires the fundamental accessory cells such as the macrophage and the complement system.

Both the T and B cells have a common origin, the lymphoid stem cell. The origins of these lymphocyte precursors are found in the yolk sac in the early embryonic period. From the yolk sac, these stem cells migrate to liver, bone marrow, and eventually thymus and other lymphoid tissues (for reviews of the developing immune system see Park and Good [1974]; Roberts and Chapman [1981]). The maturation of the accessory cells and the complement system spans much of the prenatal period and the early weeks of neonatal life.

With such complex cellular interactions it is not surprising that the developing system is sensitive to prenatal exposures to chemicals which ultimately compromise adult immunocompetence. Often these agents produce irreversible damage to the developing immune system, but do not produce permanent changes when administered to the adult. Examples of such agents are cortisone [Ioachim, 1971; Kalland et al, 1978; Ways and Bern, 1979], busulfan [Pinto-Machado, 1970], cyclophosphamide [Glick, 1971; Linna et al, 1972], estrogens [Reilly et al, 1967; Luz et al, 1969; Luster et al, 1979; Kalland et al, 1978, 1979; Kalland, 1980; Ways and Bern, 1979], and TCDD [Vos and Moore, 1974; Faith and Moore, 1977; Thomas and Hindsall, 1979].

Other agents, eg, 19-nortestosterone, do not substantially impair immunocompetence when administered to the adult but can depress the developing immune system. In 1959, Meyer and associates observed that 19-nortestosterone inhibited B-cell development in the chick when it was administered during development. More recent investigations have confirmed the delay in the appearance of immunologic abnormalities in mitogen response with DES [Kalland, 1980], or in T-cell functions with methyl mercury [Spyker, 1975].

In the presence of the wide range of immunological functions, and the length of time required for detection of abnormalities in these immune functions, the need for long-term immunological evaluation of perinatally exposed animals is obvious. Unfortunately, as with most comparative studies (Fig. 2), which species should be chosen?

In the human, most immune components are mature at birth. One of the exceptions is the production of immunoglobulins [Park and Good, 1974]. It is not until 30–45 days postpartum that the newborn can adequately produce IgG. Until that time maternal IgG is utilized by the newborn. This maternal IgG is available to the fetus or newborn via specific transport processes in the trophoblast and mammary gland [Brambell, 1970; Beck 1982; Faulk, 1981; Wild, 1981].

On the other hand, most common laboratory animals with short gestational periods are not immunocompetent at birth and therefore substantial immune development occurs postpartum. In addition, in most rodents the movement of IgG is via the yolk sac rather than the trophoblast [Brambell, 1970].

The difficulty in choosing an appropriate animal model notwithstanding, it is essential to consider many of the effects of chemical agents on the immune systems during the fetal and newborn periods, when evaluating any perinatal exposures problems because of the role of the immune system in infections, carcinogenesis, and immune diseases. For additional readings articles by Faulk [1981], Roberts and Chapman [1981], and Park and Good [1974] are suggested.

MECHANISMS FOR PERINATAL TOXICITY

In this section three chemicals will be evaluated. In all cases, these agents produce multiple effects, eg, ENU: lethality, malformation, tumors. However, the emphasis of this section will be to explore specific areas of fetal toxicity: (1) ENU (carcinogenesis), (2) DES (carcinogenesis, immune, reproduction, behavior); (3) cadmium (placental toxicity, behavior, and respiration).

Ethylnitrosourea

Ethylnitrosourea (ENU) is a direct-acting carcinogen in the adult; in the developing organism, it is a substantially more reactive carcinogen. This alkylating agent is among the most potent transplacental carcinogens studied to date (Table III). In the rat, the conceptus is 50-fold more sensitive than the adult to neural tumor induction. This increased fetal sensitivity is noted for other organ systems and species as well (Table V). Yet as noted in Figure 4, ENU is also a teratogen and a potent mutagen [Russell et al, 1979]. Such toxic potential makes ENU a powerful laboratory tool for studying perinatal carcinogenesis; however, ENU and other alkyl nitrosourea compounds can be generated as metabolic breakdown products from nitrosoamines [Kleihues, 1982] as well as formed in the GI tract from ethylurea and sodium nitrite [Fox et al, 1977; Ivankovic, 1979]. Although such chemical reactions are possible in the human environment, no human studies have examined this question.

It is noted that other substances may interact with nitrites and produce tumors following prenatal exposures. L-citrulline, an amino acid from fruit, 100 mg/kg, orally administered with sodium nitrite (50 mg/kg) during the last 11 days of gestation in the rat produced renal tumors in 6 of 22 offspring [Ivankovic, 1979]. This study coupled with the ENU data base is suggestive that gastrointestinal (bacterial) reactivity may be important in the prenatal toxicity of these types of compounds. Additional studies are needed to complement these initial L-citrulline results.

The multiplicity of ENU toxicity is significant in that tumor formation, at least in the rat, is not seen unless exposure occurred after day 11 of gestation. Thus, there is a window of sensitivity for tumor induction and malformations (Fig. 4). Before day 11 of gestation, higher doses of ENU are necessary to induce malformations when compared with ENU exposure later in gestation for both rats and mice [Rice, 1969, 1981; Napalkov, 1973; Ivankovic, 1979]. The offspring treated with ENU during late gestation appear normal at birth, and only demonstrate tumors as adults (Table IV). However, the opossum treated postnatally demonstrated embryonal tumors (Wilms type) (Table V).

Chemical carcinogenesis, especially genetoxicity, is thought to involve interactions, such as covalent binding, with DNA, RNA, or cellular proteins. Substances which form electrophilic intermediates, eg, ENU, are suspected of utilizing this mechanism. However, simply binding to DNA may not be sufficient to induce toxicity, in that rate of excision of the affected DNA may be more important.

Besides ENU, other nitrosourea and nitrosoamines have been shown to bind to DNA. Yet, where these alkylating agents bind to the DNA can be varied. The degree and type of covalent binding is dependent upon the length and nature of the alkyl side chain, as is the toxicity [Ivankovic and Druckery 1968; Kleihues et al, 1979; Kleihues, 1982].

With ENU, the formation of ethyl diazonium hydroxide, which in turn is converted to ethyl carbonium ions with the release of nitrogen, represents the reactive constituents [Ivankovic and Druckery, 1968]. Under physiological conditions in vitro, Goth and Rojewsky [1972] demonstrated that ENU has a half-life of less than ten minutes. Thus, the major research efforts have been focused on the interaction of this rapidly formed and short-lived carbonium ion with DNA.

In the late 60s and 70s, numerous adducts had been identified for ENU. In vitro, O^6 alkylation products of deoxyguanine were reported for ENU, methyl nitrosourea and ethylmethanesulphonate [Loveless, 1969]. In vivo, N^7 ethylguanine was the major alkylation product for ENU in liver, kidney, lung, and brain in the Wistar rat [Swann and Magee, 1971]. The N^7 adduct was abandoned as the carcinogenic adduct when ethylmethanesulphonate was observed to produce five times more N^7 ethylguanine than the carcinogenic dose of ENU, but failed to induce renal tumors [Kleihues and Magee, 1973].

Upon reexamination Goth and Rajewsky [1974] demonstrated that besides N^7 ethylguanine, O^6 ethylguanine, N^3 ethyladenine, and N^7 ethyladenine were present in both brain and liver of 10-day-old rats treated with ENU. Excision rates were determined in rat brain (a high-risk organ for carcinogenesis) and the liver (a lower-risk organ for carcinogenesis), and the major differences between half-lives were for O^6 ethylguanine (liver 36 hours; brain 229 hours). This observation for ENU has been also reported for methylnitrosourea [Kleihues et al, 1979]. All of these data are suggestive that while specific alkylation (O^6) of guanine may be the initiating step in the process, the excision-repair processes within cell types are also important determinants. Yet,

TABLE IV. Tumor Induction by Perinatal Exposure to Ethylnitrosourea in Various Species

Species	Dose	Days of administration	Tumor type	Reference
Hamster (Syrian)	60–80 mg/kg	10 days prenatal	Neurinomas, oligodendrogliomas, astrocytomas	Ivanokovic [1979]
Mouse	29–117 mg/kg[a]	12–18 days prenatal or within 24 h postnatal	Hepatomas, alveologenic adenomas and adenocarcinomas	Rice [1969] Vesselinovitch et al [1971, 1977]
	59–117 mg/kg[a]	15–42 days postnatal	Neurinomas, medulloblastomas, astrocytomas, meningeal tumors, lymphomas	Denlinger et al [1974] Wechsler et al [1979]
Monkey (Patas)	12 mg/kg[b]	30 days—parturition	Connective tissue	Rice et al [1979, 1981] Rice [1981]
Opossum	100 mg/kg[c]	1–20 days postnatal	Nephroblastomas, gangliogliomas, diktyoma (ocular), ameloblastomas	Jurgelski et al [1979]
Rabbit	60 mg/kg[a]	18 days prenatal	Renal tubular cyst adenocarcinomas, nephroblastomas,	Fox et al [1975]
	50 mg/kg[b]	8–10 days prenatal 21–32 days prenatal at intervals	gliomas, schwannomas, renal adenocarcino-sarcomas, nephroblastomas	Stavrou and Hanichen [1975] Stavrou et al [1977]
Rat	5–80 mg/kg[d]	12–22 days prenatal	Neurinomas, neurosarcomas, oligodendrogliomas, astrocytomas, ependymomas	Napalkov [1973] Ivankovic [1973, 1979]

[a]Single administration, IP.
[b]Multiple injections, IV.
[c]Single administraiton, per os.
[d]Single administration, IV, per os.

TABLE V. Summary of Abnormalities Following Prenatal Exposure to Diethylstilbestrol in Man and Rat/Mouse

Human[a]	Rat/mouse[b]
Female anatomical abnormalities	
Vagina	
Adenosis	Adenosis
Adenocarcinoma	Adenocarcinoma
Squamous cell dysplasia	Squamous cell carcinoma
	Hyperkeratosis
	Hypospadias
	Caculi
	Vaginitis
Cervix	
Hoods	
Stenosis	
Incompetence	
Uterus/fallopian tube	
T-shaped uterus	Endometrial hyperplasia
"Withered" fallopian tube	Uncoiled tubes
External genitalia	
	Patent vagina at birth
	Persistent cleft phallus
	Perineal excoriation
Female functional abnormalities	
Menstrual irregularities	Noncycling
Spontaneous abortions	Infertility
Ectopic pregnancy	
Premature delivery	
Male anatomical abnormalities	
Meatal stenosis	
Hypospadias	
Epididymal cysts	
Hypotrophic testes	Hypotrophic testes
Leydig cell hyperplasia	
Cryptorchidism	Cryptorchidism
Microphallus	Microphallus
Testicular agenesis	Undescended testes
Male functional abnormalities	
Altered semen analysis	
Azoospermia	Azoospermia
Decreased sperm penetration in hamster egg sperm penetration assay	

[a]Data for the human was derived from Stillman [1982]; Herbst and Bern [1981]; Kaufman [1982]; Bibbo et al [1975].
[b]Data for the rat was derived from Greene et al [1940]; Baggs et al [1980]; Miller et al [1982b]; Baggs and Miller [unpublished data]; for mouse—McLachlan [1981]; McLachlan et al [1980, 1981, in press].

many other controlling factors, eg, rate of cell replication, may be essential in this regulatory scheme. Whether one considers the adduct formation, organ-specific tumor induction, or time of appearance of tumor, ENU is the most potent perinatal carcinogen known to date.

Diethylstilbestrol

Since the 1940s, humans have been exposed to the nonsteroidal estrogen diethylstilbestrol (DES). DES was the wonder drug of the era due to its availability at reasonable cost and estrogenic potency, even when taken orally. DES became widely used as the estrogen for therapy and later for its growth promoting effects in livestock. With such wide utilization, human exposures occurred in the workplace, through food (meat), through contaminated medications and for specific drug therapy [Zaebst et al, 1980; Hertz, 1980; Knight, 1980; Hoover, 1980].

Although DES is the most widely recognized estrogenic contaminant, many other compounds with estrogenic effects have been reported: insecticides (eg, DDT and kepone), polychlorinated biphenyls (PCB), animal products (eg, zeranol, estradiol), and mycotoxins from fungi in contaminated grains (eg, zearalenone and zearalenol).

Even though Greene et al [1940] noted impaired testicular descent following maternal DES exposure in the rat and Dunn and Green [1963] reported reproductive tract tumors in mice following neonatal exposure to DES (2 gm/kg in saline), interest in DES as a teratogen and transplacental carcinogen did not peak until the reports by Herbst et al [1970, 1979] of clear-cell adenocarcinomas of the vagina in women exposed in utero to DES.

The effects associated with perinatal exposures to diethylstilbestrol (DES) range from altered liver enzyme imprinting [Lucier, 1976], to reproductive tract malformations [Kaufman, 1982; Stillman, 1982] and to adenocarcinoma of the vagina and cervix [Herbst et al, 1979; Herbst and Bern, 1981; Forsberg, 1981].

As noted in Table V a wide range of teratogenic effects have been observed in both male and female progeny exposed to DES. Infertility and poor reproductive history appear directly related to structural abnormalities of the urogenital system, eg, oviduct, ovarian, cervical, uterine, vaginal, testicular, phallic, and epididymal. Many of these same abnormalities have been noted in primates [Hendrickx et al, 1979; Johnson et al, 1981], mice [Forsberg, 1981; Bern and Talamantes, 1981; McLachlan, 1981; McLachlan et al, 1981], hamster [Rustia and Shubik, 1976], and rat [Greene et al, 1940; Vorherr et al, 1978; Napalkov and Anisimov, 1979; Boylan and Calhoon, 1979; Baggs et al, 1980; Miller et al, 1982b].

The origin of many of these urogenital abnormalities is Müllerian duct tissue. Forsberg and Kalland [1981] have reviewed the development of the genital tract in both humans and rodents. Such an understanding of the embryology is essential to identifying the mechanisms by which estrogens may affect urogenital tract development, especially the vaginal plate epithelium. One of the difficulties in utilizing animal models for reproductive tract studies is the wide variation in temporal development of the reproductive tract. Considerable development of the reproductive tract occurs postnatally in rodents compared to humans.

The incidence of vaginal adenocarcinoma in women under the age of 30 exposed in utero to DES is estimated to be between 0.01 and 0.1% of the exposed population [Herbst et al, 1979]. Such a low incidence raises many questions concerning timing of exposure, dose, genetics, metabolism/distribution, cell response, and susceptibility. To date, the human studies have certainly demonstrated a high incidence of malformations [Stillman, 1982; Kaufman, 1982]; however, an understanding of the mechanisms of the tumor induction must depend upon a combination of human and animal model evaluations. Rodents are the only animal models that demonstrate vaginal tumors. Even though a wide range of malformations (hypospadias, infertility, abnormal external genitalia, precocious development of the nipples, cervical ridges, vaginal hyperkeratosis, and adenosis) have been reported, only four studies have reported increased reproductive tumors (vagina, uterus) with appropriate controls [Dunn and Green, 1963; Napalkov and Anisimov, 1979; McLachlan et al, 1980, 1981, 1982; Miller et al, 1982]. Dunn and Green [1963] reported a vaginal tumor incidence of approximately 8% (our calculation) for female mice injected subcutaneously at birth with 2 g/kg of DES in saline. These tumors were epithelial in original and were observed between 20–26 months. There were no such tumors in the four controls. McLachlan et al [1980] and Newbold and McLachlan [1982] have reported the appearance of three vaginal adenocarcinomas in female mice prenatally exposed to DES. Adenocarcinomas of the vagina were observed at 0.005 (1/29), 0.01 (1/57) and 0.1 (1/46) mg/kg/day of DES. The overall incidence of adenocarcinoma in female mice exposed in utero to DES from days 9–16 was 2.4%. No vaginal tumors were observed in controls. Interestingly, little adenosis was observed in these prenatally exposed mice, while neonatal exposure to DES induced a high incidence of adenosis but no adenocarcinoma [Forsberg, 1969, 1975, 1981; Bern and Talamantes, 1981; Newbold and McLachlan, 1982]. These studies suggest a prenatal window of sensitivity for vaginal tumors and ovarian duct defects and postnatal window for vaginal adenosis.

The hamsters treated on day 14 and 15 of gestation with DES developed reproductive tumors as adults [Rustia, 1978; Rustia and Shubik, 1979]. A 28% overall incidence of reproductive tumors which included cervical polyps in the exposed females was described. No vaginal clear cell adenocarcinomas or squamous cell carcinomas were observed.

Vorherr et al [1978] have reported four reproductive tract tumors uterine, ovarian, and vaginal in ten Sprague Dawley rats exposed at five different dosing schedules (prenatal, postnatal, or perinatal). However, there were not controls reported to establish the background incidence in this strain.

Pregnant rats on day 19 of gestation were treated with 1.0 mg/kg DES [Napalkov and Anisimov, 1979]. The resulting female offspring had persistent estrus and 78% (14 of 18) had tumors but not of the uterus or cervix. There were two thecafolliculomas, three endometrial polyps, and one uterine fibroma. Twenty-seven percent of the controls (9/34) had tumors, but of nonurogenital tissue.

Boylan [1978] also utilized the Sprague Dawley rat and observed that treatment with 0.360 mg/kg (calculated) on day 15, 17, 19 of gestation did produce some pregnancy loss, and in the progeny, abnormal nipple development, reduced fertility, precocious opening of the vagina. Besides ovarian cysts in 6 females and salpingitis in 2 of 32 females exposed to DES, no other lesions were observed. Unfortunately, the rats were not maintained for a sufficient period to document tumorigenesis.

In the Wistar rat, two-year studies [Baggs et al, 1980; Miller et al, 1982b; Baggs and Miller, unpublished observations] have demonstrated that prenatal exposure to DES on days 18, 19, 20 of gestation produced a dose-related incidence of urogenital malformations (patent vagina, cleft phallus, hypospadias, immature ovary, vaginal hyperkeratosis, vaginal calculi), infertility, and vaginal tumors (adenocarcinoma and squamous cell carcinoma). DES exposure earlier during pregnancy resulted in abortion. The teratogenesis, in particular hypospadias and vaginal calculi, was a confounding variable for the tumors. The dose-related appearance of adenocarcinoma [4.2% and 9.5% at 0.1 mg/kg and 0.5 mg/kg DES] and squamous cell carcinoma [2.4%, 3.8% and 27.3% at 0.5, 25, and 50 mg/kg DES] were in animals that survived at least 120 days. There were no epithelial tumors of the vagina (one leiomyosarcoma, one leiomyoma) in 147 corn-oil treated controls. It now appears that for both the rat and the mouse, DES can induce adenocarcinoma and squamous cell carcinoma of the vagina at dosages that are substantially less than used for human therapy when calculated on a surface-area basis.

Since 1976, there has been a continued emphasis on the effects of nonsteroidal estrogens and their unusual pharmacokinetics compared with the natural estrogens. Prior to this time, glucuronides and sulfates were the reported metabolites of DES [Fischer et al, 1976; Shah and McLachlan, 1976]. However, since that time over 11 oxidative metabolites (DES-epoxide, DES-diol, W-OH DES, methoxy-DES, catechol-DES, β-dienestrol, W-OH-dienestrol, catechol-dienestrol, methoxy-dienestrol, 4-hydroxy propriophenone) have been described (see Fig. 5, Metzler, 1981; Miller et al, 1978, 1980, 1982a; Engel et al, 1978]. The arene oxides, epoxides, and quinones formed from DES have been the focus of concern involving the nonestrogenic toxicity of DES. Rudiger et al [1979] reported that DES-epoxide is substantially more effective in increasing the rate of exchange in the *sister chromatid exchange assay* than DES. It is 1,000 times more potent than benzo(α)pyrene, the positive control.

DES has been shown to bind to DNA cellular fractions [Blackburn et al, 1974, 1976] and covalent binding of DES/metabolites will occur both in vivo and in whole cells under in vitro conditions in both maternal and fetal tissues especially reproductive tract [Miller et al, 1982a; Henry et al, 1982]. Such metabolism has been documented in mouse, rat, man, and primate [Metzler et al, 1981].

Additional manipulations of DES metabolism have been undertaken to modify this incidence of covalent binding and production of oxidative metabolites [Miller et al, 1979, 1980, 1982a; Henry et al, 1982]. It is now apparent that antioxidants (BHA) and trichloropropene oxide (an inhibitor of epoxide hydrase and depletor of glutathione) will dramatically alter both of the above DES parameters. Also, peroxidases and prostaglandin synthetase may be involved in the metabolism of DES [McLachlan et al, 1982]. This is certainly consistent with the lack of increased DES metabolism when monoxygenases are induced by TCDD, 3-methylcholanthrene, phenobarbital, and pentobarbital. It is becoming increasingly apparent that epoxide hydrase may play an important regulatory role in this process, as has been demonstrated in adult liver systems for benzo(α)pyrene [Benson et al, 1978].

Besides the covalent binding, the distribution of metabolites and DES to the fetus is substantially different when compared to the mother. Note in Figure 6 that the fetal reproductive tract has a large amount of metabolites compared to the fetus as a whole or selected maternal tissue. Perhaps the reproductive tract enzymes (peroxidases and prostaglandin synthetases) are locally producing reactive intermediates.

It has been established that 17β-estradiol is not metabolized to similar reactive intermediates [Klopper, 1980; Miller et al, 1980; Slikker et al, 1982]. If 17β-estradiol is given to the pregnant monkey, human, or rat, practically all of the 17β-estradiol is converted to estrone by the trophoblast, whether in utero or in vitro [Slikker et al, 1982; Henry et al, 1982; Wier and Miller, unpublished observations].

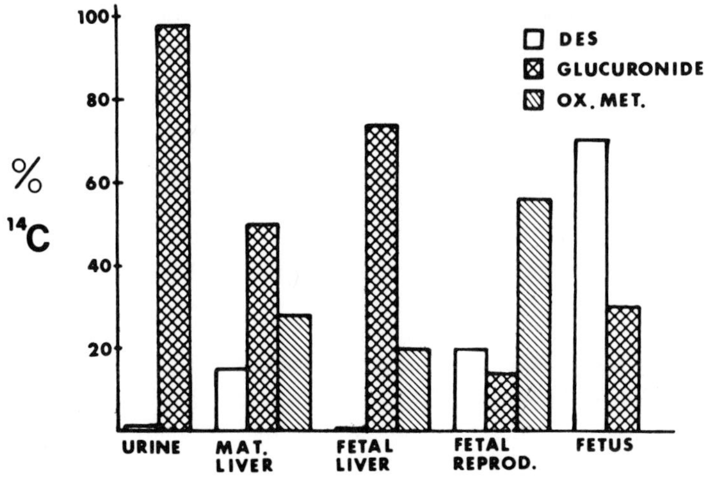

Fig. 6. Relative amounts of DES and its metabolites in selected maternal (MAT) and fetal tissues after maternal infusion of [^{14}C]DES in the Wistar rat. OX MET, oxidative metabolites (see Figure 5, especially β-dienestrol); REPROD, reproductive tract. With permission from Miller et al [1982].

If 17β-estradiol is given directly to the rat fetus, most of it is plasma protein-bound and/or metabolized to estrone [Henry et al, 1982]. Previously Sheehan et al [1980] have demonstrated that α-fetoprotein binds estradiol but not DES in the rat. Thus, plasma binding and metabolism to less active estrogens may be defense mechanisms for the conceptus to reduce the estrogenic effects of 17β-estradiol. However, if 17β-estradiol is directly injected into rat fetuses, similar urogenital malformations are produced as observed for DES. No information is available concerning the carcinogenicity of the two compounds.

Besides reproductive tract alterations, immunological deficits have been observed following perinatal exposure to DES [Kalland et al, 1978, 1979; Kalland, 1980; Luster et al, 1979]. Many of the altered immune responses have been associated with the estrogenicity of DES.

In utero exposure to DES in the human represents a cohort of extreme proportion, which provides a unique opportunity for documenting and evaluating prenatal exposures to an environmental agent. Such large numbers of well-documented exposures can give both high and low incidence of risks for malformation, functional deficits (immune, reproductive, behavioral) and carcinogenesis in the human. Thus, for perhaps the first time animal models can be compared directly with a large data base derived from human exposures.

Cadmium

Cadmium is an environmental pollutant which can be released during manufacturing processes, eg, mining, smelting of metals, manufacture of fungicides, pigments, batteries, and corrosion-resistant metals [Friberg et al, 1974]. Human exposure to cadmium may occur via water, air, food, and also tobacco., Even low-level exposures to cadmium unfortunately result in a prolonged retention of the cadmium in the body, especially in liver and kidney (Nordberg, 1974; Kjellstrom and Nordberg, 1978].

A wide variety of cardiovascular, renal, lung, and reproductive dysfunctions have been associated with cadmium exposures in man or animal [Friberg, 1974]. Cadmium exposure during gestation in animals is associated with multiple birth defects and function compromise. These teratologic effects of cadmium are dose, stage of gestation, and strain/species-specific [Ferm and Hanlon, 1983; Layton and Ferm, 1980]. The teratogenicity of cadmium is highly correlated with the movement of cadmium across the yolk sac [Ferm et al, 1969; Dencker, 1975]. In the hamster and mouse, few structural malformations are induced after day 9–9.5 of gestation, which coincides with the closure of the vitelline duct. The vitelline duct is the communication between the visceral yolk sac and the primitive gut of the embryo where cadmium is detected when administered before day 9–9.5 [Dencker et al, 1983].

In most rodents, the fetal transfer of cadmium across the placenta is low [Sonawane et al, 1975; Ahokas and Dilts, 1979; Prigge, 1978; Levin et al, in press]. These observations have been confirmed for the human under in vitro perfusion conditions [Wier and Miller, unpublished observations]. The only exception to this placental barrier equivalent for cadmium are in situ perfusion data for the guinea pig [Kelman and Walter, 1977].

Cadmium is present in the human placenta. The actual levels of heavy metals are dependent upon whether the mother is occupationally exposed, eg, potters and lead [Khera et al, 1980], and is a smoker, (cadmium, Table I). Unfortunately, most reports for human placental levels of metals have not considered these variables, and thus the degree of smoking may bias the cadmium levels reported for various regions.

In animal studies, single subcutaneous or intravenous administration of cadmium results in large concentrations in the maternal liver ≫placenta >maternal kidney [Ahokas and Dilts, 1979; Sonawane et al, 1975; Levin et al, 1983]. Even though the kidney is the primary site for adult toxicity, the placenta actually concentrates more cadmium on a weight basis. In 1964, Parizek reported that single injections of cadmium (40 μmol/kg) to the mother during late gestation will produce both fetal death and placental necrosis within 24 hours. Also during this period, there is an apparent decrease in the LD_{50} for cadmium in the mother [Samarawickrama and Webb, 1979]. As noted under Maternal Factors, this altered sensitivity may be due to the increased weight of the uterus relative to the entire weight of the mother. Thus, with poor permeability for cadmium, more cadmium is available to be materno-toxic [Magos and Webb, 1983]. The maternal toxicity has many similarities to the Schwartzman reaction and preeclampsia-eclampsia, eg, hematuria, hemorrhagic edema, swollen and hyperemic kidneys, lethargy and occasionally convulsions [Parizek, 1965; Samarawickrama and Webb, 1979].

The fetal toxicity of cadmium cannot be accounted for by the maternal toxicity due to the low incidence of renal toxicity or the direct effects of cadmium on the fetus. Levin and Miller [1980] demonstrated that direct fetal injections of 1–1 1/2 times the maternal dose of cadmium on a weight basis produced levels of 75 nmol/fetus with a fetal lethality of 12%, while maternal injections produced cadmium levels of 9 nmol/fetus and a fetal mortality of 75%. Thus, the acute fetal lethality was attributed to the placenta as suggested by Parizek and not to direct fetal toxicity [1964].

Placental injury and eventual fetal death due to cadmium is attributed to two factors: (1) direct actions of cadmium on the trophoblast, and (2) reduction of utero-placental blood flow. By 10–16 hours following maternal injection of cadmium in the day 19 pregnant rat, histologic evident necrosis was occurring in the placenta, fetal death was increasing and utero-placental blood flow was decreasing significantly [Levin and Miller, 1981; Levin et al, 1981]. These changes were coincident; however, electron microscopic changes occurred earlier (less than six hours). The earlier ultrastructural changes appear in trophoblast layer II, with lysosomal vesiculations, cytoplasmic edema, apparent mitochondrial calcium apatite [di Sant'Agnese et al, 1983]. Even when all trophoblastic layers were destroyed, the fetal endothelium was still remarkably unaffected, which certainly contrasts with the effects of cadmium on the testes. Mitochondrial calcium analysis of the placenta confirmed these earlier ultrastructural changes [Levin et al, 1981].

Thus, the fetal lethality noted for single maternal administrations of cadmium is directly associated with placental malfunction. Yet, this fetal and placental toxicity can be antagonized by zinc administration [Ahokas and Dilts, 1980]. In addition, cadmium will inhibit zinc transport across the placenta on day 12 of gestation Samarawickrama and Webb, 1981]. However, this fetal lethality was not noted for cadmium metallothionein [Plautz et al, 1980, 1981].

Besides placental toxicity, a respiratory distress-like syndrome and CNS hemorrhages have been reported in rat pups following prenatal exposure to cadmium. Administration of cadmium on days 12–15 of gestation to the pregnant rat reduced pulmonary function in the newborns, inhibited surfactant production, reduced lung weight, produced the labored breathing observed for hyaline membrane disease or the respiratory distress syndrome, and neonatal death [Chernoff, 1973; Daston and Grabowski, 1979; Daston, 1981, 1982]. The lungs of the treated fetuses demonstrated reduced cellular glycogen, fewer lamellar bodies and delayed epithelial differentiation of the alveoli [Daston, 1981]. These effects by cadmium on fetal lung development, especially type II cells, and results in an RDS like syndrome are the first reports of such an environmentally induced association. Appropriate human cases exposed to cadmium have not been reported.

Cadmium has not been considered a neurotoxin as have other heavy metals, e.g., methylmercury and lead. However, prenatal and early postnatal studies are finding permanent morphological and behavioral changes [Webster and Valais, 1981; Wong and Klassen, 1981; Infurna et al, 1982]. Cerebral hemorrhages were persistent following prenatal or postnatal exposure. Olfactory-guided behavior was disrupted following neonatal cadmium exposure and correlated with hemorrhages in the olfactory bulb [Infurna et al, 1982]. Behavioral toxicity persisted into adult life.

Besides cadmium being teratogenic in animals, fetal toxicity is readily apparent. Cadmium is concentrated by the placenta and can be placental toxic, which in turn produces fetal death. But more importantly, cadmium administered to the mother can induce severe functional deficits, respiratory distress syndrome. And finally, besides the testicular necrosis noted for cadmium, it can in the developing central nervous system produce local cerebral hemorrhages and permanent behavioral impairment.

SUMMARY

Perinatal toxicology examines the exposure of the developing organism throughout gestation and as a neonate to an agent and evaluates ensuing functional and structural deficits. Carcinogenesis, immunotoxicity, behavioral compromise, reproductive dysfunction, and endocrine disorders can all be traced to specific exposures only during the fetal or neonatal periods. As noted for ethylnitrosourea, diethylstilbestrol, and cadmium, the mechanisms for their perinatal toxicity are not identical. In particular, DES offers a unique opportunity to study a large human population. These comparisons between human and animal data will assist in developing new testing protocols for both functional and structural disorders. Postnatal functional assessment is as critical as structural evaluation when one must consider the developing organism, its mother, and her environment, whether in the workplace, home, or pharmacy. The progeny is certainly at risk throughout the perinatal period.

ACKNOWLEDGMENTS

The original research presented was supported in part by the following grants: NCI 22335, NIEHS 12407, and the R.W. and M.S. Goode Gift. The author gratefully acknowledges the critical reviews by Drs Raymond Baggs, Wendy Ng, and Ellen Henry, Roy Simmons and Patrick Wier. The typing was most capably performed by Miss Shelley DiDomenico.

REFERENCES

– (1981): Journal symposium on Transplacental carcinogenesis. Int J Biol Res Pregnancy 2:4.

Abel EL (1980): Fetal alcohol syndrome: Behavioral teratology. Psychol Bull 87:29–50.

Abel EL, Bush R, Dintcheff BA (1981): Exposure of rats to alcohol in utero alters drug sensitivity in adulthood. Science 212:1531–1533.

Abramovici AF, Svejcar J (1982): Histochemical and quantitative biochemical changes in mucopolysaccharides of the term placenta of the diabetic rat. Placenta 3:81–90.

Adams J, Buelke-Sam J (1981): Behavioral assessment of the postnatal animal: Testing and methods development. In Kimmel CA, Buelke-Sam J (eds): "Developmental Toxicology." New York: Raven Press, pp 233–258.

Ahokas RA, Dilts PV (1979): Cadmium uptake by the rat embryo as a function of gestational age. Am J Obstet Gynecol 135:219–222.

Alexandrov VA (1968): Blastomogenic effect of dimethylnitrosamine on pregnant rats and their offspring. Nature 218:280–281.

Alexandrov VA, Napalkov NP (1981): General regularities of transplacental (prenatal) carcinogenesis. Biol Res Pregnancy 2:181–187.

Althoff J, Grandjean C, Mauk S, Pour P and Takahashi M (1977): Transplacental effects of nitrosamines in Syrian hamsters II. Nitrosopipendine. Z Krebsforsch 90:71–77.

Amin-Zaki L, Elhassani S, Majeed MA, Clarkson TW, Doherty RA, Greenwood M (1974): Intra-uterine methylmercury poisoning in Iraq. Pediatrics 54:587–595.

Anders MW, Sunram JM (1982): Transplacental passage of dichloromethane and carbon monoxide. Toxicol Lett (in press).

Asmussen I (1979): Effects of maternal smoking on the fetal cardiovascular system. Cardiovasc Med 4: 777–790.

Baggs RB, Miller RK, Garman R, McKenzie RC (1980): Teratogenic and neoplastic lesions of the genitourinary system in Wistar rats exposed in utero to diethylstilbestrol. Teratology 21:26A.

Baglan RJ, Brill AB, Schubert A, Wilson D, Larsen K, Dyer N, Mansour M, Schaffner W, Hoffman L, Davies J (1974): Utility of placental tissue as an indicator of trace element exposure to adult and fetus. Environ Res 8:64–70.

Barrett JC, Wong A, McLachlan JA (1981): Diethylstilbestrol induces neoplastic transformation without measurable gene mutation. Science 212:1402–1404.

Beck F (1976): Comparative placental pathology and function. Environ Health Perspect 18:5–12.

Beck F (1982): Models in teratology research. In Snell K (ed): "Developmental Toxicology." New York: Praeger Publishers, pp 11–32.

Beckman D, Jensen M, Burman M, Koszalka TR, Brent RL (1981): Rat trophoblastic cell antigenicity. Placenta (Suppl) 3:75–84.

Beer AE, Sio JO (1982): Placenta as an immunological barrier. Biol Reprod 26:15–27.

Benson A, Batzinger R, Ou S, Bueding E, Cha Y, and Talalay P (1978): – Cancer Res 38:4486–4495.
Berkowitz RL (1980): Anti-hypertensive drugs in the pregnant parent. Obstet Gynecol Survey 35:191–204.
Berlin, M, Ullberg S (1963): The fate of cadmium 109 in the mouse: An autoradiographic study after a single intravenous injection of cadmium 109 Cl_2. Arch Environ Health 7:686–693.
Bern HA, Talamantes FJ (1981): Neonatal mouse models and their relation to disease in human female. In Herbst AL, Bern HA (eds): "Developmental Effects of DES in Pregnancy." New York: Thieme-Stratton, pp 129–145.
Berry DL, Zachariah PK, Slaga RJ, Juchau MR (1977): Analysis of biotransformation of benzo(α)pyrene in human fetal and placental tissues with high-pressure liquid chromatography. Eur J Cancer 13:667–675.
Bibbo M, Naqebb MA, Baccarini I, Gill W, Newton M, Sleeper KM, Sonek M, Wied GL (1975): Follow-up study of male and female offspring of DES-exposed mothers. J Reprod Med 15:29–32.
Bissonnette JM (1982): Membrane vesicles from trophoblast cells as models for placental exchange studies. Placenta 3:99–106.
Black IB (1982): Stages of neurotransmitter development in autonomic neurons. Science 215:1198–1204.
Blackburn GM, Thompson MH, King HWS (1976): Binding of diethylstilbestrol to DNA by rat liver microsomal fractions in vitro and in mouse foetal cells in culture. Biochem J 158:643–646.
Blackburn GM, Flavell AJ, Thompson MH (1974): Oxidative and photochemical linkage of diethylstilbestrol to DNA in vitro. Cancer Res 34:2015–2019.
Boylan ES (1978): Morphological and functional consequences of prenatal exposure to diethylstilbestrol in the rat. Biol Reprod 19:854–863.
Boylan ES, Calhoon RE (1979): Mammary tumorogenesis in the rat following prenatal exposure to diethylstilbestrol and postnatal treatment with 7,12,-dimethylbenz(a)anthracene. J Toxicol Environ Health 5:1059–1071.
Brambell FWR (1970): "The Transmission of Passive Immunity From Mother to Young." New York: American Elsevier.
Brent RL (1973): Radiation and teratogenesis in pregnancy. In Perrin EVD, Finegold MJ, Brunson JG (eds): "Pathobiology of Development—or—Ontogeny Revisited." Baltimore: Williams and Wilkins, pp 76–96.
Brent RL (1982): Computer program for searching the biomedical literature in teratology, developmental biology and reproductive biology. Teratology 25:273–275.
Brent RL, Franklin JB (1960): Uterine vascular clamping: New procedure for the study of congenital malformations. Science 132:89–91.
Brent RL, Gorson RO (1972): Radiation exposure in pregnancy. Curr Probl Diagn Radiol 2:1–48.
Brent RL, Harris ML (eds) (1976): "Prevention of Embryionic, Fetal and Perinatal Disease." Washington: DHEW Publication NIH 76-853.
Brent RL, Jensen M (1977): Immunological aspects of development. In Wilson J, Fraser C (eds): "Handbook of Teratology." New York: Plenum Press, 4:339–396.
Bruce NW (1972): Effects of temporary uterine ischaemia on the rat embryo at different maternal abdominal temperatures. J Reprod Fertil 30:63–69.
Buchet JP, Rolls H, Hubermot G, Lauwerys R (1978): Placental transfer of lead, mercury, cadmium and carbon monoxide in women. II. Environ Res 15:494–503.
Cherian NG, Panemangalore M, Banerjee D (1983): Cellular accumulation and sub-cellular localization of metallothionein in rat liver during postnatal development. In Clarkson T, Nordberg G (eds): "Developmental and Reproductive Toxicity of Metals." New York: Plenum Press.
Chernoff GF (1977): The fetal alcohol syndrome in mice: An animal model. Teratology 15:223–230.
Chernoff N (1973): Teratogenic effects of cadmium in rats. Teratology 8:29–32.
Choi BH (1983): Effects of prenatal methylmercury poisoning upon growth and development of fetal central nervous system. In Clarkson TW, Nordberg G (eds): "Reproductive and Developmental Toxicology of Metals." New York: Plenum Press (in press).
Choi BH, Lapham LW, Amin-Zaki L, Saleem T (1978): Abnormal neuronal migration, deranged cerebral cortical organization and diffuse white matter astrocytosis of human fetal brain: A major effect of methylmercury poisoning in utero. J Neuropathol Exp Neurol 37:719–733.
Clarkson T, Nordberg G (eds) (1983): "Developmental and Reproductive Toxicity of Metals." New York: Plenum Press.
Clayson DB (1981): Carcinogenesis in the developing organism: Could protocols for testing be improved? Biol Res Pregnancy 2:150–156.
Council on Environmental Quality (1981): "Chemical Hazards to Human Reproduction." Washington:

US Government Printing Office.
Cowen WM (1979): The development of the brain. Sci Am 241:113–134.
Craig JM (1966): Mechanism of serotonin induced abortion in rats. Arch Pathol Lab Med 81:257–263.
Cree JE, Meyer J, Hailey DM (1973): Diazepam in labour: Its metabolism and effect on the clinical condition and thermogenesis of the newborn. Br Med J 4:251–255.
Daston GP (1981): Effects of cadmium on the prenatal ultrastructural maturation of rat alveolar epithelium. Teratology 23:75–84.
Daston GP (1982): Toxic effects of cadmium on the developing rat lung. II. Glycogen and phospholipid metabolism. J Toxicol Environ Health 9:51–61.
Daston GP, Grabowski CT (1979): Toxic effects of cadmium on the developing lung. J Toxicol Environ Health 5:973–983.
Dawson EB, Croft HA, Clark RR, McGanity WJ (1968): Study of seasonal variation in nine cations of normal term placentas. Am J Obstet Gynecol 102:354–357.
Delivoria-Papadopoulos M, Coburn RF, Forester RE II (1972): The placental diffusing capacity for carbon monoxide in pregnant women at term. In Longo L, Baitels H (eds): "Respiratory Gas Exchange and Blood Flow in the Placenta." Washington: US Government Printing Office, pp 73–361.
Dencker L (1975): Possible mechanisms of cadmium feto toxicity in golden hamsters and mice: Uptake by the embryo placenta and ovary. J Reprod Fertil 44:461–471.
Dencker L, Danielsson B, Khayat A, Lindgren A (1983): Disposition of metals in the embryo and fetus. In Clarkson T, Nordberg G (eds): "Reproductive and Developmental Toxicity of Metals." New York: Plenum Press.
Denlinger RH, Koestner A, Wechsler W (1974): Induction of neurogenic tumors in C3HeB/FeJ mice by nitorsourea derivatives. Observations by light microscopy, tissue culture and electron microscopy. Int J Cancer 13:559–571.
di Sant'Agnese PA, Jensen K, Levin AA, Miller RK (1983): Placental toxicity of cadmium: An ultrastructural study. Placenta (in press).
Dixon RL (1980): Toxic responses of the reproductive system. In Doull J, Klaasen CD, Andur MO (eds): "Casarett and Doull's Toxicology," 2nd ed. New York: McMillan Publishing Company, Inc, pp 332–356.
Duncan R, Lloyd JB (1978): Pinocytosis in the rat visceral yolk sac. Biochim Biophys Acta 544:647–655.
Dunn TB, Green AW (1963): Cysts of the epididarysis, cancer of the cervix, granular cell myoblastoma, and other lesions after estrogen injections in newborn mice. J Natl Cancer Inst 31:425–455.
Eibs HG, Spielmann H, Jacob-Müller U, Klase J (1982): Teratogenic effects of cyproterone actate and medroxyprogesterone treatment during pre- and post-implantation period of mouse embryos. Teratology 25:291–299.
Engel LL, Marshall PJ, Orr JC, Reinhold VN, Carter P (1978): The mass spectra of diethylstilbestrol and related compounds. Biomed Mass Spectrom 5:582–587.
Eroschenko UP, Mousa (1979): Neonatal administration of insectiside chlordecone and its effects on development of the reproductive tract in the female mouse. Toxicol Appl Pharmacol 49:157–159.
Eroschenko UP, Palmiter RD (1980): Estrogenicity of Kepone in birds and mammals. In McLachlan J (ed): "Estrogens in the Environment." Holland: Elsevier, pp 305–324.
Fabro S (1973): Passage of drugs and other chemicals into the uterine fluids and preimplantation blastocyst. In Boreus L (ed): "Fetal Pharacology." New York: Raven Press, pp 443–459.
Fahim MS, Fahim Z, Hall DG (1976): Effects of subtoxic lead levels on pregnant women in the state of Missouri. Res Commun Chem Pathol Pharmacol 13:309–331.
Faith R, Moore J (1977): Impairment of thymus-dependent immune functions by exposure of the developing immune system to 2,3,7,8-tetrachlorodibenzo-p-dioxin (TCDD). J Toxicol Environ Health 3:451–464.
Farber E (1981): Chemical carcinogenesis. New Engl J Med 305:-1379–1389.
Faulk WP (1981): Trophoblast and extraembryonic membranes in the immunobiology of human pregnancy. In Miller RK, Thiede HA (eds): "Placenta: Receptors, Pathology, and Toxicology." London: W.B. Saunders Ltd, pp 3–22.
Ferm VH, Hanlon DP, Urban J (1969): Permeability of the hamster placenta to radioactive cadmium. J Embryol Exp Morphol 22:107–113.
Ferm VH, Hanlon DP, (1983): Metal-induced congenital malformations. In Clarkson TW, Nordberg GF (eds): "Reproductive and Developmental Effects of Metals." New York: Raven Press (in press).
Filler R, Lew KJ (1981): Developmental onset of mixed-function oxidase activity in preimplantation mouse embryos. Proc Natl Acad Sci USA 78:6991–6995.

Fischer LJ, Weissinger JL, Rickett DE, Hintze KL (1976): Studies on the biological disposition of diethylstilbestrol in rats and humans. J Tox Environ Health 1:587–605.
Forsberg JG (1981): Permanent changes induced by DES at critical stages in female development: 10-year experience from human and model systems. Biol Res Pregnancy 2:168–175.
Forsberg J-G, Kalland T (1981): Embryology of genital tract in humans and in rodents. In Herbst AL, Bern HA (eds): "Developmental Effects of DES in Pregnancy." New York: Thieme-Stratton, pp 4–25.
Fox RR, Diwan BA, Meier H (1975): Transplacental induction of primary renal tumors in rabbits with 1-ethyl-1-nitrosourea. J Natl Cancer Inst 54:1439–1448.
Fox RR, Diwan BA, Meier H (1977): Transplacental carcinogenic effects of combined treatment of ethylurea + sodium nitrate in rabbits. J Natl Cancer Inst 59:427–429.
Franklin J, Brent RL (1964): The effect of uterine vascular clamping on the development of rat embryos three to fourteen days old. J Morphol 115:273–290.
Freeman SJ, Beck F, Lloyd JB (1981): The role of the visceral yolk sac in mediating protein utilization by rat embryos cultured in vitro. J Embryol Exp Morphol 66:223–234.
Friberg L, Piscator N, Nordburg G, Kyellstrom T (1974): Cadmium in the environment. CRC Press.
Gellert RJ (1978a): Uterotrophic activity of polychlorinated biphenyls (PCB) and induction of precocious reproductive aging in neonatally treated female rats. Environ Res 16:123–130.
Gellert RJ (1978b): Kepone, mirex, dieldrin and aldrin: Estrogenic activity and the induction of persistent vaginal estrus and anovulation in rats following neonatal treatment. Environ Res 16:131–138.
Glick B (1971): Morphologic changes and humoral immunity in cyclophosphamide-treated chicks. Transplantation 11:433–439.
Goodman DR, James RC, Harbison RD (1982): Placental toxicology. Fd Chem Tox 20:123–128.
Goth R, Rajewsky MF (1974): Molecular and cellular mechanisms associated with pulse carcinogenesis in the rat nervous system by ethylnitosourea: Ethylaton of nucleic acids and elimination rates of ethylated bases from the DNA of different tissues. Z Krebsforsch 82:37–64.
Goth R, Rajewsky MF (1972): Ethylation of nucleic acid by ethylnitrosourea-1-^{14}C in the fetal and adult rat. Cancer Res 32:1501–1505.
Greenaway JC, Fantel AG, Shepard TH, Jachau MR (1982): The in vitro teratogenicity of cyclophasphamide in rat embryos. Teratology 25:335–344.
Greene RR, Burrill MW, Ivy AC (1940): Experimental intersexuality. Am J Anat 67:305–313.
Haddad R, Ausma R, Dumas R (1979): Neuroteratogenicity of methylizoxy methanol acetate: Behavioral deficits of ferrets with transplacentally induced lissencephaly. Neurotoxicol 1:171–190.
Harbison RD (1980): Teratogens. In Doull J, Klaasen CD, Andur MO (eds): "Casarett and Doull's Toxicology," 2nd ed. New York: McMillan Publishing Company, Inc, pp 158–178.
Hendrickx A, Benerschke K, Thompson RS, Ahern JK, Lucas WE, Oi RH (1979): The effects of prenatal diethylstilbestrol exposure on the genitalia of pubertal mucaca mulatta. J Reprod Med 22:233–240.
Henry EC, Miller RK, Baggs RB (1982): Direct fetal injections of diethylstilbestrol and estradiol: A method for investigating their teratogenicity and distribution. Teratology.
Herbst AL, Bern HA (eds) (1982): "Developmental Effects of Diethylstilbesterol in Pregnancy." New York: Thieme-Stratton.
Herbst AL, Scully RE (1970): Adenocarcinoma of the vagina in adolescence: A report of seven cases including six clear cell carcinomas (so-called mesonephromas). Cancer 25:745–751.
Herbst AL, Scully RE, Robbay SJ (1979): Prenatal diethylstilbestrol exposure an human genital tract abnormalities. In Rice JM (ed): "Perinatal Carcinogenesis." Washington: DHEW NIH 79-1633.
Hertz R (1980): Personal experiences emphasizing the environmental impact of estrogens. In McLachlan J (ed): "Estrogens in the Environment." New York: Elsevier, pp 347–352.
Honey DP, Robson JM, Sullivan FM (1967): Mechanism of inhibitory action of 5-hydroxytryptamine on placental function. Am J Obstet Gynecol 99:250–257.
Hook JB, Bailie MD (1979): Perinatal renal pharmacology. Annu Rev Pharmacol Toxicol 19:491–509.
Hoover R (1980): Association of exogenous estrogens and cancer in humans. In McLachlan J (ed): "Estrogens in the Environment." Holland: Elseiver, pp 353–376.
Hubermont G, Buchet JP, Roels H, Lauwerys R (1978): Placental transfer of lead, mercury and cadmium in women living in a rural area. Int Arch Occup Environ Health 41:117–124.
Infurna RN, Stanton M, Baggs RB, Miller RK (1982): Neonatal exposure to cadmium: Behavioral toxicity. Teratology 26:48A.
Interagency Regulatory Liaison Group (1981): "Recommended Guideline for Teratogenicity Studies in the Rat, Mouse, Hamster, or Rabbit." Washington: US Government Printing Office.

Ioachim H (1971): The cortisone-induced wasting disease of newborn rats: Histopathological and autoradiographic studies. J Pathol 104:201–205.

Ivankovic S (1973): Experimental prenatal carcinogenesis. In Tomatis L, Mohr U, Avis W (eds): "Transplacental Carcinogenesis." Lyon: International Agency for Research on Cancer #4, WHO.

Ivankovic S (1979): Teratogenic and carcinogenic effects of some chemicals during prenatal life in rats, syrian hamsters and minipigs. In Rice JM (ed): "Perinatal Carcinogenesis." Washington: NCI Monograph 51, United States Government Printing Office.

Ivankovic S, Druckrey H (1968): Transplacentore erzengung maligner tumoren des nervensystems. I Äthylnitrosoharnstoff (ÄNH) an BDIX-Ratten. Z Krebsforsch 71:320–360.

Johnson LD, Palmer AE, King NW Jr, Hertig AT (1981): Vaginal adenosis in cebus apella monkeys exposed to DES in utero. Obstet Gynecol 57:629–635.

Jones AH, Fantel AG, Kocan RA, Juchau MR (1977): Bioactivation of procarcinogens to mutagens in human fetal and placental tissues. Life Sci 21:1831–1837.

Jones KL, Smith DW, Ulleland CW, Streissguth AP (1976): Pattern of malformation in offspring of chronic alcoholic women. Lancet 1:1076–1078.

Juchau MR (1980): Drug biotransformation in the placenta. Pharmacol Ther 8:501–524.

Juchau MR (1982): The role of the placenta in developmental toxicology. In Snell K (ed): "Developmental Toxicology." New York: Praeger Publishers, pp 187–210.

Juchau MR, Namkung MJ, Jones AH, DiGiovanni J (1978): Biotransformation and bioactivation of 7,12-dimethylbenz(a)anthracene in human fetal and placentl tissues. Analyses of HPLC profiles and studies with Salmonella typhimurium. Drug Metab Dispos 6:273–281.

Jurgelski W, Hudson PM, Falk HL (1976): Embryonal neoplasm in the opossum: A new model for solid tumors of infancy and childhood. Science 193:328–332.

Jurgelski W (1974): The opossum as a biomedical model. I. Research perspective husbandry and laboratory techniques. Lab Anim Sci 24:375–403.

Jurgelski W, Forsythe W, Dahl D (1974): The opossum as a biomedical model. II. Breeding the opossum in captivity. Lab Anim Sci 24:407–411.

Jurgelski W, Porter ME (1974): The opossum as a biomedical model. III. Breeding the opossum in captivity. Lab Anim Sci 24:412–425.

Jurgelski W Jr, Hudson P, Falk HL (1979): Tissue differentiation and susceptibility to embryonal tumor induction by ethylnitrosourea in the opossum. Rice J (ed): "Perinatal Carcinogenicity." Washington: National Cancer Institute Monograph 51, pp 23–158.

Kalland T (1980): Alterations of antibody response in female mice after neonatal exposure to diethylstilbestrol. J Immunol 124:194–198.

Kalland T, Forsberg T, Forsberg J (1978): Effect of estrogen and corticosterone on the lymphoid system in neonatal mice. Exp Mol Pathol 28:76–95.

Kalland T, Strand O, Fosberg J (1979): Long-term effects of neonatal estrogen treatment on mitogen responsiveness of mouse spleen lymphocytes. J Natl Cancer Inst 63:413–421.

Karp WB, Roberstson AF (1977): Correlation of human placental enzymatic activity with trace metal concentrations in placentas from three geographical locations. Environ Res 13:470–477.

Kauffman RE, Azarnoff DL, Morris JA (1975): Placental transfer and fetal urinary excretion of gentamicin—comparison between an animal model and the human fetus. In Morselli TL, Garattini S, Sereni F (eds): "Basic and Therapeutic Aspects of Perinatal Pharmacology." New York: Raven Press, pp 75–84.

Kaufman RH, Korhonen MO, Stramm T, Adam E, Kaplan A (1982): Development of clear cell adenocarcinoma in DES-exposed offspring under observation. Obstet Gynecol 59:568–572.

Kaunitz H, Malins DC, McKay DG (1962): Studies of the generalized Shwartzman reaction by diet. II. J Exp Med 115:127–136.

Kellogg CK, Tervo D, Ison J, Miller RK (1980): Prenatal diazepam exposure alters behavioral development in rats. Science 207:205–207.

Kellogg CK, Ison J, Miller RK (1983): Auditory temporal resolution in rats: Effect of prenatal diazepam exposure. Psychopharmacologia (in press).

Kelman BJ (1979): Effects of toxic agents on movements of material across the placenta. Fed Proc 38: 2246–2250.

Kelman BJ, Walter BK (1977): Passage of cadmium across the perfused guinea pig placenta. Proc Soc Exp Biol Med 156:68–71.

Khera AK, Wibberley DG, Dathan JG (1980): Placental and stillbirth tissue lead concentrations in occupationally exposed women. Br J Ind Med 37:394–396.

Kjellström T, Nordberg GF (1978): A kinetic model of cadmium metabolism in the human being. Environ Res 16:248–269.

Kleihues P (1982): Developmental carcinogenicity. In Snell K (ed): "Developmental Toxicology." New York: Praeger Publishers, pp 211–246.

Kleihues P, Cooper HK, Bulcheler J, Kolar GF, Diessner H (1979): Mechanism of perinatal tumor induction by neuro-oncogenic alkylnitrosoureas and dialkylaryltriazenes. In Rice JM (ed): "Perinatal Carcinogenesis." Washington: US Government Printing Office, pp 79–1633.

Kleihues P, Magee DN (1973): Alkylation of rat brain nucleic acids by N-methyl-N-nitrosourea and methylmethanesulfonate. J Neurochem 20:595–606.

Kleihues P, Cooper HK, Buecheler J, Kolar GF, Dressner H (1979): Mechanism of perinatal tumor induction by neuro-oncogenic alkylnitrosoureas and dialhylaryltriagenes. In Rice JM (ed): "Perinatal Carcinogenesis." Washington: DHEW 79–1633, pp 227–231.

Klopper A (1980): Review article: The new placental proteins. Placenta 2:77–89.

Knight WM (1980): Estrogens administered to food producing animals. In McLachlan J (ed): "Estrogens in the Environment." Holland: Elsevier, pp 391–401.

Kuhnert PM, Kuhnert BR, Erhard P (1981): Comparison of mercury levels in maternal blood, fetal cord blood and placental tissues. Am J Obstet Gynecol 139:209–213.

Kuhnert PM, Kuhnert BR, Bottoms SF, Erhard P (1982): Cadmium levels in maternal blood, fetal cord blood, and placental tissues of pregnant women who smoke. Am J Obstet Gynecol 142:1021–1025.

Laerum OD, Rajewsky MF (1975): Neoplastic transformation of fetal rat brain cells in culture after exposure to ethylnitrosourea in vivo. J Natl Cancer Inst 55:1177–1187.

Laerum OD, Rajewsky MF, Schachner M, Stavrou D, Haglid KG, Haugen A (1977): Phenotypic properties of neoplastic cell lines developed from fetal rat brain cells in culture after exposure to ethylnitrosourea *in vivo.* Z Krebsforsch 89:273–295.

Laquer GL, Spatz M (1973): Transplacental induction of tumors and malformations in rats with cycasin and methoxyazomethanol. In Tomatis L, Mohr U (eds): "Transplacental Carcinogenesis." Lyon: International Agency for Research on Cancer, pp 59–64.

Lauwerys R, Bucket JP, Roels H, Hubermont G (1978): Placental transfer of lead, mercury, cadmium and carbon monoxide. I. Environ Res 15:278–289.

Layton WM, Ferm VH (1980): Protection against cadmium-induced limb malformations by pretreatment with cadmium or mercury. Teratology 21:357–360.

Lehtovista P, Forss M (1978): The acute effect of smoking on intervillous blood flow of the placenta. Br J Obstet Gynecol 85:729–732.

Leist KH, Graviler J (1974): Fetal pathology in rats following uterine-vessel clamping on day 14 of gestation. Teratology 10:55–67.

Lemaine P, Harousseau H, Borteyru JP, Mennet JC (1967): Les enfants de parents alcoholiques: Anomalies observées á propos de 127 cas. Arch Fr Pédiatr 25:820–832.

Levin AA, Miller RK (1980): Fetal toxicity of cadmium: Maternal vs. fetal injections. Teratology 22:1–5.

Levin AA, Miller RK (1981): Fetal toxicity of cadmium in the rat: Decreased utero placental blood flow. Toxicol Appl Pharmacol 58:297–306.

Levin AA, Plautz JR, di Sant'Agnese PA, Miller RK (1981): Cadmium: Placental mechanisms of fetal toxicity. Placenta (Suppl) 3:303–318.

Levin AA, Kilpper RW, Miller RK (1983): Organ specific kinetics of a fetal toxic injection of cadmium chloride in the pregnant rat. Toxicol Appl Pharmacol (in press).

Levin AA, Miller RK, di Sant'Agnese PA (1983): Heavy metal alterations of placental function: A mechanism for the induction of fetal toxicity by cadmium. In Clarkson T, Nordburg G (eds): "Developmental and Reproductive Toxicity of Metals." New York: Plenum Press.

Levitt M, Hutchings DE, Bodnarenko SR (1981): The postnatal persistence of methadone following prenatal exposure. Teratology 24:56A.

Linna TJ, Frommel D, Good RA (1972): Effects of early cyclophosphamide treatment on the development of lymphoid organs and immunological functions in the chicken. Int Arch Allergy Appl Immunol 42:20–39.

Longo LD, (1972): Disorders of placental transfer. In Assali NS (ed): "Pathophysiology of Gestation," Vol. 2. New York: Academic Press, pp 1–76.

Longo LD (1977): The biolobical effects of carbon monoxide on the pregnant woman, fetus and newborn infant. Am J Obstet Gynecol 129:69–78.

Longo LD, Hill EP (1977): Carbon monoxide uptake and elimination in fetal and maternal sleep. Am J Physiol 232:H324–329.

Loveless A (1969): Possible relevance of O-6 alkylation of deoxyguanosine to the mutagenicity and carcinogenicity of nitrosamines and nitrosamides. Nature 223:206–207.

Lucier GW (1976): Perinatal development of conjugated enzyme systems. Environ Health Perspect 18: 25–34.

Luster M, Faith R, McLachlan J, Clark G (1979): Effect of in utero exposure to diethylstilbestrol on the immune response in mice. Toxicol Appl Pharmacol 47:279–285.

Lutwak-Mann C (1973): Drugs and the blastocyst. In Boreus L (ed): "Fetal Pharmacology." New York Raven Press, pp 419–442.

Luz N, Margues M, Ayub A, Correa P (1969): Effects of estradiol upon the thymus and lymphoid organs of immature female rats. Am J Obstet Gynecol 105:525–528.

Magos L, Webb M (1983): The influence of weight and other physiological changes during pregnancy and lactation on the toxicities of mercury and cadmium. In Clarkson T, Nordburg G (eds): "Developmental and Reproductive Toxicity of Metals." New York: Plenum Press (in press).

Matsumato H, Koya G, Takeuchi T (1965): Fetal minamata disease. J Neuropathol Exp Neurol 24: 563–574.

Mattison D (1983): Ovarian toxicity: Effects on sexual maturation, reproduction and menopause. In Clarkson T, Nordberg G (eds): "Reproductive and Developmental Toxicity of Metals." New York: Plenum Press.

McKay DG, Wong TC (1962): Studies of the generalized Schwartzman reaction produced by diet. I. J Exp Med 115:1117–1126.

McLachlan JA (1980): "Estrogens in the Environment." New York: Elsevier/North Holland.

McLachlan JA (1981): Rodent models for perinatal exposure to diethylstilbestrol and their relation to human disease in the male. In Herbst AL, Bern HA (eds): "Developmental Effects of Diethylstilbestrol in Pregnancy." New York: Thieme-Stratton, Inc, pp 148–157.

McLachlan JA, Fabro SE (1982): Altered postnatal development following intrauterine exposure to hormonally active chemicals. In Yoshida H, Hagihara Y, Ebashi S (eds): "Advances in Pharmacology and Therapeutics II," Vol. 5. New York: Pergamon Press, pp 211–220.

McLachlan JA, Newbold RR, Korach KS, Lam JC IV, Suzuki Y (1981): Transplacental toxicology: Prenatal factors influencing postnatal fertility. In Kimmel CA, Buelke-Sam J (eds): "Developmental Toxicology." New York: Raven Press, pp 213–232.

McLachlan JA, Newhold RR, Bullock BC (1980): Long term effects on the female mouse genital tract associated with prenatal exposure to diethylstilbestrol. Cancer Res 40:3988–3999.

McLachlan JA, Newbold RR, Shah MC, Hogan MD, Dixon RL (1983): Reduced fertility in female mice exposed transplacentally to diethylstilbestrol. Fertil Steril (in press).

McLachlan JA, Wong A, Barrett JC (1982): Morphologic and neoplastic transformation of syrian hamster embryo fibroblasts by diethylstilbestrol and its analogs. Cancer Res (in press).

Metzler M (1981): The metabolism of diethylstilbestrol CRC. CRIT. Rev Biochem 10:171–213.

Meyer MB, Tonascia JA (1977): Maternal smoking, pregnancy complications and perinatal mortality. Am J Obstet Gynecol 128:494–498.

Meyer RK, Rao MA, Aspinall RL (1959): Inhibition of the development of the bursa of Fabricius in the embryos of common fowl by 19-nortestosterone. Endocrinol 64:890–897.

Middaugh LD, Blackwell LA, Santos CA, Zemp JW (1974): Effects of d-amphetamine sulfate given to pregnant mice on activity and on catecholamines in the brain of offspring. Dev Psychobiol 7: 429–438.

Middaugh LD, Santos CA, Zemp JW (1975): Phenobarbitol during pregnancy alters operant behavior of offspring in C57Bl/6J mice. Pharm Biochem Behav 3:1137–1139.

Miller RK (1979): Drugs as teratogens: Three common syndromes. Drug Ther 9:125–130.

Miller RK (1981): Drugs during pregnancy: A therapeutic dilemma. Ration Drug Ther 15:1–10.

Miller RK, Ng W, Levin A (1983): The placenta: Relevance to toxicology. In Clarkson T, Nordberg G (eds): "Developmental and Reproductive Toxicity of Metals." New York: Plenum Press.

Miller RK, Gardner KA (1981): Cadmium in the human placenta: Relationship to smoking. Teratology 23:51a.

Miller RK, Shaikh Z (1983): Prenatal metabolism: Metals and Metallothionein. In Clarkson T, Nordberg G (eds): "Developmental and Reproductive Toxicity of Metals." New York: Plenum Press.

Miller RK, Heckmann ME, McKenzie RC (1982a): Diethylstilbestrol: Placental transfer and metabolism, covalent binding and fetal distribution in the Wistar rat. J Pharmacol Exp Therap 220:358–365.

Miller RK, Baggs RB, Odoroff CL, McKenzie RC (1982b): Prenatal diethylstilbestrol: Vaginal tumors and malformations in the Wistar rat. Pharmacologist 24:237.

Miller RK, Koszalka TR, Brent RL (1976): Transport mechanisms for molecules across placental membranes, in cell surface in animal development 1. In Poste G, Nicolson GL (eds): "Cell Surface

Reviews." Holland: Elsevier, pp 145–223.
Miller RK, Thiede HA (1981): "Placenta: Receptors, Pathology and Toxicology." London: WB Saunders, Ltd.
Miller RW (1977): Relationship between human teratogens and carcinogens. J Natl Cancer Inst 58: 471–474.
Mirkin BL (1976): "Perinatal Pharmacology and Therapeutics." New York: Academic Press.
Mohr U, Althoff J, Authaler A (1966): Dioplacental effect of the carcinogen diethylnitrosamine in the golden hamster. Cancer Res 26:2349–2353.
Naeye RL (1981): Common environmental influences on the fetus. In Naeye RL, Kissane JM, Kaufman N (eds): "Perinatal Diseases." Baltimore: William and Wilkins, pp 52–66.
Napalkov NP (1973): Some general considerations of the problem of transplacental carcinogenesis. In Tomatis L, Mohr U (eds): "Transplacental Carcinogenesis." Lyon: WHO-IARC, pp 1–10.
Napalkov NP, Anisimov VN (1979): Transplacental effect of diethylstilbestrol in female rats. Cancer Lett 6:107–114.
Nau H, Merker HJ, Richter H (1981): Morphological alterations induced by TCED (2,3,7,8,-tetrachlorodibenzo-p-dioxin) and epithelial monolayer cultures of human and rodent fetal lung, kidney and liver. Biol Res Pregnancy 2:176–180.
Nebert DW (1981): Birth defects and the potential of genetic differences in drug metabolism. Birth Defects Original Article Series 17(1):51–70.
Nebert DW, Gelboin HV (1969): The in vivo and in vitro induction of aryl hydrocarbon hydroxylase in mammalian cells of different species. Arch Biochem 134:76–83.
New DAT, Brent RL (1972): Effect of yolk sac antibody on rat embryos grown in culture. J Embryol Exp Morphol 27:543–553.
Ng WW, Catus RG, Miller RK (1981): Macromolecule transfer in the human trophoblast: Transcobalamine. TC-II vitamin B_{34} uptake. Placenta (Suppl) 3:45–160.
Nixon RE, Koller LD, Exon JH (1979): Effect of methylmercury chloride on transplacental tumors induced by sodium nitrite and ethylruea in rats. J NCI 63:1057–1063.
Nordberg G (ed) (1974): "Effects and Dose Response Relationships of Toxic Metals." Amsterdam: Elsevier.
Oesch F (1975): Transplacental control of epoxide-forming and inactivating enzymes of rat fetal liver by clinically used drugs and by environmental chemicals. In Morselli PL, Garattini S, Sereni F (eds): "Basic and Therapeutic Aspects of Perinatal Pharmacology." New York: Raven Press, pp 853–863.
Owen JR, Trani SF, Blair AW (1972): Effect of diazepam administered to mothers during labour on temperature regulation of neonate. Arch Dis Child 47:107–110.
Page E, Villee C, Villee D (1981): "Human Reproduction: Essentials of Reproductive and Perinatal Medicine." Philadelphia: WB Saunders.
Palmer AK (1981): Regulatory requirements for reproductive toxicology: Theory and practice. In Kimmel CA, Buelke-Sam J (eds): "Developmental Toxicology." New York: Raven Press, pp 259–288.
Panigel M (1981): Placental function: Toxicology and pathology. In Miller RK, Thiede HA (eds): "Placenta: Receptors, Pathology and Toxicology." London: WB Saunders, Ltd., pp 275–287.
Papoz L, Eschivege E, Pequingnot G, Barrat J, Schwartz D (1982): Maternal smoking and birth weight in relation to dietary habits. Am J Obstet Gynecol 142:870–876.
Parizek J (1964): Vascular changes at sites of estrogen biosynthesis produced by injection of cadmium cells: A destruction of the placenta by cadmium cells. J Reprod Fertil 7:263–264.
Parizek J (1965): The peculiar toxicity of cadmium during pregnancy: An experimental toxemia of pregnancy induced by cadmium cells. J Reprod Fertil 9:111–112.
Park BH, Good RA (1974): "Principles of Modern Immunobiology." Philadelphia: Lea and Febeger.
Parrich H-J, Neubert D (1980): Significance of organ culture techniques for evaluation of prenatal toxicity. Arch Toxicol 45:161–187.
Pelkonen O (1980): Environmental influences on human fetal and placental xenobiotic metabolism. Eur J Clin Pharmacol 18:17–24.
Pelkonen O (1982): Differentiation of drug metabolism in relation to developmental toxicology. In Snell K (ed): "Developmental Toxicology." New York: Praeger Publishers, pp 165–186.
Pinto-Machado J (1970): Influence of prenatal administration of busulfan on the postnatal development of mice. Production of a syndrome including hypoplasia of the thymus. Teratology 3:363–370.
Pitkin RM, Bahns JA, Filer LJ Jr, Reynolds WA (1975): Mercury in human maternal and cord blood, placenta and milk. Proc Soc Exp Biol Med 157:565–570.
Plautz JR, Levin AA, Miller RK (1980): Fetal and maternal toxicity of cadmium, metallothionein and its distribution in the pregnant Wistar rat. Teratology 26:61.
Plautz JR, Levin AA, Miller RK (1981): Fetal placental distributions of cadmium, metallothionein and

cadmium chloride in the Wistar rat. Teratology 23:56.

Pratt RM, Solomon DS (1981): Biochemical basis for the teratogenic effects of glucocorticoids. In Iuchau MR (ed): "The Biochemical Basis of Chemical Teratogenesis." New York: Elseiver/North Holland, pp 179–193.

Prigge E (1978): Inhalative cadmium effects in pregnant and fetal rats. Toxicology 10:297–309.

Rankin JHG (1978): Role of protaglandins in the maintenance of the placental circulation. In Coceani F, Olley PM (eds): "Advances in Prostaglandins and Thromboxane Research," Vol. 4. New York: Raven Press, pp 261–269.

Rech RH, Lomuscio G, Algeri S (1980): Methadone exposure in utero: Effects on brain biogenic amines and behavior. Neurobehav Toxicol 2:75–78.

Reilly RW, Thompson JA, Eielski RK, Severson CD (1967): Estradiol-induced wasting syndrome in neonatal mice. J Immunol 98:321–330.

Reinisch JM, Simon NE, Karow WG, Gauddum R (1978): Prenatal exposure to prednisone in humans and animals retards intrauterine growth. Science 202:436–440.

Rementeria JL, Bhatt K (1977): Withdrawal symptoms in neonates from intrauterine exposure to diazepam J Pediatr 90:123–126.

Rice JM (1969): Transplacental carcinogenesis in mice by 1-ethyl-1-nitrosourea. Ann NY Acad Sci 63: 813–827.

Rice JM (1979): Perinatal carcinogenesis. Bethesda, MD: National Cancer Institute Monograph 51, DHEW Publication #79–1633.

Rice JM (1981): Effects of prenatal exposure to chemical carcinogens and methods for their detection. In Kimmel CA, Buelke-Sam J (eds): "Developmental Toxicology." New York: Raven Press, pp 191–212.

Rice JM, Sly DL, Palmer AE, London WT (1979): Transplacental effects of ethylnitrosourea in a nonhuman primate, erythrocebus patas. In Rice JM (ed): "Perinatal Carcinogenesis." Washington: US Government Printing Office, pp 79–1633.

Rice JM, Williams GM, Palmer AE, London WT, Sly DL (1981): Pathology of gestational choriocarcinoma induced in patus monkeys by ethylnitrosourea given during pregnancy. In Miller RK, Thiede HA (eds): "Placenta: Receptors, Pathology and Toxicology." London: WB Saunders, pp 223–230.

Roberts DW, Chapman JR (1981): Concepts essential to the assessment of toxicity to the developing immune system. In Kimmel CA, Buelke-Sam J (eds): "Developmental Toxicology." New York: Raven Press, pp 167–189.

Robson JM, Sullivan FM (1966): Analysis of actions of 5-hydroxytryptamine in pregancy. J Physiol 184: 717–732.

Rodier PM (1980): Chronology of neuron development: Animal studies and their clinical implications. Dev Med Child Neurol 22:525–545.

Rodier PM, Gramann WJ (1979): Morphologic effects of interference with cell proliferation in the early fetal period. Neurobehav Toxicol 1:129–135.

Rodier PM, Koeter HBWM, Aschner M, Lewis LS (1982): Response of the developing CNS after brief exposure to inhalant. Teratology 25:71A.

Rodier PM, Reynolds SS, Roberts WM (1979): Behavioral consequences of interference with CNS development in the early fetal period. Teratology 19:327–336.

Roels H, Hubermont G, Buchet JP, Lauwerys R (1978): Placental transfer of lead, mercury, cadmium and carbon dioxide. III. Environ Res 16:236–247.

Rowell PP (1981): The effect of maternal cigarette smoking on the ability of human placental villae to concentrate α-aminoisobutyric acid in vitro. Res Commun Sub Abuse 2:253–266.

Rüdiger HW, Haenisch F, Metzler M, Oesch F, Glatt HR (1979): Metabolites of diethylstilbestrol induce sister chromatid exchange in human fibroblasts. Nature 281:392–393.

Russell WL, Kelley EM, Hunsicker PR, Bangham JW, Maddux SC, Phipps EL (1979): Specific locus test shows ethylnitrosourea to be the most potent mutagen in the mouse. Proc Natl Acad Sci USA 76: 5818–5819.

Rustia M, Sherkik P (1974): Prenatal induction of neurogenic tumors in hamsters by precursors of ethylurea and sodium nitrite. J Natl Cancer Inst 52:605–608.

Sager PR, Doherty RA, Rodier PM (1982): Effects of methylmercury on developing mouse cerebellar cortex. Exp Neurol 77:179–193.

Samarawickrama GP, Webb M (1979): Acute effects of cadmium during pregnancy and embryo fetal development in the rat. Environ Health Perspect 28:345–349.

Samarawickrama GP, Webb M (1981): The acute toxicity and teratogenicity of cadmium in the pregnant rat. J Appl Toxicol 1:264–269.

Sasser LB, Levin AA, Miller RK, Kelman BJ (1980): The effect of fetal cadmium injection on metallothi-

onein and fetal and maternal rat liver. Soc Toxicol Abstr.
Scher J, Hailey DM, Beard RW (1972): The effects of diazepam on the fetus. J Obstet Gynecol Brit Comm 79:635-638.
Schwarz RH, Jaffe SJ (eds) (1979): "Drugs and Chemical Risks to the Fetus and Newborn." New York: AR Liss, Inc.
Shah HC, McLachlan JA (1976): The fate of diethylstilbestrol in the pregnant mouse. J Pharm Exp Ther 197:687-695.
Sheehan DM, Branham WS, Medlock KL, Olsen ME, Zehr D (1980): Estrogen plasma binding and regulation of development in the neonatal rat. Teratology 21:68A.
Shepard TH (1980): "Catalog of Teratogenic Agents." Baltimore: Johns Hopkins University Press.
Shields JR, Resnik R (1979): Fetal lung maturation in the antenatal use of glucocorticoids to prevent the respiratory distress syndrome. Obstetrical and Gynecologic Survey 34:343-363.
Shum S, Jensen NM, Nebert DM (1979): The murine Ah locus: In uterotoxicity and teratogenesis associated with genetic differences in benzo(α)pyrene metabolism. Teratology 20:365-376.
Simmonds RJ, Coade SB, Harkness RA, Drury L, Hytten FE (1982a): Nucleotide, nucleoside and purine base concentrations in human placentae. Placenta 3:29-38.
Simmons RD, Kellogg CK, Miller RK (1982): Prenatal diazepam: Its effects on central catecholamine neurons in the rat. Teratology 25:76A.
Simmons RD, Miller RK, Kellogg CK (1983): Prenatal diazepam: Distribution and metabolism in perinatal rats. Teratology (in press).
Slikker WJR, Hill DE, Young JF (1982): Comparison of the transplacental pharmacokinetics of 17β-estradiol and diethylstilbestrol in the subhuman primate. J Pharmol Exp Ther 221:173-182.
Snell K (ed) (1982): "Developmental Toxicology." New York: Praeger Press.
Socol ML, Manning FA, Murata Y, Druzin ML (1982): Maternal smoking causes fetal hypoxia. Am J Obstet Gynecol 142:214-218.
Sonawane BR, Nordburg M, Nordburg G, Lucier G (1975): Placental transfer of cadmium in rats. Influence of dose and gestational age. Environ Health Perspect 28:248-249.
Spatz M, Laquer GL (1967): Transplacental induction of tumors in Sprague-Dawley rats with crude cycad material. J Natl Cancer Inst 38:233-245.
Spyker JM (1975): Assessing the impact of low level chemicals on development: Behavioral and latent effects. Fed Proc 34:1835-ν844.
Spyker JM, Fernandes G (1973): Impaired immune function in offspring from methylmercury-treated mice. Teratology 7:A28.
Stave U (1978): "Perinatal Physiology." New York: Plenum Press.
Stavro D, Hanichen T (1975): Oncogene wirkung von äthylnitrasohainstoff beim kanischen während der pranatalen periode. Z Krebsforsch 84:207-215.
Stavrou D, Dahme E, Schroder B (1977): Transplacentare neuroonkognene wirkung von äthylnitrosoharnstoff beim kanichen während der fruhen graviditätsphase. Z Krebsforsch 89:331-339.
Stillman RJ (1982): In utero exposure to diethylstilbestrol: Adverse effects on the reproductive tract and reproductive performance in male and female offspring. Am J Obstet Gynecol 142:905-921.
Suzuki T, Miyama T, Katsunuma H (1971): Comparison of mercury contents in maternal blood, umbilical cord blood and placental tissues. Bull Environ Contam Toxicol 5:502-506.
Swenberg JA (1979): Incorporation of transplacental exposure into routine carcinogenicity testing. Rice M (ed): "Perinatal Carcinogenesis." Wahington: US Government Printing Office, pp 265-268.
Swenberg JA, Koestner A, Wechsler W, Denlinger RH (1972): Quantitative aspects of transplacental tumor induction with ethylnitrasourea in rats. Cancer Res 32:2656-2660.
Thieme R, Schramel P, Klose G, Waidl E (1974): The influence of regional environmental factors on the trace element composition of the placenta. Geburtsch U Frauenheilk 14:36-41.
Thomas PT, Hinsdill RD (1979): The effect of perinatal exposure to tetrachlorodibenzo-ϱ-dioxin on the immune response of young mice. Drug Chem Toxicol 2:77-98.
Thurauf J, Schaller KH, Engehardt E, Gossler K (1975): Der cadmium-gehalt der mensheishleu placenta. Int Arch Occup Environ Health 36:19-27.
Tomatis L, Mohr U (1973): "Transplacental Carcinogenesis." Lyon: International Agency for Research on Cancer, #4.
United States Public Health Service (1973): Smoking and pregnancy. In: "The Health Consequences of Smoking, A Report of the Surgeon General." Washington DHEW Publication 73-8704.
Vesselinovitch SD, Koka M, Rao KVN, Mihailovich N, Rice JM (1977): Prenatal carcinogenesis by ethylnitrosourea in mice. Cancer Res 39:4378-4380.
Vesselinovitch SD, Lombard LS, Mihailovich N (1971): Broad spectrum carcinogenicity of ethylnitrosourea in the newborn and infant mice. Proc Am Assoc Cancer Res 12:56.

Vorhees CV, Butcher RE (1982): Behavioral teratogenicity. In Snell K (ed): "Developmental Toxicology." New York: Praeger Publishers, pp 247–298.

Vorherr H (1974): "The Breast: Morphology, Physiology, and Lactation." New York: Academic Press.

Vorherr H, Messer R, Vorherr U, Jordan SO, Kamfeld M (1979): Teratogenesis and carcinogenesis in rat offspring after transplacental and transmammary exposure to diethylstilbestrol. Biochem Pharm 28:1865–1877.

Vos J, Moore J (1974): Suppression of cellular immunity in mice by maternal treatment with 2,3,7,8,-tetrachlorodibenzo-p-dioxin. Int Arch Allergy Appl Immunol 47:777–794.

Waddell WJ, Marlowe C (1981): Biochemical regulation of the accessibility of teratogens to the developing embryo. In Juchau MR (ed): "The Biochemical Basis of Chemical Teratogenesis." New York: Elsevier/North Holland, Inc, pp 1–62.

Warkany J (1978): History of teratology. In Wilson J, Fraser C (eds): "Handbook of Teratology." 1:3–46.

Ways SC, Bern HA (1979): Long-term effects of neonatal treatment with cortisol and/or estrogen in the female BALB/c mouse. Proc Soc Exp Biol Med 160:94–98.

Webb W, Samarawickrama GP (1981): Placental transport and embryonic utilization of essential metabolites in the rat at teratogenic doses of cadmium. J Appl Toxicol 1:270–274.

Webster WS (1978): Cadmium induced fetal growth retardation in the mouse. Arch Environ Health 33: 36–42.

Webster W, Valois A (1981): The toxic effects of cadmium in the neonatal mouse CNS. J Neuropathol Exp Neurol 40:247–257.

Wechsler W (1973): Carcinogenic and teratogenic effects of ethylnitosourea and methylnitrosourea during pregnancy in experimental rats. In Tomatis L, Mohr V (eds): "Transplacental Carcinogenesis." Lyon: IARC Scientific Publication No. 4.

Wechsler W, Rice JM, Vesselinovitch SD (1979): Transplacental neonatal induction of neurogenic tumors in mice: Comparison with related species and with human pediatric neoplasms. In Rice JM (ed): "Perinatal Carcinogenesis." Washington: US Government Printing Office, pp 79–1633.

Weisberger JH, Williams GH (1981): Carcinogen testing: Current problems and new approaches. Science 214:401–407.

Weiss B, Doherty RA (1975): Methyl mercury poisoning. Teratology 12:311–313.

Welch RM, Harrison JE, Gommi BW, Poppers PI, Finster M, Conney AH (1969): Stimulatory effect of cigarette smoking and the hydroxylation of benzpyrine by enzymes in human placenta. Clin Pharmacol Ther 10:100–110.

Wibberley DG, Khera AK, Edwards JH, Rushton DI (1977): Lead levels in human placentae from normal and malformed births. J Med Genet 14:339–345.

Wigglesworth JS (1964): Experimental growth retardation in the fetal rat. J Pathol Bacteriol 88:1–13.

Wild AE (1981): Endocytic mechanisms in protein transfer across the placenta. Placenta Suppl 1:165–185.

Willhite CC (1981): Congenital malformations induced by laetrile. Science 215:513–515.

Willhite CC (1982): Developmental toxicology of acetonitrile in the syrian golden hamster. Teratology (in press).

Willhite CC, Ferm VH, Smith RP (1981): Teratogenic effects of aliphatic nitrites. Teratology 23:317.

Williams KE (1982): Biochemical mechanisms of teratogenesis. In Snell K (ed): "Developmental Toxicology." New York: Praeger Publishers, pp 93–122.

Williams KE, Kidston EM, Beck F, Lloyd JB (1975): Quantitative studies of pinocytosis. I. Kinetics of uptake of (125I) polyvinyl pyrrolidone by rat yolk sac cultured in vitro. J Cell Biol 64:113–122.

Williams KE, Kidston EM, Beck F, Lloyd JB (1975): Quantitative studies of pinocytosis. II. Kinetics and protein uptake and digestion by rat yolk sac cultured in vitro. J Cell Biol 64:123–134.

Williams KE, Roberts G, Kidston ME, Beck F, Lloyd JB (1976): Inhibition of pinocytosis in rat yolk sac by tripan blue. Teratology 14:343–354.

Wilson J, Fraser C (1977): "Handbook of Teratology." New York: Plenum Press.

Woodward D, Bichett D, Chanda R (1975): Purkinje cell dendritic alteration after transient developmental injury of the external granular layer. Brain Res 97:195–214.

Wong KL, Klassen CD (1981): Toxic effects of cadmium in the brain of newborn rats. Toxicologist 1:62.

Yerusholmy J (1971): The relationship of parents' cigarette smoking to outcome of pregnancy. Am J Epidemiol 93:143–158.

Young M, Boyd RDH, Longo LD, Telegdy G (1981): "Placental Transfer: Methods and Interpretations." London: WB Saunders, Ltd.

Zaebst DD, Tanaka S, Haring M (1980): Occupational exposure to estrogens: Problems and approaches. In McLachlan J (ed): "Estrogens in the Environment." New York: Elsevier/North Holland, pp 377–401.

American Journal of Industrial Medicine 4:245–258 (1983)

The Teratogenic Risk

H. Tuchmann-Duplessis, Prof, MD, PhD

Reproduction can be impaired in animals and man by drugs and various environmental agents. Depending on the time of exposure – from fertilization through the fetal period and eventually during lactation – the consequences can range from embryotoxicity, gross malformations and a large variety of more subtle morphological, biochemical, and functional abnormalities. The high susceptibility of the embryo to exogenous agents is due to cellular multiplication and differentiation and to the lack of development of the enzyme systems necessary for the detoxification of chemicals. At present, developmental impairments represent the main cause of perinatal mortality and postnatal morbidity.

After a review of prenatal physiology and teratogenic principles, the action of selected drugs and environmental agents is analyzed. The potential danger of environmental factors during intrauterine development is of particular concern because of its irreversible nature.

Key words: teratogenicity and environment, drug teratogenicity, teratogenetic potential of exogeneous agents

INTRODUCTION

Drug teratogenicity is a relatively recent problem. Only after the thalidomide accidents was it recognized that a drug harmless to the maternal organism could impair intrauterine development, leading to congenital malformations.

Prenatal development is governed by a constant interplay of hereditary and environmental factors. The genetic information contains the program for the phenotype of the future baby; the environment supplies nutrients necessary for growth and differentiation. Therefore, exogenous agents can modify the expression of inherited genes.

The practical implications of congenital malformations are increasing. At present, developmental accidents represent the main cause of perinatal mortality and postnatal morbidity. More infants are handicapped for the rest of their lives through developmental impairments than by any other cause. The National Foundation has

Faculté de Médecine, Laboratorie d'Embryologie, Paris
Address reprint requests to Professor H. Tuchmann-Duplessis, Faculté de Médecine, Laboratoire d'Embryologie, 45, rue des Saints-Pères, 75006 Paris, France.

0271-3586/83/0401-02-0245$04.00 © 1983 Alan R. Liss, Inc.

estimated that 7% of Americans suffer from some type of congenital defect and that 700,000 infants are born defective each year. With the increasing number of women employed in industry, environmental pollution due to industrial wastes and the use of insecticides, herbicides, and defoliants, and other xenobiotics constitutes a definite potential hazard for the human embryo. It is therefore important to evaluate the hazard that occupational activities, drugs and environmental exposures may pose for the offspring.

Detection of an "occupational" teratogen is usually extremely difficult, except in the case of a highly dangerous substance that produces a high percentage of malformations of a specific type and during a specific period in the exposed group. This was the case with mercurial contamination in Minimata, Japan, or with polychlorinated biphenyls. The effect of agents with a low teratogenic potential may pass unnoticed for a long time, since the effect is masked by the normal rate of spontaneous malformations.

Since the experimental and epidemiological data on drugs are better documented than those for occupational agents, examples from therapeutic agents will be explored here. To provide an overview of the various factors involved in reproduction, the characteristics of prenatal development, and the principles of teratogenesis will be summarized.

IMPAIRMENT OF REPRODUCTION

Depending upon the time of exposure—from gametogenesis to the fetal period and eventually also during lactation—the adverse actions on the conceptus are variable. They range from infertility to embryotoxicity, gross congenital malformations, and a large variety of more subtle morphological, biochemical, and functional abnormalities. Some of these subtle defects require numerous investigations to be detected. They may be recognized a few months after birth (like heart anomalies) or several years later (like kidney and central nervous system malformations). Some anomalies are only fully recognized after puberty—like adenosis, impaired reproduction and vaginal cancer—attributed to prenatal diethylstilbestrol exposure.

Prenatal Physiology

The physiological activity of the conceptus can be divided into three main periods: (a) embryonic, (b) fetal and (c) neonatal period.

The embryo synthesizes the fundamental cell components: nucleic acids, proteins, and lipids. Nutrients necessary for embryonic synthesis are furnished by the maternal organism through placental exchange. During the early stages of development, the physiological activities of the embryo are mostly at the cellular level.

The general metabolic functions of the embryo are characterized by a predominance of anaerobic glycolysis of carbohydrate metabolism, the pentose pathway being very active. Specific metabolic functions begin to develop within embryonic cells. Differentiation is characterized by a dissimilarity in the protein content and metabolic function of different cell populations, and results from a selective activation and inactivation of genes.

The fetal period is characterized mainly by differentiation of the embryonic primordium into definitive organs, general growth, and the storage of energy substrates. Morphological maturation of organs is paralleled by an increase in functional activity with protein content increasing more than DNA content.

Compared with the neonate, functional activities of the fetus are limited. The central nervous system does not play the important integrative role it assumes later. However, very important cellular remodeling occurs in the brain during the fetal period. The development of fundamental central nervous system structures proceeds at different rates and is not completed until puberty. Periods of active growth and differentiation represent critical stages in central nervous system (CNS) development; the fine structures have the highest susceptibility to environmental and internal factors.

While many organ functions are restricted during prenatal development, the second feature of fetal physiology—growth—plays a predominant role. Increase in the rate of growth results from two phenomena: cell proliferation, and increase in cell size. Growth begins with an increase in DNA followed by accelerated protein synthesis. Formation of intercellular substances occur simultaneously, particularly in connective tissues and their derivatives. Growth is linear and continuous but its rate is not constant. After a sharp increase in the rate of growth up to the third month of gestation, the rate of growth decreases throughout the remainder of gestation.

On the cellular level, there exists a strict sequence of transformation. Growth and differentiation proceed by critical stages, starting with the early totipotent cell and leading to the highly specialized cell. When an embryonic cell loses its totipotency after the segmentation stage, the cell orients in a new direction along a particular developmental path and becomes "determined." The mechanisms underlying this new orientation remain unknown and will have to await an understanding of the basis of regulatory mechanisms.

The initiation of cellular proliferation occurs concurrently or immediately after determination. The cell divides, passing through one or several cell cycles. The most important part of the cell cycle is the G_1 phase, during which DNA synthesis is arrested while RNA synthesis proceeds at a high rate. When the cell proceeds beyond the proliferation phase into the later phases of differentiation and maturation, it usually passes through a migratory phase. The migrating phase is also critical and can be disrupted by environmental agents. After migration, the cell becomes fully differentiated and mature. Biochemically, differentiation is characterized by the appearance of distinctive molecules specific to this particular type of cell—hemoglobin for red blood cells, albumin for hepatocytes, and lactase for intestinal cells.

The neonatal period is characterized by adaptation to extrauterine life. It is important as the period in which metabolic anomalies acquired during gestation may cause a dysfunction of enzymatic systems required for the autonomic life of the baby. For example, at birth, pulmonary function is dependent on the production of surfactant, which prevents the small alveoli from collapsing after the first expiration.

The general metabolism performs a transfer from an anabolic state to a catabolic one. Before birth, the metabolic flow is oriented from the mother to the fetus, which stored glycogen and lipids. After birth, new energy needs appear for thermogenesis, respiratory work, and motility. This energy is supplied by the utilization of stored glycogen and lipids. A fault in the storage or utilization processes may have a deleterious effect on the survival of the newborn. Therefore, alterations occurring during late intrauterine stages may be expressed only postnatally.

TERATOGENIC PRINCIPLES

The action of a teratogenic agent on the conceptus depends mainly on three conditions:

1. the developmental stages of the embryo;
2. the genetic susceptibility of the embryo;
3. the physiological or pathological status of the mother.

The Developmental Stages

The period in which injurious agents can affect the development of the human embryo is very short. It is largely completed by the eighth week of pregnancy, just about the time a woman knows she is pregnant.

During the preimplantation period, when the blastocyst lies free within the uterus and depends for its nutrition on the uterine secretions, exogenous agents can kill the embryo, but there is no evidence that they would produce congenital malformations. This is the period of maximal embryolethality. However, minor injuries can be overcome without obvious harmful consequences on the growing embryo because during the segmentation stage, many blastomeres retain their totipotency, being able to replace damaged cells by newly formed cells. Once implantation has occurred in the human, at seven to eight days after fertilization, the embryo undergoes very rapid and important transformations.

The sequence of embryonic events shows that each organ and system undergoes a critical stage of differentiation at a precise moment of prenatal development. It is during this critical period that the vulnerability of the developing embryo is greatest, and that specific gross malformations can be produced. This is the susceptible period, which lasts until the 56th day in the human embryo.

The fetal period begins at the end of the eighth week, when little further differentiation of organs remain to be completed. The most important events at this stage are closure of the palate; reduction of the umbilical hernia at the end of the 9th week; differentiation of external genitalia at the 12th week; and histogenesis of the central nervous system, which lasts for the entire period of intrauterine development, and is not complete until several months after birth.

Consequently, during the fetal period, teratogenic agents do not determine major morphological malformations but can impair the differentiation of external genitalia, leading, in severe cases, to pseudo-hermaphrodism. Interferences with the histogenesis of the central nervous system can lead to various degrees of encephalopathies.

Genetic Susceptibility and Species Differences

The reaction of the embryo to exogenous agents depends upon its genetic constitution. It varies not only between different animal species but also within a given species among each strain and even between individuals of the same strain.

Mendelian susceptibility. Hybridization of inbred strains showed that the susceptibility to a drug may depend on one or several genes, the effects of which can be followed through successive generations. Several characteristics have emerged:

1. The susceptibility is not a general one: the gene action is organ specific.
2. The tendency to malformation of one organ may depend on several genes, each being affected by various teratogens. A strain susceptible to one agent may be found to be resistant if a different teratogen is used.

Recent advances in basic research furnish a biochemical explanation for strain differences in teratogenic susceptibility. Consider, for example, cortisone. It has been known for many years that certain strains of mice are more susceptible to glucocorticoids than others [Fraser, 1978]. When mice of the A/J strain are treated between days 11–14 of gestation with cortisone, they exhibit a 100% incidence of cleft palate. The same treatment produces only 20 to 25% cleft palate in the C57BL/6J strain.

Salmon and Pratt [1976], and Goldman et al [1977], have shown that these differences in susceptibility are related to cytosolic receptors for glucocorticoids. Comparing the facial mesenchyme cells of these two strains, the authors found that the cells of cortisone sensitive strains have a much greater ability to bind glucocorticoids than do embryonic facial mesenchymal cells of cortisone resistant strains. The susceptibility to cleft palate is closely related to the affinity of embryonic mesenchymal cells to bind glucocorticoids.

Physiological and Pathological Status of the Mother

Besides the developmental stage and the genetic constitution of the embryo, the action of a drug is dependent on the physiological and pathological condition of the mother. The most important physiological factors are age, diet, hormonal balance, and uterine environment. Among pathological conditions metabolic diseases, diabetes, and obesity tend to increase the action of potential teratogenic agents.

TERATOGENIC AGENTS

The tragic thalidomide accidents in several countries revealed to the medical profession and to governmental regulatory agencies that a drug harmless to the mother could impair the development of the human embryo and produce congenital malformations. Several thousand children suffered from limb deformities ranging from shortening and torsions of the upper and lower extremities, phocomelia or amelia complete absence of the arms and legs.

Research was stimulated from this period forward. Scientists of various disciplines (including toxicology, pharmacology, and biochemistry) became interested in teratology and mechanisms of abnormal growth. We will review here experimental and clinical data of drugs that are frequently used.

Psychoactive Drugs

The first cases of thalidomide embryopathies were described in West Germany [Lenz, 1961] and in Australia [McBride, 1961]. Besides the characteristic limb deformities, thalidomide also produces visceral defects: heart, kidney and gastrointestinal anomalies. The teratogenic potential of thalidomide was demonstrated experimentally in 1962 in our laboratory.

While the rat is not susceptible to thalidomide, a large variety of congenital malformations can be produced in other species: rabbits, monkeys, dogs, and pigs.

Two compounds chemically related to thalidomide— α-phenyl, and α-ethylglutaridmide (Doriden) and α-phenylbetadiethylaminoethylglutarimide (Aturban)—which had become suspect were investigated. No malformations were observed, even using very high doses, including those toxic to the mother. This is an example which demonstrates the complex interrelationship between pharmacological properties and teratogenic action.

Phenothiazines. A large number of phenothiazines have been used for long periods in pregnancy to relieve tension and nausea. Although in a few isolated cases a slight teratogenic action of chlorpromazine and prochlorpromazine has been reported in the rat, there has been no proof of any toxic effect on the human fetus, whether the drug has been taken for a short time or for several months. In a study of 50,000 subjects, Sloane et al [1977] found no differences in the rate of malformations in children whose mothers were exposed to phenothiazines as compared to children of nonexposed mothers.

Behavioral effects. In experiments where rats were exposed to phenothiazines at various periods of pregnancy, the offspring had reduced general activity as compared to control rats. Hearing deficits have also been reported [Ordy et al, 1966]. Other antipsychotic drugs, like prochlorperazine, penfluidol and trifluroperazine have similar effects. Chlorpromazine, when given to rats during the first three postnatal weeks, modifies the sensibility of the central nervous system. It is possible that these behavorial effects are related to a modification of the developing dopaminergic system.

Benzodiazepines. Several compounds of this group such as chlordiazepoxide (Librium), diazepam (Valium), and nitrazepam (Mogandon), are widely used in clinical practice. Experimental investigations in rats and rabbits did not reveal teratogenic effects of these drugs. Neither the evolution of pregnancy nor the development of the offspring was affected, even when doses 20 to 40 times higher than therapeutic doses were used.

It has been claimed that benzodiazepines produce chromosomal aberrations; however, detailed studies did not confirm this claim either by in vitro or in vivo investigations. The clinical data suggest that although the benzodiazepines cross the placenta rapidly they have no adverse effect on the fetus.

Babies of women treated up to the time of labor with benzodiazepines (chlorazepate, nitrazepam) as well as stimulants of the CNS (impramine, lithium) often have various nervous impairments: depression and somnolence, or in other cases withdrawal, hyperexcitability, and convulsions with respiratory depression. In the majority of cases these symptoms disappear after a few days and the children have a complete recovery.

Valium, which has been used for some time in threatened abortion, does not seem to increase the rate of congenital malformation. According to Wallner et al [1973], no significant differences of anomalies have been observed in women treated for threatened abortion with sex hormones alone or with a combination of sex hormones and valium. Available experimental and clinical data suggest that neither librium nor valium are teratogenic.

Milkovich and Van Den Berg [1974] claimed that there might be an association between the use of chloridazepoxide hydrochloride or other tranquilizers (like meprobamate) in early pregnancy and an increased incidence of severe congenital malformations. This assumption was based on a prospective epidemiological study that included 19,044 live births. Rates of the anomalies were four times higher in infants born to mothers who took the drugs than of mothers who did not take drugs. The heterogeneity of malformations found in this study and the lack of similar findings in other epidemiological studies led to doubts that the two drugs were potential human teratogens [Tuchmann-Duplessis, 1975].

In an independent follow up study 1,870 children exposed to meprobamate or chlordiazepoxide in the first trimester or at other times of pregnancy were compared with 48,412 children who were not [Hartz et al, 1975]. The authors concluded that there was no evidence that either of these drugs was teratogenic. No relationship with major malformations, cardiac anomalies or stillbirth was found. In addition, as judged by mental and motor scores at eight months and at four years, there was no evidence that these drugs cause brain damage.

Butyrophenones. Large doses of Haloperidol can delay implantation in the rat and in the mouse. In our experience, this action is due to a modification of the gonadotropic activity of the maternal pituitary leading to excessive releases of luteotrophic hormone. No teratogenic action was observed. Reserpine, which is used in hypertension and in psychiatric disorders, may affect the fetus. A syndrome characterized by lethargy, bradycardia, hypothermia, and nasal congestion has been described. A similar syndrome due to depletion of the catecholamine stores could be produced in guinea pigs. However, no abnormalities were observed.

Experiments with reserpine which involved three generations of rats gave interesting results. Nulliparous Wistar female rats were treated with reserpine at a dose of 100 $\mu g/kg$ daily, the drug being mixed with their food. They were mated with untreated males after three, seven and 12 months of this treatment. In the group mated at three months of treatment, fertility was practically the same as in untreated controls, but in the group mated at seven months there was a 50% reduction of fertility, and in the group mated at 12 months only 5% of the animals produced young [Tuchmann-Duplessis and Mercieri-Parot, 1962].

Lysergic acid diethylamide (LSD). Studies with ^{14}C-labeled LSD provides evidence that LSD crosses the placenta and enters the fetus during different stages of gestation in mice and hamster. Metabolic studies have measured the half-life of LSD as 175 minutes in man and only seven minutes in mice. The distribution of LSD in the rat has been determined after intravenous administration. The drug accumulates rapidly in the brain within 15 to 20 minutes after injection. Free LSD is distributed generally in the cerebellum, hippocampus choroidal plexus, pituitary, and pineal glands. Bound LSD is localized in the cortex, the midbrain, and medulla.

Analysis of various human data suggests that there is an increased risk of spontaneous abortion in users of illicit LSD. Dishotsky and coworkers [1971] concluded that pure LSD ingested in moderate doses does not produce chromosome damage detectable by available methods, does not cause detectable genetic damage, and is not teratogenic or carcinogenic in man. A similar opinion has been reached by Long [1972], who considers that it is unrealistic to ascribe to LSD the chromosome breaks found in prior LSD users, since there was no correlation between the percentage of breaks and the amount of LSD, number of doses, or time between the last dose and blood sampling. Despite several claims, no definite dysmorphogenetic effect was observed in LSD treated animals. Thus there is no strong evidence of dysmorphogenic action by LSD in animals or man.

Narcotics. Narcotics are often used as premedicants for anesthesia, and are also widely utilized in pregnancy. It is estimated that there are 100,000 to 200,000 heroin addicts in the USA, of whom approximately 25,000 females live in the metropolitan area of New York City, 85% of these women are in the childbearing age. Obstetrical complications including prematurity, toxemia, and precipitate labor are significantly higher in these pregnant women than in the general population. Half of the neonates are premature and show evidence of retarded intrauterine growth. Over 70% of them show congenital neonatal addiction with clinically recognizable withdrawal symptoms suggesting hyperactivity of the autonomic nervous system and irritability of the central nervous system. In addition, diarrhea, rhinorrhoea and shrill cry are frequently present.

The length of maternal addiction was directly correlated with the occurrence of withdrawal signs in the infants. With increased maternal intake of heroin, there was a higher incidence of neonatal withdrawal symptoms. In addition, the closer to delivery the last dose of heroin was taken, the higher and earlier was the occurrence of withdrawal signs in the neonate. The incidence of congenital malformations is not higher than in the general newborn population.

Barbiturates. Addiction symptoms have been observed in infants born of mothers taking large doses of barbiturates throughout pregnancy or in the last trimester. The symptomatology is quite similar to that observed in heroin addiction: hyperexcitability, overactivity, excessive crying, tremors, hyperphagia, and outbursts of screaming. But infants with barbiturate addiction differ from babies with congenital heroin addiction; they are fully grown and show no respiratory depression in the perinatal period. In addition, the onset of symptomatology is later (six to twelve days of age) in the barbiturate group than it is in the heroin-addicted group. Most of the symptoms of disturbed neuronal function can be sedated with phenobarbitone and chlorpromazine.

Anticonvulsants and Epilepsy

The malformation rate in the offspring of epileptic parents is two to three times higher than in the general population. The etiology of these malformations is still a matter of discussion. There are embryopathies, cleft lip, cleft palate, cardiac, and skeletal abnormalities.

Specific patterns of abnormalities have been described in infants born to epileptic mothers treated during pregnancy. Hanson and Smith [1975] described the fetal hydantoin syndrome, while Seip [1976], found evidence of a fetal trimethadione syndrome (characterized by cranio-facial malformations), developmental and psychomotor delay, pre- and postnatal growth deficiency, and impaired intellectual performance associated or not associated with cardiac, genito-urinary, and skeletal malformations. Bethenod and Frederich [1975] have described a similar clinical syndrome termed "child of epileptics."

Although a large variety of malformations could be experimentally produced in rodents with anti-convulsants, it seemed difficult to establish a direct relationship between a specific antiepileptic drug and human abnormalities. Several epidemiological studies favor drug-induced teratogenicity.

Besides congenital anomalies, mental subnormality occurred in 1.5% of the epileptic mother's children, compared with 0.2% in a control group. The perinatal mortality rate was 42.5 per 1,000 for pregnancies in which anticonvulsants were taken, which was nearly twice the regional rate. The rate for all epileptic women was 28.6 per 1,000. The increased perinatal mortality was mainly due to congenital malformations and spontaneous hemorrhage which occurred in six of 388 births.

Shapiro et al [1976], analyzing data from a collaborative perinatal project, concluded that epilepsy (maternal or paternal) is the decisive factor in producing malformations. The anticonvulsant drugs are not considered to be directly implicated in the etiology of malformations.

From a collaborative study in Japan in which use was made of prospective and retrospective cases, Yoshibumi et al [1980] concluded that anticonvulsants have a teratogenic action in humans. The malformation incidence in the non-medicated group was 2.3% as compared to 11.5% in the medicated group. When the trimetadione treated patients are excluded, the malformation rate decreased to 6.75%. Therefore the authors conclude that trimetadione represents the highest teratogenic risk, while barbiturates and their derivatives are less dangerous. In this study diphenylhydantoin appeared the least dangerous drug.

A strong indication of drug-induced teratogenesis is suggested by the observations of Finnell [1981] in three mouse strains: SWV, C_3H and C57BL/6J. This last strain had spontaneous, tonic-clonic seizure disorders which mimics the human condition. Phenytoin in doses of twenty to sixty mg/kg was given two weeks before and continued throughout gestation. Fetal weights decreased significantly in all strains as the dosage of phenytoin increased. The incidence of malformations also increased in all three strains in proportion to the concentration of phenytoin in maternal plasma. Females whose seizure disorders were untreated produced normal offspring even when seizures occurred daily through the gestation period. The author therefore concluded that the malformations are determined by the drug and not by the maternal-seizure syndrome.

Stimulants of the Central Nervous System

Among drugs used as psycholeptics, MAO inhibitors and amitriptyline derivatives have been investigated.

Nialamide, a well-known MAO inhibitor, is not teratogenic. With Hercier-Parot [1962], we investigated the influence of nialamide on pregnancy in the rat. High doses given immediately after fertilization or after implantation produce fetal resorptions but no congenital malformations.

Amphetamines. Dexamphetamine sulfate is teratogenic in A/Jax mice. Large doses administered parenterally at the onset of pregnancy caused fetal resorptions, retarded growth and malformations: 11% had anophthalmia or microphthalmia, and 18% had palatoschesis. In the rabbit no teratogenic effects were observed. This drug has been incriminated in the etiology of human malformations. Clinical data suggest the possibility of a teratogenic risk.

One of the analogues of this drug, methamphetamine hydrochloride, when given at a dose of 10 mg/kg, led to 13% of gross malformations in the mouse. In the rabbit even small doses, – 1.5 mg/kg administered between days 12–15 or between days 15–20 – result in teratogenic effect, and approximately 15% of the survivors have gross congenital malformations [Kasirsky and Tansy, 1971].

Imipramine (Tofranil). In several circumstances imipramine has been incriminated in the etiology of human malformations. It has been claimed that when this antidepressant is taken in early pregnancy it can harm morphogenesis of the fetus leading to various limb and CNS malformations [McBride, 1972].

Experimental data seemed to substantiate this hypothesis; Robson and Sullivan [1963], using high doses of imipramine, induced various malformations of the CNS in the rabbit. However, in the rat and mouse no malformations were observed when large doses of imipramine were administered throughout the entire pregnancy.

Careful epidemiological studies carried out in recent years suggest a more optimistic appreciation of this drug. Crombie et al [1972] analyzed the outcome of 19 cases in which imipramine was administered during the early weeks of pregnancy. No abnormalities in this group were reported either at birth or subsequently. In another group in which 28 prescriptions for amitriptyline were given, the issue of pregnancy was normal. Kuensberg and Knox [1972], reviewed 15,000 pregnancies in Scotland.

There were 18 mothers for whom imipramine was prescribed within the first ten weeks of pregnancy. No abnormalities were reported in 14 children, one mother aborted and two mothers gave birth to malformed children; one malformed child had defective abdominal muscles and a possible abnormality of the gut, the other had diaphragmatic hernia. Of 31 mothers for whom amitriptyline was prescribed, 28 had normal babies; 1 aborted; there was 1 still-birth; and 1 child was born with hypospadias. Thus the Scottish study suggests that this antidepressant is not associated with a high teratogenic risk.

Lithium. In animals, conflicting results have been reported. It is generally believed that lithium has only a very low teratogenic potential. The number of patients with recurrent manic-depressive disorders treated with lithium carbonate is increasing constantly. A large number of pregnant women experience this treatment. Schou et al [1973] made a retrospective investigation in 118 children born to mothers who were given lithium treatment during the first trimester of pregnancy. On the basis of these findings they concluded that the risk of teratogenic effects is lower than one might expect from certain experimental data. However, the authors still could not answer the question of whether or not lithium is teratogenic in man.

Hormones

Sex steroids. Testosterone and its analogues have a masculinizing effect in female fetuses. In girls, masculinization of the urogenital sinus takes the form of hypertrophy of the clitoris and atrophy of the labia. If the drug has been given before the 12th week of pregnancy, complete fusion of the labia may occur.

The female sex hormones, progesterone and estradiol, are considered to be devoid of adverse effects on the embryo. Diethylstilbestrol, a potent synthetic estrogen, has feminizing effects on male rat fetuses. Herbst et al [1971] reported that adenocarcinoma of the vagina occurred in relatively young girls whose mother underwent stilbestrol treatment during the first trimester of pregnancy. The causal relationship between the maternal stilbestrol therapy and virginal cancers of their daughters, although not completely demonstrated, seems highly probable. Herbst [1972] also found a high proportion of vaginal adenosis in girls of 13–14 years exposed to diethylstilbestrol in utero. The exposure was before the 12th week of pregnancy and the usual dosage was 10mg/per day.

Oral Contraceptives. In animal experiments the influence of the oral contraceptive steroid Enidrel (a combination of norethynodrel 4.925 mg and mestranol 0.075 mg) on the pituitary hypothalamic system and on general reproduction [Tuchmann-Duplessis and Mercier-Parot, 1972] was examined. The treatment during organogenesis did not impair fetal development. Examination of male and female gonads did not reveal any morphological changes. Therefore, it was concluded that even very high doses of a potent fertility control drug do not alter reproduction in the rat nor lead to congenital malformations or disturbances of postnatal development.

In humans neither mutagenic effects nor an increase of chromosomal aberration were observed in women who conceived during contraceptive treatment, or several months after withdrawal of the drug. In agreement with experimental results, clinical data indicate that contraceptive drugs have no masculinizing effects. The teratogenic danger which has been suspected for some time has not been substantiated, though a very large series of clinical observations have been made.

Corticosteroids. Cortisone produces cleft palate and cardiac abnormalities in mice and rabbits. Hydrocortisone and ACTH have similar effects in mice. The incidence of malformations varies according to the dose of cortisone and the strain of the animal that is being studied. In rats a high postnatal mortality is observed. The majority of the newborn die at an age of 8–10 days. The mechanism involved in the high postnatal mortality of cortisone-treated rats is still unknown. It has been suggested that it could be related to a depression of the thymo-lymphatic system. This hypothesis is supported by the fact that there is a depression and distortion of thymo-lymphatic differentiation in the offspring of cortisone-treated rats. The thymus is atrophic and shows a shrinkage of the lobular cortex associated with lymphocytic depletion. In addition, the serum level of IgG is decreased in prenatal cortisone treated rats. In humans, it is doubtful that cortisone has a specific teratogenic action.

Anticoagulants

Two groups of anticoagulants are available; heparin, which does not cross the placenta, and the coumarin derivatives of low molecular weight, which enter the fetal circulation. Although bishydroxycoumarin and its analogues are nearly completely bound to plasma proteins, they enter the fetal compartment and produce a more pronounced anticoagulant effect in the fetus than in the mother.

The noxious effects of bishydroxycoumarin might also be attributable to the enzymatic immaturity of the fetal liver. The detoxification mechanism, which includes conjugation with glucuronic acid, is deficient; even minute quantities of free bishydroxycoumarin are able to interfere with prothrombin synthesis in the liver.

Oral anticoagulants have been incriminated in fetal morbidity and mortality as well as bone malformations. However, this has not been substantiated. The influence of early oral anticoagulant treatment is still uncertain. Various case reports suggest no adverse effect, and our personal observations in five patients treated before conception and during the first trimester had a normal outcome of pregnancy.

Despite the theoretical risk of maternal hemorrhage, there is no evidence that anticoagulants in therapeutic doses increase the incidence of maternal hemorrhage either before, during, or after delivery.

Beta Adrenergic Drugs

Recently, Loser et al [1981] have investigated the influence of a tocolytic agent, fenoterol (Partusisten), in neonates. Among 42 children of mothers treated with this drug, 11 suffered from severe toxic myocardiopathy associated with heart failure. The mothers had been given excessive fenoterol doses for 13 weeks.

Antitumor Drugs

Embryonic cells are much more sensitive to the toxic effects of antitumor drugs than are adult cells. Therefore, most compounds at lower doses have deleterious actions on the embryo than the adult. They are embryotoxic and frequently teratogenic also. A large number of compounds induce congenital malformations in experimental animals; however, in humans, teratogenic potential could only be demonstrated for aminopterine, methotrexate, busulfan, cyclophosphamide, and chlorambucil.

Most of the antibiotics used in cancer therapy have a high degree of toxicity. They interfere with protein synthesis and nucleic acid metabolism. The most teratogenic in rodents are actinomycin-D, mitomycin-C, and streptonigrin.

Antimetabolites are the most interesting compounds in morphogenic research; they act on basic biological systems and cause functional deficiencies in the cell. The most frequently investigated compounds are 6-mercaptopurine, azathioprine, fluoropyrimidine, 5-fluoro-2-desoxyuridine, aminopterine, X-methylfolic acid (methotrexate), azaserine, and 6-aminonicotinamide. All these compounds are teratogenic.

Cancer Chemotherapy

Blatt et al [1980] examined the outcome of pregnancies of 30 patients who had chemotherapy during the time of conception and/or during the first trimester. They had 24 live infants. No births were premature and the birth weights were normal. Three children had anomalies: a hip dysplasia, febrile convulsions, and a deformity of the spinal column without spina bifida, respectively. A follow-up of the offspring until twelve years of age showed that growth, development, and school performance were normal. Therefore, the authors conclude that chemotherapy given to a woman prior to conception does not jeopardize her infants. Chemotherapy given to a male prior to conception does not result in fetal damage. However, many more children must be examined and followed for long-term to rule out potential adverse effects.

Cigarette Smoking and Nicotine

The harmful effects of cigarette smoking and nicotine have been substantiated by experimental animal and human data. Exposure to nicotine during pregnancy has deleterious effects in rodents. Offspring of nicotine-treated rats are significantly underweight compared to saline injected controls. The average number of liveborn young is lower than in the control group and parturition is delayed. The offspring whose mothers received nicotine throughout pregnancy and during the nursing period had decreased survival after birth.

In humans, the offspring of smokers are underweight; there is also a high incidence of prematurity and abortion. The lower birth weight at any given stage of gestation has been attributed to fetal malnutrition, resulting from either a smoking effect on uterine circulation or a smoking effect on maternal appetite. In pregnant smokers as much as 8.5% of the hemoglobin is saturated with carbon monoxide; as much as 9.6% of hemoglobin is saturated with carbon monoxide in red blood of infants of these mothers. It has also been suggested that chronic hypoxia of the fetus of the smoking mother might be the cause of the lower fetal weight.

There are also reports concerning biological modifications correlated with tobacco: the level of serum vitamin B_{12} is lower in smokers than in nonsmokers. This vitamin decrease is probably related to the cyanide content of tobacco smoke, since cyanide may be detoxified by a mechanism which depletes the stores of B_{12} in the body.

Alcohol

Alcohol, which is widely used, has been extensively studied in its consequences on metabolism and human reproduction. The teratogenicity of alcohol has been demonstrated by clinical investigations and in animals through controlled laboratory experiments. The human fetal alcohol syndrome is characterized by CNS dysfunction, mild to moderate mental retardation, growth deficiency, and facial deformities. The facial abnormalities include short palpebral fissures, low nasal bridge epicanthic folds, short nose, narrow upper lip, and various abnormalities of the eyes. The experimental data obtained in rodents and guinea pigs give strong evidence of the noxious effect of alcohol on prenatal and postnatal development. The results are particularly obvious in several mice strains, in which concentrations of 25 to 30% alcohol in the food or water induce CNS malformations, ie, microcephaly, hydrocephaly, and exencephaly.

Mechanisms of action. Alcohol injected into pregnant mice rapidly enters the fetal compartment. Equivalent feto-maternal concentrations are reached 40 minutes after maternal injection in the mouse. Although all fetal tissues absorb alcohol, the highest concentrations are reached in the CNS. Cell membranes and the membranes of subcellular components are considered to be critical sites of action of ethanol.

CONCLUSION

A large variety of drugs and environmental agents when administered during pregnancy can impair intrauterine development and produce congenital malformations in animals and humans. Prenatal development is characterized by the existence of critical periods during which the cellular activity can be disrupted leading to various morphogenetic accidents. The high susceptibility of the embryo to exogenous agents is also due to the lack of development of the enzyme systems necessary for detoxification of chemicals. Important species differences exist in the susceptibility to a given agent. Neither the chemical structure nor the pharmacological properties of a compound gives an indication of the possible teratogenic potential. While thalidomide is highly teratogenic, closely related compounds do not produce malformations in animals or humans.

The potential danger of environmental factors during intrauterine development is of particular concern because of its irreversible nature. While in the past the teratogenic risk to exogenous agents has been overlooked despite the experimental results, the danger has probably been overemphasized after the thalidomide accident. Excessive fear should not deprive women of the benefit of useful drugs.

Prevention of congenital malformations should be directed to the improvement of experimental and epidemiological investigations and the analysis of the mechanisms involved in abnormal development.

REFERENCES

Bethenod M, Frederich A (1975): Les enfants des antiepileptiques. Pédiatrie 30:227–248.

Blatt JJ, Mulvihill JL, Ziegler RC, Young DG (1980): Pregnancy outcome following cancer chemotherapy, Psycho-social aspects of Cancer. American Society of Clinical Oncology Abstract c/ 2 47 p. 381.

Crombie DL, Pinsent RFH, Fleming D (1972): Imipramine in pregnancy. Brit mad J 1:745.

Dishotsky NI, Loughman WD, Mogar RE, Lipcomb WR (1971): LSD and genetic damage. Science 172:431–441.

Finell RH (1981): Phenytoin induced teratogenesis: a mouse model. Science 211:483–484.

Fraser FC (1978): Prevention of birth defects. How are we doing. Teratology 17:198–202.

Goldmann AS, Katsumata S, Jaffe , Gasser DL (1977): Palatal cytosol cortisol binding protein associated with cleft palate susceptibility and H_2 genotypes. Nature 245:643–645.

Hanson JW, Smith DW (1975): The fetal hydantoin syndrome. J Pediat 87:285–290.

Hartz SC, Heinonen OP, Shapiro S, Siskind V, Slone D (1975): Antenatal exposure to meprobamate and chloridazopexide in relation to malformations, mental development and childhood mortality, New Engl J Med 292:727–728.

Herbst AL, Kurman RJ, Scully RE (1972): Vaginal and cervical abnormalities after exposure to stilbestrol in utero. Obstet Gynec 40:287–298.

Herbst AL, Ulfelder H, Poskanzer DC (1971): Andenocarcinoma of the vagina: association of maternal stilbestrol therapy with tumor appearance in young women. New Engl J Med 284:878–881.

Kuensberg EV, Knox JDE (1972): Imipramine in pregnancy. Brit Med J 2:292.

Lenz W Von (1961): Die thalidomid-embryopathie. Dtsch med Wschr 86:2555.

Long SY (1972): Does LSD induce chromosomal damage and malformations. A review of the literature. Teratology 6:75–90.

Loser H, Steinkam K, Muller M, Pfefferkorn JR, Dame WR, Hilgenberg FR: Kardiotoxische Wirkung des Toskolytikoma Fenoterol. Munch Med Woehr 123 2:49–52.

McBride WG (1961): Thalidomide and congenital abnormalities. Lancet 2:1358.

Mercler-Parot L, Tuchmann-Duplessis H (1974): The dysmorphogenic potential of phenytoin: experimental observations. Drugs 8:340–353.

Milkovich L. Van den Berg BJ (1974): Effects of prenatal meprobamate and chlordiazepoxide hydrochloride on human embryonic and fetal development. New Engl J Med 291:1268–1271.

Robson JM, Sullivan FM (1963): The production of foetal in abnormalities in rabbits by Imipramine. Lancet 638:639–000.

Salomon D, Pratt R (1976): Glucocorticoid receptors in murine embryonic facial mesenchyme cells. Nature 264:174–177.

Schou M, Goldfield MD, Weinstein MR, Villeneuve A (1973): Lithium and pregnancy I. Report from the register of lithium babies. Brit Med J 2:135–136.

Schou M, Amdisen A (1973): Lithium and pregnancy. III. Lithium ingestion by children breast-fed by women on lithium treatment. Brit Med J 2:137–138.

Seip M (1976): Growth retardation, dysmorphic facies and minor malformations following massive exposure to phenobarbitane in utero. Acta Pediatr Scan 65:617–621.

Shapiro S, Stone D, Hartz SC, Rosenberg L, Siskind V, Monson RR, Mitchell AA, Heinonen OP (1976): Anticonvulsants and prenatal epilepsy in the development of birth defects. Lancet 1:272–275.

Slone D, Siskind V, Heinonien OP, Monson RR, Kaufman DW, Shapiro S (1977): Antenatal exposure to the phenothiazines in relation to congenital malformations perinatal mortality rate, birth weight and intelligence quotient score. Am J Obstet Gynecol 128:486–488.

Tuchmann-Duplessis H, Mercier-Parot L (1962): Influence de traitements chroniques de réserpine sur la fertilite des parents et de leurs descendants. Comptes-rendus des seances la Societe de Biologie et de ses filiales. 156:587–590.

Tuchmann-Duplessis H, Mercier-Parot L (1962): Influence des traitements chroniques de réserpine sur la fertilité CRS:156–587.

Tuchmann-Duplessis H, Mercier-Parot L (1972): Action d'un stéroide anticonceptionnel sur la descendance. J Gyne Obstet Biol Repr 1:141–159.

Tuchmann-Deplessis H, Mercier-Parot L (1973): Risque teratogene de la therapetique anti-epileptique. Nelle Press Med 2:2719–2720.

Tuchmann-Duplessis H (1975): Drug effects on the fetus. Adis Press.

Wallner JJ, Waidel E, Welsh H (1973): Arch Gunak 214:83–84.

Yoshibumi N, Okuma T, Takahashi R, Sato Y, Wada T, Sato T, Fukushima Y, Kumashiro H, Ono T, Takahashi T, Aoki Y, Kazamatsuri H, Inomi K, Komai S, Seino M, Miyakoshi M, Tanimura T, Hazama H, Kawahara R, Otsuki S, Hosokawa K, Inanaga K, Nagazawa Y, and Yamamoto K (1980): Multi-institutional study on the teratogenicity and fetal toxicity of antiepileptic drugs. A report of a collaborative study group in Japan. Epilepsia 21:663–680.

American Journal of Industrial Medicine 4:259–281 (1983)

Occupationally Derived Chemicals in Breast Milk

Mary S. Wolff, PhD

Exogenously derived chemicals have been widely reported in breast milk. Chemicals typically found in occupational exposures, including trace metals, solvents, and halogenated hydrocarbons, are reviewed, in terms of milk partition factors, potential infant exposures, and possible infant health effects. In addition to ingestion of a chemical from breast milk, an infant incurs a neonatal body burden of a chemical due to transplacental migration from maternal blood. For trace metals, neonatal blood levels are similar to maternal blood levels. Partition of metals to milk is less efficient, but nevertheless can contribute significantly to an infant's body burden. For lipid-soluble pesticide residues and halogenated biphenyls, neonatal body burden is much less than that of the mother, but transfer to milk is efficient, due to the high proportion of milk fat.

It is suggested that potential organic mercury toxicity can be estimated from concentration in maternal blood or milk. For other chemicals, available data are not sufficient to evaluate short- or long-term health effects. However, for many halogenated hydrocarbons, concentrations in normal human milk would permit infant exposure above guidelines for allowable daily intake set by the World Health organization.

Key words: chemicals in breast milk, pesticides, mercury, lead, PCBs, solvents

INTRODUCTION

The issue of toxic industrial chemicals present in trace quantities in human milk poses a unique dilemma. The problem has arisen from recent recognition of two opposing aspects of human health: the beneficial nature of breast feeding and the potentially detrimental effects of industrial chemical contaminants in human milk. The focus is principally the protection of the newborn, in whom a body burden of toxic chemicals may be created by subsequent breast feeding.

The resolution of the latter problem seems obvious: Protection of infants requires removal of lactating mothers from exposure to such chemicals, through appro-

Environmental Sciences Laboratory, Mount Sinai School of Medicine of the City University of New York, New York.

Address reprint requests to Dr. Mary S. Wolff, Environmental Sciences Laboratory, Mt. Sinai School of Medicine, 1 Gustave L. Levy Place, New York, NY 10029.

0271-3586/83/0401-02-0259$06.50 **© 1983 Alan R. Liss, Inc.**

TABLE I. Components of Human Milk Beneficial to Neonates

	Specific effect	Difference in cow's milk
Proteins (0.8–0.9%) – immunological benefit		
β-Lactoglobulin (absent)	Most common infant allergin	Present
β-Casein		α-Casein
Amino acid profile		Different proportions
Lactoferrin		Absent
IGA (major)	Higher in breast-fed infants	Minor
Lysozyme		1/300 human milk
Fats (2–3%) – important to brain development		1/10 human milk
Arachidonic and decosahexaenoic acids neonate cannot synthesize		
Human lymphocytes (including T cells) and macrophages (immune function)		
Unsaturated folic acid and vitamin B_{12} (immunity)		
Factors affecting iron absorption and thus anemia (ascorbic acid, copper, vitamin E)		

From Jeliffe [1976] and Goldman [1976].

priate industrial hygiene in the workplace. The efforts of various regulatory agencies, eg, OSHA, trade unions, and active industrial hygiene divisions of major industries, have led to better protection of all workers, but not necessarily to an immediate solution to the problem of chemicals in mother's milk. Such exposures still exist. Moreover, past exposures may have produced a body burden of chemicals that are not readily excreted and which are retained in maternal tissues and partitioned into milk when lactation occurs. Individual awareness of potential risks often occurs only after conception, when it may be too late for a woman to take the initiative to reduce her exposure.

Whatever the infant's chemical exposure from maternal milk, it is superimposed on an existing neonatal body burden. Transplacental migration of toxic chemicals to the growing fetus causes a prenatal exposure, and subsequent burden, of maternally carried chemicals. But, whereas prenatal exposure cannot at present be avoided in the course of childbearing, breast feeding has an acceptable alternative. The question that parents and their physicians must address is whether or not to nurse, and it is one that cannot be answered with any clear substantiating evidence. For most chemicals, a safe level in milk cannot be assigned. Even more equivocal is the choice between avoiding exposure to chemicals in breast milk and obtaining the "beneficial factors" in human milk.

The decision to breast feed has become a popular one: 53% of mothers in 1976, compared with 25% in 1974 [Wilson et al, 1980]. Presumably, this option has been exercised in the face of a compendium of biological findings (see Table I). However, although these scientific data are impressive, no epidemiologic evidence testifies in their favor in a comprehensive way. The majority opinion has been reached by personal and medical evaluation of piecemeal data on the benefits of human milk.

The advantages of breast feeding have been summarized in many reviews [Jeliffe, 1976; Goldman, 1976]. Some of the physiological parameters are listed in Table I. Proteins, fats, and carbohydrates are actually synthesized in the mammary gland [Wilson

TABLE II. Properties of Xenobiotics That Affect Their Concentration in Human Milk

Property	Effect on milk (plasma partition by type of xenobiotic compound)	
	Ionizable	Neutral
pKa		
"Acidic" (<pH 7–7.25 of milk)	Decrease	
"Basic" (>pH 7.35 of plasma)	Increase	
Protein binding		
Plasma > milk	Decrease	Decrease
Plasma < milk	Increase	Increase
Erythrocyte bound	Decrease	Decrease
Molecular weight (>200)	Decrease	Decrease
Mammary metabolism	Decrease	Decrease
Route of absorption by milk from plasma		
Direct (intercellular cleft)	?	?
Dipolar membrane (intracellular diffusion)	?	?
Ionic strength of milk or plasma	Affects	?
Lipid solubility		Increase
Blood flow to breast and tissue storage compartments	Affects	Affects

After Wilson et al [1980].

et al, 1980; Vorherr, 1974]. Immunological factors in human milk have received particular attention, although their mode of action in the newborn is unclear. Transport of high molecular weight proteins, ingested from milk, across the infant gut is unlikely. Interaction of immunoproteins with mucous membranes, throughout the infant nasal and gastrointestinal system, may be responsible for producing immunological factors in the infant [Goldman, 1976; Hanson and Johansson, 1970; Ogra and Dayton, 1979].

Passive transfer,[1] which may account for some of the beneficial components in human milk, accounts for all of the potentially detrimental chemicals found therein. It is a process involving diffusion of a compound from the blood, within capillaries surrounding the secretory mechanism of the mammary gland, into the milk. Since it relies on relative solubility in these compartments, the extent of passive transfer of a substance is determined by its chemical properties and by the chemical constitution of blood and milk. The efficiency of the process is often depicted in the milk–plasma (M/P) ratio, indicating the relative concentration of a chemical in maternal milk and blood. The phenomenon has been elaborated in the context of a pharmacokinetic model by Wilson et al [1980]. Table II summarizes the critical biochemical parameters.

CHEMICALS IN BREAST MILK

Concern with chemicals in breast milk implies the existence of a dose response. That is, above a certain concentration in breast milk, a chemical will be ingested by an infant at a level that will be harmful. Similarly, a "no response" level is implied. That is,

[1]Active transfer may also be involved in absorption of some chemicals by milk.

TABLE III. Apparent Half-Life of Chemicals in Breast Milk

Chemicals	Half-life
Ethanol[a]	3 hr
Theophylline[b]	4–8 hr
Tc (salt or EDTA)[c]	4 hr
I (as MAA)[c]	20 hr
Prednisolone[d]	23 hr
Lead[e]	~ 13 weeks
PCBs[f,g] and p,p′-DDE[g]	5–8 months

[a]Kesaniemi [1974].
[b]Yurchak and Jusko [1976].
[c]Wyburn [1973].
[d]McKenzie et al [1975].
[e]Ryu et al [1978].
[f]Kodama and Ota [1980].
[g]Yakushiji et al [1978]; 5–6 months on fat basis; whole basis $t_{1/2}$ = 8.4 months.

below a certain concentration, the amount of chemical ingested by an infant is unlikely to be harmful. Both the acute, usually short-term, and chronic effects must be considered. Chronic effects are usually long term and often involve accumulation of toxic agents in the body. Perhaps individual chemicals should be considered within the context of the many exogenous chemicals that comprise a normal human body burden, since there may be interactive effects. However, just as there is a paucity of data to provide an upper, or danger, limit for chemicals in breast milk, there are no data to substantiate any detrimental effects of multiple chemicals at low levels.

Various agencies have published guidelines for safe levels of chemicals in foods (for example [WHO, 1973]). Recommended dietary limits are extrapolated from studies of animal toxicity and from reported poisoning episodes in humans, in circumstances of accidental or occupational exposure. Most standards are developed for adults, and their applicability to infants, who may be more or less susceptible to chemical toxicity, is a moot question. It has been suggested that women with pesticide or halogenated hydrocarbon residues at high levels, for example above the 90th percentile of the normal population, refrain from breast feeding [Barr, 1981].

Quite a large number of compounds, including drugs and environmentally derived chemicals, have been measured in breast milk [Wilson et al, 1980; Rogan et al, 1980; Yaffe and Waletsky, 1976]. The review by Wilson and co-workers [1980] is especially useful in assessing the danger of various drugs in breast milk. Their review suggests that in many cases a safe nursing period can be estimated from milk–plasma ratios and maternal clearance times for pharmacological substances.

For example, many chemicals are rapidly discharged from maternal milk (Table III), corresponding to maternal excretion rate of the chemicals. More rapid clearance of a drug from milk is proportional to (1) greater water solubility (direct urinary excretion) and (2) ease of metabolism (excretion of water-soluble metabolites) in the mother. Slower clearance occurs with drugs that (1) are not water soluble, (2) are not readily metabolized, and (3) are stored in the body (thus overriding the direct fecal excretion route). The lack of water solubility often means that a substance is lipid soluble and is thus stored in fat depots, eg, PCBs. During lactation, transfer to milk occurs readily due to its high fat content (Scheme 1). For lead, which can be water soluble, storage in the bone and slow mobilization may account for slow clearance of an accumulated body burden.

Scheme 1

	$[C_A]$	Q_A	$[C_B]$	Q_M	$[C_M]$
Compartment:	adipose tissue	$\leftrightarrows$	blood	$\rightleftarrows$	milk
Carrier:	70–90% fat		0.2–0.5% lipproteins		2–4% fat

Partition of lipid-soluble chemicals to milk
Milk–plasma ratio $= C_M/C_B$

Q = blood flow, C = chemical concentration

Of particular concern are chemicals that occur at higher concentrations in milk than in maternal plasma, ie, where the M/P ratio is high. This occurrence is made possible, even with a nominal blood-to-milk transfer rate (diffusion), since blood flow to the mammary tissue exceeds milk production 400–500 times [Wilson et al, 1980; Vorherr, 1974].

In the examples listed in Table IV, chemicals have a M/P ratio less than 1 where extensive plasma binding occurs or where the chemical has a low pKa (eg, aspirin, M/P = 0.6–1). Warfarin, an anticoagulant with a pKa equivalent to aspirin (5), binds extensively to plasma (97%) and is not transferred to milk (M/P $<$ 0.02 [Wilson et al, 1980; Yaffe and Waletzky, 1976]). The molecular weight of heparin (17,000) prevents its appearance in milk.

Theobromine (present in chocolate), caffeine, and theophylline belong to the same chemical family (xanthines). Using the model of Wilson et al [1980], the theoretical M/P ratios can be calculated from the pKa: theobromine 2.24, caffeine 1.20, theophylline 2.20, at milk pH 7.0. The differences observed in the actual partition (Table IV) can be accounted for by the extent of plasma binding, 20–25% for theobromine and caffeine vs 50% for theophylline.

Erythromycin (M/P $\sim$ 3) and penicillin (M/P $<$ 0.2) are two antibiotics whose transference to milk is quite different. Erythromycin is not ionized in the blood, and its pKa (8.8) predisposes partition to milk (see Table II). The penicillins, with low pKa (2–3), exist in an ionized form in plasma, which impedes transfer to milk [Wilson et al, 1980].

Neutral, fat-soluble, poorly metabolized chemicals, such as PBBs, PCBs, and various pesticide residues, have high M/P ratios, which typify their accumulation in the environment toward the top of the food chain (Scheme 1).

OCCUPATIONAL VERSUS BASE POPULATION LEVELS OF CHEMICALS IN HUMAN MILK

Mercury

Although the instances of mercury contamination of human milk do not constitute occupational exposure, they may be expected to represent the worst case example

TABLE IV. Ratio of Chemical Concentration in Breast Milk (Whole Basis) to That in Maternal Blood (M/P Ratio)

Chemical	M/P ratio
Salicylate[a]	0.35
Lithium[b]	0.4
Caffeine[c]	0.5
Theobromine[d]	0.8
Theophylline[e]	0.7
Phenobarbital[a]	0.7
Methadone[f]	0.8
Ethanol[g]	0.8–1
Antipyrine[h]	1
Mercury	
United States[h]	0.9
Japan[i]	0.1
Iran[d] (organic)	0.03
Lead[j]	≤ 1
Tetrachlorethylene[k]	~ 3
Polybrominated biphenyls (PBB)[l]	~ 3
Polychlorinated biphenyls (PCB)[m]	4–10
Dieldrin[n]	~ 6
BHC[o]	4–5
DDT residues[o]	6–7

[a]Berlin [1980].
[b]Schou and Amdisen [1973].
[c]Tyrala and Dodson [1979].
[d]From Wilson et al [1980].
[e]Yurchak and Jusko [1976].
[f]Blinick et al [1975].
[g]Kesaniemi [1974].
[h]Pitkin et al [1976].
[i]Fugita and Takabatake [1977].
[j]See Table VI.
[k]Bagnell and Ellenberger [1977].
[l]Brilliant et al [1978].
[m]Occupational exposure or not; Kodama and Ota [1980]; Yakushiji et al [1978, 1979]; Kuwabara et al [1978].
[n]Curley and Kimbrough [1969]; Eckenhausen et al [1981]; Siddiqui et al [1981].
[o]Curley and Kimbrough [1969].

of mercury poisoning. Such episodes occurred in Japan in 1953 and 1970 and in Iran in 1971–1972. Severe damage to the central nervous system, including the cerebral cortex, were typical effects. In Minimata, death occurred in 46/121 cases following ingestion of fish, and 23/359 children born in the vicinity had a disease resembling cerebral palsy [Nelson, 1971].

Pitkin et al [1976] in the United States reported that 100 nonexposed women had 0.9 ± 0.2 (SE) μg/liter total mercury in their milk, compared with 3.6 μg/liter reported in Japanese women [Fugita and Takabatake, 1977] (Table V). A possible explanation was the greater intake by the Japanese of fish, known to be contaminated with mercury, mainly organic mercury. As Pitkin pointed out, the discrepancies in various studies may also have been due in part to analytical technique. Of particular interest is the

TABLE V. Mercury Concentrations in Human Milk and Blood and in Children's Blood (Means)

Source	N	Breast milk μg/liter (ppb)	Mother's blood ng/gm (ppb)	Cord blood ng/gm
Minimata, newborns[a]	14		4.2	6.2
Minimata, 1968; 1973[b]		63; 500–540		
Japanese, 1977[b]	34	3.6	25	26
United States, 1976[c]	100	0.9 ± 0.2 (SE)	1.0	1.2
United States, 1974[d]	~900		8.7	11.5
Belgium, 1977[e]	440		12	14
Sweden[c]	5		6.0	7.7
Japan, 1971[f]	9		16	20
Iran[g] (organic Hg)	19	29 ± 29 (SD)	870 ± 750	
	11		740 ± 790[g]	1020 ± 1190[g]
	11		850 ± 730[h]	700 ± 350[h]

[a]Nelson [1971].
[b]Fugita and Takabatake [1977] and citations therein.
[c]Pitkin et al [1976] and citations therein.
[d]Baglan et al [1974].
[e]Buchet et al [1978].
[f]Suzuki et al [1971] and citations therein, as calculated by Pitkin.
[g]Bakir et al [1973], estimated from data reported by Wilson et al [1980]. Infants exposed by milk and in utero (mean age 3 months).
[h]Infants exposed only by maternal milk.

difference in maternal M/P ratios (Table IV). Whereas the United States findings suggested a M/P ratio of about 1 for total mercury, the ratio in the Japanese study was approximately 0.1. It is interesting to speculate that the difference is again due to the high proportion of organic mercury in the Japanese. Inorganic mercury is approximately equally distributed in plasma and red blood cells [Friberg, 1969]. But organic mercury binds very efficiently to erythrocytes (over 90% [Clarkson, 1972]), where it may be less available for transfer to milk than inorganic mercury. This speculation is further supported by data on mercury poisoning in Iran (reported by Wilson et al [1980]). Very high levels of organic mercury were found in mothers who ate contaminated grain, and the M/P ratio was about 0.03. In one individual followed over several months, the M/P ratio of total mercury was 0.1 initially and 0.3 5 months later. Here, the increased M/P ratio might be due to the marked decrease in concentration of maternal organic mercury during this period. The half-life in this woman was about 40 days for total mercury. The half-life of methyl mercury in humans has been reported as about 70 days [Nelson, 1971; Clarkson, 1972].

In contrast with milk, transplacental migration of methyl mercury is facile. Clarkson [1972] reported that fetal blood had twice the level in maternal blood. For humans, total mercury in fetal blood is evidently greater than or about equal to (1.1 times) maternal blood (Table V). However, the data of Bakir et al (reported by Wilson et al [1980]) show organic mercury in blood, from 11 infants with suspected prenatal exposure, to be 1.7 ± 1.0 (SD) times that of their mothers, close to the predicted ratio [Clarkson, 1972]. For 11 infants with only postnatal (breast milk) exposure, blood organic mercury was 0.7 ± 0.4 times the mothers' blood.

The recommended exposure limits for mercury incorporate a designated 10-fold safety factor [Clarkson, 1972; Nelson, 1971; Friberg, 1969]. Thus 20 μg/liter blood mercury (the MAC: maximum allowable concentration) or 30 μg/day intake (0.4 μg/kg body weight) is considered dangerous in an adult. The "safe" intake level is equivalent to approximately 2 μg/day for a 5-kg infant. The data in Tables V and X suggest that normal United States women have safe levels of mercury in breast milk, but that breast milk with mercury levels greater than 4 μg/liter may exceed the safe intake level for an infant.

Mercury levels in milk above 4 μg/liter would very likely be associated with maternal blood levels above 20 μg/liter (Table V). Several of the studies cited in Table V might exceed these guidepoints, *both of which also correspond to the MAC.* Mercury in the organic form has been implicated in poisoning episodes, so that a predominance of this form is assumed in designating toxicity.

The two citations, Iran and Minimata, where overt mercury toxicity was observed, exceed these allowances by 7–135 times (milk), or 40–50 times (average maternal or cord blood)—1–2 orders of *magnitude*—coincidentally the inherent safety factor. Hence, maternal and infant blood exceed the *danger* level by 4–5 times (Iran) (Table V). In the Iranian incident, application of the current United States standards for mercury intake *or* blood level would preclude breast feeding.

For both mercury and lead, newborn blood concentration is comparable to that in the mother. Since the blood volumes differ, the body burdens are not necessarily equivalent. If blood were the only compartment, the neonatal body burden would be about 1/20 that of the mother. Using a milk-blood partition of 0.01–0.1 for mercury, it can be estimated that the body burden would then double during 4–40 days of breast feeding (100% absorption). Therefore, even with a very low milk-blood partition, the cumulative dosage from breast feeding could theoretically be significant.

Lead

Lead has been reported in normal human milk at levels from 2 to 30 μg/liter (ppb) (Table VI). According to one survey [Johnson et al, 1981], an infant diet including formula or cow's milk has less than 1 ppb (μg/kg) lead. Lead in milk and maternal and infant blood has been investigated in the case of an ex-battery worker (Table 7) [Ryu et al, 1978]. The range of blood lead concentration in mother and child during the first 3 weeks after birth were similar (30–70 μg/dl), and these levels were high compared with normal infants (6 ± 3) (Table 6). The upper limit of normal in children is considered to be 30 μg/dl, a level requiring consideration of possible lead poisoning [Ratcliffe, 1981]. Evidence of subacute lead toxicity has been reported among exposed workers with blood lead below 80 μg/dl, the maximum allowable concentration in occupational exposures [Lilis et al, 1978; Ratcliffe, 1981].

In the exposed worker (Table VII), the range of lead concentrations in *milk* was similar to that in controls supplied by the author, which is surprising since her *blood* lead levels were approximately 3 times that of controls. Moreover, in the lead-exposed case, the M/P ratio was less than 0.1. The M/P ratio was not reported for the controls, but if the data provided by the authors were representative, then the M/P partition for normal persons would be around 0.3. This case study showed fluctuations in maternal blood lead during pregnancy and 8 months postpartum. Blood lead values during pregnancy ranged from 33 to 57 μg/dl, lower values occurring during the ninth month. Lead exposure (employment) had been terminated after 6 months of pregnancy. Indeed, blood lead values, following removal from exposure, predictably decrease, lead being slowly transferred to bone tissue [Ratcliffe, 1981]. Typically, the half-life of

TABLE VI. Lead in Maternal Milk, Blood, and in Their Children's Blood

Reference	N	Breast milk μg/liter (ppb) mean (range)	Maternal blood μg/dl (ppb × 10^{-1})	Children's blood μg/dl
Dillon et al [1974], United States	29	26 (6–58)		
Ryu et al [1978]				
Unexposed	9	29 (15–65)		
	900		9.9 (2–29)	
	31			6.1 ± 2.6
Ex-battery worker 1–3 wks postpartum	1	(19–63)	(33–60)	(30–70)
Buchet et al [1978]	442		10.1 (3–31)	8.3 (2.7–27)[a]
Baglan et al [1978]	84, 130		16.5 ± 14	12.3 ± 19.8
Larsson et al [1981], Sweden	41	2 (0.5–9)[b]		
WHO: Safe level[c]		51		

[a]Cord blood.
[b]Median.
[c]"Safe" upper limit for adults: less than 50 μg/kg body weight per week: 50 μg/kg/wk ÷ 7 days/wk ÷ 0.7 kg milk/day × 5 kg infant body weight = μg/kg (liter) milk allowed.

TABLE VII. Lead Concentrations in Maternal Milk, Blood, and Infant's Blood in the Case of Past Lead Exposure (Ryu et al [1978])

	Lead concentration[a]		
Weeks after birth	Maternal milk (μg/liter)	maternal blood (μg/dl)	Infant's blood[b] (μg/dl)
0–4	19–63	33–43 (39)	35–70 (50)
4–8	36–62	52–57 (55)	28–45 (32)
8–16 (11–15 for milk)	(24)	42–57 (49)	28–31 (30)
16–32	4–14	38–62 (49)	18–22 (21)

[a]Values for milk as reported. Values for blood estimated from graph. Means in parentheses.
[b]Infant was being treated with calcium gluconate.

lead in blood is 2–5 weeks [Ratcliffe, 1981], when it is being transferred to bone. However, during lactation, in the case reported by Ryu et al [1978], blood lead varied (Table VII) *above* the nadir observed during late pregnancy, suggesting mobilization of lead from tissue stores. The apparent decrease in lead concentration in milk might be interpreted as depletion of stored lead. More than 90% of the lead body burden is in bone tissues, but it has been suggested that one-fourth of the lead in blood represents resorption from bone [Ratcliffe, 1981]. Buchet et al]1978] tentatively suggested that stored lead might be mobilized during pregnancy.

The pharmacokinetic model for lead appears to be complex. While milk concentrations showed a trend of decreasing, blood lead varied between 33 and 62 μg/dl, so that the M/P ratio was decreasing during 8 months of lactation (from ca 0.1 to ca 0.02) (Table VII). Lead is strongly bound to hemoglobin, which may greatly impede partition to milk. Plasma levels of lead normally constitute a small proportion of lead in blood (~5%). Experimentally, the plasma concentration accounted for 80% of a dosage after 20 min, but only 10% after 2 hr [Ratcliffe, 1981]. The partition of lead between blood compartments may thus determine its milk partition. Determination of plasma lead levels may be important in understanding this phenomenon. Moreover, the difference between M/P ratio of lead in a worker vs the general population may be due to differences in blood binding in such different exposures.

Other Metals

Various trace metals have been reported in human milk. Cadmium (Table X) is an interesting example. Buchet et al [1978] reported cadmium and carboxyhemoglobin levels in maternal and cord blood. Cigarette smoking, by the mother, strongly influenced the levels of both cadmium and carboxyhemoglobin in mother and infant, as well as reduced fetal weight. Smoking might affect such compounds in human milk, as well. Nicotine has been detected in milk from mothers who smoked as few as 1–4 cigarettes per day [Yaffe and Waletsky, 1976].

Solvents

The pharmacokinetic behavior of solvents is similar to that of inhalation anesthetics. Hydrocarbons and halogenated hydrocarbons (eg, carbon tetrachloride, chlorinated ethane or ethylene) are absorbed, mainly through the respiratory route, with almost infinite uptake possible by adipose tissue. Upon cessation of exposure, excretion occurs (1) directly via the lungs and (2) by metabolites in urine. Depending on volatility, respiratory excretion is usually complete within 8–24 hr, although traces may remain in adipose tissue. However, compounds that are not readily metabolized remain in the body in spite of their volatility. For example, tetrachloroethylene (PCE) is detectable in breath as long as 2 weeks following exposure [Stewart et al, 1970]. This commonly used cleaning solvent was the source of a reported family exposure, where severe neonatal jaundice was attributed to ingestion of PCE in breast milk [Bagnell and Ellenberger, 1977]. The father, who worked in a drycleaning plant, had 30 μg/ml (ppm) PCE in blood. Two hours after visiting him at work, the mother had 3 μg/ml in blood. Her milk had 10 and 3 ppm PCE 1 and 24 hr after exposure. The infant recovered when breastfeeding was discontinued.

Halothane, a commonly used inhalation anesthetic, was determined in breast milk (0.8–2 ppm) of an anesthetist following exposure in an operating room, where levels were 0.27–1.5 ppm in air [Cole et al, 1976]. Carbon disulfide (CS_2) is a volatile solvent, commonly used in extrusion of viscose rayon. It is a potent neuro- and cardiovascular poison, and the recommended 8-hr exposure limit in the United States is 30 mg/M^3 (10 ppm) [ACGIH, 1979]. Concentrations in this range (37–56 mg/M^3) were recently associated with menstrual disturbances and pregnancy toxemia in Chinese women [Cai and Bao, 1981]. In one infant, whose mother had worked on the day of delivery, the cord blood had 50 μg/liter CS_2. The level of CS_2 in breast milk, during or after work, was 21–306 μg/liter, with a mean value of 85 (46 samples from 17 women). In 10 cases, milk collected in the morning before work had an average CS_2 concentration of 48.5 μg/liter (0–126), reduced from 110 μg/liter (22–306) after work the previous day. CS_2 was detected in urine in 5/10 infants being breast-fed (16–71 μg/liter). It may be significant that CS_2 was found on the hands and uniforms of the women, as well as in their milk.

TABLE VIII. General Population Versus Occupational Exposure to Typical Organochlorine Breast Milk Contaminants

	Breast milk concentration (whole basis, μg/liter)		Corresponding blood serum levels (μg/liter)		Typical occupational blood levels (μg/liter)	
	Mean	Range	Mean	Range	Mean	Range
Total DDT	~100[a]	(20–830)	15[a]	(3–60)	108[b]	(8–705)
					591[c]	(110–2200)
Dieldrin	6[d,e]	(0.5–70)	1[d]	(<0.1–6)	27[b]	(0–350)
					68[f]	(3–220)
PCBs[g]	~45[h]	(~<10–150)	9[i]	(0.4–69)	~5–500[j]	
	87[e]	(50–4091)				
PBBs	~2[k]	(~0.3–36)			43[l]	(0–1240)
	~11[m]	(~1–23)	2.1[m]	(0.2–9)	600[n]	(10–1700)
			1.1[i]	(0.2–120)	16[n]	(1–500)

[a]Composite estimate from Curley and Kimbrough [1969]: mean milk 70 (40–160); blood 10 (3–20); Siddiqui et al [1981]: mean milk 127 (20–300); blood 20 (10–60); Wilson et al [1973]: mean milk 170 (<20–830). Blood levels similar to those found for p.p-DDE by Wolff et al [1982a] for 431 Michigan females: mean blood serum 13 (1.2–69).
[b]Morgan and Roan [1974].
[c]Laws et al [1967].
[d]Curley and Kimbrough [1969].
[e]Savage [1976]. Mean dieldrin (fat basis) 164 ppb, range (whole basis) 0.5–70.
[f]Brown et al [1964].
[g]Cf. Occupational PCB exposures where milk PCB were ~400 ppb, blood ~40 ppb [Kuwabara et al, 1978] and 230–470 ppb milk, 20–57 ppb blood [Yakushiji et al, 1978].
[h]Wickizer et al [1981] for the general population of Michigan estimated from concentrations given on fat basis, mean 1.5 ppm (0.3–5.1 ppm); similar to mean reported by Polishuk et al [1977] 44 ppb whole basis, and Yakushiji et al [1979] 39 ppb whole basis.
[i]Data of Wolff et al [1982a] for the general population of Michigan. PCBs: 527 females ≥ 18 years old, median 6 ppb. PBBs: N = 622, median 0.4 ppb, in serum.
[j]Blood levels and PCB isomer constitutions vary widely. 5–500 ppb is typical for Aroclor 1254 or 1260 exposure [Wolff et al, 1982b and unpublished data]. Levels up to 2500 ppb have been reported in blood for lower chlorinated PCBs (eg, Wolff et al, [1982b]).
[k]Brilliant et al [1978]. Data for women from the lower peninsula of Michigan, which were higher than for the upper peninsula. Levels were estimated from data reported on fat basis, mean 68 μg/liter (<10–1200).
[l]Landrigan et al [1980]. Data for 216 chemical workers and their families, median 4.5 ppb.
[m]Estimated from data in [k] above, for a subset of paired milk and serum values.
[n]Wolff et al [1979]. Mean of 600 ppb for 10 PBB production workers (male), median 108. Mean of 16 for 45 nonproduction workers, median 6.

Halogenated Hydrocarbons

These compounds have been widely found in the environment, including human tissues and breast milk [Rogan et al, 1980; Savage, 1976; Yaffe and Waletsky, 1976; Kroger, 1972]. Several typical pesticide residues in breast milk, along with polychlorinated and polybrominated biphenyls (PCBs and PBBs), are summarized in Tables VIII and X.

For many of these chemicals, no association has been found between blood (or adipose or, therefore, milk) concentrations and human health effects, eg, PCBs, PBBs, and DDT residues.

In animals most of these compounds are carcinogenic and cause alterations in reproductive systems and hepatic function, which has led to concern about long-term health effects. Even with continued low-level exposure from environmental sources, these compounds accumulate in the body, since they are lipid soluble and cannot be effectively metabolized.

The amount of organochlorine residues accumulated by an infant depends upon (1) exposure in utero, (2) concentration in breast milk (both related to maternal body burden), and (3) duration of breast feeding.

Cord blood concentrations are a measure of prenatal exposure or transplacental transfer. PCB concentrations in cord blood are approximately one-fourth that in the mother (Table XI). Similar ratios have been reported for DDT residues [Eckenhausen et al, 1981; Siddiqui et al, 1981]. Somewhat higher cord-to-maternal blood ratios of aldrin, dieldrin, and lindane were found in the same studies, 0.8–1. Of interest in this regard is the fact that dieldrin, like organic mercury, binds more to the erythrocyte fraction of blood than do the less polar PCBs and DDT [Morgan et al, 1972]. This phenomenon may be implicated in greater placental transfer for such chemicals.

In any case, cord blood concentrations reflect the neonatal body burden. For a neonatal blood concentration one-fourth that of the mother, it can be estimated that the relative body burden, newborn-to-mother, is approximately 1:150 (maternal blood = 61; maternal fat = 18 kg; 3-kg infant blood = 300 ml; infant fat = 450 gm; fat-to-blood ratio = 300:1; milk-to-blood ratio = 10:1). For the case presented by Yakushiji et al [1978], the mother would thus have a total of 290 mg PCBs, the infant 2.2 mg. Her breast milk (PCB concentration 400 μg/liter) would, therefore, cause the infant's body burden to double within the first week of life (ingestion of 0.7 liter/day). Wickizer et al [1981] calculated that the infant body burden of PCBs, expressed as milligrams per kilogram, would double within 2 months, based on the average PCB level in human milk in Michigan. Intake of PCBs and organochlorine pesticide residues from synthetic infant formulae is nil [Johnson et al, 1981]. For PCBs, and probably other nonpolar compounds of this class, breast milk is potentially a much greater source of exposure than transplacental migration.

An infant's body burden increases with longer nursing. Several studies of PCBs have demonstrated this point (Table IX). Children nursed longer than 3 months had higher PCB concentrations in blood than their mothers. Conversely, nursing reduces the maternal body burden of organochlorine compounds, and hence also transplacental and breast milk exposure of successive siblings. In the example cited previously, where an unusual volume of breast milk was produced during laboratory studies, the maternal body burden eventually became less than one-third that at partition (200 mg were excreted in a total of 818 liters of milk during 16 months) [Yakushiji et al, 1978]. Multiparous women have been shown to have lower levels of PCBs and other pesticide residues in milk than primiparas [Kroger, 1972; Yakushiji et al, 1979; Savage, 1976].

The concentration of organochlorine residues in milk (see Table IV) is close to that theoretically predicted by diffusion directly from adipose tissue. That is, the level in milk fat is practically that in adipose tissue. In two studies that reported individual values, the ratio of adipose (fat basis) to milk fat concentration was 1.1 (from data in Yakushiji et al [1978] [PCBs]; Brilliant et al [1978] [PBBs]). Of course, transfer from adipose to milk occurs indirectly via the blood compartment (Scheme I). The efficiency

TABLE IX. Blood PCB Concentration, According to Duration of Breast Feeding, in Children Compared With Their Mothers

	Duration of breast feeding Blood PCB concentration, ppb mean ± SD (N)		
	< 1 month or none	1 to < 3 months	≥ 3 months
Occupational exposure[a]			
Mothers	45.3 ± 28.5 (6)	35.5 ± 19.8 (8)	21.9 ± 13.4 (5)
Children	5.8 ± 5.8 (11)[b]	12.6 ± 6.9 (13)[c]	32.8 ± 33.8 (8)[d]
No occupational exposure			
Mothers[e]	2.5 ± 0.6 (5)	3.9 ± 0.9 (3)	2.5 ± 0.4 (9)
Children[e]	0.2 ± 0.4[f]	3.3 ± 0.4[g]	6.0 ± 3.5[h]
Mothers[i]		1.8 ± 2.6 (15)	1.3 ± 3.3 (27)
Children		2.4 ± 0.9	2.5 ± 1.2
Mothers[j]	2.7 ± 1.4 (9)	2.5 ± 2.0 (46)	
Children	1.6 ± 2.2[k] (15)	3.6 ± 2.3[l] (42)	4.5 ± 3.7 (32)[m]

[a]Kuwabara et al [1978].
[b]Age of children: 4 ± 4 years.
[c]Age of children: 4 ± 2 years.
[d]Age of children: 8 ± 3 years.
[e]Kuwabara et al [1979].
[f]4/5 PCBs were not detected; 0 used as value; age of children 0.4 ± 0.1 year.
[g]Age of children: 0.9 ± 1 year.
[h]Age of children: 3 ± 2 years.
[i]Masuda et al [1978]. The two age groups of children were 2–3 and 4–12 months.
[j]Kodama and Ota [1980].
[k]Age of children: < 3 months (bottle fed).
[l]3 months.
[m]1 year—mothers same as 3 month group, not retested.

of the process is due to efficient blood flow to both adipose (12 liter/hr; Mapleson [1963]) and mammary tissue (30–40 liter/hr; Vorherr [1974]). Both greatly exceed milk production, which is about 0.03 liter/hr. Therefore, adipose tissue and milk fat essentially attain equilibrium concentrations of dissolved organochlorine residues.

Adipose-to-milk transfer may also be estimated from the M/P partition ratios. Again, the adipose–milk fat ratio approaches 1. A M/P partition of 6 (for whole milk) represents a milk fat–plasma (MF/P) partition of about 200. The M/P partition is thus the same as the adipose-to-blood (or serum) partition, which for most of the chemicals in this group has been reported in various studies as 200–300.

$$\text{Milk fat/plasma partition} = \text{adipose/plasma partition}$$
$$MF/P = A/P$$

$$M/P = 6;\ M = 0.03 \times MF;\ (MF \times 0.03)/P = 6;\ MF/P = 6/0.03 = 200$$
$$A/P = 200 \quad \therefore\ A/MF = 1$$

The idea of "direct" transfer assumes a blood carrier, common to both adipose-blood and milk-blood diffusion. There is some evidence to suggest that the blood car-

TABLE X. Breast Milk Concentrations of Chemicals and Estimated Infant Daily Intake, Compared With Recommended Limits (Based on 5-kg Infant Ingesting 0.7 liter (kg) per Day)

Chemical	Population		Reported level in milk (μg/liter)	Calculated intake (μg/day)	Allowable daily intake (μg/day)
Mercury	United States[a]		0.9 ± 0.2 (SE)	0.6	2[b]
	Japan[c]		3.6	2.5	
	Poisoning[d]		30–540	20–380	
Lead	United States[e]		26–29	18–20	< 36[f]
	1 occupational exposure[g]		63	44	
	Sweden[h]		2	1	
Chromium[i]			~ 0.4	0.27	5–15
Cadmium[j]			3–35		
PCB	Occupational exposure[k]		~ 400	~ 300	5[l]
	General population[m]	Mean	~ 45	32	
		Highest	150	107	
DDT	General population[n]	Mean	70–170	50–120	25[o]
		High	160–830	100–600	
Dieldrin	General population[p]	Mean	2–7	1–5	0.5[o]
		High	70	50	
Heptachlorepoxide	General population[q]	Median	2–9	1–6	2.5[o]
		High	39	27	

γ-BHC	General population[r]	Mean	10; 40	7–30	62[o]
Oxychlordane	General population[s]	Mean	~3	~2	5[o]
		High	37.4	26	

[a]Pitkin et al [1976].
[b]10 times less than "safe" = 0.03 mg/70-kg adult = 0.4 μg/kg body weight = 2 μg/day for 5-kg infant.
[c]Fugita and Takabatake [1977].
[d]See Table V.
[e,g,h]See Table VI.
[f]WHO [1973]. < 50 μg/kg body weight per week ÷ 7 days/week × 5-kg body weight = 36 μg/day.
[i]Kumpulainen and Vuori [1980].
[j]Schulte-Lobbert and Bohn [1977].
[k]Cf. Table VIII. Yakushiji et al [1978]; Kuwabara et al [1978].
[l]Rogan et al [1980]. 1 μg/kg body weight (for adults) × 5-kg body weight (infant).
[m]Cf. Table VIII. The mean value estimated from the data of Wickizer et al (45 μg/liter) is similar to that of Yakushiji et al [1979], 39 μg/liter for Japanese women, and Polishuk et al [1977] 44 ppb.
[n]Cf. Table VIII.
[o]WHO [1973].
[p]Cf. Table VIII. Range of means reported by Curley and Kimbrough [1969], Polishuk et al [1977], Eckenhausen et al [1981], and Savage [1976]. The latter mean (164 ppb, fat basis) corresponds to approximately 5 ppb in whole milk. The high value reported by Savage [1976] was 70 ppb on a whole milk basis.
[q]Means reported by Curley and Kimbrough [1969] and Polishuk et al [1977]. The mean (fat basis) reported by Savage [1976] was 91 ppb, corresponding to approximately 3 ppb whole basis. The high value reported by Savage was 39 ppb (whole basis).
[r]Means reported by Polishuk et al [1977] and Siddiqui et al [1981].
[s]Savage [1976]. Mean (fat basis) 96 ppb ~3 ppb whole basis. High was that reported on whole basis.

TABLE XI. PCB Concentrations in Umbilical Cord and Maternal Blood

	PCB in ppb, mean ± SD	
N	Cord blood	Maternal blood
1[a]	16	57
60[b]	0.61 ± 0.39	2.5 ± 1.1
76[c]	1.1 ± 1.0	4.5 ± 2.9

[a]Case of an occupationally exposed woman, Yakushiji et al [1978].
[b]Masuda et al [1978].
[c]Kodama and Ota [1980].

rier for nonpolar substances is the low-density lipoprotein fraction of serum [Morgan et al, 1972; Roboz et al, 1979]. The blood carrier, calculated from adipose–blood partition ratios, is equivalent to that calculated from milk–blood partition.

$$A/P = 200\text{–}300 = \frac{\%\ \text{lipid, adipose}}{\%\ \text{carrier, blood}} = \frac{60\text{–}90\%}{\%\ \text{carrier}};\ \%\ \text{carrier} = \frac{60\text{–}90}{200\text{–}300} = 0.2\text{–}0.45\%$$

$$M/P = 6\text{–}10 = \frac{\%\ \text{milk fat}}{\%\ \text{carrier, blood}} = \frac{1\text{–}4\%}{\%\ \text{carrier}};\ \%\ \text{carrier} = \frac{1\text{–}4}{6\text{–}10} = 0.1\text{–}0.7\%$$

The blood carrier, calculated from adipose–plasma partition, is very close to the normal lipoprotein component of human blood. The range calculated from M/P ratio is somewhat wider, although it encompasses the same values. Individual variations in M/P ratios reflect the marked variations in the proportion of milk fat, both in individuals and in populations. Individual milk fat level may change during a feeding, during the day, with diet, and other unknown factors [Mes and Davies, 1978]. In the various population studies cited in this review, milk fat averages ranged from 1 to 4%.

The maternal body burden of chemicals, reflected in milk concentration, is determined by accumulated dietary, as well as occupational, exposure in the mother. In a study of milk from women without occupational exposure, increasing levels of PCBs were associated with concomitantly increasing levels of other organochlorine residues, suggesting similar dietary sources for such chemicals [Yakushiji et al, 1979]. In this investigation of Japanese women, marine products had been found to be the major source for both PCB and DDT residues. A similar trend can be found for some, but not all, pesticide residues in other studies [Savage, 1976; Siddiqui et al, 1981].

Substantial data on organochlorine chemicals in breast milk from occupational exposures are absent. However, it may be helpful, in evaluating potential contamination of milk from an occupational exposure, to compare typical blood concentrations in occupational exposures with blood and milk concentrations reported in various normal populations (Table VIII). A striking feature of these data is that the upper range of chemical levels in blood, reported for so-called unexposed persons (general population), encompasses the lower range of levels reported in occupationally exposed groups.

For dieldrin, levels greater than 130 μg/liter in blood have been associated with "intoxication," seizures, and CNS changes among workers [Brown et al, 1964 and others]. The highest dieldrin level reported by Savage in 1976 (12,300 μg/liter fat basis) would correspond roughly to 40 μg/liter in blood (using 3% milk fat and fat blood partition of 300 or a M/P partition of 9). Unless infants are more susceptible to acute toxic effects of dieldrin than adults, a dangerous level in blood would apparently require marked bioaccumulation, even with high levels in milk.

DDT residues have been reported in human tissues throughout the world, including many studies of breast milk contamination. p,p′-DDE, which generally constitutes 80–90% of DDT residues, is practically ubiquitous in humans. Even though DDT usage ceased in the United States after 1972, some environmental contamination still exists. DDT residues concentrations in people have since declined in the United States and Canada, especially among youth [Kutz et al, 1978; Currie et al, 1979], but whatever body burden of DDT exists is not likely to diminish significantly during a person's lifetime. Human health effects, even at unusually high exposures, have not been definitely established. However, this chemical is representative of the class of halogenated aromatic hydrocarbons, and is generally present in humans at higher levels than any other residue of this class.

PCBs have been associated with a multisystem toxic syndrome in Japan (Yusho) where ingested rice oil had high concentrations of PCBs (≤ 1000 ppm) as well as other toxic contaminants (1–5 ppm chlorinated dibenzofurans; ≤ 1000 ppm chlorinated quaterphenyls). Patients with symptoms had ingested as little as 111 mg PCB (29 μg/kg/day) along with the other contaminants [Kuratsune, 1980]. According to one investigator, two of six breast-fed infants, with no prenatal exposure, had symptoms of Yusho disease. All six mothers, who had ingested the contaminated oil after parturition, exhibited symptoms [Yoshimura, 1974]. One mother, with symptoms, was supposed to have ingested 1.2 gm PCBs during lactation, and her infant's symptoms appeared 1–2 months later. Using a model assuming 100% absorption into adipose and 1:1 partition to milk with 3% milk fat, it can be estimated that the infant would ingest 100 mg in 1 month. Adjusted for body weight, these dosages would be comparable in mother and infant, and would approximate the minimum effective dosage reported by Kuratsune [1980].

In one case of occupational exposure to PCB, breast milk, maternal and infant blood were studied for 16 months [Yakushiji et al, 1978]. The cord blood:maternal blood ratio at birth (1:3.6) reflected the transplacentally conferred body burden. The infant, who was not breast fed, had similar blood levels at 4 and 16 months (3–4 μg/liter), compared with 16 μg/liter (ppb) in the cord blood. The decrease in the child's blood is consistent with dilution of the initial body burden by the adipose tissue, which increases with growth, although some superimposed body burden of PCB may also have occurred through the diet. The PCB level in milk declined from 0.5 ppm (14–16 ppm fat basis) to 0.1 (3.8 ppm fat basis), while the maternal blood level changed from 57 to 12 μg/liter and adipose from 16 to 4 ppm (μg/gm) over the same period. This woman was described as "a chronic case of PCB intoxication." Her estimated body burden (290 mg, see above) and that excreted in milk (200 mg) can be simply interpreted to be a dosage (a) sufficient to have caused the mother's "PCB intoxication" and (b) to therefore be potentially dangerous to an infant (who was not breast fed). A more reasonable interpretation of point (b) must take into account the unusual volume of breast milk collected during the experimental studies and the very long period of nursing. If the milk volume were reduced by one-third and the nursing period reduced to 6 months, the total amount of PCBs in milk would be closer to 40 mg (almost 20 times the same infant's neonatal body burden as estimated above). Adjusted for body weight, the total body burden would still be comparable in mother and child. Whether or not this dosage, compared with the Yusho case of 111 mg, would be toxic cannot be judged. Compared with the United States allowable daily intake (ADI, Table X), a 40-mg dose would be approximately 50 times too high. Again, the potential toxicity cannot be judged.

It has long been recognized that environmental PCB residues, especially near the top of the food chain, are qualitatively different from synthetic PCB mixtures. Biodegradation of certain PCB congeners is facile, whereas isomers with chlorine substitution of the type shown in Scheme 2 are more resistant to metabolism. These PCB congeners accumulate and constitute almost all of the PCBs found in humans, including those in human milk [Yakushiji et al, 1979].

Scheme 2

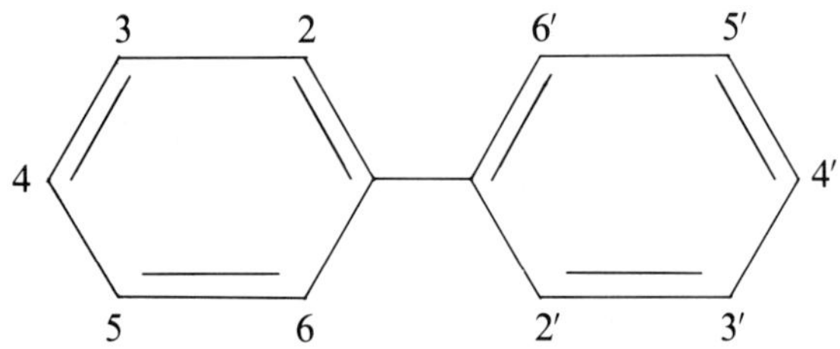

2,4–
3,4–
2,3,4–
2,3,5–
2,4,5–
2,3,4,5–

Chlorine substitution patterns that predominate in PCBs found in human tissues (1 pattern present on each biphenyl ring)

A PCB mixture, constituted to simulate the PCB congeners reported by Yakushiji et al [1979], was found to be 7 times as potent as a similar commercial PCB in its ability to induce hepatic aromatic hydrocarbon hydroxylase (AHH) [Parkinson et al, 1980], a measure related to animal toxicity in various studies. The authors suggested that current regulations might require readjustment, since they have been based on toxicological studies of commercial PCB mixtures. Inducibility of AHH in humans has been reported with extraordinary PCB and DDT exposure, although the impact on human health is unknown. It is interesting that recent evidence shows that children less than 10 years old have more drug-metabolizing enzymes proportional to body size than do older persons [C&E News, Nov. 23, 1981]. Higher activity of such enzymes (including AHH) may be protective (in accelerating removal of chemicals from the body) or may be detrimental (by activating chemicals to a more toxic form).

Polybrominated biphenyls (PBB), once used as fire retardants, are chemically similar to PCBs. The chemical widely contaminated humans and livestock in Michigan, following its accidental introduction into dairy feed [Landrigan et al, 1979; Wolff et al, 1982a]. Measurement of PBB in breast milk suggested that the chemical had been widely disseminated across the state [Brilliant et al, 1978], a finding later confirmed in a larger study [Wolff et al, 1982a]. The levels in the general population were low compared with farmers and chemical workers (Table VIII). However, health effects have been reported among persons with low PBB levels, and it has been difficult to assign a safe level of exposure.

The chemicals summarized in Table X are representative of the variety of environmental contaminants reported in breast milk, which are also commonly found in all of us. In many cases, the concentrations normally observed in human milk would cause the daily intake to exceed recommended limits, which were not specifically designed for infants. The lowest mean concentrations of organochlorine compounds in breast milk (Table X) represent 0.1 (γ-BHC) to 6 (PCBs) times the allowable daily intake. The highest values given (Table X) represent 5–100 times the suggested dietary limits. If the effects of chemicals in milk were additive, the lowest average levels would be approximately 11 and the highest 160 times the suggested daily limits. The dietary estimates were determined with the knowledge that such chemicals accumulate in the body, and attempted to incorporate a margin of safety. WHO states that the allowable daily intake (ADI) is a "...daily dosage...that, during an entire lifetime, appears to be without appreciable risk..." [WHO, 1973].

It may be construed from the mean blood concentrations in Table VIII that occupational exposures typically exceed those in the normal population by 10–300 times. Since the levels of the organochlorines in milk normally represent 2–8 times the allowable daily intake, then for average occupational exposures the milk would exceed the ADI by 20–2400 times, or 1–3 orders of magnitude. In fact, the level of PCBs in milk in a Japanese woman, occupationally exposed, was approximately 400 μg/liter, 10 times the mean level in normal Japanese women's milk, 39 μg/liter, and 60 times the ADI [Yakushiji et al, 1978, 1979]. Breast milk from women with occupational exposures represented by the highest blood values (50–1700 times normal) could exceed the ADI by 2–4 orders of magnitude. If the margin of safety of the ADI were known, or if a definite upper (danger) limit for daily exposure could be specified, an individual and her physician would have some basis for deciding whether or not to breast feed.

CONCLUSIONS

Contamination of breast milk with several chemicals of industrial origin has been described in various reports. Can such measurements be used to evaluate the safety of breast feeding? The only unequivocal example is mercury. Levels in milk above 4 μg/liter, equivalent to maternal blood above 20 μg/liter, would exceed the current United States guidelines for maternal or infant exposure (MAC or maximum allowable concentration). The MAC was designed to incorporate a 10-fold safety factor, and in two examples cited, where mercury poisoning occurred, the maternal milk and blood and infant blood were 10–100 times the MAC. Thus, the MAC would appear to provide appropriate limits for breast feeding.

In normal human milk, organochlorine pesticide and PCB residues exceeded the ADI by less than 10 times (mean values, Table X). Highest values were 5–100 times the ADI. Occupational exposures were projected to reach 70–70,000 times the ADIs. The ADI is intended to avoid long-term risks. For PCBs and dieldrin, where short-term (acutely) toxic dosages could be roughly estimated, normal human milk seemed unlikely to provide an acutely toxic level in an infant, even at the highest levels observed. However, among women occupationally exposed to PCBs, even a conservative clinical approach would dictate extreme caution in breast feeding. The decision could be greatly facilitated by measurement of maternal blood (or fat) PCB level. Further, the effects of Yusho disease became evident within a relatively short term. The long-term health effects of PCBs retained in the body, or of PCB-related contaminants, are not known. It is just that risk that the ADI seeks to protect.

In occupational circumstances, a single substance is likely to overwhelm the normal body burden of other organochlorine residues. In normal women, several different residues are present. A simple additive model (used earlier) showed a total milk concentration (organochlorine compounds in Table X) approximately 200 times the ADI at the highest reported levels in normal women. By analogy with the previous argument, this is unlikely to achieve an acutely toxic dosage. But these chemicals are likely to become a permanent body burden in the infant, and whatever originates from milk is superimposed on an infant's body burden at birth. It was estimated that breast feeding would double an infant's body burden after approximately 1 week (total residues) or 2 months (corrected for increasing body weight). Therefore, potential long-term health effects of dosages above the ADI may be the greatest concern. The margin of safety of the ADI for long-term effects seems to urgently require investigation. Alternatively, does short-term exposure, 10–100 times the ADI, merely constitute an insignificant excursion? It may incur little risk if exposure during the infant's remaining lifetime is very low.

In the absence of suitable standards, the problem of solvent exposure and breast milk contamination can be fairly readily evaluated for individual compounds. Such exposures should clearly be avoided while breast feeding. Even so, most solvents can be expected to clear maternal tissues within a few days, which can be estimated from various human and animal studies.

Finally, estimation of risk necessitates some estimate of exposure, ideally maternal body burden. Women at risk should preferably be aware of their body burden *prior* to nursing. Measurement of a chemical in blood is usually an adequate indicator of possible acutely toxic levels in milk. The matter of long-term effects, or toxicity from accumulated industrial chemicals, is at present entirely conjectural. For those organochlorine compounds that constitute a lifetime body burden, the ADI—the WHO promulgated guideline most commonly used—would render most of us, whatever our occupation, unsuitable for nursing.

ACKNOWLEDGMENTS

The excellent and patient assistance of Julia Roberts and Judy Marmor in preparation of the manuscript is gratefully acknowledged.

REFERENCES

Baglan RJ, Brill AB, Schulert A, Wilson D, Larsen K, Dyer N, Mansour M, Schaffner W, Hoffman L, Davies J (1974): Utility of placental tissue as an indicator of trace element exposure to adult and fetus. Environ Res 8:4–70.

Bagnell PC, Ellenberger HA (1977): Obstructive jaundice due to a clorinated hydrocarbon in breast milk. Can Med J 117:1047–1048.

Barr M (1981): Environmental contamination of human breast milk. Am J Public Health 71:124–126.

Berlin CM (1980): The excretion of drugs in human milk. In "Drug and Chemical Risks to the Fetus and Newborn." New York: Alan R. Liss, pp 115–127.

Blinick G, Inturrisi CE, Jerez E, Wallach RC (1975): Methadone assays in pregnant women and progeny. Am J Obstet Gynecol 121:617–621.

Brilliant LB, Amburg GV, Isbister J, Humphrey H, Wilcox K, Eyster J, Bloomer AW, Price H (1978): Breast-milk monitoring to measure Michigan's contamination with polybrominated biphenyls. Lancet 2:643–646.

Brown VKH, Hunter CG, Richardson A (1964): A blood test diagnostic of exposure to aldrin and dieldrin. Br J Ind Med 21:283–286.

Buchet JP, Roels H, Hubermont G, Lauwerys R (1978): Placental transfer of lead mercury, cadmium, and carbon monoxide in women. II. Influence of some epidemiological factors on the frequency distributions of the biological indices in maternal and umbilical cord blood. Environ Res 15:494–503.

Cai SX, Bao YS (1981): Placental transfer, secretion into mother milk of carbon disulphide and the effects on maternal function of female viscose rayon workers. Ind Health 19:15–29.

Clarkson TW (1972): Recent advances in the toxicity of mercury with emphasis on the alkylmercurials. Crit Rev Toxicol 1:203–234.

Cole CJ, Kenepp NB, Reed SB, Strobel G (1976): Trace concentrations of halothane in human breast milk. Br J Anaesth 48:541–543.

Curley A, Kimbrough R (1969): Chlorinated hydrocarbon insecticides in plasma and milk of pregnant and lactating women. Arch Environ Health 18:156–164.

Currie RA, Kadis VW, Breitkreitz WE, Cunningham GB, Bruns GW (1979): Pesticide residues in human milk, Alberta, Canada – 1966–70, 1977–78. Pestic Monit J 13:52–55.

Dillon HK, Wilson DJ, Schaffner W (1974): Lead concentrations in human milk. Am J Dis Child 128: 491–492.

Eckenhausen FW, Bennett D, Beynon KI, Elgar KE (1981): Organochlorine pesticide concentrations in perinatal samples from mothers and babies. Arch Environ Health 36:81–92.

Friberg LT, chairman (1969): Maximum allowable concentrations of mercury compounds. Arch Environ Health 19:91–905.

Fugita M, Takabatake, E (1977): Mercury levels in human maternal and neonatal blood, hair and milk. Bull Environ Contam Toxicol 18:205–209.

Goldman AS (1976): Immunologic aspects of human milk. In Waletzky LR (ed): "Symposium on Human Lactation." Arlington, VA, DHEW Publ No. 79-5107, pp 49–58.

Hanson LA, Johansson BG (1970): Immunological studies of milk. In McKenzie HA (ed): "Milk Proteins Chemistry and Molecular Biology." New York: Academic Press, Vol 1.

Jeliffe EFP (1976): Nutritional aspects of human lactation. In Waletsky LR (ed): Symposium on Human Lactation." Arlington, VA, DHEW Pub. No. HSA 79 5107.

Johnson RD, Manske DD, New DH, Podrebarac DS (1981): pesticides and other chemical residues in infant and toddler total diet samples – II – 1975–76. Pestic Monit J 15:39–50.

Kesaniemi M (1974): Ethanol and acetaldehyde in the milk and peripheral blood of lactating women after ethanol administration. J Obstet Gynecol Br Comm 81:84–86.

Kodama H, Ota H (1980): Transfer of polychlorinated biphenyls to infants from their mothers. Arch Environ Health 35:95–100.

Kroger M (1972): Insecticide residues in human milk. J Pediatr 80:401–405.

Kumpulainen J, Vuori E (1980): Longitudinal study of chromium in human milk. Am J Clin Nutr 33: 2299–2302.

Kuratsune M (1980): Yusho. In Kimbrough R (ed): "Halogenated Biphenyls, Terphenyls, Naphthalenes, Dibenzodioxins and Related Products." Amsterdam: Elsevier/North Holland Biomedical Press.

Kutz FW, Yobs AR, Strassman SC, Viar JF (1978): Effects of reducing DDT usage on total DDT storage in humans. Pestic Monit J 11:61–63.

Kuwabara K, Yakushiji T, Watanabe I, Yoshida S, Koyama K, Kunita N (1979): Levels of polychlorinated biphenyls in blood of breast-fed children whose mothers are non-occupationally exposed to PCBs. Bull Environ Contam Toxicol 21:458–462.

Kuwabara K, Yakushiji T, Watanabe I, Yoshida S, Koyama K, Kunita N, Hara I (1978): Relationship between breast feeding and PCB residues in blood of the children whose mothers were occupationally exposed to PCBs. Int Arch Occup Environ Health 41:189–197.

Landrigan PJ, Wilcox KR, Silva J, Humphrey HEB, Kauffman C, Heath CW (1979): Cohort study of Michigan residents exposed to polybrominated biphenyls: Epidemiologic and immunologic findings. Ann NY Acad Sci 320:284–294.

Landrigan P (1980): General population exposure to environmental concentrations of halogenated biphenyls. In Kimbrough R (ed): "Halogenated Biphenyls, Terphenyls, Naphthalenes, Dibenzodioxins and Related Products." Amsterdam: Elsevier/North Holland Biomedical Press.

Larsson B, Slovach SA, Hagman U, Hofrander Y (1981): WHO collaborative breast feeding study. II.

Levels of lead and cadmium in Swedish human milk, 1978–1979. Acta Paediatr Scand 70:281–284.
Laws ER, Curley A, Biros FJ (1962): Men with intensive occupational exposure to DDT. Arch Environ Health 15:766–775.
Lilis R, Fischbein A, Diamond S, Anderson HA, Selikoff IJ, Blumberg WE, Eisinger J (1977): Lead effects among secondary lead smelter workers with blood lead levels below 80 μg/100 ml. Arch Environ Health 32:56–266.
Mapleson WW (1963): An electric analogue for uptake and exchange of inert gases and other agents. J Appl Physiol 18:197–204.
McKenzie SA, Selley JA, Agnew JE (1975): Secretion of prednisolone into breast milk. Arch Dis Child 50: 894–896.
Masuda Y, Kagawa R, Kuroki H, Kuratsune M, Yoshimura T, Taki I, Kusuda M, Yamashita F, Hayashi MM (1978): Transfer of polychlorinated biphenyls from mothers to foetuses and infants. Food Cosmet Toxicol 16:543–546.
Mes J, Davies DJ (1978): Variation in the polychlorinated biphenyl and organochlorine pesticide residues during human breast feeding and its diurnal pattern. Chemosphere 44:699–706.
Morgan DP, Roan CC (1974): Liver function in workers having high tissue stores of chlorinated hydrocarbon pesticides. Arch Environ Health 29:14–17.
Morgan DP, Roan CC, Paschal EH (1972): Transport of DDT, DDE, and dieldrin in human blood. Bull Environ Contam Toxicol 8:321–326.
Nelson N (1971): Hazards of mercury. Environ Res 4:1–69.
Ogra PL, Dayton DH (eds) (1979): "Immunology of Breast Milk." New York: Raven Press.
Parkinson A, Robertson LW, Safe S (1980): Reconstituted human breast milk PCBs as potent inducers of aryl hydrocarbon hydroxylase. Biochem Biophys Res Commun 96:882–889.
Pitkin RM, Bahns JA, Filer LJ, Reynolds WA (1976): Mercury in human maternal and cord blood, placenta, and milk. Proc Soc Exp Biol Med 151:65–567.
Polishuk ZW, Ron M, Wassermann M, Cucos S, Wasserman D, Lemesch C (1977): Organochlorine compounds in human blood plasma and milk. Pestic Monit J 10:121–127.
Ratcliffe JM (1981): "Lead in Man and the Environment. New York: John Wiley and Sons.
Roboz J, Suzuki RK, Bekesi JG, Holland JF, Rosenman K, Selikoff IJ (1979): Mass spectrometric identification and quantification of polybrominated biphenyls in blood compartments of exposed Michigan chemical workers. J Environ Pathol Toxicol 3:363–378.
Rogan WJ, Bagniewska A, Damstra T (1980): Pollutants in breast milk. N Engl J Med 302:1450–1453.
Ryu JE, Ziegler EE, Fomon SJ (1978): Maternal lead exposure and blood lead concentration in infancy. J Pediatr 93:476–478.
Savage EP (1976): National study to determine levels of chlorinated hydrocarbon insecticides in human milk. EPA Contract 68-01-3190.
Schou M, Amdisen A (1973): Lithium and pregnancy-III: Lithium ingestion by children breast-fed by women on lithium treatment. Br Med J 2:138.
Schulte-Lobbert FJ, Bohn G (1977): Determination of cadmium in human milk during lactation. Arch Toxicol 37:155–157.
Siddiqui MKJ, Saxena MC, Bhargava AK, Seth TD, Krishna Murti CR, Kutty D (1981): Agrochemicals in the maternal blood, milk, and cord blood: A source of toxicants for prenates and neonates. Environ Res 24:24–32.
Stewart RD, Baretta ED, Dodd HC, Torkelson TR (1970): Experimental human exposure to tetrachloroethylene. Arch Environ Health 20:224–229.
Suzuki T, Miyama T, Katsunuma H (1971): Comparison of mercury contents in maternal blood, umbilical cord blood, and placental tissues. Bull Environ Contam Toxicol 5:502–508.
Tyrala EE, Dodson WE (1979): Caffeine secretion into breast milk. Arch Dis Child 54:787–800.
Vorherr H (1974): "The Breast." New York: Academic Press, p 71.
WHO: Pesticide residues in food (1973): WHO technical report series No. 525. Report of the 1972 joint FAO/WHO meeting. Geneva.
Wickizer TM, Brilliant LB, Copeland T, Tilden R (1980): Polychlorinated biphenyl contamination of nursing mothers' milk in Michigan. Am J Public Health 71:132–136.
Wilson DJ, Locker DJ, Ritzen CA, Watson JT, Schaffner W (1973): DDE concentrations in human milk. Am J Dis Child 125:814–817.
Wilson JT, Brown RD, Cherek DR, Dailey JW, Hilman B, Jobe PC, Manno BR, Manno JE, Redetzki HM Stewart JJ (1980): Drug excretion in human breast milk. Clin Pharmacokinet 5:1–66.
Wolff MS, Anderson HA, Selikoff IJ (1982a): Human tissue burdens of halogenated aromatic chemicals in Michigan. J Am Med Assoc 247:2112–2116.

Wolff MS, Fischbein A, Thornton J, Rice C, Lilis R, Selikoff IJ (1982b): Body burden of polychlorinated biphenyls among persons employed in capacitor manufacturing. Int Arch Occup Environ Health 49:199–208.

Wolff MS, Anderson HA, Camper F, Nikaido MN, Daum SM, Haymes N, Selikoff IJ, Aubrey BB (1979): Analysis of adipose tissue and serum from PBB (polybrominated biphenyl)-exposed workers. J Environ Pathol Toxicol 2:1397–1411.

Wyburn JR (1973): Human breast milk excretion of radionuclides following administration of radiopharmaceuticals. J Nucl Med 14:115–117.

Yaffe SJ, Waletzky LR (1976): Drugs and chemicals in breast milk. In Waletzky LR (ed): "Symposium on Human Lactation." Arlington, VA, DHEW Publ No. 79-5107, pp 77–93.

Yakushiji T, Watanabe I, Kuwabara K, Yoshida S, Koyama K, Kunita N (1979): Levels of polychlorinated biphenyls (PCBs) and organochlorine pesticides in human milk and blood collected in Osaka prefecture from 1972 to 1977. Int Arch Occup Environ Health 43:1–15.

Yakushiji T, Watanabe I, Kuwabara K, Yoshida S, Koyama K, Hara I, Kunita N (1978): Long-term studies of the excretion of polychlorinated biphenyls (PCBs) through the mother's milk of an occupationally exposed worker. Arch Environ Contam Toxicol 7:493–504.

Yoshimura T (1974): Epidemiological study on Yusho babies born to mothers who had consumed oil contaminated by PCB. Fukuoka Igka Zasshi 65:74–80.

Yurchak AM, Jusko WJ (1976): Theophylline secretion into breast milk. Pediatrics 57:518–520.

REPRODUCTIVE TOXICOLOGY SURVEILLANCE

American Journal of Industrial Medicine 4:285–291 (1983)

Surveillance of Pregnancy Loss in Human Populations

Allen J. Wilcox, MD, PhD

Surveillance of pregnancy loss as a way to detect hazardous exposures is attractive in principle. However, there are no established methods for monitoring pregnancy loss in humans. Surveillance is difficult because most loss occurs within the first 12 weeks of gestation, when pregnancy may not be documented or even recognized. Three possible approaches to surveillance of pregnancy loss are discussed here: direct surveillance of recognized loss, direct surveillance of unrecognized loss, and indirect surveillance of all loss. These approaches vary in cost, technique, power, and interpretability. While these surveillance methods appear potentially useful, their effectiveness in practice has not yet been determined.

Key words: population surveillance, spontaneous abortion, fetal death

INTRODUCTION

Exposure to toxic agents can increase the risk of pregnancy loss. Thus, the surveillance of pregnancy loss may be useful for identifying groups of persons exposed to toxic substances. Furthermore, the latency period between exposure and reproductive damage is potentially short, so that surveillance of pregnancy loss may provide earlier evidence of toxic effect than can chronic diseases such as cancer. While the surveillance of pregnancy loss is attractive in principle, there are no established methods for doing so. The purpose of this paper is to describe feasible approaches to such surveillance.

Pregnancy loss (that is, the failure of conception to produce a live birth) can occur before pregnancy is even recognized, or it can be manifest as spontaneous abortion or stillbirth. This paper describes three possible approaches to the surveillance of pregnancy loss in human populations: direct surveillance of recognized pregnancy loss, direct surveillance of unrecognized pregnancy loss, and indirect surveillance of loss

Biometry and Risk Assessment Program, National Institute of Environmental Health Sciences, Research Triangle Park, North Carolina.

Address reprint requests to Dr. Allen J. Wilcox, Biometry and Risk Assessment Program, National Institute of Environmental Health Sciences, Research Triangle Park, NC 27709.

0271-3586/83/0401-02-0285$02.50 **© 1983 Alan R. Liss, Inc.**

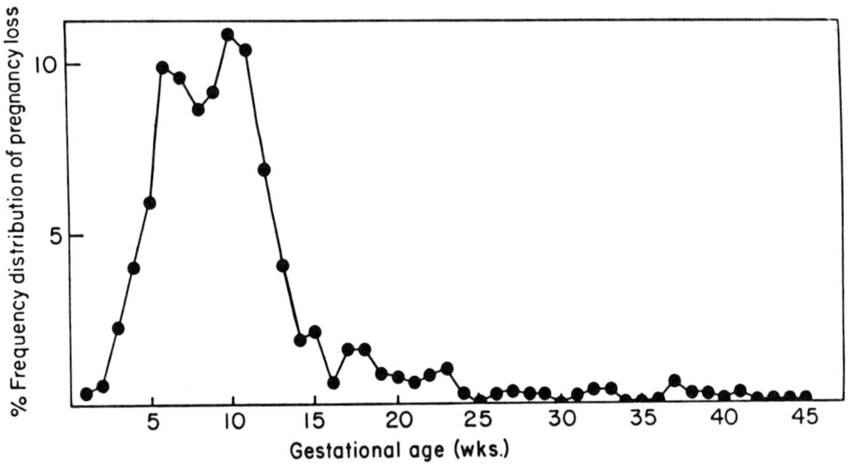

Fig. 1. Frequency distribution of 723 losses among 4351 pregnancies, by gestational age (Menstrual and Reproductive Health Study).

through the analysis of patterns of live births. While these reflect a common purpose, they represent very different research strategies.

DIRECT SURVEILLANCE OF RECOGNIZED PREGNANCY LOSS

Self-recognized pregnancy loss was recorded in a long-term prospective study of white college-educated women (Menstrual and Reproductive Health (MRH) Study) [Treloar et al, 1967; Wilcox et al, 1981]. Figure 1 shows the pattern of 723 losses occurring among 4351 pregnancies. Pregnancy losses are customarily divided into two categories: loss that occurs before the fetus is presumed mature enough to survive outside the uterus, and loss that occurs after. Losses in the first category are known as spontaneous abortions (miscarriages), and are by far the preponderance of recognized pregnancy loss. In Figure 1, spontaneous abortions before 20 weeks of gestation represent 91% of all recognized pregnancy loss. The smaller portion of pregnancy loss consists of late fetal deaths (stillbirths). In the MRH Study, which includes pregnancies over a 40-year period, the overall risk of stillbirth was 17.3/1000. Risk of stillbirth is falling steadily and stands at 9.9/1000 in the United States [Vital Statistics, 1981].

Pregnancy loss late in pregnancy is, in effect, monitored by law: report of a stillbirth is required as part of the collection of vital statistics, and summaries are routinely published at state and national levels. In contrast, the collection of spontaneous abortion data is more erratic, and accurate estimates of its occurrence have proven far more difficult.

Monitoring Spontaneous Abortion

The usual criterion for a spontaneous abortion is pregnancy loss before 20 or 28 weeks of gestation. From a statistical standpoint, the exact gestational criterion is not crucial, since most loss has occurred by 20 weeks (Fig. 1). In the MRH study, 15.2% of all recognized pregnancies with known gestational age ended before the 20th week. In other special studies, the occurrence of spontaneous abortion is typically 10 to 15% [Leridon, 1977].

One method for monitoring spontaneous abortion would be to extend vital registration by law to include every recognized pregnancy loss, regardless of gestational age. However, where it has been attempted, compulsory registration has been difficult to enforce [Tietze, 1953]. Early losses are presumably underreported in the eight United States states that require registration of all pregnancy losses: the occurrence of reported spontaneous abortions ranges from 7% to virtually zero [Vital Statistics, 1980a, Vital Statistics, 1980b].

Another approach would be to monitor medical records. Spontaneous abortion may or may not require hospitalization; thus, complete ascertainment of medically documented abortions among a group of women requires access to all medical records—private practice and clinic, as well as hospital. Where such access exists, medical records have proven useful in the study of spontaneous abortion [Shapiro et al, 1962]. However, medical records are rarely so accessible.

Furthermore, it is not known what proportion of spontaneous abortions are brought to the attention of physicians, and how that proportion might vary by socioeconomic or other factors. Perhaps a third of the spontaneous abortions recognized by women occur before the second missed menstrual period, and thus before the time when most pregnancies are medically documented (Fig. 1). Even in a special study where women were encouraged to enroll at the first suspicion of pregnancy, early recognized spontaneous abortions were underregistered [French and Bierman, 1962].

Yet another approach would be to monitor all spontaneous abortions as they are recognized by women themselves. The MRH Study [Treloar et al, 1967] offers a possible model for doing so. In this study, daily records of menstrual flow are kept on a calendar card by participants. When a woman misses a period, she records any known or suspected reason on the back of the card. As a result, all pregnancies are documented from the time of the first missed menstrual period.

Without objective criteria for pregnancy, there is a question of how reliable or valid such records can be. In the MRH Study, the pattern of occurrence of spontaneous abortion has been stable over several decades. However, in particular situations, subjective factors might disrupt this measure of spontaneous abortion. Women who fear they or their husbands have been exposed to some hazard may be more quick to interpret a late menstrual period as an early spontaneous abortion. The availability of menstrual intervals, as well as self-reported pregnancies from menstrual history records, would allow exploration of such bias.

To extend this line of reasoning, it may be possible to develop a statistical analysis of menstrual cycle variability that detects the menstrual irregularities of early pregnancy loss, regardless of women's reports of pregnancy. However, the susceptibility of menstruation to disruption from many sources (stress, illness, injury, etc) probably makes menstruation by itself an impractical outcome measure.

A crude estimate of the statistical power of menstrual history surveillance can be made by assuming that there are two groups of women being monitored: one exposed and one unexposed to a possibly hazardous substance. (This oversimplifies for the sake of illustration; the statistical analysis of surveillance data presents special problems not pursued here.) If we assume each group consists of 500 women between the ages of 20 and 40 having the age distribution and fertility of the United States population as a whole, then the surveillance of menstrual histories of these women for 1 year would yield 60 self-recognized pregnancies, and could detect a 2.8-fold increase in risk of abortion ($\alpha = 0.05$, $\beta = 0.10$) [Schlesselman, 1974]. Monitoring those two groups for 2 years would permit a 2.2-fold increase in abortion risk to be detected.

DIRECT SURVEILLANCE OF UNRECOGNIZED PREGNANCY LOSS

Pregnancy loss may occur before a woman even suspects she is pregnant. The loss may be indistinguishable from menstruation, given the natural variability in length of the menstrual cycle and duration of flow among normal women [Treloar et al, 1967]. Accordingly, surveillance of earliest pregnancy loss depends on two things: a biological assay to determine early pregnancy, and cooperation of women in obtaining that assay.

The Extent of Unrecognized Pregnancy Loss

Until recently, there has been little basis on which to estimate the extent of early fetal loss in humans. In 1959, Hertig et al described 34 fertilized ova found in the reproductive tract of women who had had unprotected intercourse prior to hysterectomy. Of these 34 ova, 10 (or 29%) were abnormal and would presumably have resulted in very early pregnancy loss. James [1970] reanalyzed the data of Hertig, and estimated that 35% of all fertilizations are lost before pregnancy is recognized. On more speculative grounds, Roberts and Lowe [1975] proposed that three-quarters of conceptions end in unrecognized loss.

By the 1960s, immunologic assays of the pregnancy hormone human chorionic gonadotropin (hCG) were developed to the point that direct epidemiologic measures of early pregnancy loss could be attempted [Udry et al, 1971]. However, the early immunologic tests cross-reacted with other glycoprotein hormones, and were relatively insensitive to low levels of hCG. In the 1970s, the more sensitive radioimmune and radioreceptor assays offered new possibilities for epidemiologic research [Kosasa et al, 1973; McCready et al, 1978; Landesman and Saxena, 1976; Saxena et al, 1977]. The radioimmune assay (RIA) is of particular interest because it is reactive to the unique beta-subunit of the hCG molecule, and therefore has little cross-reactivity with luteinizing hormone or other similar hormones.

In the first reported application of a radioimmune assay on an epidemiologic scale, Miller and his colleagues [1980] performed urine assays on 197 women beginning on the 21st day of each menstrual cycle and thereafter on alternate days. In 6 months of surveillance, they diagnosed 152 pregnancies, of which 43% were lost by the 20th week of gestation. Three-quarters of those losses occurred unknown to the women. Among all clinically recognized pregnancies, the occurrence of spontaneous abortions was 14%. Based on this study, the current best estimate is that for every recognized spontaneous abortion, there are three unrecognized early fetal losses. The observation that 33% of conceptions ended in unrecognized loss is surprisingly close to the estimates of Hertig et al [1959] and James [1970].

Recent developments in hCG assay include an RIA specific to the carboxy-terminal peptide of the beta-subunit of hCG, creating a test that has virtually no reactivity to non-hCG molecules [Wehmann et al, 1981]. Additionally, the means are available to extract and concentrate hCG, leading to a urine test that is up to 10 times more sensitive than the routine RIA blood test [Wehmann et al, 1981]. Although this new generation of tests is not yet commercially available, the tests are simple enough to be used in wide-scale studies and should be able to produce even more accurate estimates of early pregnancy loss.

With increasingly specific and sensitive assays for hCG, the limits of early pregnancy detection by these tests may be the natural production of hCG itself. HCG is thought to be produced by trophoblastic tissue beginning about 1 week after conception. The possibility of even earlier pregnancy detection has been raised with the discovery of an "early pregnancy factor," which apparently alters maternal immune response [Morton et al, 1977; Shaw and Morton, 1980]. This factor has been detected in the serum of humans and other mammals as early as 6 to 24 hr after conception. An epidemiologically useful assay of this substance has yet to be developed.

Given that the means are at hand to monitor very early fetal loss, there is a question as to whether or not it would be useful to do so. Is unrecognized early pregnancy loss an extension of spontaneous abortion, or does it represent a biologically different end point? More specifically, do the two types of early pregnancy loss behave the same or differently with respect to reproductive toxins? If there is a difference in this regard between unrecognized and recognized pregnancy loss, then the added burden of biochemical monitoring may be justifiable. This question can be addressed by descriptive epidemiologic studies of unrecognized early fetal loss.

The statistical power of monitoring early fetal loss through hCG assay can be roughly compared to the power of menstrual history surveillance. The same assumptions about fertility, sample size, and statistical test characteristics are made, and in addition it is assumed that half of the women who are trying to become pregnant volunteer for surveillance before conceiving. About 38 women would be monitored in each group in the first year, and 70 women in each over 2 years (including volunteers who fail to conceive). Based on an estimated 55 conceptions in 1 year and 102 conceptions in 2 years, this method would permit the detection of 1.9- and 1.7-fold increases in risk, respectively. Thus, compared to menstrual history monitoring, this method would involve fewer women more intensively at greater expense, with clearer documentation of outcome and somewhat more statistical power.

INDIRECT SURVEILLANCE OF PREGNANCY LOSS THROUGH THE ANALYSIS OF PARITY

While intensive surveillance may be attractive from a scientific standpoint, it is not always practical. Especially for routine, wide-scale monitoring, data collection must be simple, cheap, and quick. The need for a surveillance tool that meets these criteria has led several authors to propose the analysis of parity as a means of assessing reproductive function [Levine et al, 1980; Levine et al, 1981; Wong et al, 1979; Dobbins et al, 1978]. A couple's parity (number of live births) is a function of ability to conceive and risk of pregnancy loss, among other things. Furthermore, information on live births is perhaps the most accessible and reliable of all reproductive data [Selevan, 1980]. Thus, in theory, the surveillance of parity offers a convenient means of assessing both fertility and pregnancy loss for large groups of people at modest expense. Furthermore, it is suitable for retrospective studies as well as surveillance. In practice, it has not yet been satisfactorily established how such data should be analyzed nor how useful the results might be in screening for toxins, given the variety of other factors that influence parity.

Levine et al [1980, 1981] have proposed an analysis of parity in which expected births (based on United States rates for maternal age, birth year, parity, and race) are compared to observed births. The authors call this ratio of observed to expected the Standardized Fertility Ratio (SFR), where fertility is defined in terms of live births. In this sense of the word, fertility is affected by failure of pregnancy as well as failure of fertilization. The authors suggest that by comparing SFRs of groups exposed and unexposed to occupational (or other) toxins, it may be possible to detect exposure-related differences in reproduction.

Patterns of live births reflect a mixture of physiologic, social, and behavioral influences that typically cannot be separated. Levine et al assume all extraneous influences on fertility and pregnancy loss are independent of employment. This is unlikely to be true for female employees, whose decisions regarding pregnancy can be influenced by employment and vice versa. Other limitations of this method have been noted by Harber [1981]. Dobbins et al [1978] have taken a different approach to this problem: They propose a "survivorship" analysis of the time from one birth to the next. While this approach also appears promising, a complete presentation remains to be published. All its advantages notwithstanding, the use of parity as a reproductive outcome faces unresolved problems of analysis and interpretation.

CONCLUSION

In principle, the surveillance of pregnancy loss offers a valuable tool for the study of environmental toxins. Three potential approaches to the surveillance of pregnancy loss have been described here: surveillance of recognized pregnancies, surveillance of unrecognized pregnancies, and surveillance of live births. These differ markedly in their technique, cost, power, and interpretability. One feature these methods do have in common is that they are untested. How useful these methods will prove to be, and under what circumstances, remains to be demonstrated.

REFERENCES

Dobbins JG, Eifler CW, Buffler PA (1978): The use of parity survivorship analysis in the study of reproductive outcome. Am J Epidemiol 108:245 (abstract).

French FE, Bierman JM (1962): Probabilities of fetal mortality. Public Health Rep 77:835–847.

Harber P (1981): Some questions concerning "A method for monitoring the fertility of workers." J Occup Med 23:324–325.

Hertig AT, Rock J, Adams EC, Menkin MC (1959): Thirty-four fertilized human ova, good, bad and indifferent, recovered from 210 women of known fertility. Pediatrics 23:202–211.

James WH (1970): The incidence of spontaneous abortion. Popul Studies 24:241–245.

Kosasa T, Levesque L, Goldstein DP, Taymor ML (1973): Early detection of implantation using a radioimmunoassay specific for human chorionic gonadotropin. J Clin Endocrinol Metab 36:622–624.

Landesman R, Saxena BB (1976): Results of the first 1000 radioreceptor assays for the determination of human chorionic gonadotropin: A new, rapid, reliable, and sensitive pregnancy test. Fertil Steril 27:357–368.

Leridon H (1977): Intrauterine mortality. In: "Human Fertility: The Basic Components." Chicago: University of Chicago Press, pp 48–81.

Levine RJ, Symons JM, Balogh SA, Arndt DM, Kaswandik NT, Gentile JW (1980): A method for monitoring the fertility of workers. 1. Method and Pilot Studies. J Occup Med 22:781–791.

Levine RJ, Symons JM, Balogh SA, Milby TH, Whorton MD (1981): A method for monitoring the fertility of workers. 2. Validation of the method among workers exposed to dibromochloropropane. J Occup Med 23:183–188.

McCready J, Braunstein GD, Helm D, Wade ME (1978): Modification of the choriogonadotropin beta-subunit radioimmunassay for determination of urinary choriogonadotropin. Clin Chem 24:1958–1961.

Miller JF, Williamson E, Glue J, Gordon YB, Grudzinskas JG, Sykes A (1980): Fetal loss after implantation: A prospective study. Lancet ii:554–556.

Morton H, Rolfe B, Clunie GJA, Anderson JM, Morrison J (1977): An early pregnancy factor detected in human serum by the rosette inhibition test. Lancet i:394–397.

Roberts CJ, Lowe CR (1975): Where have all the conceptions gone? Lancet i:498–499.

Saxena BB, Saito T, Said N, Landesman R (1977): Radioreceptor assay of luteinizing hormone-human chorionic gonadotropin in urine: Detection of the luteinizing hormone surge and pregnancy. Fertil Steril 28:163–167.

Schlesselman JJ (1974): Sample size requirements in cohort and case-control studies of disease. Am J Epidemiol 99:381–384.

Selevan SG (1980): "Evaluation of Data Sources for Occupational Pregnancy Outcome Studies." Doctoral dissertation. University of Cincinnati, Department of Environmental Health, College of Medicine, Ann Arbor: University Microfilms.

Shapiro S, Jones EW, Densen PM (1962): A life table of pregnancy terminations and correlates of fetal loss. Milbank Mem Fund Q 40:7–45.

Shaw FD, Morton H (1980): The immunological approach to pregnancy diagnosis: A review. Vet Rec 106: 268–269.

Tietze C (1953): Introduction to the statistics of abortion. In Engle ET (ed): "Pregnancy Wastage." Springfield, Illinois: CC Thomas, pp 135–145.

Treloar AE, Boynton RE, Behn BG, Brown BW (1967): Variation of the human menstrual cycle through reproductive life. Int J Fertil 12:77–126.

Udry JR, Keovichit S, Burnright R, Cowgill DO, Morris NM, Yamarat C (1971): Pregnancy testing as a fertility measurement technique: A preliminary report on field results. Am J Public Health 61: 344–352.

"Vital Statistics of the United States, 1976" (1980a). Vol I—Natality, USHHS, PHS, OHRST, NCHS, Hyattsville MD.

"Vital Statistics of the United States, 1976" (1980b). Vol II—Mortality, Part A, USHHS, PHS, OHRST, NCHS, Hyattsville MD.

"Vital Statistics of the United States, 1977" (1981). Vol II—Mortality, Part A, USHHS, PHS, OHRST, NCHS, Hyattsville MD.

Wehmann RE, Harman SM, Birken S, Canfield RE, Nisula BC (1981): Convenient radioimmunoassay that measures urinary human choriogonadotropin in the presence of urinary human lutropin. Clin Chem 27:1997–2001.

Wilcox AJ, Treloar AE, Sandler DP (1981): Spontaneous abortion over time: Comparing occurrence in two cohorts of women a generation apart. Am J Epidemiol 114:548–553.

Wong O, Utidjian HM, Karten VS (1979): Retrospective evaluation of reproductive performance of workers exposed to ethylene dibromide (EDB). J Occup Med 21:98–102.

American Journal of Industrial Medicine 4:293–307 (1983)

Assessment of Methods and Results of Reproductive Occupational Epidemiology: Spontaneous Abortions and Malformations in the Offspring of Working Women

Kari Hemminki, MD, Olav Axelson, MD, Marja-Liisa Niemi, CandPolSci, and Gunnar Ahlborg, MD

Epidemiological studies relating occupational exposures of working women to spontaneous abortions and malformation are reviewed and some methodological considerations are presented. The reproductive epidemiology is less developed than epidemiology in general and seems to involve some specific problems. The ascertainment of exposure and of outcome may be problematic. The exposures may be reported differently by the women depending on the outcome of the pregnancy; thus confirmation of exposure from an independent data source would be an asset. The recognition of spontaneous abortions by women is variable and this should be considered in the study design. An access to medical records could provide objective data on the outcome, but it may introduce other types of asymmetries.

The types of occupational exposures of the women, suggested to carry a risk of spontaneous abortions, include anesthetic agents, laboratory work, copper smelting, soldering, and chemical sterilization using ethylene oxide and glutaraldehyde. Maternal employment in laboratories and exposure to solvents has been linked to a risk of congenital malformations in the offspring in five studies. Data on the teratogenic effects of anesthetic gases has been conflicting. In one study, employment in copper smelting was associated with malformations in the offspring.

Key words: spontaneous abortions, malformations, occupational hazards, anesthetic gases, solvents, smelter work, ethylene oxide

INTRODUCTION

The number of employed women and particularly working mothers has increased markedly in all industrialized countries since the Second World War, promulgated by changes in industrial structure and by need for new labor force. Today women represent over one-third of the world's economically active population. About 46% of women of working age are part of the labor force [ILO, 1980]. In the Nordic countries the percentage of employed women aged 20–44 years varied among the countries from 63% to 79% at the end of 1970s [Yearbook of Nordic Statistics, 1980].

Institute of Occupational Health, Haartmaninkatu 1, SF-00290 Helsinki 29, Finland (K.H., M.-L.N.).
Department of Occupational Medicine, University Hospital, 581 85 Linköping, Sweden (O.A.).
Department of Occupational Medicine, Regional Hospital, 701 85 Örebro, Sweden (G.A.).
Address reprint requests to Dr. Kari Hemminki, Institute of Occupational Health, Haartmaninkatu 1, SF-00290 Helsinki 29, Finland.

0271-3586/83/0401-02-0293$04.50 © 1983 Alan R. Liss, Inc.

The increasing participation of pregnant women in the labor force has raised concerns about the effects of occupational exposure on the embryo and fetus. It has been estimated that in the US 41.5% of all women who had a live birth worked during pregnancy in 1972–1973 [Hunt, 1979]. In Finland about 67% of women having a live birth in 1976 were working at the end of the year 1975 [Niemi, in press]. The potential risks of reproductive hazards have posed a challenge to the occupational health community to evaluate the magnitude of the problems involved and to provide preventive measures.

Major surveys have been carried out to investigate the types of chemicals used in various industries and the number of workers exposed to each agent [NIOSH, 1977]. Yet the information on the various exposures of women is quite meager. Women are occupationally segregated into "traditional" female jobs, and therefore tend to be exposed to different kinds of hazards than male workers [Stellman, 1978]. However, the studies on the occupational health hazards have mainly been limited to male-dominated occupations. The potential hazards faced by pregnant and nonpregnant women in their workplaces would require specific inquiries in the future.

The most important female-dominated occupations with potential chemical exposures are probably health-care professions and work tasks in the clothing and textile industry. Also laboratory workers, hairdressers, and cosmetologists are exposed to large amounts of chemicals. It may be estimated that inorganic gases, solvents, and textile dust are the most prevalent chemical exposures among working women in industrialized countries.

The origins and mechanisms of reproductive problems may be complex, as both maternal and paternal effects can be noted [Strobino et al, 1978; Sullivan and Barlow, 1979; Hemminki et al, 1979]. Increasing attention has recently been paid to the possibility of paternal effects [Haas and Schottenfeld, 1979] possibly causing spontaneous abortions [Infante et al, 1976a; Sanotsky, 1976; Kharrazi et al, 1980], malformations [Cohen et al, 1974; Corbett et al, 1974; Knill-Jones et al, 1975], and cancer in the offspring [Fabia and Thuy, 1974; Hemminki et al, 1981b]. However, the paternal effects are outside the scope of the present review and they will not be considered any further. The maternal effects may either be genetic, taking place before conception, or toxic (or genotoxic) taking place during pregnancy [Hemminki et al, 1979; Sever, 1981]. However, as no maternal preconceptional occupational chemical hazard has yet been recognized and since the ascertainment of such effects is a most laborious task, the present article focuses exclusively on the harmful agents affecting reproduction during pregnancy and reviews the literature in this respect.

INCIDENCE OF SPONTANEOUS ABORTIONS AND MALFORMATIONS

According to prospective studies, the frequency of spontaneous abortions in humans ranges between 13% and 20% of all pregnancies, whereas retrospective studies have estimated the frequency to be 10% to 15% [Carr, 1977]. The differences in these figures are mainly due to the difficulty in observing pregnancies and fetal losses during the first few weeks after conception. Fetal losses are more likely to occur in the early weeks of pregnancy, and their frequency decreases considerably after the first trimester.

Application of modern methods of pregnancy testing has provided more insight in the "true" frequency of spontaneous abortions [Miller et al, 1980]. When 197 healthy women were followed prospectively, 152 pregnancies were detected using chorion-gonadotropin determination and clinical judgment. By the 20th week of pregnancy 14 (9.2%) clinical spontaneous abortions and 87 (57%) living fetuses were observed. These figures are interesting insofar as they show that the frequency of all spontaneous abortions was over 40% and that the frequency of spontaneous abortions, diagnosed clinically, was 13.7% of the pregnancies. This figure agrees with several other clinical observations where the frequency of spontaneous abortions is given at about 10% of the clinically manifest pregnancies.

The frequency of birth defects depends on the time of observation after birth and on the types of malformations included, the figures ranging from 2% to 6% of all living children. Morphological inspection of the abortuses may be a valuable source for teratological monitoring. The specimens that are intact enough for inspection (usually over 60% of the material) may reveal fetal malformations in at least 5% to 10% of the cases [Sentrakul and Potter, 1966; Saxén and Rapola, 1969]. The reason for the concentration of malformations in the abortion material is the decreased survival of the malformed fetus. It has been estimated that fetuses with neural tube defects miscarry at a frequency of 60% to 90% [Leck, 1977].

Chromosome studies of aborted embryos have revealed that anomalies are about 60 to 100 times more frequent in spontaneous abortions than in live births, amounting to about 30–50% of all aborted embryos [Bochkov, 1977; Lauritsen, 1976; Stein et al, 1975]. The frequency of chromosomal anomalies is very dependent on the developmental age of the embryo or fetus. At 12 and 28 weeks anomalies are observed in 2.4% and 0.7% of the fetuses, respectively [Stein et al, in press].

METHODS OF RISK EVALUATION

Animal testing is one possible approach in investigations of reproductive toxicity of chemicals. A large number of commonly used industrial chemicals have been tested for teratogenicity and embryotoxicity [Wilson, 1977a,b; Shephard, 1980]. The results from animal experiments on these various chemicals have been correlated to the possible human exposures in the workplaces, regulated by hygienic norms [Hemminki, 1980]. The theoretical safety factors have been calculated to vary from about 1 to 1000. Yet as the extrapolation of teratogenic risk from experimental animals to humans has never been thoroughly validated, the calculation of safety factors has theoretical interest at best. As a consequence, epidemiological studies are of particular interest in the field of reproductive epidemiology as in other occupational-hazard evaluations. However, reproductive epidemiology tends to be less developed than epidemiology in general and also seems to involve some specific problems of study designs, including aspects of validity.

Sources of Data

In many countries, registers on congenital birth defects are maintained and may be valuable sources for the ascertainment of cases. Yet underreporting, particularly of minor malformations, may harass even an efficient register, and it has to be considered to what extent such register deficiencies might introduce any bias or otherwise hamper the particular study on occupational reproductive hazards. Generally judging, however, it seems likely that occupational studies should be less sensitive to bias of this background than, say, studies relating to side effects of drugs or so-called life-style factors like use of alcohol and smoking. The reason is that reporting might tend to be merely conditional on information about use of drugs, and, similarly, selectional forces in utilizing the health-care system might obtain in relation to life-style factors and also influence reporting and registration, particularly of minor malformations.

Spontaneous abortions are not registered, and two sources of data are available, namely, interview data or medical records. Both sources have advantages and disadvantages. Usually the interviews are more sensitive in registering early spontaneous abortions, which would not always lead to medical contacts. Medical records, on the other hand, are more reliable. Yet, hospital registration may be biased insofar as there are differential rates for hospitalization on socioeconomic or geographical grounds.

As most spontaneous abortions are not always recognized clinically even by experienced observers, as discussed above [Miller et al, 1980], it is very likely that the ability of the women to recognize their spontaneous abortions is highly variable. Therefore interview reporting may be rather sensitive to factors like social status, medical experience, concern about harmful environmental factors, etc, which all might introduce severe bias (particularly if the social background is heterogeneous in the study population) unless data acquisition is very careful. It has even been observed that hospital personnel tended to report a higher frequency of spontaneous abortions for recent pregnancies as compared with those one or two decades earlier [Hemminki et al, in press]. Whenever objective data on spontaneous abortions are available, they should be used to control at least a sample of the information collected; this is particularly compelling if the involved study population is heterogenous in factors that may influence subjective reporting.

Technical Aspects of Study Design

In traditional epidemiology, the measures of occurrence of illness have the number of cases in the numerator and the number of individuals or person-years at observation in the denominator, ie, the cumulative incidence and prevalence have count denominators, whereas incidence density (incidence rate) has person-years in the denominator. In reproductive epidemiology, measures of malformations might preferably be taken as the prevalence of defects at the time of birth (or noted during some period thereafter) per number of born children, whereas the number of abortions might ideally be referred to the number of pregnancies (ie, births + induced abortions + spontaneous abortions) to obtain the spontaneous abortion proportion (or rate of spontaneous abortions). A correspondence of the incidence density (incidence rate) would be obtained by taking spontaneous abortions per pregnancy months (or, preferably even, early pregnancy months) as the measure of occurrence. However, this possibility does not seem to have been utilized, and it may also involve specific problems, as the occurrence of spontaneous abortions rapidly declines with advancing pregnancy.

Since pregnancy denominators might be difficult to adequately assess due to unrecognized pregnancies and unreported induced abortions, births might also be considered for obtaining a denominator. However, the resulting measure of occurrence of spontaneous abortions, which has been called the spontaneous-abortion ratio [Hemminki et al, 1980a], is conceptually less clear, but might nevertheless be justified as a supplementary measure, being more robust and seemingly also more sensitive than the spontaneous-abortion proportion, especially if the agent under study might cause (preconceptional) infertility together with an increased frequency of abortions.

Malformation data might simply be looked upon as of a cross-sectional character, anchored at the time of birth, and no specific difficulties would be involved in this view. For example, since birth defects might be taken as rare conditions, especially when considering any specific malformation entity, a case-referent (case-control) approach could be taken by estimating the exposure frequency of the source population from a sample of noncase births as referents (or controls). Preferably, however, the exposure frequency should be obtained through a sample from all births, ie, including the cases as well (ie, a cross-sectional view).

Studying abortions, the cohort view is similarly straightforward, comparing the spontaneous abortion proportions among the exposed and nonexposed. However, since spontaneous abortions are quite common, an unbiased case-referent approach would involve somewhat more complicated considerations in referent (control) selection. Thus, the referents cannot simply be taken from among the successful (noncase) pregnancies, due to the fact that spontaneous abortion is a rather frequent outcome of pregnancy, and consequently, there would be a considerable influence on the ratio of exposed and nonexposed successful pregnancies, if there would be an abortion effect from the exposure. Instead, there are some other options: one possibility would be to estimate the relative occurrence of exposed and nonexposed pregnancies by taking a sample of pregnancies, irrespective of outcome (as abortions or births), which would be close to a cohort view although not permitting the absolute rates to be assessed in contrast to the rate ratio. Another way would be to give the study a prospective character as based on incidence density, namely, by drawing the sample of referents from ongoing pregnancies, ie, from the pool of candidates for abortion [cf Miettinen, 1976], and, preferably, allowing for the order of pregnancy month in the study design (matching or restriction) or in the analysis, (eg, by stratification).

It is worth noticing perhaps that the quotient of the aforementioned spontaneous-abortion ratios is easily obtainable as an odds ratio in a retrospective case-referent approach, taking the referents as a sample of the births in the study population for spontaneous abortions. The term "abortion-birth's odds ratio" might be suggested for this quotient.

Some Validity Aspects

The difficulties involved in recognizing the event of an early pregnancy and/or early abortion have already been indicated. Therefore, should the exposure cause early abortions in particular, there is an obvious risk of underestimating its effect. Moreover, in some countries at least, there might be a tendency to induce abortion if the mother is exposed to, say, chemical agents with suspected adverse effects. Thus, some rather particular selective forces might sometimes be at hand. Regarding malformations, and particularly the severe ones, the situation is somewhat more robust, although there could be tendencies to hide away malformed children, eg, in particular ethnic groups, but registration should be expected to be rather complete in countries where the births take place at hospitals.

The observation of exposure tends to be a difficult issue in reproductive epidemiology if information has to be obtained through questionnaires. The exposure reporting might therefore be influenced by the pregnancy outcome, ie, mothers with normal pregnancies might be apt to forget and falsely underreport exposure, whereas exposure might even be exaggerated by those having an abortion or a malformed child. However, concerns of this sort might not be totally relevant since questionnaire information sometimes seems quite reliable, at least in terms of employment and as to the general character of the job (cf experiences in cancer studies, Pershagen and Axelson [1982]) but the situation might be different with regard to use of drugs, alcohol and other life-style factors during pregnancy [Klemetti and Saxén, 1967].

Objective exposure information is often possible to obtain in occupational health epidemiology through the employment registers held by the company. If there are antenatal care units one might also try to achieve exposure information in very early pregnancy. There is also a possibility to evaluate observational bias through introducing a supporting exposure variable as has been pointed out in the context of case-referent studies [Axelson, 1980a], but the principle would apply also in cross-sectional and in cohort studies [cf also Axelson, in press].

The aspects of comparability between groups is another matter of general concern in epidemiology. Should the exposed group be compared to another group with some other type of adverse exposure, there is obviously the risk of overlooking an effect if both groups have increased numbers of abortions or malformations. Office workers or housewives may be thought of as a suitable comparison group, but there might be generally operating risk factors for abortions among employed women, eg, due to unspecific stress, work in a standing position, traveling to the workplace, etc; and there might also be social differences influencing nutritional status, use of cosmetics, and other ill-defined circumstances that might influence pregnancy outcome. To some extent, such problems of comparability might be overcome by restriction of the study to a particular social, or better still, vocational group of women, or by comparisons within a single group of women, since some of their pregnancies might be exposed, others not, thus permitting suitable contrasts to be created.

If several determinants of unfavorable pregnancy outcome tend to be associated with each other, a confounded study situation results. Detailed discussions of the issues involved in allowing for confounding can be found elsewhere [eg, Miettinen, 1974; Axelson, 1980b], and it might suffice here to indicate that the strength of confounding in reproductive epidemiology seems to be greater than in many other areas of epidemiology, eg, in occupational health. For example, age of the mother is a rather strong determinant of pregnancy outcome, and various types of exposure might be positively or negatively associated with age through occupational career. Attitudes toward use of alcohol and tobacco might change relatively quickly in certain social groups, and over a period of time; this might lead to an association by change, positive or negative, between such risk factors for abortions and specific occupational exposures.

The various potential confounding factors have to be identified from the literature, but the data, representing the population base of the study, should be checked to determine whether there really is an association at hand between exposure and specific, already known, or suspected risk factors; otherwise confounding is not present. Parity and previous successful pregnancies are of particular interest with regard to confounding since there is likely to be a tendency for women with multiple abortions to remain in work, whereas successful pregnancies tend to force a woman into a housewife status, where she remains nonexposed during subsequent pregnancies. In some industries, where women may subjectively experience risk factors, eg, the rubber industry, an opposite situation may prevail: women with successful pregnancies stay on working. This has some analogies with the "healthy worker effect" in occupational mortality studies [Hemminki et al, 1982].

Due to the relatively complex patterns of confounding, which often are present in reproduction epidemiology studies, the analyses might be rather demanding and require, say, multivariate procedures. From this point of view, a rather attractive possibility could be to apply the Miettinen multivariate confounder score technique [Miettinen, 1976], particularly since it tends to permit a better insight into the data than regular multivariate methods. It is furthermore important to evaluate the strength of various confounding factors, since it might well happen that some of these exert a rather small effect and might therefore turn out to be of less relevance than is usually assumed; particularly, the weak negative confounders might then be neglected in some of the subsequent analyses. Such negligence would make a study slightly conservative, which could even be somewhat beneficial in meeting the criticism, given a positive association between exposure and outcome (should the study be nonpositive or negative, the negative confounders, especially, have to be carefully allowed for; whereas the slightly positive ones, ie, "exaggerating" factors, might be more or less neglected, without attracting too much criticism).

Confounders should not be mixed up with effect-modifying factors, ie, the effect of an exposure might be increased or decreased in the presence of a modifying factor. Taking smoking as an example, an excess of abortions might be found to have occurred among the exposed smokers, even when the exposed and nonexposed group have identical smoking frequencies. Then this is an inherent biological phenomenon that might be important to consider in preventive efforts, and decrease of smoking would then be adjuvant to decreasing the specific hazardous exposure. It should be recognized as well that an effect found in a certain population, where an effect-modifier is widely operating, might be difficult to reproduce in another population more or less without this factor.

Evaluating Methodological Issues

In the near future, there is a need not only to delineate specific industrial risk factors for malformations and abortions but also to focus on the methodological issues involved in these studies, since there is still rather little methodological experience of reproductive epidemiology relating to occupational hazards. Thus, explicit evaluation and reporting of various types of bias in these studies is desirable as one such methodological aspect. For example, it might be mentioned here that when the Finnish Register of Congenital Malformations was used to study the association with congenital malformations of about 60 variables, almost all collected by a retrospective interview, it turned out that the strongest association was found with the woman's statement that she had felt depressed during the pregnancy [Hemminki et al, 1981a]. Reporting bias has also been detected in a Swedish postal interview study, with an 80% response rate, on personnel exposed to anesthetic gases [Axelsson and Rylander, 1982]. Through checking various records it was found that among the exposed nonrespondents no spontaneous abortions had occurred, while among the control nonrespondents a large number of spontaneous abortions were identified.

OCCUPATIONS AND EXPOSURES WITH REPORTED INCREASE IN RISK

Data on the distribution of spontaneous abortions in various socioeconomic groups of women is meager, but some information is available. For example, in a Finnish study, information on the social class and occupation of women was retrieved from the hospital discharge register from 1973 to 1975. The analysis showed an increasing gradient of spontaneous abortions from social class 1 (higher officials, employers) to social class 4 (unskilled workers). Analysis by occupation revealed that industrial and construction workers (rate 8.11%) and students and trainees (rate 8.37%) had more spontaneous abortions than all women combined (rate 7.43%). Housewives had a rate of 6.85%, significantly lower than that of all women [Hemminki et al, 1980d]. A study from Iran has also noted a higher rate for spontaneous abortions among working as opposed to nonworking women, 12% and 8%, respectively [Kavoussi, 1977].

Studies relating the socioeconomic status of the family to the prevalence of malformations have shown a gradient in Great Britain and the United States, while no such difference was detected in Hungary or among Jews in Israel and the United States [Leck, 1977]. A socioeconomic risk factor has also been noted in the British mortality study [Registrar General, 1978] and the Norwegian occupational study [Bjerkedal and Lund, 1978]. The Norwegian survey has detected more malformed children born to the employed and the less-educated women. The Finnish Register of Congenital Malformations, a case-referent register, has been used to analyze the associations between the parental occupation and children born with malformations. The analysis of 3,600 case-referent pairs indicated that the women working in industrial and construction occupations had more children with central nervous system (CNS) and musculoskeletal malformations than the referent mothers [Hemminki et al, 1980b]. However, no statistically significant odds ratios were found when the material was analyzed according to the paternal occupation. Some associations were found between low social class and some types of malformations. In a subsequent study the association with maternal occupation in industry and children born with malformations was controlled with regard to over 20 possible confounding factors using multivariate analysis. The possible confounding factors were taken as characteristics of the mother, the child, and the family; maternal illnesses and medication at the first trimester of pregnancy were also considered. The association with oral clefts remained and that with central nervous system malformation was strengthened in the multivariate analysis [Hemminki et al, 1981a].

Metals and the Metal Industry

Many metals have been found to be teratogenic in experimental animals [Wilson, 1977a; Hemminki, 1980], but the information on human effects is limited. However, methyl mercury has been shown to cause severe intrauterine toxicity in humans as a result of nonoccupational poisoning in Japan and in Iraq [Wilson, 1977a]. Historical data and case studies are also available on the induction of spontaneous abortions in humans by lead [Rom, 1976; Chang et al, 1980]. This literature shows that ingestion of high concentrations of lead causes spontaneous abortions. Unfortunately, no recent information is available on the effects of maternal occupational exposures. However, occupational lead poisoning has been reported to decrease the fertility of men [Lancranjan et al, 1975]. Occupational exposure of the pregnant women to lead may be problematic from another point of view, because of the reported impairment of mental functions in children exposed to lead environmentally [Beattie et al, 1975].

In a Swedish series of studies, the reproductive health of employees of a copper smelter was investigated. The workers were potentially exposed to a number of metals such as copper, lead, arsenic, and cadmium in addition to sulfur dioxide. An increased frequency of spontaneous abortions and malformations in the offspring was reported for women employed in the smelter [Nordström et al, 1979a,b]. The reliability of the data on spontaneous abortions was somewhat unclear, however, as they were collected over several decades, while the health-care system was continuously changing and developing.

In a Finnish study, the hospitalized spontaneous abortions of the female members of the Metal Workers' Union were analyzed. Among these metal workers, 195 spontaneous abortions were recorded in the hospital discharge register. The proportion (rate) of spontaneous abortions was 7.82% (spontaneous abortions/pregnancies) and the abortion-birth's ratio 13.79% (spontaneous abortions/births), of which the latter figure was significantly higher than the average for all Finnish women [Hemminki et al, 1980c]. Both the proportion and the ratio were significantly increased upon joining the union, implicating risk factors relating to metalwork. A particularly high-risk industry appeared to be the production of radio and television sets. An exposure to solder fumes was suggested to explain the increased risk. The implications for the validity of using the hospital discharge register have been reported elsewhere [Hemminki et al, 1982; Niemi et al, in press].

Laboratory Work and Solvents

Several recent studies conducted in Sweden and Finland have provided evidence that employment of women in chemical laboratories may sometimes affect the outcome of pregnancy. In a study based on questionnaires, Strandberg et al [1978] found that the women who had been employed both in hospital laboratories and elsewhere in the hospital had more spontaneous abortions while working in laboratories than in their unexposed pregnancies. The data were adjusted for several confounding factors. Similarly increased frequency of spontaneous abortions has also been reported on two other groups of laboratory personnel in Sweden as detected in retrospective interview studies [Kolmodin-Hedman and Hedström, 1978; Hansson et al, 1980]. In these three studies, solvent exposure in particular has been suggested to contribute to the adverse effects.

Three other Swedish studies have found an increase in major malformations in the children of women who were working in chemical laboratories during pregnancy [Meirik et al, 1979; Hansson et al, 1980; Blomqvist et al, 1981]. The data on exposure were collected from company records and by questionnaires. The authors paid attention to the presence of gut atresias and oral clefts in the small number of malformations recorded. In a Finnish case-control study an excess of central nervous system malformations was noted among women who were exposed to solvents during their pregnancy [Holmberg, 1979]. All four cited studies relied on registered cases of malformations adding weight to the findings.

Chemical, Plastics, and Rubber Industries

Vinyl chloride has attracted particular concern and adverse effects on reproduction have been reported. Infante et al [1976a] found more spontaneous abortions among the wives of vinyl chloride polymerization workers as compared to the controls or to the same workers prior to the employment at the vinyl chloride plant. Two studies have also suggested that the population living in the vicinity of a vinyl chloride plant tend to have an increased frequency of malformed children [Infante, 1976b; Edmonds et al, 1978]. In a Finnish case-control study on central nervous system defects three mothers with exposure to styrene were found while less than one was expected [Holmberg, 1977].

Phthalate plasticizers are used as additives in plastics. Aldyreva et al [1975] have reported that exposure to phthalates might cause spontaneous abortions and disturbances in the ovulatory cycles.

In Finland a study encompassing 9,000 women belonging to the Union of Chemical Workers in 1973–1976 showed significantly more hospitalized spontaneous abortions among the union members than among all women in Finland [Hemminki et al, 1980a]. When the union members were divided according to the main branch of employment, the occurrence of spontaneous abortions was more common in the plastics industry, particularly styrene workplaces; viscose rayon industry; laundries; and pharmaceutical industry. However, the results should be interpreted cautiously as only 52 spontaneous abortions were enrolled in total, and therefore the number of spontaneous abortions in each branch was small. Furthermore, the referent group might be less than adequate, taken as the total Finnish population. Additionally, both spontaneous abortions, contributing to the numerator, and births, contributing to the denominator, were defined by fewer International Classification of Diseases (ICD) codes than necessary [cf Hemminki et al, 1980c,d]. It is of interest in this context that an earlier survey suggested that exposure to carbon disulfide increased the frequency of spontaneous abortions [Ehrhardt, 1967].

A further register-based study on spontaneous abortions has been carried out in collaboration with the Union of Rubber and Leather Workers including about 10,000 women [Hemminki et al, 1982]. Again the occurrence of spontaneous abortions was slightly increased for all union members as compared to the national average. The age-standardized frequencies were higher (nonsignificant statistically) when the pregnancy occurred during the union membership as compared to pregnancies before or after the membership. These observations concerned rubber and leather industries. Rubber work did not distinguish itself from other types of work within the union members. The study revealed some interesting relations between the length of employment and the frequency of spontaneous abortion, as discussed in the section on validity aspects.

Anesthetic Gases and Sterilization Media

The first suggestion of possible reproductive hazards in the operating rooms came from a Russian questionnaire study [Vaisman, 1967]. Only seven of 31 pregnancies among the doctors surveyed were found to have been trouble-free. This report launched a number of similar studies in other countries. The standard study protocol has been to identify a group of exposed persons from some sort of a register and to send out a questionnaire inquiring about spontaneous abortions and malformed offspring.

Fairly consistent conclusions have been drawn by various reviewers of the literature on operating theatre personnel and there is some agreement on an increased risk of spontaneous abortions in operating-room staff [Spence and Knill-Jones, 1978; Vessey and Nunn, 1980; Edling, 1980]. A recent study done among dentists and chairside assistants also supports the idea that a risk is present [Cohen et al, 1980]. Furthermore, this study together with the evidence from experimental animals suggested that high concentrations of nitrous oxide may be the primary causative factor [Lane et al, 1980; Vessey and Nunn, 1980].

However, all the studies were carried out using subjective information on spontaneous abortion and response rates were relatively low, varying from 56% to 72% in the large studies. In some studies the response rates have varied between the exposed and the nonexposed. Therefore, several of the studies have a somewhat questionable validity, particularly in view of the observations made on the rather different occurrence of abortions among the nonrespondents with and without an exposure as referred to in the section on evaluating methodological issues.

In several studies, eg, those by Cohen et al [1974], Knill-Jones et al [1972], Göthe et al [1976] and Tomlin [1979], a higher risk of malformations in children born by women exposed to anesthetic gases during pregnancy was also reported. No increase in congenital malformations was observed among the children born to dentists exposed to inhalation anesthetics [Cohen et al, 1980] or in two Swedish studies on female operating-room personnel [Ericson and Källen, 1979; Baltzar et al, 1979a]. The existing evidence of an increased risk of malformations caused by operating-room exposure therefore seems to rest on a weaker basis than that on spontaneous abortions [Spence and Knill-Jones, 1978; Vessey and Nunn, 1980].

Chemical sterilization, using gaseous or liquid media, is carried out extensively in the hospitals and the pharmaceutical industry. A number of chemical sterilants have been used, eg, ethylene oxide, formaldehyde, propylene oxide, glutaraldehyde, methyl bromide, β-propiolactone, etc. Furthermore, a number of antimicrobial agents, which inhibit the growth of the bacteria, are used. These include hexachlorophene and sodium azide.

Mutagenicity and carcinogenicity in experimental animals has been reported for ethylene oxide, formaldehyde, propylene oxide, and β-propiolactone. Few reproductive studies have been carried out with regard to exposure to chemical sterilization media, even though many of them would be highly suspect because of their mutagenic and carcinogenic properties.

In a Swedish study, based on a cluster of malformations in the offspring of hospital personnel, hexachlorophene was suggested to be a risk factor for malformations [Halling, 1977]. This suggestion was not confirmed by other Swedish workers, however, correlating the hospital use of hexachlorophene and the occurrence of malformations in the offspring of the personnel [Baltzar et al, 1979b]. Using questionnaire and register information a study was carried out in Finland among chemical sterilizers in hospitals exposed to ethylene oxide, glutaraldehyde, and formaldehyde. The results of this study suggest an association between an exposure to ethylene oxide and a risk of spontaneous abortions [Hemminki et al, in press]. In this study no effect was seen within the small group of women using formaldehyde sterilization. Yet formaldehyde exposure has been linked to an increased frequency of spontaneous abortions in a Russian study [Shumlina, 1975].

Pesticides

The use of pesticides has grown all over the world during the past 30 years, and the number of persons occupationally exposed to them amounts to millions. Many pesticides also leave residues in the plants treated whereby a large proportion of the world's population is exposed to small concentrations of pesticide residues in food. The possible effects of pesticides on human reproduction are not well established, and more studies are urgently needed in this area. In an American study an association was found between oral clefts in the offspring and a possible maternal exposure to pesticides [Gordon and Shy, 1981]. A Finnish study on the occupation of mothers bearing children with musculoskeletal malformations showed an overpresentation of gardeners among the case mothers [Hemminki et al, 1980b].

Environmental and Medicinal Risk Factors With Occupational Relevance

Some nonoccupational chemical exposures have been suggested to carry reproductive risks. As such compounds may also cause occupational exposures, it is reasonable to mention them here. These include methyl mercury and lead, which are linked to spontaneous abortions, microcephaly, as well as neurological and mental disorders (see section on the metals industry); polychlorinated biphenyls, ubiquitous environmental contaminants, causing retardation of growth, dark pigmentation of skin, and other defects [Miller, 1971; Yamashita, 1977]; and dioxins, extremely potent teratogens in experimental animals [Neubert et al, 1973], with probable effects on spontaneous abortions in Seveso [Bisanti et al, 1980; Remotti, personal communication], and spontaneous abortions and malformations in Vietnam [Laporte, 1977]. Spray adhesives, of unspecified composition, have been suggested to cause malformations [Silberg et al, 1979], and many medicines such as antimetabolites, alkylating agents, and anticonvulsants have also been associated with malformations [Wilson, 1977b].

CONCLUSIONS

Occupational epidemiology on spontaneous abortions and malformations is still young, and many selection mechanisms and risk factors have to be recognized and evaluated before far-reaching causal conclusions can be drawn. The basic study designs, particularly on spontaneous abortions, have not been thoroughly evaluated, and some methodological progress might be anticipated. Studies on malformations tend to be more clear in design and usually resort to case registers, ie, objective sources of cases, which makes the evaluation easier. With an increasing female workforce in industry and with certain reproductive hazards having been more or less convincingly identified, further investment in occupational reproductive epidemiology seems to be urgently needed.

REFERENCES

Aldyreva MV, Slimona TS, Izyumova AS, Timofievska LA (1975): The influence of phthalate plasticisers on the generative function. Gig Tr Prof Zabol 19:25–29.

Axelson O (1980a): A note on observational bias in case-referent studies in occupational health epidemiology. Scand J Work Environ Health 6:80–82.

Axelson O (1980b): Aspects of confounding and effect modification in the assessment of occupational cancer risk. J Toxicol Environ Health 45:125–126.

Axelsson G, Rylander R (1982): Exposure to anaesthetic gases and spontaneous abortions: response bias in a postal questionnaire study. Int J Epidemiol 11:250–256.

Baltzar B, Ericson A, Källen B (1979a): Pregnancy outcome among women working in Swedish hospitals. N Engl J Med 300:627–628.

Baltzar B, Ericson A, Källen B (1979b): Delivery outcome in women employed in medical occupations in Sweden. J Occup Med 21:543–548.

Beattie AB, Moore MR, Goldberg A, Finlayson MJW, Graham JF, Mackie EM, Main JC, McLare DA, Murdoch RM, Stewart GT (1975): Role of chronic lowlevel lead exposure in the aetiology of mental retardation. Lancet i:589–592.

Bisanti L, Bonetti F, Caramaschi F, Del Corno G, Favaretti C, Giambelluca SE, Marni E, Montesarchio E, Puccinelli V, Remotti G, Volpato C, Zambrelli E, Fara GM (1980): Experiences from the accident of Seveso. Ann Morphol Acad Sci Hung 28:139–157.

Bjerkedal T, Lund TE (1978): Yrke og fødsel. Hygienisk Institutt, Universitetet i Oslo, pp 1–116.

Blomqvist U, Ericson A, Källen B, Westerholm P (1981): Delivery outcome for women working in the pulp and paper industry. Scand J Work Environ Health 7:114–118.

Bochkov NP (1977): Monitoring of human populations in connection with environmental pollution by the evaluation of chromosome anomalies. In Böhme H, Schöneich J (eds): "Environmental Mutagens." Berlin: Akademie-Verlag, pp 167–179.

Carr DH (1977): Detection and evaluation of pregnancy wastage. In Wilson JG, Fraser FC (eds): "Handbook of Teratology," Vol. 3. New York: Plenum Press, pp 189–213.

Chang LW, Wade PR, Pounds JG, Reuhl KR (1980): Prenatal and neonatal toxicology and pathology of heavy metals. Adv Pharmacol Chemother 17:195–231.

Cohen EN, Brown BW, Bruce DL, Cascorbi HF, Corbett TH, Jones TW, Whitcher CE (1974): Occupational disease among operating room personnel: A national study. Anesthesiology 41:321–340.

Cohen EN, Brown BW, Wu ML, Whitcher CE, Brodsky JB, Gift HC, Greenfield W, Jones TW, Driscoll EJ (1980): Occupational disease in dentistry and chronic exposure to trace anesthetic gases. J Am Dent Assoc 101:21–31.

Corbett TH, Cornell RG, Lieding K, Endres JL (1974): Birth defects of children among Michigan nurse-anesthetists. Anesthesiology 41:341–344.

Edling C (1980): Anesthetic gases as an occupational hazard—A review. Scand J Work Environ Health 6: 85–93.

Edmonds LD, Anderson CE, Flynt JW Jr, James LM (1978): Congenital central nervous system malformations and vinyl chloride monomer exposure: A community study. Teratology 17:137–142.

Ehrhardt W (1967): Experiences with the employment of women exposed to carbon disulphide. In "International Symposium on Toxicology of Carbon Disulphide, Prague 1966." Amsterdam: Excerpta Medica Foundation, p. 240.

Ericson A, Källén B (1979): Survey of infants born in 1973 or 1975 to Swedish women working in operating rooms during their pregnancies. Anesth Analg 58:302–305.

Fabia J, Thuy TD (1974): Occupation of father at time of birth of children dying of malignant diseases. Br J Prev Soc Med 28:98–100.

Gordon JE, Shy CM (1981): Agricultural chemical use and congenital cleft lip and/or palate. Arch Environ Health 36:213–221.

Göthe C-J, Dahlgren B-E, Hulten B, Olander L, Övrum P, Westerholm P (1976): Narkosgaser som yrkesrisk. Läkartidningen 73:2553–2563.

Haas JF, Schottenfeld D (1979): Risks to the offspring from parental occupational exposures. J Occup Med 21:607–613.

Halling H (1977): Misstänkt samband mellan hexaklorofenexposition och missbildningsbörd. Läkartidningen 74:542–546.

Hansson E, Jansa S, Wande H, Källen B, Östlund E (1980): Pregnancy outcome for women working in laboratories in some of the pharmaceutical industries in Sweden. Scand J Work Environ Health 6: 131–134.

Hemminki K (1980): Occupational chemicals tested for teratogenicity. Int Arch Occup Environ Health 47: 191–207.

Hemminki K, Sorsa M, Vainio H (1979): Genetic risks caused by occupational chemicals. Scand J Work Environ Health 5:307–327.

Hemminki K, Franssila E, Vainio H (1980a): Spontaneous abortions among female chemical workers in Finland. Int Arch Occup Environ Health 45:123–126.

Hemminki K, Mutanen P, Luoma K, Saloniemi I (1980b): Congenital malformations by the parental occupation in Finland. Int Arch Occup Environ Health 46:93–98.

Hemminki K, Niemi M-L, Koskinen K, Vainio H (1980c): Spontaneous abortions among women employed in the metal industry in Finland. Int Arch Occup Environ Health 47:53–60.

Hemminki K, Niemi M-L, Saloniemi I, Vainio H, Hemminki E (1980d): Spontaneous abortions by occupation and social class in Finland. Int J Epidemiol 9:149–153.

Hemminki K, Mutanen P, Saloniemi I, Luoma K (1981a): Congenital malformations and maternal occupation in Finland: Multivariate analysis. J Epidemiol Community Health 35:5–10.

Hemminki K, Saloniemi I, Salonen T, Partanen T, Vainio H (1981b): Childhood cancer and parental occupation in Finland. J Epidemiol Community Health 35:11–15.

Hemminki K, Niemi M-L, Kyyrönen P, Kilpikari I, Vainio H (1982): Spontaneous abortions and reproductive selection mechanisms in rubber and leather industry in Finland. Br J Ind Med (in press).

Hemminki K, Mutanen P, Saloniemi I, Niemi M-L, Vainio H (1982): Spontaneous abortions in hospital staff engaged in sterilising instruments with chemical agents. Br Med J (in press).

Holmberg PC (1977): Central nervous defects in two children of mothers exposed to chemicals in the reinforced plastics industry. Scand J Work Environ Health 3:212–214.

Holmberg PC (1979): Central nervous system defects in children born to mothers exposed to organic solvents during pregnancy. Lancet i:177–179.

Hunt VR (1979): "Work and the Health of Women." Boca Raton, Florida: CRC Press.

ILO (1980): Equal opportunities and equal treatment for men and women workers: Workers with family responsibilities. Report VI(1): International Labour Conference 66th Session, Geneva.

Infante PF, Wagoner JK, McMichael AJ, Waxweiler RJ, Falk H (1976a): Genetic risks of vinyl chloride. Lancet i:734–735.

Infante PF (1976b): Oncogenic and mutagenic risks in communities with polyvinyl chloride production facilities. In Saffiotti U, Wagoner JK (eds): "Occupational Carcinogenesis." Ann NY Acad Sci 271: 49–57.

Kavoussi N (1977): The effect of industrialization on spontaneous abortion in Iran. J Occup Med 19:419–423.

Kharrazi M, Potashnik G, Goldsmith JR (1980): Reproductive effects of dibromochloropropane. Isr J Med Sci 16:403–406.

Klemetti A, Saxen L (1967): Prospective versus retrospective approach in the search for environmental causes of malformations. Am J Public Health 57:2071–2075.

Knill-Jones RP, Moir DD, Rodrigues LV, Spence AA (1972): Anesthetic practice and pregnancy: Controlled survey of women anesthetists in United Kingdom. Lancet i:1326–1328.

Knill-Jones RP, Newman BJ, Spence AA (1975): Anesthetic practice and pregnancy: Controlled survey of male anaesthetists in the United Kingdom. Lancet ii:807–809.

Kolmodin-Hedman B, Hedstrom L (1978): Enkätundersökning has kemikalie exponerad laboratorie per sonal rörande spontanabortes. Läkartidningen 75:3044–3045.

Lancranjan I, Posocu H, Galvenescu O, Klepsch I, Serbanescu M (1975): Reproductive ability of workmen occupationally exposed to lead. Arch Environ Health 30:396–401.

Lane GA, Nahrwold ML, Tait AR, Taylor-Busch S, Cohen PJ, Beaudoin AR (1980): Anesthetics as teratogens: Nitrous oxide is fetotoxic, xenon is not. Science 899–901.

Laporte J-R (1977): Effects of dioxin exposure. Lancet 1:1049–1050.

Lauritsen JG (1976): Aetiology of spontaneous abortion. Acta Obstet Gynecol Scand (Suppl) 52:1–29.

Leck I (1977): Correlations of malformation frequency with environmental and genetic attributes in man. In Wilson JG, Clark FC (eds): "Handbook of Teratology," Vol 3. New York: Plenum Press, pp 243–324.

Meirik O, Källén B, Gauffin U, Ericson A (1979): Major malformations in infants born to women who worked in laboratories while pregnant. Lancet i:91.

Miettinen OS (1974): Confounding and effect-modification. Am J Epidemiol 100:350–353.

Miettinen OS (1976): Stratification by a multivariate confounder score. Am J Epidemiol 104:609–620.

Miller RW (1971): Cola-colored babies. Chlorobiphenyl poisoning in Japan. Teratology 4:211–212.

Miller JF, Williamson E, Glue J, Gordon YB, Grudzinskas JG, Sykes A (1980): Fetal loss after implantation. A prospective study. Lancet ii:554–556.

NIOSH (1977): National occupational hazard survey, Vol. I-III, National Institute for Occupational Safety and Health, DHEW (NIOSH) Publ. no. 78–114, Washington DC.

Neubert D, Zens P, Rothenwallner A, Merker H-J (1973): Survey of the embryotoxic effects of TCDD in mammalian species. Environ Health Perspect 5:67–79.

Niemi M-L, Hemminki K, Sallmén M (In Press): Application of hospital discharge register for studies on spontaneous abortions. In Hemminki K, Sorsa M, Vainio H (eds): "Occupational Hazards and Reproduction." Washington DC: Hemisphere Publishing Co.

Nordström S, Beckman L, Nordenson I (1979a): Occupational and environmental risks in and around a smelter in northern Sweden. V. Spontaneous abortions among female employees and decreased birth weight in their offspring. Hereditas 90:291–296.

Nordström S, Beckman L, Nordenson I (1979b): Occupational and environmental risks in and around a smelter in northern Sweden. VI. Congenital malformations. Hereditas 90:297–300.

Pershagen G, Axelson O (1982): A validation of questionnaire information on occupational exposure and smoking. Scand J Work Environ Health 8:24–28.

Registrar General (1978): "Occupational Mortality 1970–1972." The Registrar General's Decennial Supplement for England and Wales. London: Her Majesty's Stationery Office, pp 1–224.
Rom WN (1976): Effects of lead on the female and reproduction: A review. Mt Sinai J Med 43:542–552.
Sanotsky IV (1976): Aspects of the toxicology of chloroprene: Immediate and longterm effects. Environ Health Perspect 17:85–93.
Saxén L, Rapola J (1969): "Congenital defects." New York: Holt, Rinehart and Winston Inc., p 247.
Sentrakul P, Potter EL (1966): Pathologic diagnosis on 2681 abortions at the Chicago lying-in hospital, 1957–1965. Am J Public Health 56:2083–2092.
Sever LE (1981): Reproductive hazards of the workplace. J Occup Med 23:685–689.
Shephard TH (1980): "Catalog of Teratogenic Agents," 4th ed. Baltimore: Johns Hopkins University Press.
Shumlina AV (1975): Menstrual and child-bearing functions of female workers occupationally exposed to the effects of formaldehyde. Prof Gig Zabol 19:18–21.
Silberg SL, Ransom DR, Lyon JA, Anderson PS (1979): Relationship between spray adhesives and congenital malformations. South Med J 72:1170–1173.
Spence AA, Knill-Jones RP (1978): Is there a health hazard in anaesthetic practice? Br J Anaesth 50:713–719.
Stein Z, Susser M, Warburton D, Wittes J, Kline J (1975): Spontaneous abortion as a screening device: The effect of fetal survival on the incidence of birth defects. Am J Epidemiol 102:275–290.
Stein Z, Warburton D, Kline J (In Press): Epidemiology of chromosomal anomalies and other malformations in spontaneous abortions. In Hemminki K, Sorsa M, Vainio H (eds): "Occupational Hazards and Reproduction." New York: Hemisphere Publ. Co.
Stellman JS (1978): Occupational health hazards of women: An overview. Prev Med 7:281–293.
Strandberg M, Sandbäck K, Axelson O, Sundell L (1978): Spontaneous abortions among women in hospital laboratory. Lancet i:384–385.
Strobino BR, Kline J, Stein Z (1978): Chemical and physical exposures of parents: Effects on human reproduction and offspring. Early Hum Dev 1:371–399.
Sullivan FM, Barlow SM (1979): Congenital malformations and other reproductive hazards from environmental chemicals. Proc R Soc Lond [Biol] 205:91–110.
Tomlin PJ (1979): Health problems of anaesthetists and their families in the West Midlands. Br Med J 1:779–784.
Vaisman AI (1967): Working conditions in surgery and their effect on the health of anaesthesiologists. Eksp Khir Anesteziol 3:44–49.
Vessey MP, Nunn JF (1980): Occupational hazards of anaesthesia. Br Med J 281:696–698.
Wilson JG (1977a): Environmental chemicals. In Wilson JG, Fraser FC (eds): "Handbook of Teratology," Vol. 1. New York and London: Plenum Press, pp 357–385.
Wilson JG (1977b): Embryotoxicity of drugs in man. In Wilson JG, Fraser FC (eds): "Handbook of Teratology," Vol. 1. New York and London: Plenum Press, pp 309–355.
Yamashita F (1977): Clinical features of polichlorbiphenyls (PCB)-induced fetopathy. Paediatrician 6: 20–27.
Yearbook of Nordic Statistics 1980 (1981): The Nordic Statistical Secretariat (eds): Stockholm, The Nordic Council.

American Journal of Industrial Medicine 4:309–328 (1983)

Birth Defects Monitoring

Marcus A. Klingberg, MD, Cheri M. Papier, MS, MPH, and Jacob Hart, MD

Population monitoring of birth defects provides a means for detecting relative changes in their frequency. Many varied systems have been developed throughout the world since the thalidomide tragedy of the early 1960s. Although it is difficult to pinpoint specific teratogenic agents based on rises in rates of a particular defect or a constellation of defects, monitoring systems can provide clues for hypothesis testing in epidemiological investigations. International coordination of efforts in this area resulted in the founding of the International Clearinghouse for Birth Defects Monitoring Systems (ICBDMS) in 1974. In this paper we will describe the functions and basic requirements of monitoring systems in general, and look at the development and activities of the ICBDMS. A review of known and suspected environmental teratogenic agents (eg, chemical, habitual, biological, physical, and nutritional) is also presented.

Key words: birth defects monitoring, monitoring, birth defects, congenital malformations, teratogens

INTRODUCTION

In this chapter a review of known and suspected environmental human teratogenic and embryolethal agents will be presented as well as a discussion of birth defects monitoring.

Despite the vast literature relating to this topic, Wilson has estimated that only about 10% of all congenital malformations have been definitely attributed to known environmental factors; about 25% are the result of genetic transmission and chromosomal abnormalities; and the remainder are of unknown origin [Wilson, 1977]. This may be explained, in part, by the tremendous difficulties involved in detecting specific etiologic factors when the outcome is a rare event—as are most congenital malformations. Environmental teratogens may be regarded in the following classification scheme: chemical (including drugs), habitual, biological, physical, and nutritional.

Department of Preventive and Social Medicine, Tel-Aviv University Sackler School of Medicine, Israel.
Address reprint requests to Prof. Marcus A. Klingberg, MD, Chairman, Department of Preventive and Social Medicine, Tel-Aviv University Sackler School of Medicine, P.O.B. 30940, Tel-Aviv 69978, Israel.

0271-3586/83/0401-02-0309$05.50 **© 1983 Alan R. Liss, Inc.**

DRUGS

The potential for drug teratogenicity is dependent upon a number of determinants. The first of these is the nature of the drug. Some embryonic tissues are more sensitive to certain categories of drug compounds because of the chemical structure, mode of action, or biochemical peculiarities. The second deals with the dosage, timing, and mode of drug administration. Some drugs exert their teratogenic action if given in low doses over an extended period of time; others, at higher doses for a short period of time; still others (such as thalidomide) are known to have produced malformations after only a single exposure. In general, the critical teratogenic period in humans is the 3rd–12th weeks of pregnancy. Before the third week there is usually an "all or none" response to teratogens; ie, either embryolethality or complete cellular recovery occurs. For example, with thalidomide, the critical period is 33–51 days LMP (last menstrual period) and for aminopterin, 40–54 days postconception. The mode of drug administration (ie, oral, injected, or topical) may also have an effect on teratogenic outcome. The third determinant in drug teratogenicity is the interactive effect if more than one drug is taken. The four possible effects are: none, antagonistic, additive, or synergistic. Finally, the fourth factor deals with genetic susceptibility. The fact that not all fetuses exposed to a specific agent develop abnormally implies an individual selective feature, probably inherent susceptibility.

Due to these determinants and the fact that most congenital malformations are relatively rare phenomena, it is not surprising that only a few drugs have been definitely found to be teratogenic. These include thalidomide, androgenic steroid hormones, diethylstilbestrol (DES), folic acid antagonists (eg, aminopterin and other cytotoxic drugs), anticoagulants (warfarin), and some anticonvulsants. In addition, some maternal metabolic disorders such as diabetes, epilepsy, phenylketonuria, and thyroid disorders are associated with higher rates of various malformations; it is not known for certain, however, whether the metabolic disorders themselves or the drug(s) used in treating them are responsible.

Other categories of drugs for which teratogenic activity has been suspected include the following (in alphabetical order):

Alkylating agents

Alkylating agents taken during pregnancy (such as 6-mercaptopurine, busulfan, chlorambucil, and cyclophosphamide) for the treatment of neoplasms have been implicated in increased spontaneous abortions and various malformations [Diamond et al, 1960; Nicholson, 1968; Shotton and Monie, 1963; Greenberg and Tanaka, 1964].

Antibiotics

Some studies have suggested low-level teratogenic or fetotoxic effects of penicillines, tetracyclines, and/or chloramphenicol [Carter and Wilson, 1965; Carter, 1965; Saxén, 1975]; others have not [Nelson and Forfar, 1971; Richards, 1969; Heinonen et al, 1977]. It has been recommended that tetracyclines be avoided in early pregnancy as they are known to have an affinity for teeth and bones [Heinonen et al, 1977]. In utero exposure to tetracyclines often results in staining of the deciduous teeth [Cohlan, 1977]. Others consider tetracyclines to be contraindicated only from the fourth month of pregnancy.

Anticonvulsants

Malformations found to be associated with anticonvulsants (including phenytoin, phenobarbitone, sulthiame, phensuximide, pheneturide, trimethadione (troxidon), paramethadione, exthosuximide, and carbamazepine) are oral clefts, cardiovascular, skeletal, digital,, central nervous system, and mental abnormalities [Visser et al, 1976; Speidel and Meadow, 1974; Annegers et al, 1978; Loughnan et al, 1973; Zachai et al, 1975]. Other studies have implicated the disease being treated (eg, either maternal or paternal epilepsy) rather than the drugs as the primary teratogenic determinant [Shapiro et al, 1976].

Antidiabetic Drugs

In the Collaborative Perinatal Project (CPP), a large-scale prospective investigation of the relationship between drug consumption during pregnancy and congenital malformations, a strong association was found between insulin use and congenital malformations; however, it was impossible to determine whether the association was due to the insulin therapy or to the disease being treated [Heinonen et al, 1977]. Epidemiological studies have found higher rates of central nervous system defects, cardiac defects, and caudal regression among offspring of diabetic women [Kučera, 1971; Chung and Myrianthopoulos, 1975]. Treatment with oral hypoglycemics such as chlorpropamide has been reported to incur a threefold excess risk of intrauterine fetal death than either insulin or diet therapy [Babson and Benson, 1971]. In addition, tolbutamide use by pregnant diabetics has been implicated in thrombocytopenia and hypoglycemia in the neonate [Babson and Benson, 1971].

Antinauseants and Antihistamines

A great deal of controversy has recently been generated regarding the possible teratogenicity of the antinauseant Bendectin (also known as Debendox). The components of this preparation originally included doxylamine succinate (an antihistamine), dicyclomine hydrochloride (an antispasmodic), and pyridoxine hydrochloride (a vitamin), although dicyclomine hydrochloride was removed as of 1976 in the US due to ineffectiveness. A number of studies have suggested a slightly increased risk of various malformations following exposure to the drug in early pregnancy [Paterson, 1977; Dickson, 1977; Frith, 1978; Menzies, 1978; Henderson, 1977; Golding and Baldwin, 1980; Rothman et al, 1979], but the lack of a characteristic pattern of defects weakens the possibility of its teratogenicity. In addition, studies which carefully controlled for confounding variables and biased recall did *not* reveal any increased malformation risk with this preparation [Mitchell et al, 1981; Smithells and Sheppard, 1978; Cordero et al, 1981; Gibson et al, 1981; Milkovich and van den Berg, 1976; Shapiro et al, 1977]. Therefore, in our opinion, there is *no convincing evidence* that Bendectin is teratogenic. Other antinauseant/antihistamines, including meclizine and cyclizine, were similarly found not to exhibit teratogenic effects [Shapiro et al, 1978; Mellin, 1963; Smithells and Chinn, 1964; Yerushalmy and Milkovich, 1965].

Antithyroid Drugs

Drugs such as iodine, iodides, uracils, perchlorates, and imidazoles have been found to produce fetal goiters [Babson and Benson, 1971; Wolff, 1969; Croughs and

Visser, 1965]. Thyroid suppressants were suggestive of possible teratogenicity in the Collaborative Perinatal Project [Heinonen et al, 1977], consistent with the findings of Milham and Elledge [1972].

Antituberculosis Drugs

In a review of antituberculosis drugs, Warkany [1979] concluded that (1) streptomycin and dihydrostreptomycin seem to cause damage to the eighth nerve resulting in hearing loss in approximately 10% of prenatally exposed children; other potentiating factors may include exposure during the first trimester, high doses, prolonged treatment, and individual susceptibility; and (2) other antituberculosis drugs such as rifampicin, ethionamide, isoniazid, aminosalicylic acid and ethambutol have not shown conclusive evidence of human teratogenicity to date, although a slight risk of ototoxic and possibly cerebral damage may occur. The author suggests that the risks would be reduced by treating the women after the first trimester and by prescribing minimal effective doses.

Hormones

In a recent critical review of the possible teratogenicity of female sex hormones, a WHO Scientific Group [1981] concluded that "there was *no* evidence for any adverse effects on the fetus" for oral contraceptives taken *before* pregnancy. Inadvertent use during early pregnancy may increase the rate of spontaneous abortions, but there is no indication of increased risk of congenital malformations. Regarding hormonal pregnancy tests, no definite association with increased risk of birth defects has been demonstrated. However, it has been recommended that their use be discontinued since other safe means of pregnancy testing are available. Hormonal drugs used in treatment of threatened abortion have not been shown to be effective, so their discontinuance has also been recommended. Diethylstilbestrol (DES), widely used in the 1940s and 1950s has recently been found to be teratogenic [Kaufman et al, 1977; Berger and Goldstein, 1980; Barnes et al, 1980; Gill et al, 1979] in addition to its action as a transplacental carcinogen [Herbst and Scully, 1970; Herbst et al, 1977].

Immunizing Agents

Active immunization with a vaccine, whether it be a live-attenuated or a killed-inactivated product, should, as a general rule, be initiated *before* the onset of pregnancy. The object is to render a woman immune so that if viral exposures occur during pregnancy, the fetus will be protected. Another general principle is that all live vaccines are contraindicated in pregnancy. On the other hand, killed vaccines or toxoids are not contraindicated.

Rubella vaccines are contraindicated throughout the entire pregnancy because the vaccine strains, although attenuated, must be assumed to be potentially teratogenic to the fetus. Pregnancy should be avoided for at least eight weeks after vaccination. Rubella vaccines should not be administered to any woman of childbearing age unless her prior immune status is determined. Although smallpox has been successfully eradicated and there is no longer a need for vaccination, it should be remembered that primary maternal vaccination in the first or second trimesters may lead to fetal infection and possibly stillbirth or abortion [Horstman, 1969]. Vaccination of the pregnant mother at the later stage of pregnancy seems relatively safe to the fetus. Oral poliovaccine is contraindicated in pregnancy on the general principle that all live vaccines are best avoided in pregnancy. If immunization is considered necessary, the inactivated Salk-type vaccine can be used. Measles and mumps vaccines are contraindicated during pregnancy.

Neurotropic-Anorectics

Amphetamines and phenmetrazine, which are sometimes used to control weight during pregnancy, have been implicated as being possibly teratogenic in several studies [Nelson and Forfar, 1971; Nora et al, 1970; Levin, 1971; Lenz, 1962; Powell and Johnstone, 1962]. No such associations were found in the Collaborative Perinatal Study [Heinonen et al, 1977]

Sedatives, Neuroleptics, and Antidepressants

As a class, these drugs exhibited no overall association with congenital malformations in the Collaborative Perinatal Study [Heinonen et al, 1977]. A controlled study by Milkovich and van den Berg [1974] showed significantly higher rates of defects and increased rates of fetal death for those exposed during early pregnancy to the tranquilizers meprobamate and chlordiazepoxide, although no definite pattern of defects was discerned. No such associations were observed in the CPP.

Lithium carbonate, used in treating manic-depressives, has recently become suspect in the rare cardiac defect Ebstein's anomaly [Schou et al, 1973; Nora et al, 1974] as well as in other cardiovascular defects [Mignot et al, 1978; Weinstein and Goldfield, 1975].

OTHER CHEMICAL AGENTS

Apart from drugs, other chemical exposures in the environment and workplace have become suspect as possible teratogenic agents.

Anesthetic Gases

Although numerous retrospective studies have suggested the possibility of increased risk of spontaneous abortion, stillbirths and/or congenital malformations among operating room personnel (and even among wives of male anesthesiologists) [Askrog and Harvald, 1970; Vaisman, 1967; Pharoah et al, 1977], Vessey [1978] concluded that due to methodological problems and bias in many of the studies, such inferences are unjustified except for the possible slightly increased risk of spontaneous abortion. One other study, conducted in Sweden, in which possible confounding factors were taken into consideration and bias reduced to a minimum found *no* differences in incidence of threatened abortions, birth weight, perinatal death rate, or congenital malformations [Ericson and Källén, 1979].

Vinyl Chloride

Wives of vinyl chloride workers have been reported to exhibit excess fetal loss, possibly due to paternal germ cell damage [Infante et al, 1976]. Rates of congenital abnormalities were reported to be higher in three Ohio communities in which a polyvinyl chloride plant is located, using the state rates for comparison [Infante, 1976a], although such indirect evidence is far from satisfactory. Case-control studies conducted by the Center for Disease Control (CDC), on the other hand, did not show an association between vinyl chloride exposure (either occupational or environmental) and congenital malformations [Edmonds et al, 1975; Edmonds et al, 1978]. Therefore no definite conclusions can be drawn.

Organic Solvents

Organic solvents apart from volatile anesthetics have been found to be associated with CNS defects in a case-control study in Finland [Holmberg, 1979]. Significantly more mothers of affected babies than control mothers had been exposed to organic solvents at their place of work during the first trimester of pregnancy. Another Finnish study dealing with female chemical workers linked information on 9,000 workers with data from the National Board of Health Hospital Discharge Registry. Rates of spontaneous abortions were compared with national rates and found to be significantly higher [Hemminki et al, 1980]. In addition, exposure to laboratory chemicals among women in hospital laboratories has been implicated in increased rates of spontaneous abortions [Strandberg et al, 1978]. Confirmation of these studies is urgently needed.

Methyl Mercury

One of the already "classic" cases of environmental contamination occurred in Japan, where mercury compounds were dumped into the waters of Minamata Bay by a chemical company known as the Chisso Corporation. Congenital Minamata disease, first described in 1965 [Matsumoto et al, 1965; Takeuchi and Matsumoto, 1969], was caused by maternal ingestion of contaminated fish mainly during the second and third trimesters of pregnancy and resulted in babies born with severe neurological disorders similar to cerebral palsy.

PCBs, PBBs, and TCDD

Exposure to polychlorinated biphenyls (PCBs), polybrominated biphenyls (PBBs) and to 2,3,7,8-tetrachlorodibenzo-p-dioxin (TCDD) due to environmental contamination has recently become the focus of many epidemiological investigations; however, to date, no teratogenic outcomes have been definitely implicated in humans.

Lead

Both maternal and paternal exposure to lead at the workplace have been known to result in pregnancy wastage (spontaneous abortions, stillbirths, and neonatal deaths) for more than a century [Paul, 1860]. This has been reconfirmed more recently [Lund, 1936; Infante, 1976b; Angle and McIntire, 1964], although no evidence of teratogenic effects has been observed.

Pesticides and Herbicides

Manufacture of DBCP (1,2-dibromo-3-chloropropane) has been implicated in male infertility in studies in California [Whorton et al, 1977] and in Israel [Kharrazi et al, 1980]. Women vineyard workers in Crimea exposed to DDT, sulfur, methyl parathion, and copper sulfate had higher frequencies of miscarriage than unexposed controls [Nikitina, 1974]. Central nervous system disturbances, histological changes in the placenta, and low birth weight were also reported in that study — one of the few such studies of occupationally exposed women. There is little evidence to date of human embryotoxic or teratogenic effects resulting from pesticide use at the current levels.

Herbicides have also been claimed to be associated with various adverse reproductive outcomes; however, no proof of these claims has been forthcoming. The herbicide Agent Orange used in Vietnam has increasingly become the focus of lawsuits by US servicemen, claiming that the birth of their babies with defects was due to their Agent Orange exposure in Vietnam. TCDD has been found as an impurity in the manufacture of Agent Orange [Shapley, 1973; Cutting et al, 1970]. A major epidemiological study of this problem currently being conducted on the outcome of pregnancy in Vietnam has shown inconsistent results, and to date no correlation has been apparent between rates of stillbirths or malformations and intensity of spraying.

Toxic Waste Dumps

Love Canal, located in the city of Niagara Falls, New York, has become the most renowned toxic waste dump site in the US, although thousands of others are scattered throughout the country. A major epidemiological study conducted by the New York State Health Department [1981] examined three adverse pregnancy outcomes – spontaneous abortions, congenital malformations, and low birth weight. Two control groups were used for comparison, one from a Canadian population and the other from a neighborhood just beyond the Love Canal district. Results showed a slight increase in spontaneous abortions and low birth weight among women living closest to the canal and those living in wetland areas of the neighborhood; no significant increase in congenital malformations was found in the areas under study, with the exception of a slight increase in rates for the wetland areas. This study, however, has been questioned on methodological grounds.

HABITUAL PRACTICES

The category called "habitual practices," including smoking, drinking and use of addictive drugs, could easily have been included in the previous chemical/drug section, although it was thought appropriate to discuss separately.

Smoking

To date, no studies have presented convincing evidence of any teratological effects due to maternal smoking, except for the well-known and consistently found effect of low birth weight. Increased rates of spontaneous abortion have also been implicated [Kline et al, 1977], as have rates of abruptio placentae and perinatal deaths [Surgeon General, 1981].

Drinking

The "fetal alcohol syndrome" has recently been quite well delineated [Jones et al, 1976; Clarren and Smith, 1978; Streissguth, 1978]. Typically associated defects include CNS deficiencies, characteristic facial abnormalities, variable major and minor malformations and growth deficiencies. Moderate drinking has been suggested to be associated with increased rates of spontaneous abortions, after controlling for smoking and other risk factors [Harlap and Shiono, 1980; Kline et al, 1980].

Addictive Drugs

In a review of over 100 publications on the possible cytogenic and teratogenic effects of LSD (lysergic acid diethylamide), the compound in question could not be directly attributed to increased mutagenic or teratogenic effects in humans [Dishotsky et al, 1971]. Similarly, marijuana used during pregnancy has not been determined to be teratogenic [Wilson, 1976], nor have narcotics such as morphine, heroin, or methadone.

BIOLOGICAL HAZARDS

Biological factors representing known or potential teratogenic hazards to the developing fetus include a number of viral, bacterial, and protozoan infections. Among them are the following:

Rubella Virus

The Australian ophthalmologist, Gregg, noted a higher-than-usual frequency of congenital cataracts occurring among children born in 1941, and suggested rubella infection as the causal factor. This represented the first time that an environmental agent was found to produce congenital malformations in humans. Subsequent case-control studies confirmed his hypothesis, and later prospective studies established rubella as a teratogenic agent beyond doubt. The period of greatest risk to the fetus seems to be in the first eight to ten weeks of pregnancy [Rawls et al, 1968], although the first four months are considered as susceptible ones. Other associated defects that may occur include persistent ductus arteriosus, glaucoma, microphthalmia, skin and skeletal lesions, speech and hearing disorders, brain damage, and growth retardation [Smithells, 1976].

Cytomegalovirus

Although this viral infection is the one most commonly affecting the human fetus, it is estimated that only about 10% of the infected actually develop symptoms [Hanshaw et al, 1973]. As opposed to the many possible malformations resulting from rubella infection, cytomegalovirus tends to cause damage primarily to the central nervous system. It is difficult to pinpoint the critical period of infection, as the disease may pass asymptomatically in the mother.

Herpes Simplex Virus

Case studies have been reported of both cervical and transplacental herpes virus infections (types 1 and 2) in which outcomes of microcephaly, microphthalmia, cerebral calcification, retinal dysplasia, patent ductus arteriosus, skin lesions, short digits, and psychomotor retardation have occurred [South et al, 1969; Montgomery et al, 1973].

Varicella and Herpes Zoster

Under certain conditions, these viruses are thought to be capable of producing congenital malformations, although in a study of 293 varicella infections during pregnancy only 7 were born with major malformations [Manson et al, 1960]. Congenital heart disease, limb reduction deformities, cutaneous scars, ocular and CNS lesions, encephalomyelitis, and low birth weight have been reported to be associated [Dodion-Fransen et al, 1973; Savage et al, 1973].

Venezuelan Equine Encephalitis

Wenger [1967] has implicated Venezuelan equine encephalitis in the production of various CNS disorders, including encephalitis, hydroencephaly, microcephaly with microphthalmia, and cerebral cysts following an epidemic of the viral disease.

Influenza and Respiratory Viruses

The possible teratogenic effects of common cold viruses and influenza are inconclusive. Although some case-control studies have shown higher rates of malformations among babies whose mothers had influenza during pregnancy than among controls [Coffey and Jessop, 1955; Hardy et al, 1961; Hakosalo and Saxén, 1971] and cohort studies have indicated higher incidence rates following influenza epidemics [Coffey and Jessop, 1959], other studies have not [Manson et al, 1960; Leck, 1963]. As findings have not been consistent, the causal nature of any association is put into doubt.

Other Viral Infections

Viral infections that may produce fetal disease but not necessarily result in teratogenic outcomes include mumps, rubeola, enteroviruses, polio, variola, vaccinia, hepatitis, and Western equine encephalitis [Kleinman et al, 1962; Landsman et al, 1964].

Bacterial Infections

The spirochete Treponema pallidum, found in syphilitic infections, causes stillbirths in nearly a quarter of all affected pregnancies. Those born alive may exhibit a variety of defects including CNS damage; osteochondritis of joints and periostitis of bones; hepatosplenomegaly; lymphadenopathy; rhinitis; and mouth, anal, and genital lesions [Sever et al, 1976]. However, it should be noted that congenital syphilis is a very rare phenomenon. The annual number of cases in the US has fallen from 451 in 1971 to 111 in 1980 [CDC Morbidity and Mortality Weekly Report, 1981].

Listeria bacillus infection may result in stillbirths, spontaneous abortions, and neonatal death [Sever et al, 1976].

Protozoan Infections

The protozoan parasite *Toxoplasma gondii*, transmitted through improperly cooked mutton and pork or through contaminated soil or cat feces, may result in a 20% malformation rate in exposed fetuses, with the mother asymptomatic. Defects thought to be associated include microcephaly or hydrocephaly, chorioretinitis, anemia, cerebral palsy, splenomegaly or hepatomegaly, epilepsy, lymphadenopathy, and mental retardation [Sever et al, 1976; Remington, 1968].

PHYSICAL AGENTS

One "physical" teratogen that has been extensively studied is radiation. Background and ionizing radiation (X-rays) will be discussed here. Noise is another physical agent that has been considered as a potential teratogenic agent, but too little information is available to date for comment.

Background Radiation

Studies on the effects of background radiation on the human fetus have been controversial. Some have indicated increased rates of congenital malformations (based on average exposure per pregnancy of 0.1 rad), while others, based on the same average exposure, have not [Brent, 1977]. "Although the background radiation may contribute a small part (less than 1%) to the spontaneous mutation rate in the prenatal as well as the postnatal period in humans, both the magnitude and rate are so low that background radiation absorbed by the embryo during gestation is thought not to be a factor in the normal incidence of congenital malformations, intrauterine or extrauterine growth retardation, or embryonic death" [Brent, 1977].

X-rays

As to the threshold dose levels necessary to produce serious effects, "from a clinical point of view, an absorbed dose of 10 rads to the fetus at any time during gestation can be considered as a practical threshold for the induction of congenital defects, below which probability of producing adverse effects becomes vanishingly small" [Brent, 1976]. Exposure during the first three weeks of pregnancy more often results in spontaneous abortion than congenital malformations [Swartz and Reichling, 1978]. A recent report of a long-term cohort study dealing with the reproductive experience of black females exposed in utero to low doses of X-ray (from 1–5 rads), revealed higher rates of stillbirths compared to an unexposed group matched for parity, hospital of birth, and birthdate [Meyer and Tonascia, 1981].

For pregnant women working as X-ray technicians, radiologists, or other professions in which exposure to radiation is likely to occur, it has been recommended by the US National Council on Radiation Protection that the dosage equivalent of 0.5 rem to the fetus should not be exceeded during the entire gestational period [Parker and Taylor, 1971].

Further information on the subject of radiation teratogenesis can be found in Brent [1980].

NUTRITIONAL FACTORS

The final category of potential teratogenic agents to be discussed is nutritional.

Malnutrition

Maternal malnutrition during pregnancy is known to result in fetal growth retardation, affecting both cell number and cell size [Winick, 1970]. During periods of famine, as occurred during World War II, rates of spontaneous abortions, stillbirths, neonatal deaths and congenital malformations were found to be higher than in prewar years [Smith, 1947a; Smith, 1947b; Antonov, 1942].

Folate Deficiency

Administration of folate antagonists such as aminopterin and other cytotoxic drugs (see also alkylating agents) have been associated with congenital malformations as early as 1952 [Thiersch, 1952]. A prospective study of 900 women whose red cell folic acid levels were measured during the first trimester of pregnancy revealed significantly lower levels among those subsequently giving birth to babies with CNS defects than among controls [Smithells et al, 1976]. Depression of folic acid levels has been noted following the ingestion of alcohol, suggesting a possible mechanism for the fetal alcohol syndrome [Eichner and Hillman, 1973].

Not all studies, however, have demonstrated a teratogenic effect of folate deficiency, which led Smithells to remark: "It seems improbable that folate deficiency is ever an isolated cause of congenital defect in the human being, but it may prove to be one correctable factor involved in malformations of multifactorial origin" [Smithells, 1976].

Other Vitamins and Minerals

Although much experimental animal evidence exists with regard to the possible teratogenicity of both deficiencies *and* excesses of most vitamins and minerals, very little information is available for humans. A recent study by Smithells et al [1980] suggests the possible prevention of CNS defects by periconceptional vitamin supplementation.

BIRTH DEFECTS MONITORING

Our search for teratogenic agents is a long and difficult one. Simple, fast techniques for detecting teratogens, similar to the Ames test for carcinogenicity/mutagenicity, have not been developed. In a recent article, Barrach and Neubert [1980] discuss progress in this area. The authors point out, however, that in vitro techniques, such as organ culture, cultivation of preimplantation embryos, or whole embryo culture, "provide little, if any advantage over *in vivo* experiments if a 'mass screening' of possible teratogenic potential of chemicals (hazards for the human population) is attempted." In fact, they think in vitro systems used separately may be leading us astray in our search for teratogenic and embryolethal agents.

Discussion of the most desirable approach to use in identifying environmental teratogens is worthless. We should employ whatever sources are available, evaluating and comparing their usefulness with the understanding that each may make a specific contribution. It should be remembered that even the potent teratogen, thalidomide, with its characteristic malformations, required six years from the time it was marketed in West Germany and the birth of over 6,000 deformed babies before the drug-defect relationship was established.

One or perhaps the most practical means to reveal the etiology of birth defects and discover associations of malformations with potential teratogens is by monitoring birth defects in the newborn population. Monitoring provides an early-warning system through which possible teratogenic agents can be brought under investigation and eventually controlled. Therefore, on the heels of the thalidomide experience, birth defects monitoring was initiated in various countries as one of the principal means of preventing the "yet-unborn" from becoming the victims of our negligence [Klingberg and Papier, 1979a].

In 1963, monitoring systems on community, regional and national levels were proposed in order to alert physicians to sudden rises in frequency of specific congenital malformations and to the presence of new teratogens in the environment [Ingalls and Klingberg, 1963, 1965].

Population monitoring is clearly an indirect and difficult way of detecting environmental hazards. But obvious ethical and practical limitations preclude attempts to investigate human teratogenic effects directly. And despite recent progress in experimental teratology, it still appears a long way off before all environmental agents to which we are exposed can be investigated by approaches in the laboratory that are both ethically acceptable and clinically relevant.

Birth defects monitoring, as considered in this chapter, refers to monitoring the incidence of birth defects in the population to detect possible introduction of teratogens in the environment.

The main functions of a birth defects monitoring system can be summarized as follows:

1. An index facilitates follow-up investigations. It is essential when the frequency of a specific defect or combination of defects in live and stillbirths exceeds a critical level, possibly suggesting the presence of a new teratogenic agent.
2. Incidence rates are generated from reliable diagnostic input so that divergences from the baseline rates can be measured with confidence.
3. A monitoring system provides a base for epidemiologic investigations and research into etiology. The extent of data usage depends upon the nature of the information available to the program. In order to meet the demands of rapid monitoring, this information may be quite limited, although accessible in other forms available to the monitoring system.
4. In addition, a monitoring program can have very useful educational and public relations purposes in order to justify the existence of the system.

The basic requirements for operating a birth defects monitoring system are—

1. consistency in terms of accuracy, completeness, and uniformity of recording events;
2. minimal delay in forwarding reports;
3. rapid processing of data and communication of results;
4. appropriate analysis of output;
5. mechanism for follow-up of an apparently significant increase.

It should be stressed that although the principal feature of a monitoring system is urgency, both in accumulating data and reporting rates and trends, this must not be accomplished at the expense of accurate diagnosis.

With such programs, one of the most important aims is to ensure comparability of the data from different centers, and within the same center over a period of time by standardization of techniques. In the present situation, there is much diversity in the methods of data collection. The forms used for recording information vary considerably and are related to the availability of special data-handling equipment. Differences in classifications of malformations as well as differential age at diagnosis tend to reduce or even prohibit valid comparison of data. The fundamental prerequisite, of course, is good, reliable baseline data. Without knowledge of baseline rates for the specific malformations being monitored, there would be no way to determine changes in frequency by comparison of observed versus expected rates.

In both animals and humans, teratogenic agents tend to produce multiple malformations. Therefore, it is essential that multiple as well as single malformations be recorded and analyzed. A sudden rise in frequency of a specific malformation or an unusual syndrome may be the first sign that a new teratogenic substance has been introduced.

In order to be effective, monitoring programs require prompt reporting, and frequent review and analysis of recorded data. But there are some critical points in statistical analysis and methods used in measuring rates and trends. The whole success of the operation depends on sounding a warning only when a genuine increase in the rate of one or more of the defects being monitored occurs, and not when an apparent increase due to random fluctuation occurs. Some false alarms can be tolerated, but too many will produce dissatisfaction on the part of those whose cooperation is necessary for the success of the program, especially if specific types of malformations are considered individually, so one type is likely to be small unless the population base is very large. Long-term studies on incidence clearly show considerable annual variation. A surveillance system must be able to detect this normal random variation and differentiate it from that produced by a new teratogenic agent. The agent may be introduced gradually and cause a steady rather than a sharp increase. Therefore, most programs use multiple tests to detect both types of trends. A more detailed description of the various statistical methods for analyzing birth defects data used in the different systems can be found in reports by Hill et al [1968], Weatherall and Haskey [1976], Bjerkedal and Bakketeig [1975], and Chen [1978].

Finally, a monitoring program must have a well-established mechanism for immediate follow-up, for a teratoepidemiologic investigation when a genuine increase in rate is detected. It is also essential to remember that the reporting card or forms have sufficient identifying information to enable tracing the original hospital records of the mother and child for facilitating the follow-up.

After the thalidomide tragedy, a number of countries developed birth defects monitoring programs; however, there was a need for a systematic means of exchanging information and collaboration among the operating systems. As a result, several persons already responsible for monitoring congenital malformations met in Helsinki in 1974 at the invitation of the March of Dimes Birth Defects Foundation (formerly The National Foundation — March of Dimes) to discuss international cooperation in monitoring congenital malformations. As a result of the Helsinki meeting, the International Clearinghouse for Birth Defects Monitoring Systems (ICBDMS) was created. Its main goal is international collaboration to facilitate the routine exchange of information among the existing monitoring programs throughout the world by means of quarterly reports and annual workshops of the representative countries. It thus promotes international cooperation, encouragement and consultation to others interested in developing such programs. The International Clearinghouse provides the means for obtaining data on trends in incidence rates of specific malformations and syndromes, and exchanging ideas on various experiences such as improvements in identification and classification of congenital malformations (single as well as multiple), coding, reporting, and follow-up investigations.

One of the most important aims of the International Clearinghouse is to provide expert training for members and others on methods of detection of birth defects and on the investigation of suspected outbreaks.

During the year following the establishment of the International Clearinghouse, nine countries participated and exchanged data: Canada, England and Wales, France Hungary, Israel, Norway, Sweden, and the United States. To date, 20 monitoring programs in 24 countries are collaborating in this organization: Canada, Czechoslovakia, Denmark, England and Wales, Finland, France, Hungary, Israel, Italy, Japan, Mexico, New Zealand, Northern Ireland, Norway, South America (7 countries), Spain, Sweden, and the United States. Some of the programs are nationally based, some cover a defined population in a state or province, while others are based upon hospital reporting without definable population bases. The size of the programs, in terms of annual number of births, ranges widely from 10,000 to more than 1 million. Several other countries are in the process of preparing their programs for participation by establishing the basic requirements such as two-year baseline data. These are Australia and the Netherlands with nationwide programs and Switzerland with Kanton-based system.

During the working conference in 1981, criteria for two levels of membership were adopted: full and associate membership. A monitoring system with a minimum of 20,000 annual births and at least 100,000 births accumulated during at least two years may apply for full membership. One with a minimum of 5,000 annual births may apply for associate membership. Although associate members cannot participate in the regular quarterly exchange of data, they may take part in joint activities of the Clearinghouse and communicate on unusual events [ICBDMS, 1982].

From among the many conditions monitored in the various individual programs, 11 malformations were selected by the program directors for use in the regular exchange of incidence data: anencephaly; spina bifida; hydrocephaly; cleft palate; total cleft lip; tracheoesophageal fistula, esophageal atresia, and stenosis; rectal and anal atresia and stenosis; hypospadias, reduction deformity of limbs; omphalocele and Down's syndrome. These malformations were selected for a variety of reasons, including ease and uniformity of recognition and diagnosis, discrete code number, and preexisting concern about observed changes in incidence. Of course, the programs are not restricted to monitoring these particular malformations only; they continue to monitor as many as are feasible and possible (whether major, minor, or multiple). Increases in rates of malformations other than the 11 listed are also reported to the Clearinghouse for distribution among the participating countries. It must be emphasized that even minor malformations which may be of no serious medical or cosmetic consequence to the patient may reflect some environmental insult to the developing fetus. In addition, thorough examination of stillbirths as well as karyotype analysis of abortuses is desirable in any monitoring system.

Each quarterly report, which is sent in confidence to the members of the Clearinghouse, contains a section devoted to brief comments from the participants about observations or inquiries regarding possible teratogens, reports of new studies, recent publications, and other pertinent information.

The quarterly report contains individual tables for each of the 11 selected congenital malformations during the period concerned, showing the baseline rates, the expected and observed number of cases, the observed rate per 10,000 births and the ratio of observed to expected cases. It also includes a series of graphs of incidence trends for each of the 11 malformations over the past five years. The last section of the report is a series of tables showing, for each monitoring program, the number of births and cases of the selected malformations by quarter and year for the past five years.

The Clearinghouse members, who form a society of experts in the field of recording, interpreting, and testing specific hypotheses about a suspected causal agent, have divided their responsibilities and planning efforts into small working subcommittees. One deals with improving the classification of congenital malformations and making recommendations for the tenth revision of the International Classification of Diseases (ICD). Another deals specifically with the problems of classifying, coding, and reporting multiple malformations. There is a group of experts which is setting up a system for reporting intrauterine growth disturbances. The epidemiology committee is developing procedures and methodologies for the prompt investigation of temporal and/or spatial clusters of specific malformations of high incidence. This group is also considering the feasibility of international cooperation in case-control studies of rare malformations, developing studies of fetal effects of maternal ingestion of rarely-prescribed drugs and the possible adverse effects of industrial exposure during pregnancy on the embryo.

At this point it must be stressed that the field of occupational teratology is still a neglected one. Physicians, epidemiologists and teratologists should be encouraged to conduct studies on accidental or industrial exposures of workers to potential teratogens. Combined exposures which are commonly encountered at the workplace must be taken into consideration in these studies.

During the last few years a number of unusual events have been reported to the International Clearinghouse. In Norway and Sweden, for example, an increase in incidence of hypospadias was reported [Bjerkedal and Bakketeig, 1975; Bjerkedal, 1977]; in Israel, a steady rise in rates of congenital dislocation of the hip was encountered from 1966 to 1975, followed by a sharp downward trend [Klingberg and Papier, 1979b]. The current first Annual Report for 1980 issued by the International Clearinghouse [1982], provides a table of reported unexplained increases in incidence or localized clusters of one or more congenital malformations during the period of January 1980 to June 1981. Five monitoring systems reported increased incidence of omphalocele, two of which experienced higher rates for two or three consecutive quarters.

Of special interest is the outbreak of a specific and previously rare type of malformation — unilateral femur aplasia or hypoplasia — reported by the monitoring system in the region of Rhone Alps, France. Until 1979, only one case of such femur malformation had occurred in that area, in which there are approximately 75,000 annual births; in 1980, 11 cases were reported, and in the first half of 1981 an additional 6 cases were noticed [Robert et al, 1981]. In response to this event, the International Clearinghouse collected all available information from the various monitoring programs to estimate the prevalence at birth of femur reduction malformations. The outbreak in the Rhone Alps is "a model for what can be found in a monitoring system. Some help in evaluation of the observed accumulation of cases was obtained by the international contribution of data..." [ICBDMS, 1982]. Thus, the monitoring programs are sensitive enough to pick up departures from the "normal," allowing subsequent follow-up.

In conclusion, we should like to mention that during the past 20 years — since the thalidomide era — a gradual shift in the attitude of the medical profession has occurred. Whereas in the 1960s and 1970s all chemicals (including drugs) were incriminated as being harmful to the embryo until proven otherwise, in the last few years, the trend has been to consider them harmless until proven otherwise. Both extremes, as in other fields, are inappropriate.

REFERENCES

Angle CR, McIntire MS (1964): Lead poisoning during pregnancy. Am J Dis Child 108:436–439.

Annegers JF, Hauser WA, Elvebach LR, Anderson VE, Kurland LT (1978): Congenital malformations and seizure disorders in offspring of parents with epilepsy. Int J Epidemiol 7:241.

Antonov AN (1942): Children born during the siege of Leningrad in 1942. J Pediatr 30:250.

Askrog V, Harvald B (1970): Teratogen effekt of inhalation-sanaestetika. Nord Med 83:498–500.

Babson SG, Benson RC (1971): "Management of High-Risk Pregnancy and Intensive Care of the Neonate." St. Louis: C.V. Mosby.

Barnes AB, Colton T, Gurdersen J, Noller KL, Tilley BC, Strama T, Townsend DE, Hatab P, O'Brien OC (1980): Fertility and outcome of pregnancy in women exposed in utero to diethylstilbestrol. New Engl J Med 302:609–613.

Barrach HJ, Neubert D (1980): Significance of organ culture techniques for evaluation of prenatal toxicity. Arch Toxicol 45:161–187.
Berger MJ, Goldstein DP (1980): Impaired reproductive performance in DES-exposed women. Obstet Gynecol 55:25–27.
Bjerkedal T (1977): Monitoring of birth defects as a method in the study of etiologic factors of congenital malformations. Paper presented at 2nd International Congress of Human Reproduction, Tel Aviv, Israel.
Bjerkedal T, Bakketeig LS (1975): Surveillance of congenital malformations and other conditions of the newborn. Int J Epidemiol. 4:31–36.
Brent RL (1976): Environmental factors. In Brent RL, Harris MI (eds): "Prevention of Embryonic, Fetal and Perinatal Disease, Vol 3." Bethesda: USDHEW, p 182.
Brent RL (1977): Radiations and other physical agents. In Wilson JG, Fraser FC (eds): "Handbook of Teratology, Vol I. General Principles and Etiology." New York: Plenum, pp. 153–223.
Brent RL (1980): Radiation teratogenesis. Teratology 21:281–298.
Carter MP (1965): Discussion of a paper by B.C.S. Slater. In Robson JM, Sullivan FM, Smith RL (eds): "Embryopathic Activity of Drugs." Boston: Little, Brown, p 253.
Carter MP, Wilson F (1965): Antibiotics in early pregnancy and congenital malformations. Dev Med Child Neurol 7:353.
CDC Morbidity and Mortality Weekly Report (1981): Syphilis trends in the United States. 30:441–444.
Chen R (1978): A surveillance system for congenital malformations. J Am Stat Assoc 73:323–327.
Chung CS, Myrianthopoulos NC (1975): Factors affecting risk of congenital malformations. II. Effect of maternal diabetes on congenital malformations. Birth Defects Original Articles Series 11:23.
Clarren SK, Smith DW (1978): The fetal alcohol syndrome. N Engl J Med 298:1063–1067.
Coffey VP, Jessop WJE (1955): Congenital abnormalities. Ir J Med Sci 349:30–46.
Coffey VP, Jessop WJE (1959): Maternal influenza and congenital deformities. A prospective study. Lancet ii:935–938.
Coffey VP, Jessop WJE (1955): Congenital abnormalities. Ir J Med Sci 30.
Coffey VP, Jessop WJE (1959): Maternal influenza and congenital deformities. A prospective study. Lancet ii:935.
Cohlan SQ (1977): Tetracycline staining of teeth. Teratology 15:127–130.
Cordero JF, Oakley GP, Greenberg F, James LM (1981): Is Benedectin a teratogen? J Am Med Assoc 245: 2307–2310.
Croughs W, Visser HKA (1965): Familial iodide-induced goiter. Evidence for an abnormality in the pituitary-thyroid homeostatic control. J Pediatr 67:353.
Cutting RT, Phuoc TH, Ballo JM, Benenson MW, Evans CH (1970): Congenital malformations, hydatidiform moles and stillbirths in the Republic of Vietnam 1960–1969. Washington, D.C.: Government Printing Office.
Diamond I, Anderson MM, McCreadie SR (1960): Transplacental transmission of busulfan (Myleran) in a mother with leukemia: Production of fetal malformation and cytomegaly. Pediatrics (Springfield) 25:85.
Dickson JH (1977): Congenital deformities associated with Bendectin. Can Med Assoc J 117:721.
Dishotsky NI, Loughman WD, Mogar RE, Lipscomb WR (1971): LSD and genetic damage. Is LSD chromosome damaging, carcinogenic, mutagenic or teratogenic? Science 172:431–440.
Dodion-Fransen J, Dekegel D, Thiry L (1973): Congenital varicella-zoster infection related to maternal disease in early pregnancy. Scand J Infect Dis 5:149.
Edmonds LD, Anderson CE, Flynt JW, James LM (1978): Congenital central nervous system malformations and vinyl chloride monomer exposure: A community study. Teratology 17:137–142.
Edmonds LD, Falk H, Nissim JE (1975): Congenital malformations and vinyl chloride. Lancet ii:1095.
Eichner ER, Hillman RS (1973): Effect of alcohol on serum folate level. J Clin Invest 52:584.
Ericson A, Källén B (1979): Survey of infants born in 1973 or 1975 to Swedish women working in operating rooms during their pregnancies. Anesth Analg (Cleve) 58:302–305.
Frith K (1978): Fetal malformation after Debendox treatment in early pregnancy. Br Med J 1:925.
Gibson GT, Colley DP, McMichael AJ, Hartshorne JM (1981): Congenital anomalies in relation to the use of doxylamine dicyclomine and other antenatal factors. An ongoing prospective study. Med J Aust 1:410–414.
Gill WB, Schumacher GF, Bibbo M, Straus FH, Schoenberg HW (1979): Association of diethylstilbestrol exposure in utero with cryptorchidism, testicular hypoplasia and semen abnormalities. J Urol 122: 36–39.

Golding J, Baldwin JA (1980): Clefts of lip and palate and maternal drug consumption: I. Antinauseants. Read before the FDA Fertility and Maternal Health Drugs Advisory Committee, Washington, D.C., Sept. 15, 1980.

Greenberg LH, Tanaka KR (1964): Congenital anomalies probably induced by cyclophosphamide. J Am Med Assoc 188:423.

Hakosalo J, Saxén L (1971): Influenza epidemic and congenital defects. Lancet ii:1346.

Hanshaw JB, Schultz FW, Melish MM, Didgeon JA (1973): Congenital cytomegalovirus infection. In Elliot K, Knight J (eds): "Intrauterine Infections." Ciba Foundation Symposium n.s. no. 10, Associated Scientific Publishers, Amsterdam and London, pp 23, 32.

Hardy JMB, Azarowicz EN, Mannini A, Medearis DN, Jr, Cooke RE (1961): The effect of Asian influenza on the outcome of pregnancy, Baltimore, 1957–1958. Am J Public Health 51:1182

Harlap S, Shiono PH (1980): Alcohol, smoking and incidence of spontaneous abortions in the first and second trimester. Lancet ii:173–176.

Heinonen OP, Slone D, Shapiro S (1977): "Birth Defects and Drugs in Pregnancy." Littleton, Mass.: PSG Publishing Co.

Hemminki K, Franssila E, Vainio H (1980): Spontaneous abortions among female chemical workers in Finland. Int Arch Occup Environ Health 45:123–126.

Henderson IWD (1977): Congenital deformities associated with Bendectin. Can Med Assoc J 117:721–722.

Herbst AL, Cole P, Colton T, Robboy SJ, Scully RE (1977): Age-incidence and risk of diethylstilbestrol-related clear cell adenocarcinoma of the vagina and cervix. Am J Obstet Gynecol 128:43.

Herbst AL, Scully RE (1970): Adenocarcinoma of the vagina in adolescents - A report of 7 cases including 6 clear-cell carconomas (so-called mesonephromas). Cancer 25:745.

Herbst AL, Scully RE (1970): Adenocarcinoma of the vagina in adolescents—A report of 7 cases including 6 clear-cell carcinomas (so-called mesonephromas). Cancer 25:745.

Hill GB, Spicer CC, Weatherall JAC (1968): The computer surveillance of congenital malformations. Br Med Bull 24:215–218.

Holmberg PC (1979): Central nervous-system defects in children born to mothers exposed to organic solvents during pregnancy. Lancet ii:177–179.

Horstman DM (1969): Viral infections in pregnancy. Yale J Biol Med 42:99–112.

Infante PF (1976a): Oncogenic and mutagenic risks in communities with polyvinyl chloride production facilities. In Saffiotti U, Wagoner JK (eds): "Occupational Carcinogenesis." Ann NY Acad Sci 271: 49–57.

Infante PF (1976b): The effect of lead on reproduction. In "Proceedings of Conference on Women and the Workplace." Washington, D.C., June 1976.

Infante PF, Wagoner JK, McMichael AJ, Waxweiler RJ, Falk H (1976): Genetic risks of vinyl chloride. Lancet i:734–735.

Ingalls TH, Klingberg MA (1963): Implications of epidemic embryopathy for public health. In: "Proceedings of the 91st Annual Meeting APHA." Kansas City, November 1963.

Ingalls TH, Klingberg MA (1965): Implications of epidemic embryopathy for public health. Am J Public Health 55:200.

International Clearinghouse for Birth Defects Monitoring Systems. 1980 Annual Report. Stockholm, 1982.

Jones KL, Smith DW, Ulleland CN, Streissguth AP (1973): Pattern of malformation in offspring of chronic alcoholic mothers. Lancet i:1267–1271.

Kaufman RH, Binder GL, Gray PM, Adam E (1977): Upper genital tract changes associated with exposure in utero to diethylstilbestrol. Am J Obstet Gynecol 128:51–59.

Kharrazi M, Potashnik G, Goldsmith JR (1980): Reproductive effects of dibromochloropropane. Isr J Med Sci 16:403–406.

Kleinman H, Prince JT, Mathey WE, Rosenfeld AB, Berman JE, Syverton JT (1962): ECHO 9 virus infection and congenital abnormalities: A negative report. Pediatrics (Springfield) 29:261.

Kline J, Stein Z, Shrout P, Susser M, Warburton W (1980): Drinking during pregnancy and spontaneous abortion. Lancet ii:176–180.

Kline J, Stein ZA, Susser M, Warburton D (1977): Smoking: A risk factor for spontaneous abortion. New Engl J Med 297:793.

Klingberg MA, Papier CM (1979a): Environmental teratogens. Their significance and epidemiologic methods of detection, In Lingberg MA, Weatherall JAC (eds): "Contributions to Epidemiology and Biostatistics, Vol I. Epidemiologic Methods for Detection of Teratogens." Basel: Karger, pp 1–22.

Klingberg MA, Papier CM (1979b): Teratoepidemiology. J Biosoc Sci 11:233–258.

Kučera J (1971): Rate and type of congenital anomalies among offspring of diabetic women. J Reprod Med 7:73.
Landsman JB, Grist NR, Ross CAC (1964): ECHO 9 virus infection and congenital malformations. Br J Prev Soc Med 18:152.
Leck I (1963): Incidence of malformations following influenza epidemics. Br J Prev Soc Med 17:70.
Lenz W (1962): Drugs and congenital abnormalities. Lancet ii:1332.
Levin JN (1971): Amphetamine ingestion with biliary atresia. J Pediatr 79:130
Loughnan PM, Gold H, Vance JC (1973): Phenytoin teratogenicity in man. Lancet i:70.
Lund C (1936): The effect of chronic lead poisoning on reproductive capacity. Nord Hyg Tidski 18:12.
Manson MM, Logan WPD, Loy RM (1960): Rubella and other virus infections during pregnancy. In: Ministry of Health Reports on Public Health and Medical Subjects no. 101. HMSO, London, 1960.
Matsumoto HG, Koya G, Takeuchi T (1965): Fetal Minamata disease. A neuropathological study of two cases of intrauterine intoxication by a methylmercury compound. J Neuropathol Exp Neurol 24: 563–574.
Mellin GW (1963): Fetal life study: A prospective epidemiologic study of prenatal influences on fetal development: Meclizine and other drugs. Bull Soc R Belge Gynecol Obstet 33:79.
Menzies CJG (1978): Fetal malformation after Debendox treatment in early pregnancy. Br Med J 1:925.
Meyer MB, Tonascia J (1981): Long-term effects of prenatal X-ray of human females. I. Reproductive experience. Am J Epidemiol 114:304–316.
Mignot G, Devie M, Dumont M (1978): Lithium of grossesse. J Gynecol Obstet Biol Reprod (Paris) 7: 1303–1317.
Milham S, Elledge W (1972): Maternal methimazole and congenital defects in children. Teratology 5:125.
Milkovich L, van den Berg BJ (1974): Effects of prenatal meprobamate and chlordiazepoxide on human embryonic and fetal development. New Engl J Med 291:1268–1271.
Milkovich L, van den Berg BJ (1976): An evaluation of the teratogenicity of certain antinauseant drugs. Am J Obstet Gynecol 125:244–248.
Mitchell AA, Rosenberg L, Shapiro S, Slone D (1981): Birth defects related to Bendectin use in pregnancy. I. Oral clefts and cardiac defects. J Am Med Assoc 245:2311–2314.
Montgomery JR, Flanders RW, Yow MD (1973): Congenital anomalies and herpes virus infection. Am J Dis Child 126:364.
Nelson MM, Forfar JO (1971): Associations between drugs administered during pregnancy and congenital abnormalities of the fetus. Br Med J 1:523.
New York State Department of Health: Love Canal. A special report to the Governor and Legislature, New York, April 1981.
Nicholson HO (1968): Cytotoxic drugs in pregnancy. J Obstet Gynaec Br Commonw 75:307.
Nikitina YI (1974): Course of labor and puerperium in the vineyard workers and milkmaids in Crimea. Gig Tr Prof Zabol 18:17–20.
Nora JJ, Nora AH, Toews WH (1974): Lithium, Ebstein's anomaly and other congenital heart defects. Lancet ii:594–595.
Nora JJ, Vargo TA, Nora AH, Love KE, McNamara DG (1970): Dexamphetamine: A possible environmental trigger in cardiovascular malformations. Lancet i:1290.
Parker HM, Taylor LS (1971): Basic radiation protection criteria. Report No. 39, National Council on Radiation Protection and Measurements, Washington, D.C.
Paterson DC (1977): Congenital deformities associated with Benedectin. Can Med Assoc J 116:1348.
Paul C (1860): Mémoires originaux. Archives générales de médecine 1:513.
Pharoah POD, Alberman E, Doyle P, Chamberlain G (1977): Outcome of pregnancy among women in anaesthetic practice. Lancet i:34–36.
Powell PD, Johnstone JM (1962): Phenmetrazine and foetal abnormalities. Br Med J 2:1327.
Rawls WE, Desmyter J, Melnick JL (1968): Serologic diagnosis and fetal involvement in maternal rubella: Criteria for abortion. J Am Med Assoc 203:627.
Remington JS (1968): Intrauterine infections: Birth defects. Birth Defects Original Articles Series 4:47.
Richards IDG (1969): Congenital malformations and environmental influences in pregnancy. Br J Prev Soc Med 22:218.
Robert JM, Guibaud P, Robert E (1981): A local outbreak of femoral hypoplasia or aplasia and femoral fibula-ulnar-complex. J Genet Hum 29:379–394.
Rothman KJ, Fyler DC, Goldblatt A, Kreidberg MD (1979): Exogenous hormones and other drug exposures of children with congenital heart disease. Am J Epidemiol 109:433–439.

Savage MO, Moosa A, Gordon RR (1973): Maternal varicella infection as a cause of fetal malformations. Lancet i:352.
Saxén I (1975): Associations between oral clefts and drugs taken during pregnancy. Int J Epidemiol 4: 37–44.
Schou M, Goldfield MD, Weinstein MR, Villeneuve A (1973): Lithium and pregnancy. I. Report from the register of lithium babies. Br Med J 2:135–136.
Sever JL, Fuccilla DA, Bowes WA, Jr (1976): Environmental factors: Infection and immunizations. In Brent RL, Harris MI (eds): "Prevention of Embryonic, Fetal and Perinatal Disease, Vol. 3." Bethesda: USDHEW, pp 199–209.
Shapiro S, Heinonen OP, Siskind V, Kaufman DW, Monson RR, Slone D (1977): Antenatal exposure to doxylamine succinate and dicyclomine hydrochloride (Bendectin) in relation to congenital malformations, perinatal mortality rate, birth weight and intelligence quotient scores. Am J Obstet Gynecol 128:480–485.
Shapiro S, Kaufman DW, Rosenberg L, Slone D, Monson RR, Siskind V, Heinonen OP (1978): Meclizine in pregnancy in relation to congenital malformations. Br Med J 1:483.
Shapiro S, Slone D, Hartz SC, Rosenberg L, Siskind V, Monson RR, Mitchell AA, Heinonen OP, Idanpaan-Heikila J, Haro S, Saxén L (1976): Anticonvulsants and parental epilepsy in the development of birth defects. Lancet i:272.
Shapley D (1973): Herbicide:AAAS study finds dioxin in Vietnam fish. Science 180:285.
Shotton D, Monie IW (1963): Possible teratogenic effect of chlorambucil on a human fetus. J Am Med Assoc 186:74.
Smith CA (1947a): Effects of maternal undernutrition upon the newborn infant in Holland (1944–1945). J Pediatr 30:229.
Smith CA (1947b): The effects of wartime starvation in Holland upon pregnancy and its product. Am J Obstet Gynecol 53:599.
Smithells RW (1976): Environmental teratogens of man. Br Med Bull 32:27–33.
Smithells RW, Chinn ER (1964): Meclozine and fetal malformations: A prospective study. Br Med J 1: 217–218.
Smithells RW, Sheppard S (1978): Teratogenicity testing in humans: A method demonstrating the safety of Bendectin. Teratology 17:31–36.
Smithells RW, Sheppard S, Schorah CJ (1976): Vitamin deficiencies and neural tube defects. Arch Dis Child 51:944–950.
Smithells RW, Sheppard S, Schorah CJ, Seller MJ, Nevin NC, Harris R, Read AP, Fielding DW (1980): Possible prevention of neural tube defects by periconceptional vitamin supplementation. Lancet i: 339–340.
South MA, Tompkins WAF, Morris CR, Rawls WE (1969): Congenital malformation of the central nervous system associated with the genital type (type 2) herpes virus. J Pediatr 75:13.
Speidel BD, Meadow SR (1974): Epilepsy, anticonvulsants and congenital malformations. Drugs 8:354.
Strandberg M, Sandback K, Axelson O, Sundell L (1978): Spontaneous abortions among women in hospital laboratory. Lancet i:384–385.
Streissguth AP (1978): Fetal alcohol syndrome: An epidemiologic perspective. Am J Epidemiol 107: 467–478.
Surgeon General (1981): "The Health Consequences of Smoking. The Changing Cigarette." Rockville: DHHS No. 81-50156, pp 157–172.
Swartz HM, Reichling BA (1978): Hazards of radiation exposure for pregnant women. J Am Med Assoc 239:1907–1908.
Takeuchi T, Matsumoto H (1969): Minamata disease of human fetuses. In Nishimura H, Miller JR (eds): "Methods for Teratological Studies in Experimental Animals and Man." Tokyo: Igaku Shoin, p 280.
Thiersch JB (1952): Therapeutic abortions with a folic acid antagonist, 4-aminopteroylglutamic acid (4-amino-P.G.A.) administered by the oral route. Am J Obstet Gynecol 63:1298–1304.
Vaisman AI (1967): Work in surgical theatres and its influence on the health of anaesthesiologists. Eksp Khir Anesteziol 3:44–49.
Vessey MP (1978): Epidemiological studies of the occupational hazards of anaesthesia-A review. Anaesthesia 33:430–438.
Visser GHA, Huisjes HJ, Elshove H (1976): Anticonvulsants and fetal malformations. Lancet i:970.
Warkany J (1979): Antituberculous drugs. Teratology 20:133–138.

Weatherall JAC, Haskey JC (1976): Surveillance of malformations. Br Med Bull 32:39–44.
Weinstein MR, Goldfield M (1975): Cardiovascular malformations with lithium used during pregnancy.
1259–1261.
Am J Psychiatry 132:529–531.
Wenger F (1967): Necrosis cerebral masiva del feto en casos de encefalitis equina Venezuelana. Investigation Clinica 21:13.
Whorton D, Krauss RM, Marshall S, Milby TH (1977): Infertility in male pesticide workers. Lancet ii:
Wilson JG (1976): Environmental factors: Teratogenic drugs. In Brent RL, Harris MI (eds): "Prevention of Embryonic, Fetal and Perinatal Disease, Vol 3." Bethesda: USDHEW, p 147–161.
Wilson JG (1977): Embryotoxicity of drugs in man. In Wilson JG, Fraser FC (eds): "Handbook of Teratology, Vol I." New York: Plenum Press, p 309–356.
Winick M (1970): Fetal malnutrition. Clin Obstet Gynecol 13:526.
Wolff J (1969): Iodide goiter and the pharmacologic effects of excess iodide. Am J Med 47:101.
WHO (1981): The effect of female sex hormones on fetal development and infant health. WHO Tech Rep Ser 657, World Health Organization, Geneva.
Yerushalmy J, Milkovich L (1965): Evaluation of the teratogenic effect of meclizine in man. Am J Obstet Gynecol 93:553–562.
Zachai EH, Millman WJ, Neiderer B, Hanson JW (1975): The fetal trimethadione syndrome. J Pediatr 87: 280.

American Journal of Industrial Medicine 4:329–348 (1983)

Uses of Available Record Systems in Epidemiologic Studies of Reproductive Toxicology

Anthony P. Polednak, PhD, and Dwight T. Janerich, DDS, MPH

The uses of available record systems in epidemiologic studies of reproductive toxicology are described with reference to New York State. The available record systems (and relevant reproductive end points) described include: a newborn screening program for metabolic diseases and hemoglobinopathies (relevant to point mutations); chromosome registries and prenatal cytogenetics (for chromosome anomalies); live birth certificates (for birth defects, birthweight, sex ratio, etc); fetal death certificates (for spontaneous fetal deaths); and a statewide cancer registry (for childhood cancers and transplacental carcinogenesis). The uses and limitations of these record systems are discussed, along with examples of their use in descriptive and analytic epidemiologic studies. Descriptive studies outlined include investigations of temporal and geographic trends in birth defects, birth weight, and fetal deaths, with reference to environmental questions (eg, Love Canal, nuclear power plants). Analytic studies described concern parental occupation in relation to specific birth defects (neural tube defects and Down syndrome) and maternal use of contraceptive drugs.

Key words: epidemiology, birth defects, birth weight, childhood cancer, fetal death, occupational exposure, reproductive toxicology, vital records

INTRODUCTION

Surveillance systems that have been developed for monitoring adverse reproductive outcomes (mainly birth defects, but also fetal deaths) and various characteristics of live births (eg, sex ratio and birth weight) may be used in epidemiologic studies of reproductive toxicology. The advantages and disadvantages or limitations of such usage will be outlined, and examples involving environmental or occupational data will be provided. Our discussion and examples are mainly from the vital record system of Upstate New York (ie, New York State exclusive of New York City); birth defect monitoring systems in other states have been described elsewhere [Flynt, 1974; Edmonds et

Division of Community Health and Epidemiology, New York State Department of Health, Albany.
Address reprint requests to Anthony P. Polednak, Division of Community Health and Epidemiology, New York State Department of Health, Albany, NY 12237.

0271-3586/83/0401-02-0329$05.50 © 1983 Alan R. Liss, Inc.

TABLE I. Outcomes or End Points of Interest in Reproductive Toxicology, and Available Record Systems in New York State

Outcomes or end points of interest	Sources of data (available record systems)
Point mutations (amino acids, hemoglobins)	Newborn Screening Program
Cytogenetics (chromosome abnormalities)	Chromosome Registry, including prenatal diagnosis (amniocentesis)
Birth defects, sex ratio, birth weight, gestation length, multiple births	Birth certificates (live births)
Fetal death (spontaneous abortion), early and late	Fetal death certificates
Childhood cancer Mutagenesis in parent(s) Transplacental carcinogenesis	New York State Cancer Registry

al, 1981]. The end points of interest and the available record systems in New York State are listed in Table I.

The use of these record systems is mainly in descriptive studies of temporal or geographic variations in rates of adverse reproductive outcomes. Examples are "ecologic" studies that examine correlations with selected environmental variables and investigations of apparent "clusters" of events to explore possible occupational or environmental factors. The second major use involves the identification of cases (fetal deaths, birth defects, low birth weight live births, etc) and controls, for analytic epidemiologic studies of possible risk factors by examination of parental occupation and other factors. Examples of these uses of available record systems in New York State will be provided.

Reproductive toxicology will be discussed in terms of mutagenesis, teratogenesis, and carcinogenesis; although these processes may be related (eg, the same agent may produce multiple effects), the timing of the exposure and the outcome are of importance. For example, exposure of parental germ cells to a physical or chemical agent prior to conception may produce mutations expressed as congential malformations and/or as childhood cancer, whereas maternal exposure early in pregnancy may produce teratogenic effects on the fetus or transplacental carcinogenesis expressed in childhood.

DESCRIPTION OF AVAILABLE RECORD SYSTEMS

Newborn Screening and Prenatal Cytogenetics

Available record systems for newborn screening for inborn metabolic diseases and prenatal cytogenetics (Table I) have been described elsewhere [Porter, 1979; Hook et al, 1981]; their use in studies of environmental mutagens has been limited.

Since 1977, the New York State Department of Health (Birth Defects Institute) has operated a Newborn Screening Program, legally mandated to include screening for several genetic metabolic diseases, along with abnormal hemoglobins and hypothyroidism [Porter, 1979]. Since autosomal recessive genes are involved in the genetic metabolic diseases screened, their relevance to studies of reproductive toxicology is limited. Neel et al [1979] have advocated biochemical approaches to monitoring human populations for changes in point mutation rate, involving collection of placental cord blood and venous samples from parents, and have suggested that the costs are justified in terms of measuring the impact of industrialization and pollution.

In lieu of sophisticated methods that may detect heterozygotes, dominant disorders (eg, achondroplasia and polydactyly) have been proffered as markers for monitoring mutations [Anon, 1971; Hook, 1978], but their use has not received enthusiastic support [Bloom and Paul, 1981]. With regard to polydactyly, ascertainment through birth records alone (discussed below) would be incomplete; that is less than about 40–45% of cases, on the basis of a New York State study [Gittelsohn and Milham, 1965] and a study of cases ascertained from a variety of sources in another state [Hay, 1971]. Recently available data on hospital discharges in New York State, however, could significantly improve these figures for monitoring purposes (see below) and for use in in-depth investigations.

As noted previously, newborn screening includes abnormal hemoglobins (Table I). Studies of hemoglobins are of potential value in detecting point mutations, since most globin gene structural mutations involve 20 to 50% abnormal protein, and a number of variants have been used; the limitations of this system for studying induced (vs spontaneous) mutagenesis have been discussed by Stamatoyannopoulos [1979]. Methods involved in newborn screening for abnormal hemoglobins in New York State have been described by Warren et al [1982], who also assessed the views of parents of children with hemoglobinopathies on the value of newborn screening for sickle hemoglobin. To be useful in studies detecting possible induced mutations, available record systems would have to be expanded considerably, including studies of hemoglobins of parents of newborns (with the attendant problem of paternity determination).

Use of "dominant lethals," by definition expressed as early fetal death, has been advocated as markers for monitoring mutations in human populations. However, since rates of fetal wastage may vary greatly due to many factors (including environmental but nonmutagenic agents), monitoring fetal loss is more relevant to investigation of embryotoxins rather than assessment of induced mutations [Hook, 1978]. Surveillance of reported fetal deaths and epidemiologic investigations using fetal death records are discussed below.

Fabricant and Legator [1981] have reviewed current cytogenetic techniques and their use in monitoring for industrial and environmental hazards. In New York State, the selective nature of amniocentesis results in greater detection of trisomies relative to sex-chromosome abnormalities [Hook et al, 1981]. Theoretically, these cases from amniocentesis services could be useful in analytic studies of parental (occupational or other environmental) factors. In case-control studies, controls would have to be matched for variables (such as maternal age) that are related to occupation. Due to their previous reproductive history, women using amniocentesis services may be at high risk for induced mutations related to environmental factors. Since some factors may have operated in women some years before the conception of births with chromosomal anomalies [Alberman et al, 1972], such studies may be difficult.

The present paper focuses on the use of certificates of birth and fetal death, and the New York State Cancer Registry, in studies of reproductive toxicology.

Reproductive Outcomes From Vital Records

Birth weight. Birth weight is recorded on birth certificates and fetal death records in Upstate New York, and distributions among live births are reported annually by maternal age and ethnic groups. At the county level (or in smaller geographic units), such data have been used in ecologic studies—eg, of possible effects of environmental radiation exposure (discussed below). Advantages of the use of these data, as noted elsewhere [Bloom and Paul, 1981], include their completeness and accuracy of reporting and their correlation with concurrent or subsequent health outcomes. Among Upstate New York births, 91.7% agreement within 28 gm (about 1 oz) between birth certificates and hospital records has been reported [Carucci, 1979]; similar findings were reported from the National Natality Survey [Querec, 1980].

Limitations of the use of birth weight include the necessity of taking into account many other factors known to be associated with birth weight. Some of these variables (eg, maternal age, parity, ethnic group) are available from birth records, along with maternal and paternal education and occupation (relevant to assessment of socioeconomic level), outcomes of previous pregnancies, and prenatal care. Other important factors such as dietary habits, and consumption of cigarettes and alcohol, can be evaluated only by additional data collection or analytic studies (examples of which are provided below).

Studies involving analyses of birth weight among live births obviously exclude examination of fetal deaths (spontaneous abortions), and may result in a biased assessment of toxicologic (or other) effects of agents. Inclusion of weight of fetal deaths would be desirable, but in Upstate New York in 1980 only 11.0% of certificates of spontaneous fetal deaths (vs 99.8% of live births) included valid information on birth weight.

It should be noted that induced mutations in parental germ cells may also be expressed as disturbances in birth weight of offspring; for example, anomalies of X chromosomes and autosomes have been shown to be related to reduced birth weight [Robson, 1978].

Despite the limitations of using birth weight outlined above, studies of birth weight (in conjunction with birth defects and abortions) may be useful in epidemiologic investigations of environmental pollutants [Beckman et al, 1979]. Surveillance of birth weight distributions by geographic area may identify populations requiring more intensive study by analytic methods to explore the role of occupational or other environmental factors.

Birth defects. Surveillance of birth defects in Upstate New York has been based mainly on birth and death records [Gittelsohn and Milham, 1965; Milham, 1971]. Currently, birth defects reported on birth certificates are coded according to the classification of Hay and Tonascia [1968] and tabulated as frequencies and rates (per 10,000 live births) by county of residence of mother in Upstate New York.

Completeness and accuracy of reporting of birth defects on birth certificates have been evaluated in a number of studies. In New York State, Gittelsohn and Milham [1965] compared reports on birth and death certificates with data from a State Aid Program; vital record reporting was most complete for spina bifida (79.9%) and for oral clefts (74.8%), less complete for hypospadias (44.3%), and least complete for circulatory defects (8.4%). In a comprehensive study of birth defects among live births and fetal deaths in Iowa in 1963 [Hay, 1971], birth and fetal death certificates recorded nearly all cases of neural tube defects ascertained from hospital records of live births and fetal deaths. Oral clefts were also well reported. Among live births alone, reporting was nearly complete for neural tube defects and oral clefts, but about 50% for Down syndrome and 26% for hypospadias; consistent with the findings of Gittelsohn and Milham [1965], congenital heart defects were reported mainly from hospital records of live births and from the State Crippled Children's Service; only 23% of total cases were recorded on birth records.

A comparison of more recent data from hospital live birth records and birth certificates in Upstate New York in 1972 [Carucci, 1979] also shows nearly complete reporting for neural tube defects (7/9 cases or 77.8%), lower reporting of oral clefts (4/9 or 44.4%) hypospadias (5/24 or 20.8%), and lowest reporting for cardiovascular defects (0/20 or 0%). The single case of achondroplasia, a condition mentioned above with reference to monitoring for mutations, was reported on both the birth certificate and the hospital record. For Down syndrome, 4 of 8 cases (50.0%) were reported on both documents [Carucci, 1979], in agreement with the findings of Hay [1971]. Thus, despite the increased use of chromosome analysis since 1959, the reporting of Down syndrome on birth records does not appear to have improved in recent years. Hook and Cross [1981] reported a decrease in completeness of recording of Down syndrome on birth certificates, as a percentage of total cases from cytogenetic laboratories in New York State and other sources, from 1965–1969 to 1970–1974. The need for improved methods of clinical diagnosis of Down syndrome in the newborn is recognized [Anon, 1976].

In conclusion, use of both fetal death and live birth certificates is adequate for studies of neural tube defects, whereas use of birth records alone is suitable for certain externally obvious defects associated with apparently low fetal mortality (eg, oral clefts). For Down syndrome, results of epidemiologic studies using birth certificates alone (exemplified below) must be interpreted with caution. Nevertheless, epidemiologic studies based on birth records may result in the formulation of hypotheses to be explored by more intensive studies, even for underreported defects such as hypospadias (see Polednak and Janerich [1982].

The effects of biases resulting from the use of birth records in case ascertainment may be evaluated by special studies of samples of cases ascertained from multiple sources, for comparison of characteristics vs birth certificate cases, as accomplished by Hay [1967] for oral clefts. For spina bifida, completeness of reporting on birth certificates appears to be lower for blacks than for whites [Hook et al, 1980]. Such information on biases may be of value in analyzing and interpreting the results of studies involving vital records.

Among those defects that are reported on vital records, incomplete reporting of associated defects and lack of specificity of the morphological description of the defect may preclude exploration of specific etiologic hypotheses. Studies of oral clefts using multiple sources of ascertainment, for example, have shown distinctive epidemiologic associations for specific categories [Emanuel et al, 1973]. Isolated vs multiple defects may also differ in their descriptive epidemiologic characteristics, suggesting etiologic heterogeneity [Khoury et al, 1982]. Specific categories of hypospadias should be distinguished in studies of the timing of teratogenic effects, as suggested by data on use of exogenous hormones [Aarskog, 1979]. Reporting of limb malformations on birth records is imprecise [Piper, 1973], whereas distinguishing unilateral from bilateral limb reduction defects may be of importance in evaluating teratogens [Wilson and Brent, 1981]. Thus, epidemiologic studies of broad defect categories ascertained from vital records may need to obtain additional data from other sources (eg, records of physicians and hospitals).

Use of multiple record systems for ascertainment of birth defects would reduce the possibility of significant biases in the results of studies of reproductive toxicology, and enhance the generalizability of the findings. Hospital records may not include many defects reported on birth records, as shown for Down syndrome [Carucci, 1979] and oral clefts [Gittelsohn and Milham, 1965], whereas others will be missed by both systems. Thus, the recently established New York State Statewide Planning and Research Cooperative System (SPARCS), which provides patient statistics on all general care hospitals in the State (including birth defect codes), should improve monitoring for birth defects when used in conjunction with birth certificate data. Ideally, however, other record systems (eg, cytogenetic screening, State Aid Program, fetal death records) should also be integrated to ensure more complete reporting. The importance of including data on birth defects among spontaneous abortions in environmental monitoring has been emphasized by Stein et al [1975] and Kline et al [1977].

Fetal deaths (spontaneous abortions). As noted by others [Kline et al, 1977; Bloom and Paul, 1981], surveillance systems based on spontaneous abortions (vs birth defects) offer several advantages, including increased detection of anomalies (structural or chromosomal) and increased statistical power to detect smaller possible effects of environmental mutagens and teratogens.

Vital certificate data generally provide an incomplete source of ascertainment of abortions, and reporting of fetal deaths before 20–34 weeks of gestation is not required in some states [Querec, 1980]. Since 1967, however, New York State Public Health Laws have required registration of all fetal deaths irrespective of the length of gestation; prior to that year, reporting was mandated only deaths of gestation 20 weeks or longer. Spontaneous fetal deaths are reported separately from induced abortions, and the latter are of interest here only with regard to accuracy of reporting of spontaneous fetal deaths. That is, regional differences in rates of spontaneous fetal deaths may be due in part to differences in reporting by hospitals and physicians; induced abortions may be reported as spontaneous fetal deaths.

The extent of underreporting of spontaneous abortions on vital records in New York State is difficult to evaluate. Estimates of recognized spontaneous abortions from hospital-based and prospective studies range from about 9 to 25%, with highest proportions among older women [Warburton and Fraser, 1964; Miller et al, 1980; Wilcox et al, 1981]. Using data from Upstate New York, in 1980, reported spontaneous fetal deaths of any gestation length accounted for 6.5% of total live births and spontaneous fetal deaths (ie, 9396/145,534). This suggests considerable underreporting, especially since some reported spontaneous abortions are in fact induced; however, this proportion has increased in recent years (ie, 5.5% in 1978, and 6.3% in 1979), suggesting improvement in reporting. In addition, use of statistical data on hospitalizations from the Statewide Planning and Research Cooperative System (SPARCS, as mentioned above) should facilitate screening for spontaneous abortions by geographic area.

With respect to potential confounding variables in comparisons of spontaneous abortion rates across geographic areas, the fetal death certificate provides information on maternal age (99.5% of certificates in Upstate New York) and on previous spontaneous abortions (item completed on 84.5% of spontaneous fetal death certificates). Examination of other confounding variables (ie, maternal smoking and alcohol consumption), of course, requires more intensive investigation.

For the select group of women who have at least one live birth, the item on previous spontaneous abortions is completed on 97.4% of live birth records. Studies comparing birth certificate recording of number of previous fetal deaths with data from hospital questionnaires in 42 states [Querec, 1980] and from hospital records in New

York [Carucci, 1979] have shown good agreement (over 80%). Thus, the advantages of this record source (birth certificates) for monitoring fetal deaths are the completeness of records and the accuracy of reporting. Disadvantages include the select nature of the group, and the fact that the recorded event(s) occurred some time in the past. Thus, if an increase in reporting of previous fetal deaths was detected in a given area, the exact times of occurrence would need to be obtained from fetal death records (when available) and direct interview.

Childhood cancers. The source of information on cases of childhood cancer in New York State is the Cancer Registry. Briefly, state law requires the reporting of all cancer cases diagnosed in the State (including New York City since 1973).

Statistical reports are published periodically [Greenwald et al, 1976]. Cancers are coded to the ICD 9th Revision, and histologic type is coded according to the Manual of Tumor Nomenclature and Coding [American Cancer Society, 1968].

Temporal and geographic patterns of childhood cancer rates may be examined for clues to environmental factors. As noted by Bloom and Paul [1981], childhood cancers are infrequent, and may not be as useful in detecting transplacental carcinogens as regional comparisons of adult cancers have been useful in identifying carcinogenic agents (eg, asbestos and lung cancer in World War II shipyard workers in coastal southeastern states). Data from the New York State Cancer Registry (Table II), however, indicate that average annual numbers of childhood cancers (age at diagnosis 0–14 years) may be sufficient for surveillance purposes in certain counties, especially urbanized areas where occupational or other environmental exposures are of particular interest.

These kinds of data (Table II) may be analyzed by various statistical procedures. The scan statistic [Wallenstein, 1980] may be useful in detecting clustering of cases over time if the population at risk remains fairly constant. Sequential analysis of cases or rates within a geographic area over time may also be useful in monitoring temporal changes [Howe and Hoff, 1981] that may be related to environmental factors.

The Registry is used extensively as a source for cases in analytic epidemiologic studies of childhood (and other) cancers of specific sites. Although occupation of the case is included on the Registry report form, occupations of parents of childhood cancer cases must be obtained by record linkage (with birth records) and preferably by direct interview of parents (discussed below). Environmental exposures of children must also be considered in such studies; for example, a relationship between childhood cancer and electric power line configurations has been suggested, although the findings of studies have not been consistent [Fulton et al, 1980].

Descriptive Epidemiologic Studies: Investigations of Temporal Trends and Geographic Patterns

Birth defects and birth weight. The first system for monitoring birth defects was established at the New York State Health Department in 1962, after the removal of thalidomide from the British Market [Flynt, 1974]. The surveillance system based on vital records (live births and fetal deaths) had actually been planned prior to knowledge of the thalidomide incident, and the system detected (in 1962) evidence of the thalidomide phocomelia episode as well as the effects of an epidemic of rubella (in 1964) on eye malformations and fetal deaths [Milham, 1971]. Subsequently, birth defect rates have been monitored in several surveillance systems in an attempt to detect other environmental teratogens.

TABLE II. Average Annual Numbers (1976–1979) of Childhood Cancers (Age 0–14 Years at Diagnosis) in Selected Counties* in New York State

	Males						Females					
	All cancers		Leukemia		Lymphoma		All cancers		Leukemia		Lymphoma	
County	(No.)	(Rate per 100,000)[a]	(No.)	(Rate per 100,000)[a]	(No.)	(Rate per 100,000)[a]	(No.)	(Rate per 100,000)[a]	(No.)	(Rate per 100,000)[a]	(No.)	(Rate per 100,000)[a]
Albany	4.8	17.2	1.3	4.6	0.8	2.9	2.8	10.5	1.8	6.8	0.3	1.1
Broome	3.3	14.7	1.5	6.7	0.3	1.3	3.5	16.6	1.3	6.2	0.0	0.0
Erie	15.5	14.3	5.3	4.9	3.0	2.8	13.5	12.9	4.5	4.3	1.8	1.7
Monroe	14.3	18.3	3.3	4.2	3.0	3.8	11.8	15.1	4.8	6.4	0.8	1.1
Nassau	20.3	15.3	5.5	4.1	4.0	3.0	18.0	14.3	5.8	4.6	1.0	0.8
Niagara	5.3	20.6	2.5	9.7	0.8	3.1	2.0	8.1	0.5	2.0	0.3	1.2
Onondaga	7.3	14.2	1.5	2.9	1.3	2.5	5.8	11.6	2.0	4.0	0.0	0.0
Orange	3.8	11.6	1.8	5.5	0.3	0.9	1.8	5.8	0.3	1.0	0.0	0.0
Rockland	3.5	10.8	0.8	2.5	0.8	2.5	5.0	16.3	0.8	2.6	0.5	1.6
Saratoga	4.3	21.9	0.8	4.1	1.3	6.6	3.3	17.7	1.0	5.4	0.5	2.7
Schenectady	3.8	24.4	1.3	8.3	0.5	3.2	2.5	16.9	1.0	6.8	0.3	2.0
Suffolk	25.8	15.7	8.0	4.9	4.0	2.4	19.5	12.5	6.5	4.2	3.0	1.9
Westchester	13.8	16.0	4.0	4.6	2.8	3.2	11.0	13.3	3.5	4.2	1.8	2.2
Total Upstate	173.3	14.7	53.5	4.5	31.0	2.6	138.3	12.3	47.8	4.3	14.8	1.3
Bronx	14.3	10.2	4.5	3.2	3.0	2.1	15.8	11.7	4.3	3.2	1.0	0.7
Kings	28.3	10.9	9.5	3.7	3.3	1.3	27.0	10.6	8.3	3.3	2.0	0.8
Queens	29.3	16.5	11.5	6.5	5.0	2.8	22.3	13.0	6.8	4.0	2.0	1.2
New York	17.3	16.7	3.5	3.4	2.8	2.7	11.3	11.3	2.5	2.5	2.0	2.0
Richmond	4.5	10.6	1.0	2.4	1.0	2.4	2.8	6.9	0.3	0.7	0.3	0.7
Total New York City	93.5	12.9	30.0	4.1	15.0	2.1	79.0	11.3	22.5	3.2	7.2	1.0

*Counties with three or more average annual male childhood cancers.

[a]Rates are based on 1980 census populations (0–14 years). Data are from the New York State Cancer Registry (see text).

Temporal trends (seasonal or long-term) in the frequency of birth defects have been studied to provide evidence for operation of environmental factors, to be investigated more precisely by analytic studies. Epidemics or secular trends in neural tube defects reported in New York State and other areas are most likely related to environmental factors [MacMahon and Yen, 1971; Janerich, 1973, 1979], but these factors have yet to be identified; studies on parental occupation are discussed below. Recently, concern has been expressed about mutagens in drinking water, and short-term tests for evaluating mutagenicity of municipal water have been developed [DeMarini et al, 1982]. Recent studies on anencephaly and various elements in water, however, have disclosed no strong associations [Elmwood and Coldman, 1981]. The possible association between secular and seasonal trends in neural tube defects and herbicide use [Field and Kerr, 1979] requires further investigation.

Temporal changes in the frequency of Down syndrome have been reported in some studies. In Upstate New York, an increase among live births to women 35 and over (but not under 35) has been reported [Hook and Cross, 1981]. It has been suggested that reported seasonal and geographic variations in Down syndrome may be related to exposure of paternal germ cells to environmental mutagens [Bennett and Abroms, 1979]. More intensive studies, preferably by cytogenetic methods, are needed to investigate this hypothesis. Presently, the proportion of Down syndrome cases that involve paternal nondisjunctional events is uncertain [Jacobs and Meyer, 1981].

Temporal and geographic variations in birth defects, fetal deaths, and low birth weight have been examined in "ecologic" epidemiologic studies, including those concerned with possible effects of emissions from nuclear power plants as well as exposure to natural background radiation. In Upstate New York, reports have compared health statistics (including rates of birth defects, fetal deaths, and prematurity) in counties with and without nuclear facilities [Burometto et al, 1977]. Counties with over half their area within a 30-mile radius of a nuclear facility, and four counties (three with commercial nuclear power plants and one with a nuclear fuel reprocessing plant) were also analyzed separately. Although the results of these and other similar analyses are inconclusive [Tokuhata and Smith, 1981], they illustrate the use of available record systems in considering a subject of considerable public concern. In examining temporal and geographic variation in rates of birth defects and fetal deaths, however, consideration must be given to differences in reporting and diagnostic practices by physicians; more detailed studies may be needed to assess these problems.

For the four counties with nuclear power or reprocessing plants, birth defects, prematurity, and fetal death rates were compared over time, before and after plant operation had begun [Burometto et al, 1977], as well as with other counties from 1960 to 1975. There was little evidence for any increases in these rates attributable to operation of the plants. Table III shows data for more recent years (1975 to 1979) on total congenital malformation rates in these four counties. These are overall rates, with no adjustment for differences in maternal age or other potential confounding variables, and some rates (based on small numbers) fluctuate considerably over time. The high rates for Wayne County in some years (Table III) also occurred in earlier years prior to the operation of the plant [Burometto et al, 1977; see also Tokuhata and Smith, 1981]. Rates for Westchester county are consistently higher (Table III); this reflects a recent decline in Upstate rates (since about 1968) that has not occurred in Westchester county [Burometto et al, 1977]. On the basis of our knowledge of the genetic effects of radiation [Kato, 1975], and the very low levels of radiation exposure (relative to back-

TABLE III. Congenital Malformation Rates Among Live Births to Residents of Counties With Nuclear Power Plants in Upstate New York, 1975–1979*

County	1975		1976		1977		1978		1979	
	(No.)	(Rate per 1,000)	(No.)	(Rate per 1,000)	(No.)	(Rate per 1,000)	(No.)	(Rate per 1,000)	(No.)	(Rate per 1,000)
Cattaraugus[a]	12	9.8	21	17.3	7	5.5	17	14.3	10	7.8
Oswego	3	1.8	13	7.8	10	5.5	8	4.3	16	8.8
Wayne	27	22.0[b]	25	20.6	22	17.2	18	13.4	12	8.9
Westchester	101	13.1	121	15.8	103	13.3	109	14.3	114	12.4
Total Upstate	1250	10.0	1380	11.2	1293	10.2	1367	11.0	1296	9.7

*Data derived from vital statistics reports from the New York State Department of Health.

[a]A nuclear reprocessing plant was operated in this county.

[b]Rates were also high in years preceding the operation of a nuclear power plant (see text).

ground) to populations around nuclear facilities, the power to detect even large effects by such studies is poor. Nevertheless, monitoring systems may indicate areas for further investigation, and further analyses of vital record data are in progress.

Preliminary studies of reproductive histories among residents in the area of Love Canal dumpsite have disclosed no significant excess of congenital defects; some analyses suggested a possible association with low birth weight in an area adjacent to the Canal, but there was only one low birth weight infant among the 65 births among residents living directly adjacent to the Canal [NY State Department of Health, 1981]. Further analyses are in progress.

Spontaneous fetal deaths. As noted above, spontaneous fetal deaths may be an important screening device in detecting environmental mutagens and teratogens. Table IV shows spontaneous fetal death rates (20+ weeks gestation) by maternal age among residents of selected counties in Upstate New York, as reported on fetal death records in 1980. This table shows regional variability in rates, which may be due to several factors, including small numbers, differences in completeness of reporting, probable misreporting of induced abortions, and socioeconomic differences. Thus, these data may be useful only as a crude screening device, and provide a resource for more intensive studies to address inaccuracies and explore the possibility of environmental influence.

As a surveillance method, monitoring within a defined region should be more effective, since reporting biases may be standardized, although practices may change over time. Sequential analysis has been applied to monitoring fetal deaths and other reproductive-outcome variables in Upstate New York [Hoff, 1975]. Numbers of fetal deaths and births are accumulated over time, and may be charted graphically. Parallel lines are drawn on the chart, their location defined by statistical and other considerations (ie, probabilities of Type I and Type II errors, the expected rate, and the amount of change selected arbitrarily as worthy of attention and possible investigation).

Spontaneous fetal death rates, as reported on fetal death records, have been examined in Upstate New York in relation to residence near nuclear facilities by Burometto et al [1977]. Their analyses included fetal mortality rates (20+ weeks of gestation) for 1960 to 1975 in three counties (Oswego, Wayne, and Westchester) with nuclear power plants, and one county (Cattaraugus) with a nuclear fuel reprocessing plant which operated from 1966 to 1972. Briefly, rates for Cattaraugus and Wayne counties fluctuated considerably, with some yearly rates being higher than those for were lower than or similar to those for other counties [Burometto et al, 1977]. High rates for Wayne county are also evident in 1980 (Table IV), but are based on small numbers and confidence limits for rates in each maternal age category overlapped with those for the total Upstate. Table IV also shows that rates for Westchester county are similar to those for Upstate New York as a whole.

We have examined all spontaneous fetal deaths (of any gestation) as reported on fetal death records, for these same four counties for 1978–1980 (Table V). There was no evidence for higher rates of total spontaneous fetal deaths (of any gestation) among residents of these counties relative to those of the entire Upstate region. For the subcategory of spontaneous fetal deaths of 20+ weeks gestation, rates were high for Wayne county (as noted above), but rates for all fetal deaths (of any gestation) were not high relative to Upstate rates. This suggests possible differences in reporting practices on fetal death certificates and requires further examination.

TABLE IV. Fetal Death Rate by Selected County* of Residence and Age of Mother, Upstate New York, 1980 (Gestation 20 Weeks or Longer, Excluding Induced Abortions)

	Mother's age											
	10–19		20–24		25–29		30–34		35–39		40 +	
County	No. fetal deaths	(Rate per 1,000)[a]	No. fetal deaths	(Rate per 1,000)	No. fetal deaths	(Rate per 1,000)	No. fetal deaths	(Rate per 1,000)	No. fetal deaths	(Rate per 1,000)	No. fetal deaths	(Rate per 1,000)
Albany	2	5.5	7	7.5	14	11.5	11	16.0	3	17.5	1	41.7
Chatauqua	2	7.0	6	8.1	8	11.8	6	20.2	0	0.0	0	0.0
Chemung	3	12.0	10	19.6	4	8.1	2	9.4	1	18.5	0	0.0
Erie	23	14.8	27	6.3	40	8.6	26	11.6	6	12.1	5	55.6
Jefferson	4	16.9	5	9.3	8	18.1	3	18.0	3	81.1	0	0.0
Monroe	10	9.0	38	12.9	32	8.5	14	7.7	3	7.0	2	35.1
Nassau	12	17.3	24	9.1	48	8.8	54	14.1	19	20.1	4	29.2
Niagara	7	16.7	21	17.0	26	23.6	9	17.4	2	18.3	3	157.9
Oneida	8	10.2	13	10.8	17	14.6	9	16.1	3	21.1	2	80.0
Onondaga	11	13.3	18	9.0	25	10.5	12	9.4	4	14.3	4	81.6
Rockland	2	14.3	2	2.9	16	13.5	4	4.3	3	11.6	1	25.6
St. Lawrence	5	19.0	7	11.2	4	7.8	0	0.0	2	35.1	1	125.0
Suffolk	15	12.2	35	8.1	56	9.0	44	11.5	15	15.7	4	31.0
Wayne	4	19.1	5	9.8	9	20.3	2	10.1	3	56.6	0	0.0
Westchester	7	9.5	22	10.5	24	7.0	26	9.6	11	14.2	2	18.2
Total Upstate	168	11.8	348	8.9	437	9.1	284	10.5	98	14.6	35	33.9

*Counties with 20 or more spontaneous fetal deaths reported in 1980.

[a]Calculations are based on data obtained from vital statistics reports from New York State. The rate is the number of spontaneous fetal deaths (20 + weeks) divided by the sum of the live births plus spontaneous fetal deaths (20 + weeks).

TABLE V. Spontaneous Fetal Death Rates* Among Residents of Counties With Nuclear Power Plants in Upstate New York, 1978–1980

		Year					
		1978		1979		1980	
County	Fetal death category	(No.)	(Rate per 1,000)	(No.)	(Rate per 1,000)	(No.)	(Rate per 1,000)
Oswego	All S.F.D.[a]	20	11.0	70	37.0	77	41.6
	20+ wks. only[b]	(7)	3.9	(12)	6.3	(8)	4.3
Wayne	All S.F.D.	79	57.7	86	63.1	95	63.8
	20+ wks. only	(18)	13.2	(20)	14.7	(8)	15.4
Westchester	All S.F.D.	531	54.6	519	53.5	595	57.3
	20+ wks. only	(95)	6.7	(115)	11.9	(92)	8.9
Cattaraugus	All S.F.D.	94	72.3	82	62.7	104	72.7
	20+ wks. only	(30)	23.1	(21)	16.1	(14)	9.8
All Upstate	All S.F.D.	7676	55.5	8547	60.3	9396	64.6
New York	20+ wks. only	(1327)	9.6	(1403)	9.9	(1380)	9.5

*Rates are spontaneous fetal deaths divided by the sum of live births and spontaneous fetal deaths, based on data from fetal death certificates (see text).
[a]All spontaneous fetal deaths (of any gestation).
[b]Spontaneous fetal deaths of 20 or more weeks gestation.

Although increased fetal death rates in offspring is an expected effect of radiation exposure [Kato, 1975], no association between fertility and radiation dose has been reported in Japanese female adult atomic bomb survivors [Blot and Sawada, 1972], and large radiation doses (or very large populations) may be required to produce detectable effects on fertility [Lushbaugh and Casarett, 1976; Polednak, 1980].

Another example of the use of fetal death rates from vital records involves a reported increase in spontaneous abortion rates (all gestational ages) in a four-county area in western New York State in the year (1973) following the 1972 floods [Janerich et al, 1981]. Using data from the New York State Cancer Registry (described above), the investigators also confirmed the occurrence of high rates of leukemia and lymphoma in river valleys in the four-county area, and suggested that stress was a possible explanation for the findings on abortions and cancer.

Analytic Epidemiologic Studies

Birth defects and parental occupational and environmental exposures. Exposure of parents of children with birth defects to various chemical and physical agents is a subject of recent investigation. Interest has focused on agricultural chemicals (pesticides) and organic solvents [Wilson, 1977b; Sternberg, 1979; Longo, 1980]. Using data from the Finnish Register of Congenital Malformations along with information on exposures (at home and at work) to organic solvents, Holmberg [1979; Holmberg and Nurminen, 1980] reported an association between maternal first-trimester exposure to organic solvents and central nervous system defects in a study of 120 case–control pairs.

On the basis of comparison with data obtained from work history questionnaires, occupations were accurately recorded on birth certificates in Upstate New

TABLE VI. Occupation of Fathers of Cases (With Anencephaly and Spina Bifida) and Controls, From Birth Records (171 Case-Control Pairs)

Father's occupation	Case + / control −	Case − / control +	Case + / control +	Relative risk[a]
Farmer	5	1	0	5.00
Machinist, machine operator, metal worker	8	7	0	1.14
Bus or truck driver	7	5	0	1.40
Auto mechanic; service station	5	2	1	2.50
Carpenter, lumber	4	8	0	0.50
Printer (except salesman)[b]	1	2	0	0.50
Electrician	3	5	0	0.60
Electrical engineer, electronics technician	4	1	0	4.00

[a]Relative risk for matched case-control analysis, unadjusted for other variables; none of the marginal chi-square tests for paired samples was statistically significant (see text).
[b]Excludes one case father who was a salesman in the printing industry.

York. Overreporting of "housewife" is considerable, however, and must be reduced in order to enhance the usefulness of these data in occupational surveillance systems [Carucci and Prasad, 1979]. Using data from a case-control study of neural tube defects (ASB or anencephaly and/or spina bifida) in Upstate New York, aimed mainly at examining contraceptive patterns [Janerich et al, 1980], we report here on parental occupations as recorded on birth records of 171 case-control pairs (matched on maternal age and residence). The majority of mothers were listed as housewives, whereas six case and eight control mothers were nurses (relative risk 0.75). Nurses are of interest in view of reports of possible teratogenic effects of anesthetic gases, including an animal study showing encephalocele in offspring of pregnant rats exposed to nitrous oxide [Lane et al, 1980]. Aside from one case mother who was a clerk in a fabric dye factory, none of the occupations was suggestive of exposure to chemical or physical agents.

Paternal occupations are of interest with respect to possible germ cell mutations, and also because of potential contamination of family members. Table VI shows some suggestive findings (based on small numbers) for the categories of farmer, automobile mechanic, or service station worker, and electrical engineer or technician. The lower mean paternal education in cases than in controls (12.9 ± 2.1 SD years and 13.4 ± 2.3 SD years, respectively; $p < 0.05$, t test, two tailed), however, could indicate confounding if socioeconomic-nutritional factors are associated with the risk of ASB (cited above). Further studies on the association with farming appear warranted; although birth defects are rare events, cohort studies of the reproductive experience of various types of farmers may be in order.

Some 20–25% of cases with Down syndrome may involve nondisjunctional events in the father [Mattei et al, 1979], although another report suggests that the proportion may be lower and require further examination [Jacobs and Mayer, 1981]. In any case, environmental-occupational exposures of fathers of Down syndrome cases should be examined. Exposure of wives of male workers may also occur when toxic substances are inadvertently brought home (eg, via contaminated clothing) or may represent a common environmental factor (eg, agricultural chemicals).

Although paternal medical radiation exposure prior to conception of the child has not been associated with Down syndrome [Alberman et al, 1972; Cohen et al, 1977], and parental occupational and education did not differ between cases and controls, military service prior to birth of the index child was more frequent in case vs control fathers in the Hopkins study [Cohen et al, 1977]. Information on parental occupations obtained from birth certificates was examined from a case control study of Down syndrome in Upstate New York; other findings (on maternal contraceptive use) have been presented elsewhere [Janerich et al, 1976, 1980]. Cases and controls (103 pairs) were matched on maternal age (± 2 years), and mean maternal and paternal education did not differ significantly between cases and controls. For five cases, and no controls, paternal occupation was listed as soldier (active duty), retired soldier, or disabled veteran; albeit based on small numbers, the marginal chi-square for paired samples ($\chi^2 = 5.00$) was statistically significant. The results of this study will be reported in more detail elsewhere. Thus, further studies are needed to search for paternal exposures or other variables that are associated with military service [Cohen et al, 1977]. Aside from interest in herbicides such as "Agent Orange," paternal exposure to infectious agents during military service should also be considered. Infectious agents (particularly viruses) have been implicated in chromosomal aberrations, and maternal pregestational contact with viral diseases may be involved in trisomies in the offspring [Ross, 1981].

Maternal contraceptive agents and adverse reproductive outcomes. Certificates of Upstate New York live births have been used to identify cases with specific defects and controls for interview in case-control studies of maternal contraceptives, and to identify cohorts of births without defects for studies of birth weight, sex ratio, and other variables.

Studies on the possible teratogenic and mutagenic effects of exposure to female sex hormones have been reviewed extensively elsewhere [Wilson, 1977; Schardein, 1980; WHO, 1981; Wilson and Brent, 1981]. In New York State, case-control studies of oral contraceptives (OC) and birth defects have indicated that the association with OC use until last menstrual period (LMP) or after conception is not large, but a causal explanation cannot be ruled out; the association may be stronger for certain major structural malformations and among older women [Janerich et al, 1980]. For Down syndrome, there was no evidence of an association with OC use, either during pregnancy or in the year prior to pregnancy [Janerich et al, 1976].

Using New York State birth records, we have examined the relationship between maternal spermicide use after LMP and during the year prior to LMP, and reproductive outcome (birth weight, sex ratio, and birth defects) [Polednak et al, 1982]. There was no evidence for a significant increase in total malformations associated with spermicide use (pre- or post-LMP); associations with specific defects were not statistically significant. The findings are, in general, consistent with those of another recent study [Shapiro et al, 1982]. In analyses of a cohort of live births (without major defects) whose mothers had used spermicides before or after LMP, post-LMP use was associated with an apparent reduction in birth weight among female (but not male) births, with no significant change in the sex ratio [Polednak et al, 1982]. These associations should be examined further in independent investigations.

The studies discussed above involved collection of information on maternal contraceptive use by interview some time after birth of the child. Although Jick et al [1981] examined retrospectively records of prescriptions for spermicides among women enrolled in a cooperative health plan, actual patterns of use and time of use in relation to conception could not be determined. In Britain, Skegg and Doll [1981] have described a feasibility study of record linkage for monitoring the effects of prescription drugs on the fetus (as well as other health effects); the feasibility of such surveillance systems should also be examined in other populations.

Childhood cancers. As Bloom and Paul [1981] have observed, detection of cancers related to transplacental chemical pollutants have been accomplished by clinicians and confirmed by other studies. Thus, diethylstilbestrol (DES) was confirmed as a transplacental carcinogen by data from the New York State Cancer Registry [see Greenwald et al, 1971; Greenwald and Nasca, 1974]. In response to public concern about DES-related cancer, a statewide DES screening program and registry were established in 1978 [Glebatis and Janerich, 1981].

Another method of detecting environmental carcinogens that may act transplacentally involves investigation of regional increases in reporting of childhood cancers, or apparent clusters. As noted above (see Table II), the New York State Cancer Registry provides a resource for such investigations. For relatively frequent cancers, such as leukemia, however, apparent clusters will occur by chance alone, or intensive studies may fail to reveal an environmental cause [eg, Halperin et al, 1980].

Recent research has focused on occupations of parents of children with various types of cancer. As with birth defects (discussed above), children themselves may be exposed to contaminants carried home by parents; other mechanisms, however, involve mutations in parental germ cells, or transplacental exposure from mothers (related to their own occupations or to contaminants derived from their husbands). Kantor et al [1979] observed that paternal mutations due to chemical (eg, lead) exposure around the time of conception could be a prezygotic event in the etiology of Wilms' tumor, in accordance with Knudson and Strong's [1972] two-step mutation theory of childhood cancer. In Kantor et al's [1979] case-control study, information on paternal occupation was obtained exlusively from birth certificates, and cases of Wilm's tumor were ascertained from the Connecticut Tumor Registry. These authors noted the nonspecificity of recorded prenatal occupations and the poor reporting of maternal occupations on birth records (as discussed above with reference to birth defects).

Kwa and Fine [1980] reported another case-control study using death records (childhood cancer deaths) and parental occupations from birth records. Although the overall distribution of occupations (including hydrocarbon-related occupations) of fathers did not differ between cases and controls, the mechanic-machinist category was significantly more frequent in urinary tract cancer cases than in controls. These findings were compared with results from other studies on hydrocarbon-related occupations, which were negative with regard to Wilms' tumor [Kwa and Fine, 1980]. Kantor et al's [1979] findings on Wilm's tumor, however, suggest the need for more rigorous case-control studies involving direct interview of parents, verification of occupational histories (with industrial hygiene data, where available), and control for potential confounding variables, along with cases ascertainment from cancer registries. Some of these desiderata have been satisfied in a case-control study of 92 childhood brain cancers ascertained from the Los Angeles County Cancer Surveillance Program, involving interviews on parental occupations and on potential confounding variables. Peters et al [1981] reported associations with maternal occupations involving chemical exposure, and paternal occupations in the aircraft industry and jobs involving solvents. The authors noted a lack of detail in occupational histories as in other studies [Hemminki et al, 1981], and further studies of cases involving specific plants or industries appear warranted. As noted above, identification of apparent space–time clusters of childhood cancers may also provide clues. Case–control studies of childhood brain cancers in Upstate New York are also in progress [Nasca, 1982].

Associations between maternal use of certain drugs during pregnancy and childhood cancer has been reported, using data on childhood cancer deaths from a British cancer registry and interview data (along with information from antenatal clinics) on drug intake [Kinnier-Wilson et al, 1981]. Although use of "hormones" was not significantly associated with cancer risk in this study, more detailed studies are needed in view of the findings on DES in humans and in animal studies [Rustia, 1979]. Although maternal DES exposure in humans has not as yet been shown to produce tumors other than adenocarcinoma of the vagina in girls [Greenwald and Nasca, 1974; Cosgrove et al, 1977], larger data bases and continued follow-up are needed to evaluate this possibility. Beral and Colwell [1981] found no malignant tumors in a small group of 70 children whose mothers had high doses of stilbestrol and ethisterone therapy during pregnancy. We are conducting a follow-up study of children whose mothers used female sex hormones (ie, oral contraceptives, supportive hormones, or hormone pregnancy tests), involving record linkage with the New York State Cancer Registry.

CONCLUSIONS

The uses and limitations of several available record systems relevant to studies of reproductive toxicology have been assessed and exemplified. The major uses are in monitoring for temporal and/or spatial changes in rates that may indicate effects of occupational or other environmental exposures, and in the identification of cases (eg, fetal deaths, birth defects, low birth weight live births, childhood cancers) for analytic epidemiologic studies of parental occupation and other potential risk factors.

Improvements in these record systems are needed, including more complete and accurate reporting of adverse reproductive outcomes and improved reporting of occupation (eg, on birth records and on cancer registry reports). Further, various types of record linkage need to be developed. Examples include: linkage of environmental data on contaminants in air, water, and soil; linkage of records on parental use of prescription drugs with vital records on adverse reproductive outcomes; and linkage of data from studies on prenatal exposures with records of childhood cancers from cancer registries. Although information may be incomplete and imprecise, studies based on existing record systems may indicate directions for further investigation and hypotheses may be developed for testing by more intensive studies. Results of preliminary studies on parental occupation in relation to birth defects are discussed in this light.

REFERENCES

Aarskog D (1979): Maternal progestins as a possible cancer of hypospadias. N Engl J Med 300:75–78.

Alberman E, Polani PE, Roberts JAF, Spicer CC, Elliott M, Armstrong E (1972): Parental exposure to x-irradiation and Down's syndrome. Ann Hum Genet 36:195–206.

American Cancer Society (1968): "Manual of Tumor Nomenclature and Coding, 1968 Edition." American Cancer Society.

Anon (1971): Report of the Committee for the study of monitoring of human mutagenesis. Teratology 4: 103–108.

Anon (1976): Diagnosis of Down's syndrome at birth. Br Med J 2:835.

Beckman G, Beckman L, Nordenson I, Nordstrom S (1979): Studies of spontaneous abortion, malformations and birth weight in populations exposed to environmental pollutants: Theoretical considerations and empirical data. In Berg K (ed): "Genetic Damage in Man Caused by Environmental Agents." New York: Academic Press, pp 317–326.

Bennett JW, Abroms KI (1979): Gametogenesis and incidence of Down syndrome. Lancet 2:913.

Beral V, Colwell L (1981): Randomised trial of high doses of stilbestrol and ethisterone therapy in pregnancy: Long-term follow-up of children. J Epidemiol Community Health 35:155–160.

Bloom AD, Paul NW (1981): "Guidelines for Studies of Human Populations Exposed to Mutagenic and Reproductive Hazards." White Plains, New York: March of Dimes Birth Defects Foundation.

Blot WJ, Sawada H (1972): Fertility among female survivors of the atomic bombs of Hiroshima and Nagasaki. Am J Hum Genet 24:613–622.

Burometto E, Therriault G, Logrillo V (1977): "Summary of Selected Health Statistics for Counties With Nuclear Facilities, New York State excluding New York City, 1960–1975." Albany, New York: New York State Department of Health, Monograph No. 13.

Carucci PM (1979): "Reliability of Statistical and Medical Information Reported on Birth and Death Certificates." Albany, New York: New York State Department of Health Monograph No. 15.

Carucci PM, Prasad S (1979): A comparison of mothers' occupation reported on live birth certificates and on a survey questionnaire. Public Health Rep 94:432–437.

Cohen BH, Lilienfeld AM, Kramer S, Hyman LC (1977): Parental factors in Down's syndrome—Results of the second Baltimore case-control study. In Hook EB, Porter IH (eds): "Population Cytogenetics. Studies in Humans." New York: Academic Press, pp 301–352.

Cosgrove MD, Benton B, Henderson BE (1977): Male genitourinary abnormalities and maternal diethylstilbestrol. J Urol 117:220–222.

DeMarini DM, Pleva MJ, Brockman HE (1982): Use of four short-term tests to evaluate the mutagenicity of municipal water. J Toxicol Environ Health 9:127–140.

Edmonds LD, Layde PM, James LM, Flynt JW, Erickson JD, Oakley GP (1981): Congenital malformation surveillance: Two American systems. Int J Epidemiol 10:247–252.

Elmwood JM, Coldman AJ (1981): Water composition in the etiology of anencephalus. Am J Epidemiol 113:681–690.

Emanuel I, Culver BH, Erickson JD, Guthrie B, and Schuldberg D (1973): The further epidemiologic differentiation of cleft lip and palate: A population study of clefts in King County, Washington, 1956–1965. Teratology 7:271–282.

Fabricant JD, Legator MS (1981): Etiology, role and detection of chromosomal aberrations in man. J Occup Med 23:617–625.

Field B, Kerr C (1979): Herbicide use and incidence of neural-tube defects. Lancet 1:1341–1342.

Flynt JW Jr (1974): Trends in surveillance of congenital malformations. In Janerich DT, Skalko RG, Porter IH (eds): "Congenital Defects. New Directions in Research." New York: Academic Press, pp 118–128.

Fulton MP, Cobb S, Preble L, Leone L, Forman E (1980): Electrical wiring configurations and childhood leukemia in Rhode Island. Am J Epidemiol 111:292–296.

Gittelsohn AM, Milham S Jr (1965): Vital record incidence of congenital malformations in New York State. In Neel JV, Shaw MW, Schull WJ (eds): "Genetics and the Epidemiology of Chronic Diseases." Washington, DC: US Public Health Service Public No. 1163, pp 305–319.

Glebatis DM, Janerich DT (1981): A statewide approach to diethylstilbestrol—The New York Program. N Engl J Med 304:47–50.

Greenwald P, Nasca PC (1974): Stilbestrol exposure in utero: Long term effects. In Janerich DT, Skalko RG, Porter IH: "Congenital Defects. New Direction in Research." New York: Academic Press, pp 149–160.

Greenwald P, Barlow JJ, Nasca PC, Burnett WS (1971): Vaginal cancer after maternal treatment with synthetic estrogens. N Engl J Med 285:390–392.

Greenwald P, Burnett WS, Kirmss V, Brennan K, Nasca PC (1976): "Cancer Incidence and Mortality in New York State (Exclusive of New York City). Albany, New York: Bureau of Cancer Control, NY State Department of Health.

Halperin W, Altman R, Stemhagen A, Iaci AW, Caldwell G, Mason T, Bill J, Abe T, Clark JF (1980): Epidemiologic investigation of cluster of leukemia and Hodgkin's disease in Rutherford, New Jersey. J Med Soc NJ 77:267–273.

Hay S (1967): Variables in under-reporting of clefts on birth certificates. Public Health Rep 82:985–993.

Hay S (1971): Incidence of selected congenital malformations in Iowa. Am J Epidemiol 94:572–584.

Hay S, Tonascia S (1968): "A Classification of Congenital Malformations." US Department of Health, Education and Welfare, Public Health Service.

Hemminki K, Saloniemi I, Salonen T, Partanen T, Vainio H (1981): Childhood cancer and parental occupation in Finland. J Epidemiol Community Health 35:11–15.

Hoff ML (1975): "Sequential Analysis – A Tool for Monitoring Health Status." Albany, New York: Vital Statistics of New York State, No. 36, pp. 4–13.

Holmberg PC (1979): Central-nervous-system defects in children born to mothers exposed to organic solvents during pregnancy. Lancet 2:177–179.

Holmberg PC, Nurminen M (1980): Congenital defects of the central nervous system and occupational factors during pregnancy. A case-referent study. Am J Indust Med 1:167–176.

Hook EB (1978): Monitoring human mutations and consideration of a dilemma posed by an apparent increase in one type of mutation rate. In Morton NE, Chung CS (eds): "Genetic Epidemiology." New York: Academic Press, pp 483–523.

Hook EB, Albright, Cross PK (1980): Use of Bernoulli census and loglinear methods for estimating the prevalence of spina bifida in live births and the completeness of vital record reports in New York State. Am J Epidemiol 112:750–758.

Hook EB, Cross PK (1981): Temporal increase in the rate of Down syndrome live births to other mothers in New York State. J Med Genet 18:29–30.

Hook EB, Schreinmachers DM, Cross PK (1981): Use of prenatal cytogenetic diagnosis in New York State. N Engl J Med 305:1410–1413.

Howe HL, Hoff MB (1981): Sequential analysis. A tool for monitoring program delivery. Eval Health Profess 4:129–143.

Jacobs PA, Mayer M (1981): The origin of human trisomy: A study of heteromorphisms and satellite associations. Ann Hum Genet 45:357–365.

Janerich DT (1973): Epidemic waves in the prevalence of anencephaly and spina bifida in New York State. Teratology 8:253–256.

Janerich DT (1979): Environmental hazards and birth defects. Ann NY Acad Sci 329:10–13.

Janerich DT, Flink EM, Keogh MD (1976): Down's syndrome and oral contraceptive usage. Br J Obstet Gynaecol 83:617–620.

Janerich DT, Piper JM, Glebatis DM (1980): Oral contraceptives and birth defects. Am J Epidemiol 112: 73–79.

Janerich DT, Stark AD, Greenwald P, Burnett WS, Jacobson HI, McCusker J (1981): Increased leukemia, lymphoma, and spontaneous abortion in western New York following a flood disaster. Public Health Rep 96:350–356.

Jick H, Walker AM, Rothman KJ, Hunter JR, Holmes LB, Watkins RN, D'Ewart DC, Danford A, Madsen S (1981): Vaginal spermicides and congenital disorder. JAMA 245:1329–1332.

Kantor AF, Curnen MGM, Meigs JW, Flannery JT (1979): Occupations of fathers of patients with Wilms's tumour. J Epidemiol Community Health 33:253–256.

Kato H (1975): A review of thirty years study of Hiroshima and Nagasaki atomic bomb survivors. II. Biological effects. B. Genetic effects. 1. Early genetic surveys and mortality study. J Radiat Res (Tokyo) 16 (Suppl.):67–74.

Khoury MJ, Erickson JD, James LM (1982): Etiologic heterogenicity of neural tube defects: Clues from epidemiology. Am J Epidemiol 115:538–548.

Kinnier-Wilson LM, Kneale GW, Stewart AM (1981): Childhood cancer and pregnancy drugs. Lancet 2: 314–315.

Kline J, Stein Z, Strobino B, Susser M, Warburton D (1977): Surveillance of spontaneous abortions. Power in environmental monitoring. Am J Epidemiol 106:345–350.

Knudson AG, Strong LC (1972): Mutation and cancer: A model for Wilms' tumor of the kidney. J Natl Cancer Inst 48:313–324.

Kwa S-L, Fine LJ (1980): The association between parental occupation and childhood malignancy. J Occup Med 22:792–794.

Lane GA, Nahrwold ML, Tait, AR, Taylor-Bush M, Cohen PJ, and Beaudoin AR (1980): Anesthetics as teratogens: Nitrous oxide is fetotoxic, xenon is not. Science 210:899–901.

Longo LD (1980): Environmental pollution and pregnancy: Risks and uncertainties for the fetus and infant. Am J Obstet Gynecol 137:162–173.

Lushbaugh CC, Casarett G (1976): The effects of gonadal irradiation in clinical radiation therapy: A review. Cancer 37:1111–1120.

MacMahon B, Yen S (1971): Unrecognized epidemic of anencephaly and spina bifida. Lancet 1:31–33.

Mattei JF, Mattel MG, Ayme S, Girard F (1979): Origin of the extra chromosome in trisomy 21. Hum Genet 46:107–110.

Milham S Jr (1971): Experience with malformation surveillance. In Hook EB, Janerich DT, Porter IH (eds): "Monitoring, Birth Defects and Environment." New York: Academic Press, pp 137–142.

Miller JF, Williamson E, Glue J, Gordon YB, Grudzinskas JG, Sykes A (1980): Fetal loss after implantation. A prospective study. Lancet 2:554–556.

Nasca PC (1982): Personal communication.

Neel JV, Mohrenweiser H, Satoh C, Hamilton HB (1979): A consideration of two biochemical approaches to monitoring human populations for a change in germ cell mutation rates. In Berg K (ed): "Genetic Damage in Man Caused by Environmental Agents." New York: Academic Press, pp 29–47.

New York State Department of Health (1981): "Love Canal. A Special Report to the Governor and Legislature." Prepared by Axelrod D, Haughie GE: Albany, New York: New York State Department of Health, Health Communications Group.

Peters JM, Preston-Martin S, Yu MC (1981): Brain tumors in children and occupational exposure of parents. Science 213:235–236.

Piper J (1973): The use of vital records in the Birth Defects Institute. Q Vital Stat Rev, New York State Health Department, Oct–Dec 1973, pp 20–23.

Polednak AP (1980): Fertility of women after exposure to internal and external radiation. J Environ Pathol Toxicol 4:457–470.

Polednak AP, Janerich DT (1982): Maternal characteristics and hypospadias: A case-control study. Submitted for publication.

Polednak AP, Janerich DT, Glebatis DM (1982): Birth weight and birth defects in relation to maternal spermicide use. Teratology 26:27–38.

Porter IH (1979): Genetic services within a state health department. In Porter IH, Hook EB (eds): "Service

Querec LJ (1980): "Comparability of Reporting Between the Birth Certificate and the National Natality Survey." National Center for Health Statistics, Data Evaluation and Methods Research, Series 2, No. 83. DHEW Public No. (PHS) 80-1357. Hyattsville, Maryland: U.S. Public Health Service.

Robson ER (1978): The genetics of birth weight. In Falkner F, Tanner JM (eds): "Human Growth. 1. Principles and Prenatal Growth." New York: Plenum Press, pp 285–297.

Ross LJ (1981): Trisomy after pregestational contact with communicable disease. Am J Epidemiol 115: 529–532.

Rustia M (1979): Role of hormone imbalance in transplacental carcinogenesis induced in Syrian golden hamsters by sex hormones. Natl Cancer Inst Monogr 51:77–87.

Schardein JL (1980): Congenital abnormalities and hormones during pregnancy: A clinical review. Teratology 22:251–270.

Shapiro S, Slone D, Heinonen OP, Kaufman DW, Rosenberg L, Mitchell AA, Helmrich SP (1982): Birth defects and vaginal spermicides. JAMA 247:2381–2384.

Skegg DCG, Doll R (1981): Record linkage for drug monitoring. J Epidemiol Community Health 35:24–31.

Stamatoyannopoulos G (1979): Possibilities for demonstrating point mutations in somatic cells, as illustrated by studies of mutant hemoglobins. In Berg K (ed): "Genetic Damage in Man Caused by Environmental Agents." New York: Academic Press, pp 49–62.

Stein Z, Susser M, Warburton D, Wittes J, Kline J (1975): Spontaneous abortion as a screening device. The effect of fetal survival on the incidence of birth defects. Am J Epidemiol 102:275–290.

Sternberg SS (1979): The carcinogenesis, mutagenesis, and teratogenesis of insecticides: Review of studies in animals and man. Pharmacol Ther 6:147–166.

Tokuhata GK, Smith MW (1981): History of health studies around nuclear facilities: A methodological consideration. Environ Res 25:75–85.

Wallenstein S (1980): A test for detection of clustering over time. Am J Epidemiol 111:367–372.

Warburton D, Fraser FC (1964): Spontaneous abortion risks in man: Data from reproductive histories collected in a medical genetics unit. Hum Genet 16:1–15.

Warren NS, Carter TP, Humbert JR, Rowley PT (1982): Newborn screening for hemoglobinopathies in New York State: Experience of physicians and parents of affected children. J Pediatr 100:373–377.

World Health Organization (1981): The effect of female sex hormones on fetal development and infant health. WHO Technical Report Series No. 657. Geneva: World Health Organization.

Wilcox AJ, Treloar AE, Sandler DP (1981): Spontaneous abortion over time: Comparing occurrence in two cohorts of women a generation apart. Am J Epidemiol 114:548–553.

Wilson JG (1977a): Embryotoxicity of drugs in man. In Wilson JG, Fraser FC (eds): "Handbook of Teratology. I. General Principles and Etiology." New York: Plenum Press, pp 309–355.

Wilson JG (1977b): Environmental chemicals. In Wilson JG, Fraser FC (eds): "Handbook of Teratology. I. General Principles and Etiology." New York: Plenum Press, pp 357–385.

Wilson JG, Brent RL (1981): Are female sex hormones teratogenic? Am J Obstet Gynecol 141:567–580.

American Journal of Industrial Medicine 4:349–364 (1983)

Problems and Pitfalls in Conducting Epidemiological Research in the Area of Reproductive Toxicology

Stephen M. Levin, MD

Defective reproductive outcome continues to represent a considerable public health burden, as well as a source of intense personal tragedy and enormous expense in our society. Evidence suggests that the rate of reproductive failure may be modifiable, in part by limiting human exposure to environmental agents toxic to the reproductive process. Methodological problems posing difficulties to researchers investigating the association between exposure to such agents and adverse reproductive outcome are discussed, including inadequacies in existing data bases available for analysis; bias in reporting by or in the collection of information from affected families; and weaknesses in the selection of study populations, "control" groups, and outcomes chosen for observation. Suggestions for possible resolution of some of these difficulties are discussed.

Key words: reproductive toxicology, study methodology, environmental toxins, human population studies, birth defects

INTRODUCTION

The epidemiologist must of necessity conduct his research on human populations in an effort to assess the possible harmful effects of toxic exposures in the human environment. Unfortunately, the problems and pitfalls involved in research of this kind have led to unease among scientists, prompting Richard Doll to refer to the "haunting fear of the epidemiologist that his conclusions might not always be correct" [Doll, 1973].

This paper is a discussion of some of the methodological difficulties encountered by epidemiologists attempting to gather meaningful information about adverse effects on the reproductive system as the result of exposures to possibly toxic environmental agents. Some suggestions for possible resolutions of these problems are also discussed.

Environmental Sciences Laboratory, Mount Sinai School of Medicine of the City University of New York, New York.
Address reprint requests to S.M. Levin, MD, Environmental Sciences Laboratory, Mount Sinai School of Medicine, New York, NY 10029.

0271-3586/83/0401-02-0349$04.50 © 1983 Alan R. Liss, Inc.

SCOPE OF THE PROBLEM

In the United States, between 5 and 7% of all newborns are born with birth defects. Over 560,000 lives are lost each year to defective fetal development resulting in infant death, spontaneous abortion, and stillbirths [National Foundation/March of Dimes, 1975]. Some 2% of all newborns have a demonstrable genetic defect [Stein et al, 1981]. Moreover, an estimated 10–15% of couples in the United States who want children are infertile [Stein et al, 1981]. Ten to twenty percent of recognized pregnancies end in spontaneous abortion [Stein et al, 1981].

The cause of 60–70% of developmental defects is at present unknown [National Foundation/March of Dimes, 1975]. One example is congenital heart disease, which occurs at a rate of approximately 5 per 1000 live births. Yet of these cases, only 5% have a known, identifiable genetic or environmental etiology. Fully 95% have no known cause [Higgins, 1965].

Recent information suggests, however, that the incidence of reproductive failure is modifiable, since improvements have been noted in stillbirth rates [Doll, 1973], for example. This fall in rates is in part the result of improved public health measures and antenatal care, as well as more effective treatment of infectious diseases in pregnant women. More aggressive measures to manage fetal distress have been developed. But it also points to the possibility that potentially avoidable, environmental agents which operate on either or both of the parents prior to conception or on the conceptus during embryonic or fetal development may be causal.

In 1973, Doll published a review of vital records in England and Wales for the 39-year period between 1931 and 1970. Figure 1 shows that the stillbirth rate has been reduced by 68% during this period, the infant mortality rate by 71%, and the mortality of children aged 1–9 by 87%. The odds of a 28-week fetus living 10 years after birth had been improved from 6.5:1 to 27:1. Figure 2 shows that the contribution made by congenital anomalies to the stillbirth rate fell between 1961 and 1970.

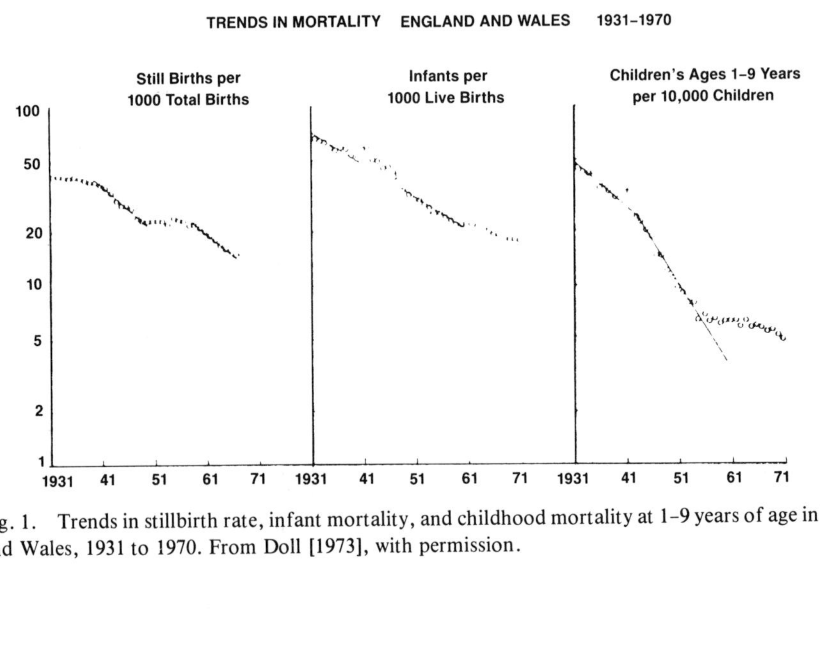

Fig. 1. Trends in stillbirth rate, infant mortality, and childhood mortality at 1–9 years of age in England and Wales, 1931 to 1970. From Doll [1973], with permission.

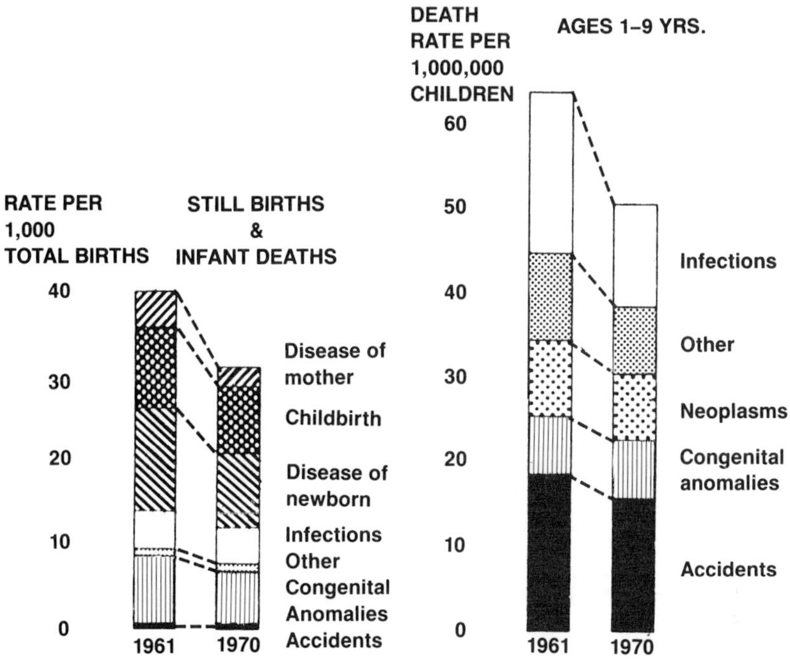

Fig. 2. Causes of (i) stillbirths and infant deaths and (ii) childhood deaths at 1–9 years of age in England and Wales in 1961 and 1970. From Doll [1973], with permission.

In Denmark, the stillbirth rate is approximately 65% that of England and Wales, as shown in Table 1, again suggesting the possible action of environmental agents among other potential factors.

AVAILABLE SOURCES OF INFORMATION

Table II lists some of the reproductive endpoints which can indicate reproductive dysfunction and is taken from Warburton's excellent discussion of data sources [Warburton, 1981]. Each of these can be measured, and each, in one study or another, has been shown to vary with exposure to environmental agents. The data from which information on these outcomes is drawn come from a number of different sources, some of which are readily available and require no further contact with the study population.

As Warburton indicates, vital statistics and other record systems are available from the National Center for Health Statistics. From these sources, one can obtain maternal age-specific birth rates, death rates including late fetal deaths, and birth weight. Some information is also available on birth certificates about malformations. Screening programs offer statistics on the incidence rates of certain conditions; for example, Down's syndrome. Surveys have been conducted on neural tube defects and their association with alpha-fetoprotein. There are some registries of malformations, chromosome abnormalities, and other genetic diseases, using vital records and/or hospital records for data sources. One such registry is maintained by the Centers for Disease

TABLE I. Countries With Low Stillbirth and Infant Mortality Rates, 1969

	Rate per 1000 live births	
Country[a]	Stillbirths[b]	Infants under one year
England and Wales	13.3	18.0
Australia	—	17.8
New Zealand	10.0	16.9
Switzerland	9.4	15.4
Denmark	8.6	14.8
Finland	—	14.4
Norway	11.2	13.8
Netherlands	11.1	13.2
Iceland	11.2	11.7
Sweden	—	11.7

[a]Ireland, infant mortality rate, 20.6

[b]Stillbirths recorded per 1000 *live* births (Demographic Yearbook, U.N.).

From Doll [1973], with permission.

TABLE II. Reproductive Endpoints to Indicate Reproductive Dysfunction

Sexual dysfunction: decreased libido; impotence.
Sperm abnormalities: decreased number; decreased motility; abnormal morphology.
Sub-fecundity: abnormal gonads, ducts of external genitalia; abnormal pubertal development; infertility of male or female origin; amenorrhea; anovulatory cycles; delay in conception.
Illness during pregnancy and parturition: toxemia; hemorrhage.
Early fetal loss (to 28 weeks).
Late fetal loss (after 28 weeks) and stillbirth. } Perinatal death
Intrapartum death. } Perinatal death
Death in first week. } Perinatal death
Decreased birth weight.
Change in gestational age at delivery: prematurity; post-maturity.
Altered sex ratio.
Multiple births.
Birth defects, major and minor.
Chromosome abnormalities, in fetal deaths, at amniocentesis, in perinatal deaths, in livebirths.
Infant death.
Childhood morbidity.
Childhood malignancies.
Age at menopause.

From Warburton [1981], with permission.

Control (CDC) in Atlanta [CDC, 19]. New York State maintains a Northeast Chromosome registry in the New York State Department of Health. Other states with similar registries include Georgia, Washington, California, and Florida [Flynt, 1974].

In addition, there are hospital and physician records which can yield important information on infertility, spontaneous abortions, congenital malformations, and infant and childhood morbidity, especially from records of prepaid insurance programs, health maintenance organizations (HMOs), etc. Although working with data from these sources often involves considerable tracing efforts, better and more useful information can be obtained than from vital statistics, and one can examine a wider range of outcomes. However, both this source and vital statistics sources share one limitation, namely, the paucity of information about exposures to toxic substances.

Studies which rely on hospital and physicians' records carry within them a potential for bias, in that one must know the pattern of hospital and physician use for both the study and the control populations. Such service usage varies, for example, with socioeconomic status, rural vs urban residence, race, and geographic area of the country. It is also necessary to look for change in patterns of use as new diagnostic procedures or screening programs are introduced, since such changes may be expected to increase the number of cases recorded.

Warburton [1981] offers an example of such an artifact in the initial reports of an increased frequency of congenital hypothyroidism among children born to residents in the area surrounding the Three Mile Island nuclear power plant, following the recent mishap there. Further investigation of this association, however, revealed that a screening program to assess thyroid function among newborns had been introduced at the time that the nuclear accident occurred. Thus, the increase in the number of cases reported was falsely attributed to the accident.

Causal inferences derived from studies which involve no contact with the study populations are likely to be weak. In addition, exposure information is usually very poor and the outcome range limited. However, such studies are less expensive to conduct, usually less time-consuming, and can offer opportunities for a first look at the possibility of an association between a given exposure and a possible adverse reproductive outcome.

STUDIES INVOLVING INTERVIEWING OF EXPOSED AND UNEXPOSED POPULATIONS

In order to obtain better information on the association between a given exposure and reproductive dysfunction, interviewing is generally necessary. Interviews are virtually indispensable for collecting data on very early abortions, duration of involuntary sterility, duration of exposure, time-relation to reproductive events, and exposures to other independently acting agents, such as smoking and alcohol, which may cause reproductive dysfunction [Warburton, 1981].

Most studies which involve interviewing exposed and unexposed populations use one of two strategies to estimate the degree of association between exposure and outcome. The first is the cohort study, which compares the frequency of adverse reproductive outcome among exposed and unexposed groups. Observation may be prospective from the point at which the baseline data are collected. One knows or determines the exposure status of the individual and follows these individuals over time, looking for selected reproductive outcomes. It is also possible to design a historical cohort study in which one establishes a cohort of exposed and unexposed groups at one point in time in the past, on the basis of earlier recorded information or interview data. Then one "follows" the study groups for a defined past period of time to evaluate the occurrence of selected reproductive outcomes.

The second study strategy, the case control study, involves comparing the frequency of exposure among people who have experienced reproductive dysfunction with the frequency of exposure among people without reproductive dysfunction, ie, beginning with cases and non-cases. Then, on the basis of interview information or other data sources, one can determine the relative frequency of exposure of the two groups.

This study design usually relies on interview data, which carries with it the potential for recall bias. Individuals who have experienced adverse reproductive outcomes may over-report or report an exposure more accurately than those with no reproductive dysfunction.

For example, a couple who have recently gone through the experience of a stillbirth or who have a congenitally malformed child are frequently more inclined to search their past histories for potential toxic exposures which could have given rise to the abnormal outcome, seeking explanations for themselves. Under-reporting can also occur, if guilt is an aspect, eg, when alcohol use is the exposure in question. The exposed group may over report adverse outcomes in prospective cohort studies, especially if they are aware of a possible association between this exposure and reproductive dysfunction. This would hold true more for reporting early spontaneous abortions and the more subtle behavioral abnormalities, than for such readily measurable adverse outcomes as severe congenital malformations.

This phenomenon is described by Doll [1973] in his discussion of the early studies of the effect of exposure *in utero* to X-rays, reported by Steward et al [1956]. The original studies relied on a comparison of the histories given by mothers whose children had died of cancer with those given by mothers of living children of the same sex who were born in the same area and at the same time period as the children who had died. It was argued that the results were an artifact due to differential recall of past events by the mothers who had different experiences. That this was not the case was ultimately demonstrated by detailed examination of the medical records of the mothers who had given a positive history of irradiation and of a sample of mothers who did not give such a history. These results were further confirmed when MacMahon, in 1962, published a study of children born in selected hospitals in the United States in which records of X-ray examination during pregnancy had been maintained. X-ray exposures were determined solely by reference to hospital records compiled before the children had been born, thus excluding the opportunity for recall bias.

To indicate how complicated this problem is, however, one can still claim that, despite the demonstrated association, it was not the radiation which caused the disease in the offspring, since some women who had been X-rayed during pregnancy were X-rayed routinely, while others were X-rayed for medical reasons. It is possible to argue, therefore, that the carcinogenic factor could be associated with the reason for the X-rays and not with the radiation itself. However, when MacMahon analyzed the effect of economic status, age, religion of the mother, birth order of the child, Casearean section versus vaginal delivery, spontaneous or foreceps delivery, placenta previa or toxemia of pregnancy, no associations were found.

Another potential source of bias in studies involving the collection of data by interview can be found in the interviewers themselves. Many studies have demonstrated the need for interviewers to be "blind," or unaware of the reproductive outcome status of the interviewees. Otherwise the interviewers might tend to pursue exposure information more vigorously for known cases than for the controls who show no abnormality. Information gathered by trained interviewers is more reliable than information collected by self-administered questionnaires, since accurate interpretation of the questions asked is important.

A number of investigators have attempted to avoid these difficulties by turning to laboratory tests to determine exposure and/or potential effect on reproductive out-

TABLE III. Sexual Dynamics and Decrease in Libido in Workmen

Group	Libido decrease, %	Pathologic erections, %	pathologic ejaculation, %	Orgasm decrease, %
A. (a) Lead-poisoned workmen, 23	21	48	30	3
(b) Lead workmen with moderate increased absorption, 42	33	33	38	4
(c) Lead workmen with slight increased absorption, 35	28	22	40	5
B. Men with physiologic absorption of lead working in a polluted environment, 50	16	14	16	2

From Lancranjan et al [1975], with permission.

come. Examples of such studies include sperm analysis and mutagenicity testing of urine, blood and other body fluids. For many agents, however, there has been a poor correlation of such laboratory findings with adverse reproductive outcome, with the exception of extremes such as azoospermia [McLeod and Ying, 1979].

Lancranjan et al [1975] utilized both laboratory data and interview information to measure reproductive outcome in a group of male workers employed at a lead storage battery plant. The workers were divided into exposure categories of lead poisoned, moderately exposed, slightly exposed and physiologic absorption, using blood lead, urinary lead, delta-aminolevulinic acid and coproporphyrin levels. A questionnaire for fertility assessment and sexual complaints at the time of the exam was also completed by the workers. Table III presents information on sexual dysfunction and decreased libido for the different worker groups as derived from the questionnaires. If one compares the frequency of decreased libido and sexual dysfunction among those workers in the lead-poisoned group with those in the lowest category of lead absorption, the frequencies of reproductive dysfunction are considerably higher in the first group. Distribution of these disorders, however, was fairly uniform across the "exposed" subgroups, with a poor relationship with dose.

Lancranjan et al [1975] expressed doubts about the accuracy of the information obtained by interview and relied therefore on semen analysis as "the only objective method to estimate the productive ability of the workmen exposed to lead." Table IV shows the frequency of alterations in spermatogenesis in the lead-exposed workmen. The increase in the frequency of abnormal spermatogenesis with increasing lead exposure was statistically significant. Figures 3, 4, 5, 6 all show the same correlations and contrast these frequencies with those for unexposed controls. Lancranjan's study [1975] demonstrates the value of finding objective measures of reproductive function and dysfunction as an indicator of an agent's toxicity to the reproductive process.

The disciplines of occupational and environmental medicine will face increasingly frequent challenges to provide information about the potential reproductive toxicity of agents to which human exposure may occur. If mistakes are made on either side in answering that question, considerable social harm may result.

If it is concluded incorrectly that a given agent or exposure is toxic to reproduc-

TABLE IV. Frequency of Alterations in Spermatogenesis in Lead-Poisoned Workmen

Group	Alterations in Spermatogenesis, no. (%)	Asthenospermia, no. (%)	Hypospermia, no. (%)	Teratospermia, no. (%)
A. (a) Lead-poisoned	15 (93)	8 (50)	8 (50)	14 (86)
(b) Lead workmen with moderately increased absorption, 29	22 (68)	15 (51)	13 (44)	17 (58)
(c) Lead workmen with slightly increased absorption, 19	12 (63)	8 (42)	8 (42)	6 (31)
B. Men with physiologic absorption of lead, working in a polluted environment, 25	7 (28)	6 (24)	7 (28)	4 (16)

From Lancranjan et al [1975], with permission.

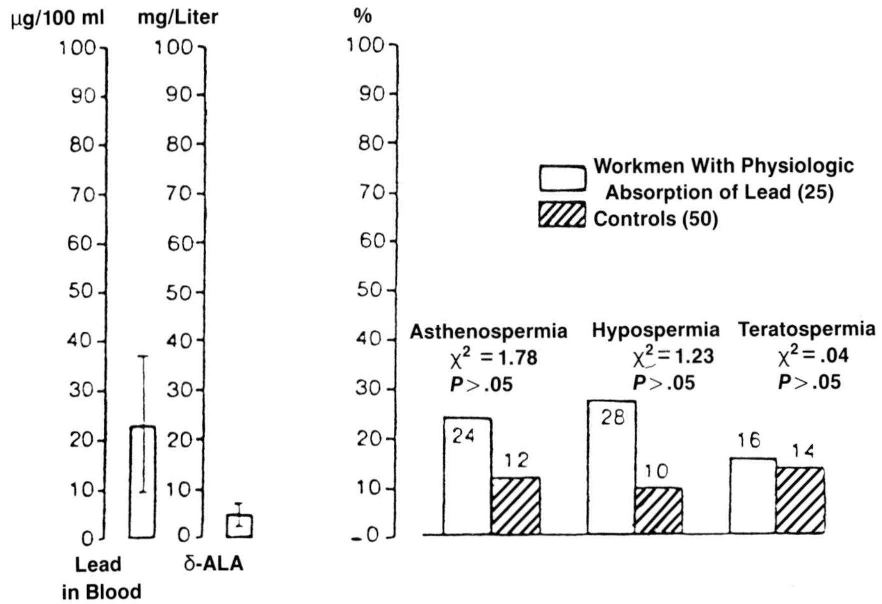

Fig. 3. Incidence of spermatogenesis alterations in workmen with physiologic absorption of lead. From Lancranjan et al [1975], with permission.

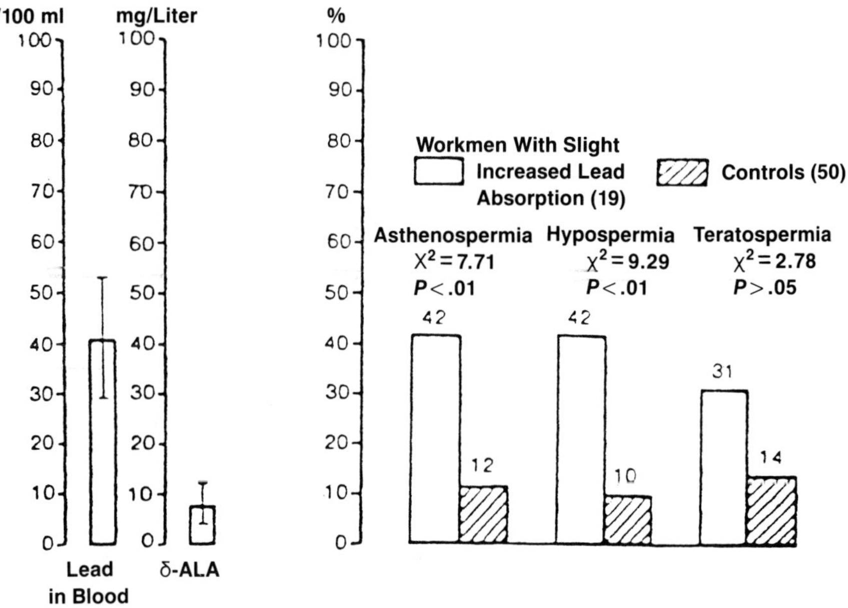

Fig. 4. Incidence of spermatogenesis alterations in workmen with slight increased lead absorption. From Lancranjan et al [1975], with permission.

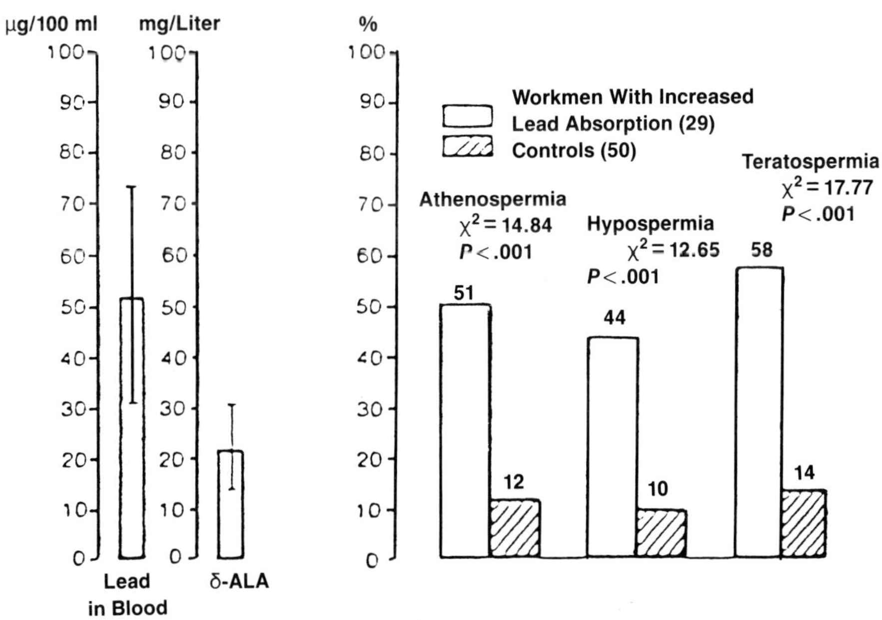

Fig. 5. Incidence of spermatogenesis alterations in workmen with increased lead absorption. From Lancranjan et al [1975], with permission.

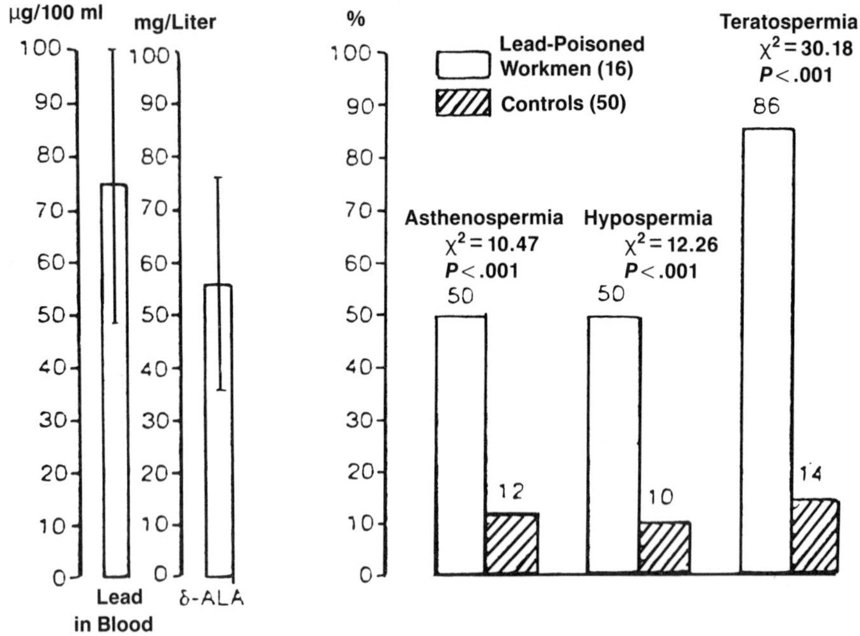

Fig. 6. Incidence of spermatogenesis troubles in lead-poisoned workmen. From Lancranjan et al [1975], with permission.

tive function, unnecessary worry and concern are caused, as well as job or residence dislocation among individuals who will then believe themselves to be exposed to a harmful material. Such an error may have occurred with the widely reported and possibly erroneous conclusions regarding the frequency of chromosome abnormalities among residents of the area surrounding Love Canal. Science can ill afford to lose the public's confidence in its conclusions, especially when the conclusions bear on issues of regulatory control of potentially hazardous exposures.

A more serious error, however, is a failure to detect the toxic effect of any given agent or exposure when in fact such toxicity exists. This kind of mistake disarms us in our attempts to eliminate, or at least reduce, exposures to substances and agents toxic to reproductive function, and there are aspects of study design which can influence our ability to demonstrate the association between exposure and subsequent reproductive dysfunction.

STUDY DESIGN

Stein et al [1981] have cited three aspects of study design which bear on the ability of investigators to detect a given agent's potential to cause adverse reproductive outcome: 1) Definition of exposure; 2) selection of outcome(s) to be observed; 3) comparability of study groups.

Definition of Exposure

Often epidemiological investigators have little data on exposure, or data of poor quality. As a result, individuals included in the category of "exposed" may in fact not be exposed, thus diluting the genuinely exposed populations and weakening our ability to detect differences between the exposed and unexposed groups. On the other hand, our definitions of exposure may be overly restrictive, in which case genuinely exposed individuals will not be included in the study population or may erroneously be included in the unexposed group, again weakening our ability to detect an effect.

As Stein points out, the more we know about the biological effects of an agent or exposure, the better we are able to define exposure. For example, if we are studying the potential toxicity of an agent which we suspect to be disruptive of chromosome replication in the gametes, then we should consider only those exposures occurring prior to conception or at the time of fertilization. If we include in the exposed group all individuals ever exposed to this agent, rather than limiting our definition of exposure to those who had contact with the agent at the critical time, we may dilute the truly exposed study population with a number of effectively unexposed individuals. While it is generally true that the greater the number of individuals in the study sample the greater our ability to detect effects, in this case increasing the number in the exposed group who are in fact not exposed reduces rather than increases the study's ability to detect the potential effect.

In cases where we have no information or suspicions regarding the timing or other critical factors regarding exposure, several dimensions of exposure may be examined: Which parent was exposed, what dose, what timing, what duration. Each of these aspects can be looked at separately or in combination with others. Using this multiple-dimension approach, one may uncover a relevant definition of exposure.

There are problems with this approach, however. The likelihood of demonstrating an association which occurs by chance alone, and not because a genuine association exists between exposure and outcome, increases with the number of such analyses that are conducted on the same body of data. If computers are used to analyze data, the number of such associations or correlations that one can readily look for can be quite large.

If one sets one's level of significance at 0.05, which is to say that one will accept as a statistically significant result an association which has a 1:20 probability of having arisen by chance alone, then if one does 20 such analyses, there is an irrefutable possibility that one of those analyses will demonstrate a statistically significant correlation by chance alone.

To circumvent this, a method has been suggested by Stein et al [1981] using the Bonferoni technique. The significance level for acceptance, let us say 0.05 in our example, can be divided by the number of comparisons one carries out. Thus, if one performs five comparisons, one divides the 0.05 level by 5, yielding a significance level of 0.01 for any one test. A 2-tailed test requires doubling the number of comparisons.

The difficulty with this procedure is that one sets such stringent levels of statistical significance that one runs the danger of missing the effect. An association which is significant, for example at the 0.03 level, may be rejected only because the criterion for statistical significance has been reduced to 0.01 because five separate analyses have been performed on the same data.

Selection of Outcome(s) To Be Observed

Some of the same problems emerge in selecting meaningful reproductive outcomes. There are times when one has a strong suspicion regarding a particular outcome of an exposure, in which case that particular outcome is selected for study. However, usually at the outset of a study, it is not clear which outcome is most likely. The decision on which outcome to observe will depend on availability of data, validity of measurement of outcome, frequency of the outcome event, and biological plausibility of the association between outcome and exposure. As a rule of thumb, for any fixed sample size, it is easier to detect changes in the frequency of more common outcomes than in the frequency of rare outcomes. For example, if neural tube defects is a rare outcome, a doubling of the frequency of this rare event is more likely to arise by chance than a doubling of the frequency of a more common event, for example, spontaneous abortion. Therefore, our power to detect a real change in frequency of outcome as a result of a harmful exposure will depend on the frequency of occurrence of that outcome. All other things being equal, the study design should thus select those outcomes with the greatest frequency in the unexposed population [Stein, 1981].

The validity of measurement of reproductive outcome, for example information through interview alone or individual recall as compared with measuring outcome by more objective techniques, has been referred to above. An increasing number of investigators are turning their attention to spermatogenesis as a sensitive and more objective, quantifiable indicator of male reproductive toxicity and perhaps the genotoxicity and/or gametotoxicity of environmental agents more generally.

A number of different aspects of spermatogenesis have been examined including ejaculate volume, sperm counts as expressed as numbers per ml, sperm motility and sperm morphology. Table V lists the criteria for normal fertility which measure aspects of spermatogenic function [McLeod and Ying, 1979], and Table VI gives a listing of the agents which have been demonstrated to produce reduced sperm counts, including the earliest detected and best documented agents.

Wyrobek et al [1981a,b] have applied an additional tool for studying toxicity to spermatogenesis, namely, nondisjunction of Y chromosomes during the process of reduction of chromosome number in half, when diploid primary spermatocytes divide to yield haploid cells with a single set of chromosomes, during spermatid formation. Nor-

TABLE V. Fertility Criteria

Motility: Greater than 40% motile sperm.
Sperm count: Greater than 20 million sperm/ml of semen.
Morphology: Greater than 70% normal forms.
Normal sperm count is generally accepted to be between 40–60 million/ml of semen.

TABLE VI. Environmental Agents With Hypospermatogenic Effects

Ionizing radiation
Lead
Ethylene dibromide
1,2-Dibromo-3-chloropropane
Cadmium
Organotin
Carbon disulfide
Glycidyl ether (in epoxy resins)
Kepone

mally, the two Y chromosomes present in the diploid primary spermatocyte separate during the process of meiosis, resulting in haploid spermatids containing one Y chromosome each. In the case of nondisjunction, this process does not occur and two Y chromosomes can be detected in the spermatozoa, using a fluorescent dye which visualizes Y chromosomes under the ultraviolet microscope. Wyrobek found that the sperm cells of 15 normal donors showed two Y chromosomes in 0.6% of donors. He also demonstrated a good correlation between the frequency of double Y spots and exposure to a number of agents toxic to spermatogenesis [Wyrobek, 1981].

Table VII shows the results of Kapp's [1979] earlier study of nondisjunction of Y chromosomes in DBCP-exposed workers. The number of DBCP-exposed individuals with more than 2% of their sperm containing two Y chromosomes was 16 out of 18, while none of the nonexposed individuals (N = 15) had more than 2% of sperm cells showing nondisjunction.

Wyrobek [1981] also reported that for any given individual over time, there is little variation in the percentage and type of morphologically abnormal sperm. Exposure to spermiotoxic agents, however, produced a change in the morphology pattern and frequency of abnormality for that individual. Thus, the techniques for detecting nondisjunction and morphology patterns appear to be more objective criteria for reproductive outcome than such measures as ejaculate volume and sperm count, for example, which may be influenced by collection techniques.

For example, most studies suggest a period of abstention for two to three days prior to sperm collection, since sexual activity within 24–48 hours prior to collection may influence ejaculate volume and sperm counts. These parameters may also be influenced by the way in which the specimen is obtained, whether by coitus interruptus, by masturbation, or by masturbation with the use of erotic reading materials.

Another problem, discussed above, concerns the choice of a number of different reproductive outcomes within the same study. The more endpoints one examines, the more likely one is to find a positive association with exposure by chance alone. On the other hand, if one sets more stringent levels of statistical significance, one increases the chance of missing an effect that is really there. It is therefore important to define the relevant endpoint where possible, using the best available biological and clinical evidence.

Comparability of Study Groups

In setting up a study to describe a relationship between exposure and outcome which is unlikely to be explainable by differences between the two study groups other than in the exposure in question, two basic factors must be controlled: variation in the

TABLE VII. Distribution of Nonexposed and DBCP-Exposed Individuals as a Function of the Percentage of YFF Sperm

	Number of individuals with YFF sperm	
Group	0–2%	>2%
Nonexposed individuals	15	0
DBCP-exposed individuals	2	16

$\chi_1^2 = 22.5$; $P < 0.001$.
From Kapp et al [1979], with permission.

characteristics in the study groups which relate either to the risk of having a history of exposure, or the risk of the adverse outcome. In a case control study, for example, cases and controls should be comparable in their opportunities for exposure as well as in characteristics other than exposure which relate both to the outcome under study and the exposure variable [Stein, 1981].

As an example of this difficulty, Doll's [1973] discussion of the relationship of birth weight of infants to the mothers' smoking during pregnancy is of interest. The birth weights of infants of mothers who smoked during pregnancy was lower, on the average, than those of infants whose mothers did not smoke. When this association was first reported in 1957 by Simpson no information regarding the socioeconomic status of the study population was given. In 1969, when Butler and Alberman included this in their study, they found that mothers who smoked in pregnancy tended to come from a lower socioeconomic background. They also were older and had had more children than the mothers who did not smoke.

Thus, low socioeconomic status, parity, and maternal age all were shown to be associated with low birth weight outcomes, and were themselves correlated with the presence or absence of smoking. This points to the importance of matching cases and controls for socioeconomic status, maternal age and other factors in order to conclude that a given agent is responsible for producing the reproductive dysfunction. Successful matching for these confounding factors is often very difficult to achieve, given the long list of agents already identified as toxic to the reproductive process. The problem is more easily handled in a cohort study, where one establishes an exposed and an unexposed group and follows both over time. Even here, however, the investigator's ability to document and record all confounding factors is limited. If one considers that many of these agents may act in combination (with as yet unstudied effects), and that considerable delay may occur before the effects of such exposures become evident, it is easy to understand Doll's [1973] statement that "it is therefore not surprising if the epidemiologist has a haunting fear that his conclusions may not always be correct." In his view, however, this should not be allowed to delay action interminably. "The best sort of experiment," he states, "is an experiment in prevention and it is not a bad last resort even if we are unable to carry it out in a properly controlled way. Proof in the strict logical sense may not have been obtained but does this matter if the disease has disappeared?" [Doll, 1973].

RECOMMENDATIONS

Infante and Legator suggested the following recommendations for conducting epidemiological investigations on assessing reproductive hazards, summarizing the deliberations of the Committee who participated in a Workshop on Methodology for Assessing Reproductive Hazards in the Workplace [Infante and Legator, 1979]:

1. That a short occupational history form be developed for use by the medical community and by existing data-gathering organizations, including registries on both the national and international level.

2. That the comprehensive NIOSH Reproductive History Questionnaire be used in collecting data for studying possible reproductive effects of specific hazardous substances on work environment.

3. That the administration of this questionnaire be standardized, with the use of an accompanying interview instruction booklet.

4. That all stillbirths (defined as 20 or more weeks of gestation or 500 or more grams) should be certified.

5. That occupational histories of both parents should be recorded on all a) birth certificates, b) stillbirth certificates, and c) death certificates for children to the age of 16 years.

6. That epidemiological assessment of a larger spectrum of endpoints should be made to measure transplacental effects.

7. That data sources, such as the Oxford Survey of Childhood Cancer and the Collaborative Perinatal Project, should be more fully utilized and supported.

8. That other existing data sources should be surveyed and assessment of potential for determining reproductive hazards in relation to occupational exposure be performed.

9. That the existing birth defects registries in the Scandinavian countries should be encouraged to coordinate their activities and be supported, so that larger data bases may be generated to permit greater statistical power (that is, greater ability to detect an effect with our studies).

The committee made other useful recommendations, such as suggesting the use of early warning cytogenetic tests with human occupationally exposed populations and the establishment of surveillance programs using population numbers of exposed individuals as a criterion to determine which groups should be monitored. Any evidence of adverse effects should lead to control of exposure, and, for all new products, analysis of body fluids of exposed workers should be conducted early on, before wide use of the product occurs.

While implementing Infante and Legator's recommendations would result in a considerable enhancement of the quantity and quality of the data available for analysis, it is clear that we also need to improve our research methodologies to improve the quality of information obtained from epidemiological studies of potential reproductive toxicity of environmental and occupational agents.

Such improvements are developed primarily in the course of carrying out the research itself. In order to conduct such studies, the financial support of Government and other agencies is indispensable. This work is essential in order to prevent the human disease and suffering associated with adverse reproductive outcomes.

ACKNOWLEDGMENTS

The author wishes to thank Mark Goldberg for his help in the preparation of this review.

REFERENCES

Centers for Disease Control: Congenital Malformation Registry, Atlanta.

Doll R (1973): Hazards of the first nine months: An epidemiologist's nightmare. J Irish Med Assoc 66: 117–126.

Flynt JW (1974): Trends in surveillance of congenital malformations. In Janevich et al (eds): "Congenital Defects—New Directions in Research." New York: Academic Press.

Higgins ITT (1965): The epidemiology of congenital heart disease. J Chron Dis 18:699.

Infante P, Legator M (1979): Workshop on methodology for assessing reproductive hazards in the workplace: Recommendations for future research. Environ Res 20:217–223.

Kapp RW, Picciano P, Jacobson C (1979): Y-Chromosomal nondisjunction in dibromochloropropane-exposed workmen. Mutat Res 64:47–51.

Lancranjan R, Popescu H, Găvănescu O, Klepsch D, Serbănescu M (1975): Reproductive ability of workmen occupationally exposed to lead. Arch Environ Health 30:396–401.

McLeod J, Ying W (1979): Male fertility potential in terms of semen quality: A review of the past, a study of the present. Fertil Steril 31(2):103–116.

National Foundation/March of Dimes: Facts, National Foundation. New York: 1975.

Stein Z, et al (1981): Epidemiologic considerations in assessing health effects at toxic waste sites. In "Symposium on Criteria for Assessment of Health Effects at Chemical Disposal Sites." New York: Rockefeller University.

Warburton I (1981): Measurement of reproductive effects in human populations: Selected outcomes for study. In "Symposium on Criteria for Assessment of Health Effects at Chemical Disposal Sites." New York: Rockefeller University.

Wyrobek AJ (1981): Human sperm assays in toxic chemical exposures. Presented at a conference on "Health Hazards of Toxic Waste Dumps," Research Triangle Park, North Carolina.

Wyrobek AJ, Watchmaker G, Gordon R, Wong K, Moore D, Whorton D (1981a): Sperm shape abnormalities in carbaryl-exposed employees. Environ Health Persp 40:225–265.

Wyrobek AJ, Brodsky J, Gordon L, Moore D, Watchmaker G, Cohen E (1981b): Sperm studies in anesthesiologists. Anesthesiology 55:527–532.

American Journal of Industrial Medicine 4:365–379 (1983)

Environmental Factors in the Etiology of Carcinoma of the Human Ovary: A Review

Carolyn H. Lingeman, MD

The woman at highest risk for epithelial neoplasia of the ovary is a nulliparous, caucasian woman of northern European descent residing in an industrialized country. She is at even higher risk if her sister, mother, maternal aunt or grandmother have had carcinomas of the breast or ovary or if she herself has had a previous carcinoma of the breast. The roles of hormonal drugs and other environmental chemical compounds in the etiology of this form of cancer are not clear but are the subjects of many ongoing investigations.

Key words: ovary, carcinoma, epithelial neoplasms, estrogens, diet

INTRODUCTION

A 1974 review revealed little activity in cancer research directed toward seeking causes of malignant neoplasms of the ovary [Lingeman, 1974]. Subsequently, a number of epidemiologic and experimental studies have contributed significant information about risk factors [Editorial, 1977; Miller, 1978]. At a 1979 workshop in Geneva, participants reviewed the embryology, biology, pathology, epidemiology, and endocrinology of neoplasia in the ovary [Weiss, 1980]. The group also discussed animal models, hormone receptors, and immunologic markers and made recommendations for future investigations [UICC, 1980]. The 1981 Directory of On-going Research in Cancer Epidemiology, published by the International Agency for Research on Cancer [World Health Organization, 1981], lists 28 epidemiologic studies of neoplasms of the ovary. These studies will further analyze risk factors including the effects of environmental chemical compounds and hormonal drugs. Other publications have described chemical markers that may be useful for development of new techniques for detection or therapy.

Because this subject was thoroughly reviewed in a previous publication (Lingeman, 1974), this paper will emphasize the most recent investigations.

Chemical Pathology Branch, National Institute of Environmental Health Sciences, Bethesda, Maryland
Address reprint requests to C.H. Lingeman, MD, Room 3A06, Landow Building, National Institute of Environmental Health Sciences, Bethesda, MD, 20205.

0271-3586/83/0401-02-0365$04.50 **© 1983 Alan R. Liss, Inc.**

EPIDEMIOLOGY

According to some estimates, cancer of the ovary will affect 1 to 2% of caucasian women and will cause death in most of them [Weiss, 1980]. In Britain, cancer of the ovary is responsible for approximately 4000 deaths each year [Editorial, 1978]. Women in other countries of northern Europe also experience a high incidence and mortality from this form of cancer. The highest *incidence* rates are reported in Sweden and other populations composed of women of northern European descent (Table I). Migrants from low incidence to high incidence countries usually develop the incidence rates nearer to those of the new host country than those of the country of birth within a generation or two. In high risk countries, epithelial cell types account for the majority of malignant neoplasms of the ovary and the epidemiologic characteristics are similar to those of carcinomas of breast, endometrium, colon, and prostate. All of these neoplasms are rare in non-industrialized countries of the world, but are very frequent in many highly industrialized countries. An exception is Japan, which though highly industrialized, continues to maintain a unique pattern of cancer incidence, with a persistently high frequency of carcinoma of the stomach and lung but relatively low incidences of the other types of epithelial carcinomas that are so prevalent in other industrialized countries of the world. Malignant neoplasia of the ovary is infrequent in the People's Republic of China. It is not listed among the most frequent sites for cancers in Chinese women (the 7 most frequent are stomach, uterine cervix, esophagus, liver, lung, colon-rectum, and breast) [Li et al, 1980]. Women of Chinese ancestry living in the United States also experience a lower in-

TABLE I. Age Standardized* Average Annual Incidence Rates Per 100,000 Population for all Malignant Neoplasms of the Ovary, Fallopian Tube, and Broad Ligament in Selected Countries With Population-Based Tumor Registries [Waterhouse et al, 1976]. (Neoplasms of the fallopian tube and broad ligament are so infrequent that they can be disregarded for statistical purposes.)

Sweden	14.9	U.K. (Birmingham)	11.1
Israel (Jews born in U.S. or Europe)	14.3	Switzerland	10.6
Norway (urban)	14.2	U.S. (Detroit, black)	9.9
Canada (British Columbia)	13.8	Canada (Alberta)	9.5
U.S. (San Francisco Bay area, white)	13.3	U.S. (Utah)	9.4
New Zealand (Maori)	12.8	Canada (Quebec)	9.4
U.S. (Iowa)	12.7	Israel (Jews born in Israel)	8.7
Norway (rural)	12.7	U.S. (San Francisco, black)	8.6
Israel (all Jews)	12.6	Finland	7.9
U.S. (Connecticut)	12.2	U.S. (New Mexico, American Indian	7.5
U.S. (New Mexico, white)	12.1	U.S. (San Francisco, Chinese)	6.7
U.S. (Detroit, white)	12.0	Brazil (São Paulo)	6.1
German Democratic Republic	11.8	Israel (Jews born in Africa or Asia)	5.8
New Zealand (non-Maori)	11.3	India (Bombay)	4.6
Poland (Warsaw City)	11.2	Japan, Miyagi	2.7

*Age-standardized rates were calculated by comparing age-specific rates to those of theoretical "world" population as described in the text (Waterhouse et al, 1976). The authors caution that these figures should be interpreted as only a rough estimate for possible differences between populations.

cidence of malignant neoplasia of the ovary than other Americans [Fraumeni et al, 1974].

Within the United States, caucasian women have higher *incidence* rates of epithelial neoplasms of the ovary than black women, whereas rates for nonepithelial types tend to be the same in both races [Weiss, 1980]. In one report it was stated that the risk of epithelial neoplasia of the ovary in American Jewish women is 1.7 times that of non-Jewish women (95% confidence limits 0.8 to 3.7) [Hildreth et al, 1981]. In the years 1950 to 1969, *deaths* in white women from all types of malignant neoplasms of the ovary were more frequent in the northeastern parts of the United States, a pattern that is similar to those of cancers of the breast, urinary bladder, kidney, rectum, and large intestine [Mason et al, 1975][1]. This distribution suggests effects of environmental carcinogens, perhaps products of industrialization such as hydrocarbons produced by combustion of fuels burned for residential heating, or chemicals produced during manufacturing processes.

Age

Malignant neoplasms of the ovary occur at all ages from infancy to old age [Lingeman, 1974]. Nonepithelial neoplasms account for most of these neoplasms in younger individuals whereas adenocarcinomas predominate in women 25 years or older in the United States and countries of northern Europe. The increase in incidence that occurs in women in their twenties continues to rise steeply into old age. The shape of the age incidence curve is similar to those of other common epithelial cancers in caucasian women including those of breast, colon, and endometrium.

Genetic and Familial Factors

Several reports have described families in which more than one member developed a malignant neoplasm of the ovary [Fraumeni et al, 1975; Lynch et al, 1981]. The majority, however, occur in women without family histories of ovarian neoplasia. It has not been possible to determine whether the occasional familial clustering is a result of genetic influences, common environmental exposures, or both. Exceptions are the clearly defined genetic syndromes such as the Peutz-Jeghers syndrome, the basal cell nevus syndrome, and 46 XY mosaics that predispose affected women to neoplasms of the ovary of distinctive histopathologic types [Lingeman, 1974].

While most evidence from epidemiologic studies is consistent with the hypothesis that environmental factors are significant risk factors in women with epithelial neoplasia (Lingeman, 1974), it is also clear that there is a variation in susceptibility, probably dependent on genetically controlled factors such as enzymes that control the metabolism of environmental chemical compounds, including those that activate precarcinogens. Genetic differences in such enzymes have been observed in mice [Mattison et al, 1979]. The report that women with carcinoma of the ovary may be deficient in serum alpha-L-fucosidase is discussed elsewhere in this paper [Barlow et al, 1981]. Two recent reports suggest that specific chromosomal markers may be present in women with adenocarcinomas of the ovary [Wake et al, 1980; Whang-Peng et al, 1982].

[1]It should be noted that these *death* rates are based on diagnoses taken from death certificates and no attempts were made to verify the diagnoses. Accuracy of death certificate diagnoses vary considerably from place to place so that conclusions drawn from these data should only be used to suggest hypotheses of causality.

Risk Factors

Carcinoma of the ovary and breast share risk factors such as reproductive experience and familial tendencies (Table II). Both are more frequent in nulliparous women than in those who have borne children. The prominent familial influences so firmly established for breast cancer are not so obvious with reference to the ovary. The fact that cancers of both sites occur more frequently than expected as multiple primary cancers in certain women (Fig. 1A, B), particularly in premenopausal ages, is evidence for common etiologic factors but their nature is yet to be clarified. Evidence obtained from studies of women with primary cancers of both sites suggests that the common etiologic event or events that initiate the neoplastic processes in one or both organs may occur before or near the time of puberty [Prior et al, 1981]. Hypotheses whereby factors designated as "initiator" and "promotor" may act to cause neoplasia of the ovary and other organs are discussed in another section. Carcinomas of the ovary and breast also share risk factors with carcinoma of the endometrium, but it is not known whether the etiologic factors act independently on each of the three target organs or whether the neoplastic processes in all three are the result of malfunction of a single target organ such as the ovary or pituitary, in which abnormalities in hormones manufactured and released may act on cells in each of the three target organs to result in neoplasia.

Hormones

The precise role of hormones in induction or promotion of neoplasms of the ovary is far from clear. Pituitary gonadotropins, estrogens, progesterones, androgens, corticosteroids, and thyroid hormones have all been implicated in or suspected of influencing the induction, growth, and clinical behavior of benign and malignant neoplasms of human or animal ovaries. Cytoplasmic receptors for 17β-estradiol (ER), 5α-dihydrotestosterone (AR), progesterone (PR), and cortisol (CR) have been identified in disease-free ovaries and in benign and malignant neoplasms in ovaries [Galli et

TABLE II. Risk Factors for Epithelial Neoplasms of the Ovary

Caucasian race[a]
Nulliparity, especially in married women who fail to conceive.[b,c,d]
Older age at first pregnancy.[d]
Residence in United States, countries of northern Europe, or Israel if born in Europe or the United States.[e]
Family history of cancer of the ovary.[f,g]
Previous breast cancer, especially in young women.[h]
Hypertension (endometrioid carcinoma).[i]
Overweight (endometrioid carcinoma).[i]
Endometriosis (clear cell carcinoma).[j]
Underactive thyroid.[d]
Heavy coffee consumption.[k]
Rubella exposure ages 12 to 18 years.[c]
Negative history of mumps parotitis.[l]

[a]Weiss, 1980
[b]Joly et al, 1974
[c]McGowan et al, 1979
[d]Hildreth et al, 1981
[e]Waterhouse et al, 1976
[f]Fraumeni et al, 1975
[g]Lynch et al, 1981
[h]Prior et al, 1981
[i]Szamborski et al, 1981
[j]Shevchuk et al, 1981
[k]Trichopoulos et al, 1981
[l]Menczer et al, 1979

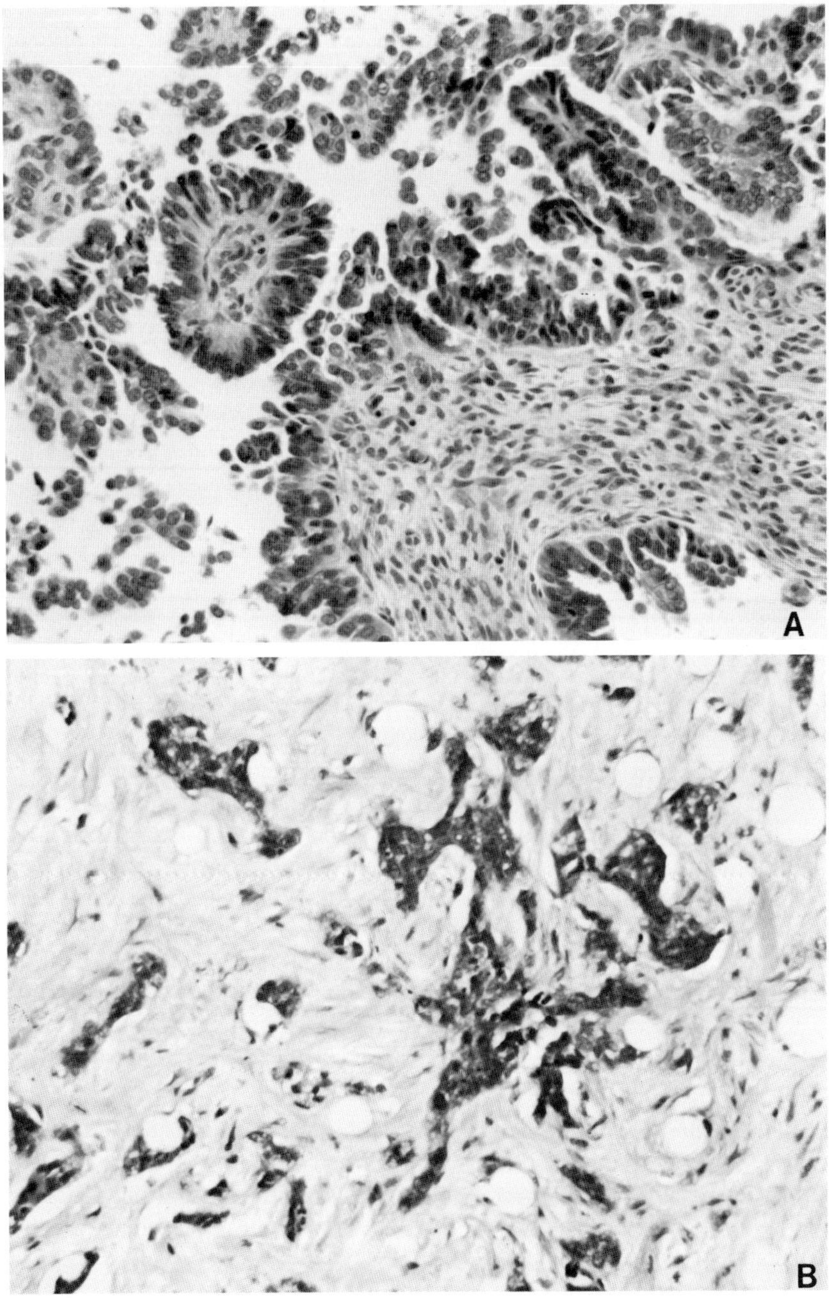

Fig. 1. Photomicrographs of tissues removed from a 36-year-old Caucasian woman with synchronous carcinomas of the ovary and breast. A diagnosis of papillary cystadenocarcinoma of the right ovary (A) was made in March, 1972. A diagnosis of infiltrating duct cell carcinoma of the right breast (B) was made in April, 1972. There was nothing abnormal in the reproductive history of this woman. Her only two pregnancies had occurred at ages21 and 26, resulting in normal full term infants. She had used Enovid approximately one year beginning at age 18. Her consumption of coffeefor contraception from 1962 to 1967. She had smoked cigarettes for was not recorded. The only known occupational exposure to chemicals had occurred during employment in a photographic printing shop in which there was exposure to hardeners and developers but the dates of employment were not recorded in the history. When questioned aboutfamilial factors, this woman reported that her father had died of "stomach cancer" at age 63 and that a maternal aunt had had bilateral cancers of the breast. A paternal uncle was also reported to have had a "cancer of the throat."

al, 1981]. Binding sites for both FSH and hCG have also been demonstrated [Kammerman et al, 1981].

The higher frequency of cancer of the ovary in nulliparous women, particularly those married women who have failed to conceive [Joly et al, 1974; McGowan et al, 1979, Hildreth et al, 1981], indicates that *endogenous* hormones are involved in the induction or growth of these neoplasms. Some investigators have postulated that gonadal failure may underlie both the neoplasia and infertility. If pregnancy per se confers some protection against cancer of the ovary, as it is believed to do with cancer of the breast, the "protective" factors have not been identified. Beral et al [1978] suggest several possible factors, including suppression of ovulation during both pregnancy and lactation, immunologic effects, or production of a chemical substance specific to pregnancy. Whatever the mechanism, the protection is only relative and most women with cancer of breast or ovary have borne children, although such women tend to have had fewer pregnancies and are frequently older than controls at the time of the first pregnancy [Hildreth et al, 1981].

One problem in attempts to define risk factors for cancer of the ovary is the failure of most investigators to recognize that different histopathologic types may have different etiologies. The inclusion of all histopathologic types in epidemiologic studies has sometimes obscured risk factors. When Hildreth et al [1981] restricted their case-control study to epithelial cancers in women ages 45 to 74, they detected the fact that women with these carcinomas are more freqeuntly caucasian, Jewish, nulliparous, that they may have had a late age of menopause and frequently a family history of cancer of the ovary or endometrium. Furthermore when histopathologic subtypes of adenocarcinomas are analyzed separately, it is found that women with the endometrioid type, which is similar histologically to primary adenocarcinoma of the uterine endometrium, tend to be overweight and hypertensive, as are women with primary adenocarcinoma of the endometrium [Szamborski, 1981]. Shevchuk et al [1981] noted the association of the clear cell variant of carcinoma of the ovary with clinical endometriosis and other evidence of unopposed estrogen on the uterine endometrium in which a spectrum of lesions occurred, ranging from cystic glandular hyperplasia progressing through adenomatous hyperplasia, to adenocarcinoma.

Exogenous Hormones

With the exception of one report that suggested an association of neoplasia of the ovary with use of exogenous estrogens [Hoover et al, 1977], most studies have failed to support such an association [Annegers, 1979; Weiss et al, 1981; Willett, 1981; Weiss et al, 1982]. In the report by Hoover et al [1977] it should be noted that the association was with only one drug, diethylstilbestrol (DES), not estrogens generally. Studies by McLachlan et al [1980, 1981], Newbold et al [1982] and Boylan [1978] have shown, in animals, an association of prenatal exposure to DES and subsequent abnormalities in ovaries, including neoplasia. This is an area that should be investigated in future epidemiologic studies of human neoplasia of the ovary. DES, a nonsteroidal estrogen, may have different properties than steroidal estrogens, and its effects should be analyzed separately from other estrogenic compounds in epidemiologic studies. Interestingly, one report suggested that use of oral contraceptives may offer some protection from cancer of the ovary, possibly by suppression of ovulation, [Willett et al, 1981; Weiss et al, 1981] but, in one study, this protective effect was observed only in women over age 40 at diagnosis [Cramer, 1982].

Viruses

Previous reports have suggested a role of mumps virus in cancer of the ovary because of its propensity for infecting the gonads [Lingeman, 1974]. Ovary cancer patients in one study appeared to be more likely than controls to have developed subclinical mumps as evidenced by a lower rate of clinical mumps history with the presence of positive serologic evidence, and they tended to have lower persistent mumps CF antibody than controls [Menczer et al, 1979]. If mumps virus is involved in the etiology of cancer of the ovary, the incidence of this neoplasm should decrease dramatically over the next several decades as vaccination for mumps becomes widespread in the population.

With regard to viruses other than mumps, Joly reported a statistically significant lower frequency of ever having had influenza or pneumonia among ovary cancer patients than controls [Joly et al, 1974], whereas McGowan et al [1979] found that patients reported having had rubella infection between the ages of 12 to 18 years significantly more frequently than controls.

Other Environmental Factors

The high incidence of epithelial cancers of the ovary, breast, endometrium, prostate and colon in certain highly industrialized countries and the changes in incidence in migrants and their offspring following migration from low to high risk countries has resulted in considerable speculation and research aimed at identifying the responsible carcinogens of these common epithelial neoplasms. Most hypotheses have focused on ble mechanisms whereby carcinogens of exogenous origin may act to induce or promote the neoplastic processes. Major efforts have centered on diet, products of industry, and extrinsic hormones acting on hosts with varying degrees of susceptibility. So far, diet has received the most serious consideration as a vehicle for the causative carcinogens of these common epithelial neoplasm. Most hypotheses have focused on diets with high intake of meats and animal fats as the most likely candidates to explain the international differences in incidence. The fact that Seventh Day Adventists, who as a group consume less meat and animal fat than other Americans, also experience lower incidences of these neoplasms has also been cited as evidence supporting this hypothesis [Phillips et al, 1980a,b]. On the other hand, reports that Utah Mormons who *do* eat meat and animal fats also experience lower frequencies of most common epithelial cancers, including those of the ovary, than non Mormon residents of Utah [Enstrom, 1980; Lyon et al, 1980] indicates that meats and animal fats may not, after all, be the most likely vehicles for these carcinogens. A report describing cancer studies of woman of two religious orders whose members ate little or no meat, yet experienced the same rates of cancers of the ovary and breast as other British women, also points to factors other than dietary meats as major causative factors [Kinlen, 1980, 1982]. These nuns did not use alcohol or cigarettes; they did not consume coffee or tea until recently [Kinlen, 1980]. The author pointed out, however, that these women did not join the religious orders until adulthood; thus effects of exposures to carcinogens may be the result of experiences prior to the entering of the religious orders.

Alcohol has not been implicated in epidemiologic studies of cancer of the ovary [Rosenberg et al, 1982]. Case-control studies of cancer of the ovaries have failed to show an association with cigarette smoking [Newhouse et al, 1977]. Women who smoke cigarettes experience decreased fertility and early menopause [Mattison and Thorgeirsson, 1978; Mattison and Thorgeirsson, 1979] and they have lower urinary estrogens than non-smokers [McMahon, 1982]. The majority of Adventists and Mormons refrain from the use of alcohol and cigarettes [West, 1980]. Most, but not all members of both religious groups use neither coffee or tea; in fact, 40% of Mormon women drink coffee [West, 1980] and 16% of Mormon men and women drink alcoholic beverages [West, 1980].

It is interesting that a dietary item as widely used as coffee has received so little attention as a possible vehicle for carcinogenic chemicals. Coffee consumption has been implicated in the etiology of cancer of the ovary in two reports. One report [Stocks, 1970] linked international mortality rates for cancers of various sites with per capita consumption of coffee. The highest incidence of and mortality from cancer of the ovary is reported from Sweden which also has the world's highest per capita coffee consumption (Table III). The other report was a case-control study in which it was found that ovary cancer patients consumed significantly more coffee than matched controls [Trichopoulos, 1981].

Coffee has been implicated as a possible causative factor for carcinoma of the pancreas [Stocks, 1970; MacMahon et al, 1981] and urinary bladder [Howe et al, 1980] although such associations have been questioned. In a report on carcinoma of the urinary bladder, coffee drinking was a risk factor for bladder cancer in men who consumed both brewed and instant coffee, and in women who used instant coffee, but no dose-response relationships were found. There was also an increased risk of bladder cancer in women reporting consumption of nondiet cola beverages which also contain caffeine. There was no excess risk of bladder cancer associated with the drinking of tea. Morrison et al [1982] concluded from their case-control studies that if there is an association between coffee drinking and bladder cancer it is likely to be weak.

Although there is no evidence to suggest that caffeine itself is carcinogenic, it is currently under test in the National Toxicology Program. Caffeine is reported to

TABLE III. Estimated Per Capita Consumption of Foods, Beverages, and Cigarettes in Populations With Low (L), Intermediate (I), and High (H) Incidence of Epithelial Cancers of the Ovary

	Incidence of epithelial cancers of ovary	Red meat	Dairy products	Coffee	Tea	Alcohol	Cigarettes
United States							
Mormons[a,b]	I	+ + +	+ + +	+		+	
Adventists[c]	I	+	+ +				
Other	H	+ + +	+ + +	+ + +	+ +	+ + +	+ + +
Great Britain							
Vegetarian or near-vegetarian religious orders[d]	H			+	+		
Other	H	+ + +	+ + +	+	+ + +	+ + +	+ + +
Sweden[e,f]	H	+ + +	+ + +	+ + +	+	+ +	+ +
Japan[f]	L	+	±	±	+ + +	+ +	+ + +
China[g]	L	±	±		+ + +	±	±

+, Estimated low per capita consumption.
+ +, Estimated intermediate per capita consumption.
+ + +, Estimated high per capita consumption.
±, Unknown per capita consumption.
[a]Enstrom, 1980
[b]Lyon et al, 1980
[c]Phillips et al, 1980a,b
[d]Kinlen, 1980, 1982
[e]Stocks, 1970
[f]Waterhouse et al, 1976
[g]Li et al, 1980

enhance DNA initiation [Bowden et al, 1982] and it potentiates the cytotoxicity of many DNA-damaging agents by inhibiting DNA repair processes [Lau et al, 1982]. Caffeine in high doses has been reported to cause fetal deaths and malformations in laboratory animals, but when coffee was given to rats in amounts equivalent to human consumption of 50 to 60 cups per day, teratogenic effects were not observed [Nolen, 1981]. Teratogenic effects have not been proven to occur in human infants whose mothers consume large amounts of coffee during pregnancy [Linn et al, 1982; Kurppa et al, 1982].

Minton [1979] reported a dose-dependent relation between cystic breast disease and consumption of methylxanthenes that was reversible after cessation of the use of beverages and medications containing these chemicals. According to Minton's hypothesis, methylxanthenes inhibit metabolism of cyclic AMP (adenosine monophosphate) and guanosine monophosphate (GMP) leading to an accumulation of these nucleotides in breast tissue, thus stimulating production of cellular products. He did not consider alternative hypotheses such as direct effects of xanthines on the ovaries with secondary effects on the breast. Other studies have failed to implicate xanthenes as a causative factor in cystic disease of the breast, partly because of the difficulty in defining "cystic disease" [Ernster et al, 1982]. Nevertheless, xanthine restriction is being advised for women with clinically significant nodular breast disease, along with tocopherol, [London et al, 1982] and the antigonadotropin danazol [Mansel et al, 1982].

In addition to caffeine, coffee contains many other naturally-occurring and additive chemical compounds that have not been tested for carcinogenicity. It is possible that prolonged storage of brewed coffee at high temperatures might create or concentrate carcinogenic compounds. The process of decaffeination formerly involved the use of trichloroethylene, but its use for this purpose was discontinued since 1976 when it was reported to be carcinogenic in mice and rats (National Cancer Institute). Subsequently untested chemicals, including methylene chloride, have been substituted for trichloroethylene in the decaffeination process; methylene chloride is undergoing testing by the National Toxicology Program at the present time. Sweeteners, cream or artificial nondairy cream substitutes might also contribute to carcinogenic effects of coffee drinking. It is conceivable that fats in added cream could selectively extract fat-soluble carcinogens from the coffee and could transport them into the blood from the intestine.

Products of Industrialization

Despite epidemiologic evidence that products of industrialization may be involved in the etiology of epithelial neoplasms of the ovary, there is little evidence to incriminate specific materials. Particulate materials of the talc-asbestos group have received serious consideration as possible etiologic factors because of reports that particles with morphologic characteristics of talc were observed in sections of ovaries containing malignant neoplasms [Graham et al, 1967] and one epidemiologic study of female asbestos workers in which there appeared to be slight excess of epithelial neoplasia in the ovaries [Newhouse, 1977]. Subsequent investigations have not confirmed the suspected associations [Henderson et al, 1979; Longo and Young, 1979; Roe, 1979], but additional investigations will be required to evaluate the possible effects on the ovaries of these ubiquitous materials. A recent case control study has suggested an association of epithelial cancers of the overy with the use of talc [Cramer et al, 1982].

There is little evidence to incriminate chemical carcinogens in the workplace as etiologic factors in neoplasia in the ovaries. One report, however, indicated that 12 women of a cohort of 78 Swedish women exposed to an antirust oil in a factory, mainly by skin contact, developed malignant neoplasms, whereas only 3.9 cases were expected; two of these neoplasms were located in the ovary, a number too small to permit significance testing [Jarvholm and Lavenius, 1981]. The composition of the oil was changed from time to time during the exposure period and among other components the oil contained N-phenyl-1-naphthylamine, a material that is widely used in cutting oils, hydraulic oils, and sometimes in the rubber industry. It is similar chemically to 2-naphthylamine, a bladder carcinogen. (One of the other 12 women with neoplasms in this cohort developed a carcinoma of the bladder.) It was postulated that perhaps the N-phenyl-1-naphthylamine reacted with sodium nitrite in the wrapping paper used to package the bearing rings covered with the antirust oil to form N-nitroso-N-phenyl-1-naphthylamine.

Questions about occupation or household exposure to chemical compounds are rarely asked during a medical examination. Women are very likely to be designated "housewife," despite the fact that many married women have worked at some time in their lives, possibly in occupations that permit exposure to toxic or carcinogenic materials. A complete occupational history should be a part of every medical examination, particularly in patients with the epithelial neoplasms that are so frequent in industrialized countries.

The case history of the woman whose simultaneous carcinomas of the ovary and breast are shown in Figure 1 illustrates the complexity of attempts to analyze the interacting environmental and genetic (familial) etiologic factors in a specific patient in whom risk factors are only relative. The familial factor appears to have been dominant in this women whose maternal aunt had had bilateral breast cancers. The familial pattern is consistent with familial patterns of cancers of both breast and ovary which tend to occur at ages younger than those in women without familial histories. The other factors in the environmental history did not contribute any information about the etiologic events in this woman's life that might have contributed to the neoplasia. Nevertheless, environmental (including occupational) histories should be part of every medical examination of patients with the common epithelial cancers that are believed to result from carcinogens in the environment.

In the case of epithelial neoplasms of the ovary, as those in other hormone-dependent organs, it is likely that the neoplasia is a multistage process involving factors in addition to hormones in the initiation or promotion phases. Sex hormones can act as cancer promotors in animals, following an initiation stage by other chemical compounds [Williams, 1982]. The concept of cancer initiation and promotion is based on studies in animals in which it can be shown that chemical carcinogenesis can be a multistage process in which a single exposure to a chemical compound (the initiator) can in some way alter the genome so that later exposures to a compound which is not carcinogenic (the promotor), usually in multiple exposures, can result in proliferation of the altered cells [Becker, 1981]. In one experimental system, rats given a single dose of diethylnitrosamine (the initiator) followed by estrogens (the promotor) developed neoplasms in the liver [Wanless and Medline, 1982]. Experiments in mice have shown that certain chemical compounds can result in destruction of primordial oocytes leading to early ovarian failure and subsequent tumor development [Jull, 1973]. The subsequent neoplasia may depend on the "promoting" effects of hormones or other chemical compounds in diet, or encountered elsewhere in the environment.

It is likely that similar phenomena can occur in humans. Inquiries into occupation and other environmental experiences may provide some clues as to which of the many possible factors may be involved in one or more stages of the development of cancers of hormone-dependent organs. If hormones do act as promotors in human ovary neoplasia, there are a number of possible candidates for initiators. It has been reported that 7,12-dimethylbenz(a)anthracene destroys primordial oocytes and reduces fertility in mice [Mattison and Thorgeirsson, 1979; Krarup, 1970]. The many opportunities for human exposure to polycyclic hydrocarbons and other toxic chemicals suggests that chemical damage to human ovaries might account for some infertile marriages. Cigarette smoke is only one of the many possible sources of exposure to such compounds. The result of epidemiologic studies showing the incidence of epithelial cancers of the ovary to be higher in married women who have failed to conceive than in unmarried women suggests that both the neoplasia and the infertility may be the result of gonadal failure in the ovary. Damage to cells in the ovary by environmental carcinogens might be one cause of such underlying gonadal failure. It is possible that chemical damage to the oocyte may result in failure of conception and also changes in DNA that might make certain cells susceptible to promoting influences of compounds such as hormones. Viruses and irradiation may act as initiators (direct-acting mutagens).

Biologic Markers

The high mortality in patients with malignant neoplasms of the ovary is mainly a result of failure of diagnosis in curable stages. Because the ovaries are relatively inaccessible for accurate evaluation during physical examination, demonstration of unique immunologic or chemical markers in blood or urine might permit development of screening procedures, particularly in high risk women.

Barlow [1981] has reported that patients with epithelial neoplasms of the ovary have deficient levels of serum alpha-L-fucosidase, an enzyme controlled by two alleles with autosomal inheritance; it is deficient in approximately 10% of healthy people and in its most severe form is associated with fucosidosis, a rare inborn error in metabolism.

Stenman et al [1981] reported that a urinary peptide of apparent molecular weight of 7,000, demonstrated by immunoadsorption techniques, was present at higher levels of serum and urine of patients with cancers of the ovary than in healthy subjects.

deBoer et al [1981] demonstrated by immunofluorescence, three intestine-associated antigens in mucinous neoplasms of the ovary which were absent in serous or mesonephroid neoplasms; in one of two endometrioid neoplasms the carcinoembryonic antigen (CEA) was present but not the other two antigens, small intestine mucin antigen (SIMA) and large intestine mucin antigen (LIMA). Thus mucinous neoplasms of the ovary contain mucins that cross-react with those of intestinal origin, a phenomenon that suggests that during the process of malignant transformation, the columnar epithelium of the ovarian cystoma forms inappropriate intestine-associated mucins.

Cauchi et al [1981] have also demonstrated by immunoassay, higher levels of CEA, α fetoprotein (AFP) and β subunit of human chorionic gonadotropin (hCG) in sera of patients with malignant neoplasms of the ovary than in controls. Because of some overlapping with values obtained in controls, and the problems encountered with frequent false positive values, it was necessary to establish cut-off points in serum values and some loss in sensitivity. The possibility that AFP levels might be higher in patients with neoplasms of germ cell and stromal origin than from those of epithelial origin, that levels of hCG might be higher with serous and epithelioid neoplasms than those of other origins and the differences in CEA levels in different histopathologic types, discussed above [deBoer et al, 1981], indicates varying degrees of specificity of these antigens that may ultimately provide useful tools for diagnosis, prognostication and immunotherapy.

Bast et al [1981] developed a murine monoclonal antibody (OC125) that reacts with 6 epithelial ovary carcinoma cell lines and with cryopreserved malignant neoplastic overy tissue from 12 of 20 patients, but not with a variety of nonmalignant tissues including adult and fetal ovary and sections of 12 nonovarian carcinomas. These authors point out that circulating CEA, although most frequently associated with mucinous carcinomas, is also associated with neoplasms of other histopathologic types. The future usefulness of immunologic markers will depend on development of improved methods for demonstrating specificity and sensitivity of these antigens.

ACKNOWLEDGMENTS

The author would like to thank Julia Cohan for her assistance in the preparation of this manuscript.

REFERENCES

Annegers JF, Strom H, Decker DG, Dockerty MB, O'Fallom WM (1979): Ovarian cancer. Incidence and case-control study. Cancer 43:723–729.

Barlow JJ, DiCioccio RA, Dillard PH, Blumenson LE, Matta KL (1981): Frequency of an allele for low activity of alpha-L-fucosidase in sera: possible increase in epithelial ovarian cancer patients. J Natl Cancer Inst 67:1005–1009.

Bast RC, Feeney M, Lazarus H, Nadler LM, Calvin RB, Knapp RC (1981): Reactivity of a monoclonal antibody with human ovarian carcinoma. J Clin Invest 68:1331–1337.

Becker FF (1981): Recent concepts of initiation and promotion in carcinogenesis. Am J Pathol 105:3–9.

Beral V, Fraser P, Chilvers C (1978): Does pregnancy protect against ovarian cancer? Lancet 1:1083–1086.

Bowden GT, Rosenthal H, Fetty M (1982): Does caffeine block DNA elongation or enhance DNA initiation in benzo(a)pyrene-diol epoxide I damaged cells? Proc AACR 23:66.

Boylan ES (1978): Morphological and functional consequences of prenatal exposure to diethylstilbestrol in the rat. Biol Reprod 19:854–863.

Cauchi MN, Koh SH, Lim D, Hay DL (1981): Oncofoetal antigens in cancer of the cervix and ovary. Br J Cancer 44:403–409.

Cramer DW, Hutchinson GB, Welch WR, Scully RE, Knapp RC (1982): Factors affecting the association of oral contraceptives and ovarian cancer. N Engl J Med 307:1047–1051.

Cramer DW, Welch WR, Scully RE, Wojciechowski CA (1982): Ovarian cancer and talc. A case-control study. Cancer 50:372–376.

deBoer WG, Ma J, Nayman J (1981): Intestine-associated antigens in ovarian tumours: an immunohistological study. Pathology 13:547–555.

Editorial (1978): Epidemiology of ovarian cancer. Br Med J 1:198.

Editorial (1977): Aetiological factors in ovarian cancer. Lancet 1:1062–1063.

Enstrom JE (1980): Health and dietary practices and cancer mortality among California Mormons. In Cairns J, Lyon JL, et al (eds): "Cancer Incidence in Defined Populations." New York: Cold Spring Harbor Laboratory, pp 69–92.

Ernster VL, Mason L, Goodson WH, Sickles EA, et al (1982): Effects of caffeine-free diet on benign breast disease: A randomized trial. Surgery 91:263–267.

Fraumeni JF Jr, Grundy GW, Creagan ET, Everson RB (1975): Six families prone to ovarian cancer. Cancer 36:364–369.

Fraumeni JF, Mason TJ (1974): Cancer mortality among Chinese Americans, 1950-1969. J Natl Cancer Inst 52:659–665.

Galli MC, DeGiovanni C, Nicoletti G, Grilli S, Nanni P, Prodi G, Gola G, Rocchetta R, Orlandi C (1981):

The occurrence of multiple steroid hormone receptors in disease-free and neoplastic human ovary. Cancer 47:1297–1302.

Graham J, Graham R (1967): Ovarian cancer and asbestos. Environ Res 1:115–128.

Henderson WJ, Hamilton TC, Griffiths K (1979): Talc in normal and malignant ovarian tissue. Lancet 1:499.

Hildreth NG, Kelsey JL, LiVolsi VA, Fischer DB, Holford TR, Mostow ED, Schwartz PE, White C (1981): An epidemiologic study of epithelial carcinoma of the ovary. Am J Epidemiol 114:398–405.

Hoover R, Gray LA Sr, Fraumeni JF Jr (1977): Stilboestrol (diethylstilbestrol) and the risk of ovarian cancer. Lancet 2:533–534.

Howe GR, Burch JD, Miller AB, Cook GM, Esteve J, Morrison B, Gordon P, Chambers LW, Fodor G, Winsor GM (1980): Tobacco use, occupation, coffee, various nutrients and bladder cancer. J Natl Cancer Inst 64:701–713.

Jarvholm B, Lavenius B (1981): A cohort study on cancer among workers exposed to an antirust oil. Scand J Work Environ Health 7:179–184.

Joly DJ, Lilienfeld AM, Diamond EL, Bross IDJ (1974): An epidemiologic study of the relationship of reproductive experience to cancer of the ovary. Am J Epidemiol 99:190–209.

Jull WJ (1973): Ovarian tumorigenesis. Methods Cancer Res 7:131–186.

Kammerman S, Demopoulos RI, Raphael C, Ross J (1981): Gonadotropic hormone binding to human ovarian tumors. Hum Pathol 12:886–890.

Kinlen LJ (1982): Meat and fat consumption and cancer mortality: A study of strict religious orders in Britain. Lancet 2:946–949.

Kinlen LJ (1980): Mortality in relation to abstinence from meat in certain orders of religious sisters in Britain. In Cairns J. Lyon JL, et al (eds): "Cancer Incidence in Defined Populations." New York: Cold Spring Harbor Laboratory, pp 135–143.

Krarup T (1970): Oocyte survival in the mouse after treatment with 9,10-dimethyl-1,2-benzanthracene. J Endocrinol 46:483–495.

Kurppa K, Holmberg PC, Kuosma E, Saxen L (1982): Coffee consumption during pregnancy. N Engl J Med 306:1548.

Lau CC, Das SK, Pardee AB (1982): Comparison of the effects of 3-aminobenzamide and caffeine as DNA repair inhibitors. Proc AACR 23:68.

Li FP, Shiang EL (1980): Cancer mortality in China. J Natl Cancer Inst 65:217–221.

Lingeman CH (1974): Etiology of cancer of the human ovary: A review. J Natl Cancer Inst 53:1603–1618

Linn S, Schoenbaum SC, Monson RR, Rosner B, Stubblefield PG, Ryan KJ (1982): No association between coffee consumption and adverse outcomes of pregnancy. N Engl J Med 306:141–145.

London RS, Sundaram GS, Goldstein PJ (1982): Medical management of mammary dysplasia. Obstet Gynecol 59:519–523.

Longo DL, Young RC (1979): Cosmetic talc and ovarian cancer. Lancet 2:349–351.

Lynch HT, Albano W, Black L, Lynch JF, Recabaren J, Pierson R (1981): Familial excess of cancer of the ovary and other anatomic sites. JAMA 245:261–264.

Lyon JL, Gardner JW, West DW (1980): Cancer risk and life-style: cancer among Mormons from 1967-1975. In Cairns J, Lyon JL (eds): "Cancer Incidence in Defined Populations." New York: Cold-Spring Harbor Laboratory, pp 3–30.

MacMahon B, Yen S, Trichopoulos D, Warren K, Nardi G (1981): Coffee and cancer of the pancreas. N Engl J Med 304:630–633.

MacMahon B, Trichopoulos D, Cole P, Brown J (1982): Cigarette smoking and urinary estrogens. N Engl J Med 307:1062–1065.

Mansel RE, Wisbey JR, Hughes LE (1982): Controlled trial of the antigonadotropin danazol in painful nodular benign breast disease. Lancet 1:928–930.

Mason TJ, McKay FW, Hoover R, Blot WJ, Fraumeni JF Jr (1975): Atlas of Cancer Mortality for U.S. Counties: 1950-1969. DHEW Publication No. (NIH) 75–780, NIH, Bethesda, MD, pp 27; No. 76–1204, pp 103.

Mattison DR, Thorgeirsson SS (1978): Smoking and industrial pollution and their effects on menopause and ovarian cancer. Lancet 1:187–188.

Mattison DR, Thorgeirsson SS (1979): Ovarian aryl hydrocarbon hydroxylase activity and primordial oocyte toxicity of polycyclic aromatic hydrocarbons in mice. Cancer Res 39:3471–3475.

McGowan L, Parent L, Lednar W, Norris HJ (1979): The woman at risk for developing ovarian cancer. Gynecol Oncol 7:325–344.

McLachlan JA, Newbold RR, Bullick BC (1980): Long-term effects on the female mouse genital tract associated with prenatal exposure to diethylstilbestrol. Cancer Res 40:3988–3999.

McLachlan JA, Newbold RR, Korach KS, Lamb JC, IV, Suzuki Y (1981): Transplacental toxicology: Prenatal factors influencing postnatal fertility. In: Developmental Toxicology, Kimmel CA, Buelke-Sam J (eds). New York, Raven Press, pp 213–232.

Menczer J, Modan M, Ranon L, Golan A (1979): Possible role of mumps virus in the etiology of ovarian cancer. Cancer 43:1375–1379.

Miller AB (1978): An overview of hormone-associated cancers. Cancer Res 38:3985–3990.

Minton JP, Foecking MK, Webster DJT, Matthews RH (1979): Caffeine, cyclic nucleotides and breast disease. Surgery 86:105–109.

Morrison AS, Buring JE, Verhoek WG, Aoki K, Leck I, Ohno Y, Obata K (1982): Coffee drinking and cancer of the lower urinary tract. J Natl Cancer Inst 68:91–94.

National Cancer Institute (1976): Carcinogenesis Bioassay of Trichloroethylene. CAS No. 79-01-6. NCI-CG-TR-2. TR #243.

Newbold RR, McLachlan JA (1982): Vaginal adenosis and adenocarcinoma in mice exposes prenatally to diethylstilbestrol. Cancer Res 42:2003–2011.

Newhouse ML, Pearson RM, Fullerton JM, Boeson EAM, Shannon HS (1977): A case control study of carcinoma of the ovary. Br J Prev Soc Med 31:148–153.

Nolen GA (1981): The effect of brewed and instant coffee on reproduction and teratogenesis in the rat. Toxicol Appl Pharmacol 58:171–183.

Phillips RL, Kuzma JW, Lotz TM (1980a): Cancer mortality among comparable members versus non-members of the Seventh-day Adventist Church. In Cairns J, Lyon JL (eds): "Cancer Incidence in Defined Populations." New York: Cold Spring Harbor Laboratory, pp 93–108.

Phillips RL, Garfinkel L, Kuzona JW, Beeson WL, Lotz T, Brin B (1980b): Mortality among California Seventh-Day Adventists for selected cancer sites. J Natl Cancer Inst 65:1097–1107.

Prior P, Waterhouse JAH (1981): Multiple primary cancers of the breast and ovary. Br J Cancer 44:628–636.

Roe FJC (1979): Controversy: cosmetic talc and ovarian cancer. Lancet 2:744.

Rosenberg L, Slone D, Shapiro S, et al (1982): Breast cancer and alcoholic-beverage consumption. Lancet i:267–271.

Shevchuk MN, Winkler-Monsanto B, Fenoglio CM, Richart RM (1981): Clear cell carcinoma of the ovary: A clinicopathologic study with review of the literature. Cancer 47:1344–1351.

Stenman UH, Huhtala ML, Koistinen R, Seppala M (1981): Isolation of a urinary peptide associated ovarian cancer (Meeting Abstr). Oncodev Biol Med; Abstracts Book 1:57.

Stocks P (1970): Cancer mortality in relation to national consumption of cigarettes, solid fuel, tea and coffee. Br J Cancer 24:215–225.

Szamborski J, Czerwinski W, Gadomska H, Kowalski M, Wacker-Pujdak B (1981): Case control study of high-risk factors in ovarian carcinomas. Gynecol Oncol 11:8–16.

Trichopoulos D, Papapostolou M, Polychronopoulou A (1981): Coffee and ovarian cancer. Int J Cancer 28:691–693.

Union Internationale Contre le Cancer (UICC) (1980): Murphy ED, Beamer WG (eds): Biology of Ovarian Neoplasia, Report #11, UICC Technical Reports, volume 50, Geneve UICC.

Wanless IR, Medline A (1982): Role of estrogens as promotors of hepatic neoplasia. Lab Invest 46:313–320.

Wake N, Hreshchyshyn MM, Piver SM, Matsui SI, Sandberg AA (1980): Specific chromosome change in ovarian cancer. Cancer Genet Cytogenet 2:87–88.

Waterhouse J, Muir C, Correa P, Powell J (1976): Cancer Incidence in Five Continents. Vol 3, Lyon: International Agency for Research on Cancer, pp 453–485.

Weiss NS, Young JL Jr, Roth GJ (1977): Marital status and incidence of ovarian cancer: The U.S. third national cancer survey, 1969-71. J Natl Cancer Inst 58:913–915.

Weiss NS (1980): Epidemiology of ovarian cancer. In Murphy ED, Beamer WG (eds): "Biology of Ovarian Neoplasia." Geneva: UICC Technical Report Series, Vol 50, pp 34–49.

Weiss NS, Lyon JL, Liff JM, Vollmer WM, Daling JR (1981): Incidence of ovarian cancer in relation to the use of oral contraceptives. Int J Cancer 28:669–671.

Weiss NS, Lyon JL, Krishnamurthy S, Dietert SE, Liff JM, Daling JR (1982): Noncontraceptive estrogen use and the occurrence of ovarian cancer. J Natl Cancer Inst 68:95–98.

West DW (1980): An assessment of cancer risk factors in Latter-day Saints and non-Latter-day Saints in Utah. In Cairns J, Lyon JL, et al (eds): "Cancer Incidence in Defined Populations." New York: Cold Spring Harbor Laboratory, pp 31–45.

Whang-Peng J, Knutsen T, Douglass EC, Chu E (1982): Is 6q–a specific chromosomal marker for ovarian cancer? Proc AACR 23:34.

Willett WC, Bain C, Hennekens CH, Rosner B, Speizer FK (1981): Oral contraceptives and risk of ovarian cancer. Cancer 48:1684–1687.

Williams GM (1982): Sex hormones and liver cancer. Lab Invest 46:352–353.

World Health Organization, International Agency for Research on Cancer (1981): Muir CS, Wagner G (eds): Directory of On-going Research in Cancer Epidemiology. Lyon: IARC Publication No. 38.

SOCIAL ASPECTS OF REPRODUCTIVE TOXICOLOGY

American Journal of Industrial Medicine 4:383–390 (1983)

Occupational Reproductive Hazards: Necessary Steps to Prevention

Peter F. Infante, DDS, DrPH, and Theodora A. Tsongas, PhD

Although some attention has recently been given to the study of occupational reproductive hazards, little is known about measures being taken to prevent exposure to substances having the potential to cause such problems. In the past, OSHA has promulgated a standard to lower the permissible exposure level to a reproductive toxin as the major focus only after experimental evidence was confirmed in humans. The method of selection of appropriate substitutes is also cause for concern. A chemical known to cause testicular atrophy, cancer, and gene mutations in subhuman test systems and to cause sterility in male workers was replaced with a substance known from experimental studies to cause testicular toxicity, cancer of multiple sites, and gene mutations. Experimental test results need to be addressed from the standpoint of determining presumptive risk to humans. A scheme for setting priorities for reduction of exposure or for consideration of regulation of occupational reproductive hazards also needs to be developed. A policy addressing occupational reproductive hazards may serve as a stimulus for protecting workers from these hazards and for improving scientific research protocols.

Key words: reproductive hazards, occupation, prevention, policy

Society is increasingly creating most of its environment. With the intention of upgrading our standard of living, it is estimated that approximately 1000 new chemical compounds are being developed and introduced into the marketplace each year. Among these chemicals are those that by their usage touch on many aspects of our lives. They include metal compounds, dyes, industrial solvents, pesticides, and other petrochemicals.

Office of Carcinogen Identification and Classification, Health Standards Programs, Occupational Safety and Health Administration, U.S. Department of Labor, Washington, D.C.

This paper was presented at the International Course on Occupational Hazards and Reproduction, August 1981, Kaupunkiopisto, Espoo, Finland. The views expressed herein are those of the authors and do not necessarily reflect those of the Occupational Safety and Health Administration.

Address reprint requests to Dr. T.A. Tsongas, Room N3718, OSHA, U.S. Department of Labor, 200 Constitution Ave., NW, Washington, DC 20210.

0271-3586/83/0401-02-0383$02.50 © 1983 Alan R. Liss, Inc.

The prevention of employee exposures to chemical substances that may be associated with occupational reproductive hazards is of primary importance in light of our inability to cure or adequately treat many of the adverse health consequences resulting from exposure to these substances. This is particularly true in the case of congenital anomalies, where the infant's deficit may be protracted over the course of a lifetime and cause debilitating consequences to family resources, both economically and psychologically, and impair the integrity of the family as a functional unit in society.

Three major approaches to the prevention of occupational illness are identification of the hazard, education of the workforce about the hazard, and control of the hazard which usually implies some form of regulation.

Shortly after the enactment of the Occupational Safety and Health Act of 1970 [U.S. Congress, 1970], the Occupational Safety and Health Administration (OSHA) adopted the American Conference of Governmental Industrial Hygienists (ACGIH) "TLVs" as Federal standards [ACGIH, 1968]. Since that time, OSHA has completed the promulgation of standards for occupational exposure to 21 substances. Of these standards, 17 have been based primarily upon information bearing on carcinogenicity, while four have been based on toxic effects other than cancer. These latter standards pertain to cotton dust, lead, asbestos, and dibromochloropropane (DBCP). The standard for DBCP exposure is the only one that has been based primarily on the substance being a reproductive hazard, although the standard for lead exposure addressed its potential for harm to reproductive capability among other forms of toxicity. One may question why there has been only one standard to control a substance that is primarily a reproductive hazard.

Unfortunately, in the past only women have been associated with reproductive hazards, and women in the U.S. have not moved into the industrial setting in great numbers or into blue collar jobs traditionally held by men until relatively recently. The traditional approach to the prevention of reproductive hazards has been to assure that women are not employed in occupations where these hazards exist [Hunt, 1979]. When the hazards of lead exposure became known, in lieu of containment of this toxic substance the tendency was, and for the most part still is, to remove fertile women from high exposure workplace situations. Because there has been relatively little concern in the past about reproductive hazards in the workplace, we now find ourselves in the difficult situation of not knowing with an adequate degree of certainty how to predict occupational reproductive hazards to humans from experimental study results and of not having sensitive enough epidemiologic methods to detect and evaluate these hazards in humans. Of even greater concern is the past approach taken by government, academia, and industry subsequent to the identification of hazards associated with a reproductive toxin.

As an example of the past approach, let us look at the historical development of the data base demonstrating the toxicity of DBCP and society's response to that data. In 1961, Torkelson and co-workers [Torkelson et al, 1961] published a study demonstrating sterility in rats and testicular atrophy in male rats, guinea pigs, and rabbits exposed to levels of DBCP ranging between 5 and 15 parts per million (ppm). As a result of this study, Torkelson et al [1961] concluded that occupational exposure to DBCP should be maintained below 1 ppm. However, there was little response to this interesting publication, and no apparent government action was based on this finding. Through 1976, there was no federal standard for occupational exposure to DBCP in the U.S.

In 1973, Olson et al reported on early results of experiments in the NCI Bioassay

Program demonstrating that exposure to DBCP by oral gavage caused cancer in experimental animals. What was the response of the public health community to the evidence of carcinogenicity of DBCP? Some argued that the cancers of the forestomach induced by administration of DBCP by oral gavage may have been irritant induced, and as such were no greater evidence of carcinogenicity than the observation of an injection site sarcoma. Therefore, the observations were not acceptable as primary evidence of carcinogenicity. These commenters apparently ignored the observation that DBCP also induced rat mammary adenocarcinoma. (Interestingly enough, when the results of the NCI Bioassay on ethylene dichloride (EDC) [NCI, 1978a] were reported, claims were made that the positive results may have been due to low level compound contamination with DBCP and other carcinogens or room air contamination with toxic substances among which was DBCP [Ames et al, 1980; Shell Oil Co., 1980]).

In 1975, DBCP was found to be mutagenic in experimental test systems [Rosenkranz, 1975]. What was society's response to this information? Experimentalists obtained samples of DBCP for evaluation in various test systems, to determine if there was concordance in mutagenicity test systems with a now "known" carcinogen. Although concordance between test systems was reported, the results apparently were meaningless to the regulatory and industrial health communities. To our knowledge no formal action was requested or taken to assess the need for regulatory action on the basis of study results available through 1976.

During 1977, DBCP workers in four industrial facilities in the United States were identified as suffering from sterility or low sperm counts; many individuals had no sperm at all [Whorton, 1980]. Who had identified the problem? An astute clinician? A clever epidemiologist? No, rather, a group of industrial workers became concerned when they realized that none of their wives had had children once they began working in the division of a chemical plant where DBCP was formulated. These men had had intermittent atmospheric exposures to DBCP that were estimated to have ranged between 0.29 to 0.43 ppm, approximately fifty times lower than the levels causing sterility in experimental animals. What was society's response? The Occupational Safety and Health Administration was petitioned in August, 1977, for an Emergency Temporary Standard (ETS). An ETS was published in September, 1977, and within a year (March, 1978) OSHA promulgated a 1 part per billion, 8-hour time-weighted-average (TWA) standard for occupational exposure to DBCP [OSHA, 1978]. In spite of the scientific evidence, the toxicity of DBCP was still not fully appreciated. During the OSHA rulemaking proceedings of 1977, the National Peach Council (NPC) wrote to the Assistant Secretary for OSHA, stating that she and the Secretary of Labor appeared to be overreacting in proposing a standard for worker exposure to this material (NPC, 1977). The NPC stated:

> "While involuntary sterility caused by a manufactured chemical may be bad, it is not necessarily so. After all, there are many people who are now paying to have themselves sterilized to assure they will no longer be able to become parents. . . . If possible sterility is the main problem, couldn't workers who were old enough that they no longer wanted to have children accept such positions voluntarily? They would know the situation, and it wouldn't matter. Or could workers be advised of the situation, and some might volunteer for such work posts as an alternative to planned surgery for a vasectomy or tubal ligation, or as a means of getting around religious bans on birth control when they want no more children." (NPC, 1977)

The fact that these statements could be made after data demonstrated that DBCP was also mutagenic and carcinogenic is difficult to comprehend.

With the promulgation of the new standard, the manufacture of DBCP in the U.S. was discontinued. Ethylene dibromide (EDB), a substance known to cause sperm abnormalities in bulls [Amir and Volcani, 1965, 1967; Amir, 1973, 1975] was used as a replacement for DBCP as a soil fumigant. Previous to the substitution, EDB administered by gavage had induced cancer of the stomach in experimental animals [Olson et al, 1973]. (The final report on this bioassay was published in 1978 [NCI, 1978b]). A more recent NCI inhalation study has demonstrated that exposures as low as 10 ppm (the lowest dose tested) induced cancers of multiple sites in a high proportion of both mice and rats [NCI, 1980]. EDB also had been found to be mutagenic in several test systems [IARC, 1977]. From a public health standpoint, the substitution of EDB for DBCP makes little sense.

For the purpose of further discussion, let us assume that there are *no* data demonstrating that DBCP causes sterility in humans. If today we found that a substance was mutagenic in several test systems, carcinogenic in experimental animals, caused sterility in male animals at five to 15 ppm, or caused adverse effects on sperm, what would be our response in terms of prevention? In retrospect, we would probably conclude that we would immediately begin to entertain feasible means to reduce exposure. This hypothetical question is important as the following information is presently available on another substance, namely ethylene oxide (ETO). ETO has been shown to be mutagenic in multiple test systems including the dominant lethal assay [Embree et al, 1977], with an increased frequency of heritable translocations [Generoso et al, 1980], sex-linked recessive lethal mutations [Bird, 1952; Fahmy and Fahmy, 1956], and alkylation of sperm DNA [Cumming et al, 1981]. An elevated incidence of mononuclear cell leukemia (MNL) has been observed in rats exposed by inhalation to ETO levels as low as 10 ppm [Snellings et al, 1981].

In humans, exposures below 50 ppm ETO has been associated with chromosomal damage; 9 of 73 exposed workers had two or more quadriradial chromosomes per 100 cells analyzed compared with none in the non-exposed controls. Increased sister chromatid exchange was also reported among these workers [Abrahams, 1980].

Epidemiologic evidence suggests an increased risk of leukemia among workers sterilizing hospital equipment and exposed to an estimated TWA of 20 ppm ETO [Hogstedt et al, 1979a]. Individuals employed in production of ETO also have been reported to have an excess risk of developing leukemia and stomach cancer [Hogstedt et al, 1979b].

The present standards for occupational exposure to ethylene oxide averaged over an 8-hour day are 50 ppm in Finland, 10 ppm in Sweden, and 50 ppm in the U.S. Current exposure levels in the U.S. are thought to range between 1 and 10 ppm in the industrial setting and to range higher in hospitals.

The question before us is whether we have enough scientific evidence to merit control of ETO as an occupational reproductive hazard or as a carcinogen. If not, what additional evidence is needed? If the evidence is sufficient, what level of exposure in the workplace should be recommended? On the basis of what scientific evidence? Is such a recommended exposure level feasible to measure and is it affordable by industry? On the basis of existing data is it possible to quantify the risks of cancer or adverse reproductive outcome? Can such risks be significantly reduced by achieving an exposure level considered feasible? How should a regulatory agency make such determinations? How should it set priorities for choosing the next substance to regulate in light of so many substances needing lower permissible exposure levels? Table I lists some additional questions that need to be answered in order to identify, classify, and prevent exposure to occupational reproductive hazards. Tables II through V list questions raised by the Interagency Regulatory Liaison Group (IRLG) Reproductive Toxicity Risk Assessment Task Group for discussion at its 1981 workshop. The workshop was intended to develop information for use in a document on risk assessment for reproductive toxicity. Table II lists questions addressing endpoints that might be useful in determining reproductive risk assessment. Table III raises questions concerning qualitative and quantitative risk, while Tables IV and V address questions related to pharmacokinetics and mechanisms, respectively.

It is clear that a policy needs to be developed for determining acceptable evidence for use toward the prevention/regulation of reproductive hazards in the workplace. Considerable effort already has been expended toward the development of guidelines for studies of human populations exposed to these hazards [Bloom, 1981]. Only through such efforts will we have peer review of the types of studies acceptable for determining presumptive risk. The development of a policy and implementation of study guidelines will clearly result in less argument after studies are completed and promote the prevention of reproductive hazards.

On the basis of the U.S. Supreme Court decision on benzene, OSHA must attempt to determine the significance of risk at current exposure levels and the reduction in risk to be achieved from lowering the exposure. This requirement has been interpreted to include the development of quantitative risk assessments when they can be ap-

TABLE I. Some Questions to Resolve in Order to Identify, Classify, and Prevent Exposure to Potential Occupational Reproductive Hazards

1. When are epidemiological studies useful in evaluating reproductive effects? And when are they not?
2. Can we relate data from animal tests to human populations in determining safe exposures that will assure insofar as practicable that no employee will suffer diminished health or functional capacity as a result of his or her work experience?
3. Can we detect toxic substances that exert a weak effect? Can methodologies used in testing safety of pharmaceutical drugs be used in assessing workplace and other environmental exposures? Or do we need new testing protocols?
4. Should teratogenicity tests consider toxic exposures prior to conception?
5. What do we need to know about the synergistic and additive effects of multiple toxic exposures?
6. Is there a need for a tiered, or hierarchical, approach in evaluating genetic and other reproductive hazards?
7. What more do we need to know about mutagens? We know that certain substances are mutagenic, but we have not yet been able to trace a specific birth defect in a human population to a specific mutagenic substance.
8. What additional information do we need about the effects of toxic substances on male and female physiology under stress?
9. In terms of social responsibility, at what point should industry, in its testing programs, inform employees and regulatory agencies of abnormal test results? At what point should government inform workers and society in general?
10. What amount of information and analysis is necessary in order to promulgate a regulation for a substance known to be associated with toxicity to the reproductive system?
11. Given the number of chemicals which still need to be evaluated for reproductive toxicity, what should be the method for assessing priorities?

TABLE II. Which Animal and Human Endpoints are Useful for Human Reproductive Risk Assessment?

1. How much should the endpoints produced by teratogens in animals be expected to mimic the human effects? How comparable are endpoints in animals and humans?
2. Are various endpoints in developmental toxicity (eg, fetal death, growth retardation, malformations, functional impairment) a continuum of responses or separable? What is the basis for concluding one or the other?
3. What is the significance of transient effects (eg, transient growth retardation of hyperactivity)?
4. What is the relationship between maternal toxicity and fetal toxicity? Should fetal effects that occur at maternal toxic doses (especially for chemicals currently in the environment) be ignored?
5. How should increases in spontaneously occurring defects be interpreted? How should variations (eg, skeletal) be interpreted?
6. Which endpoints should be evaluated in animals postnatally? Should lifespan studies be included in teratology studies?
7. Which endpoints are sensitive for assessing effects on fertility and reproduction resulting from either male or female exposures?
8. Which reproductive toxicity endpoints can be monitored adequately in epidemiologic studies? How many cases would be needed to detect a change in each of these endpoints?

Adapted from IRLG Reproductive Toxicity Risk Assessment Workshop, 1981 (Unpublished).

TABLE III. Which Procedures can be Used for Qualitative and/or Quantitative Risk Assessment of Potential Human Reproductive Hazards With Available Evidence?

1. Which statistical techniques, if any, are optimal for the analysis of teratology, infertility, and other data related to adverse pregnancy outcome?
2. How are inferences drawn from experiments consisting of multiple doses, multiple endpoints, and multiple generations?
3. Which qualitative factors contribute to identifying a compound as a potential reproductive hazard? ie,
   a. Internal consistency
   b. Reproducibility in multiple species
   c. Evidence of a dose-response relationship
   d. Molecular structure
   e. Lack of overt toxicity
4. Which mathematical models can be identified for high-to-low-dose extrapolation of both threshold and nonthreshold response data?
5. What are the pros and cons of using qualitative safety factors and which factors should be used for each type of endpoint?

Adapted from IRLG Reproductive Toxicity Risk Assessment Workshop, 1981 (Unpublished).

propriately performed. (However, OSHA has stated, "When data are not available to perform a formal quantitative risk assessment, qualitative evidence, expert testimony, and other evidence may be appropriately utilized to base a determination of significance of risk" [OSHA, 1981].) In consideration of risk assessment as it relates to establishing priorities for regulation, how does one interpret the observation of a twofold risk of spontaneous abortion or gene mutations, or adverse effects on chromosomes and on sperm as compared to cases of cancer that might be estimated to result from chemical exposure? Until the scientific and public health community begins to answer

TABLE IV. What Evidence can be Gained From Pharmacokinetic Studies and how can it be Used for Estimating Potential Risk to Fetus, Gametes, or Parents?

1. Can pharmacokinetics assist in the choice of the most appropriate test animal species, based on similarities in absorption, excretion, metabolism, placental transfer, etc? Can dissimilarities be qualified to an extent that would make data collected in a "less-appropriate" species useful for estimating potential risks?
2. Can pharmacokinetics help establish a dosing regimen in a test system that will approximate the effective dose of a compound to which the human fetus is exposed?
3. Can pharmacokinetics define the similarities and differences in fetal exposure relative to differences in critical (sensitive) periods in development of test animals and humans?
4. What should be defined as the endpoint for evaluation (target of exposure), eg, the conceptus, whole embryo/fetus, embryo/fetal blood, specific organ/tissues? Are embryo/fetal values any more predictive than maternal values alone?
5. Can pharmacokinetics assist in predicting thresholds?

From IRLG Reproductive Toxicity Risk Assessment Workshop, 1981 (Unpublished).

TABLE V. How can Information on Mechanisms be Used for Interpretation of Experimental Data and Risk Assessment?

1. What can be considered an appropriate definition of "mechanisms of action," especially as it relates to reproductive/developmental toxicity?
2. What information on mechanisms of action can help to explain qualitative or quantitative interspecies variation in response to reproductive toxins?
3. What information on mechanisms of action will allow the development of more appropriate models for extrapolation of animal data to humans?
4. What information on gene/toxicant interaction can be used for predicting risk? What markers of susceptibility to toxicants might be used for predicting risk?
5. What early indicators of toxicant effect might be used for predicting risk?

From IRLG Reproductive Toxicity Risk Assessment Workshop, 1981 (Unpublished).

this and other questions raised above, it is unlikely that government will ever promulgate a standard for a reproductive hazard unless evidence of sterility, as was the case with DBCP, or adverse pregnancy outcome is confirmed in humans.

REFERENCES

Abrahams RH (1980): Recent studies with workers exposed to ethylene oxide. In Jorkasky FJ (ed): "The Safe Use of Ethylene Oxide: Proceedings of the Education Seminar." HIMA Report No. 80-4. Washington DC: Health Industry Manufacturers Association, pp 27–38, 211–220.

American Conference of Governmental Industrial Hygienists. (ACGIH) (1968): "TLVs. Threshold Limit Values for Chemical Substances and Physical Agents in the Workroom Environment." Cincinnati.

Ames B, Infante P, Reitz R (1980): "Ethylene Dichloride: A Potential Health Risk?" Banbury Report 5. New York: Cold Spring Harbor Laboratory, pp 50.

Amir D (1973): The sites of spermicidal action of ethylene dibromide in bulls. J Reprod Fertil 35:519–525.

Amir D (1975): Individual and age differences in the spermicidal effect of ethylene dibromide in bulls. J Reprod Fertil 44:561–565.

Amir D, Volcani R (1965): Effect of dietary ethylene dibromide on bull semen. Nature 206:99–100.

Amir D, Volcani R (1967): The effect of dietary ethylene dibromide (EDB) on the testes of bulls. Fertil Steril 19:144–148.

Bird MJ (1952): Chemical production of mutations in Drosophila: Comparison of techniques. J Genet 50: 480–485.

Bloom AD (1981): "Guidelines for Studies of Human Populations Exposed to Mutagenic and Reproductive Hazards." White Plains, NY: March of Dimes Birth Defects Foundation.

Cumming RB, Michaud TA, Lewis LR, Olson WH (1981): Inhalation mutagenesis in mammals: Patterns of unscheduled DNA synthesis induced in germ cells in male mice by exposure to ethylene oxide in air. Mutat Res (in press).

Embree JW, Lyon JP, Hine CH (1977): The mutagenic potential of ethylene oxide using the dominant-lethal assay in rats. Toxicol Appl Pharmacol 40:261–267.

Fahmy OG, Fahmy MJ (1956): Cytogenetic analysis of the action of carcinogens and tumour inhibitors in Drosophila melanogaster. Part V. Differential genetic response to the alkylating mutagens and x-radiation. J Genet 50:146–164.

Generoso WM, Cain KT, Krishna M, Shea CW, Gryder RM (1980): Heritable translocation and dominant-lethal mutation induction with ethylene oxide in mice. Mutat Res 73:133–142.

Hogstedt C, Malmqvist N, Wadman B (1979a): Leukemia in workers exposed to ethylene oxide. JAMA 241:1132–1133.

Hogstedt C, Rohlen O, Berndtsson BS, Axelson O, Ehrenberg L (1979b): A cohort study of mortality and cancer incidence in ethylene oxide production workers. Br J Ind Med 36:276–280.

Hunt VR (1979): "Work and the Health of Women." Boca Raton: CRC Press.

International Agency for Research on Cancer (IARC) (1977): Ethylene dibromide. In "Some Fumigants, the Herbicides 2,4-D and 2,4,5-T, Chlorinated Dibenzodioxins and Miscellaneous Industrial Chemicals." Volume 15, IARC Monographs on the Evaluation of the Carcinogenic Risk of Chemicals to Man. Lyon: IARC, pp 195–209.

National Cancer Institute (NCI) (1978a): "Bioassay of 1,2-dichloroethane for possible carcinogenicity." Technical report series No. 55. DHEW Publication No. (NIH) 78-1361. Washington, D.C.: U.S. Government Printing Office.

National Cancer Institute (NCI) (1978b): "Bioassay of 1,2-dibromoethane for possible carcinogenicity." Technical report series No. 86. DHEW publication No. (NIH) 78-1336. Washington, D.C.: U.S. Government Printing Office.

National Cancer Institute (1980): "Bioassay of 1,2-dibromoethane for possible carcinogenicity (inhalation study)." DHHS Publication No. (NIH) 80-1766. Washington, D.C.: U.S. Government Printing Office.

NPC (National Peach Council) (1977): Letter (12 September) from Robert K. Phillips to Dr. Eula Bingham, Occupational Safety and Health Administration.

Occupational Safety and Health Administration (OSHA) (1978): Occupational exposure to 1,2-dibromo-3-chloropropane (DBCP). Fed Reg 43:11514–11533.

Occupational Safety and Health Administration (1981): Identification, classification and regulation of potential occupational carcinogens; conforming deletions. Fed Reg 46:4889–4893.

Olson WA, Habermann RT, Weisburger EK, Ward JM, Weisburger JH (1973): Induction of stomach cancer in rats and mice by halogenated aliphatic fumigants. JNCI 51:1993–1995.

Rosenkranz HS (1975): Genetic activity of 1,2-dibromo-3-chloropropane, a widely used fumigant. Bull Environ Contam Toxicol 14:8–12.

Shell Oil Company (1980): Comments on OSHA Candidate Carcinogen List – Response for ethylene dichloride (November 25) from H.L. Kusnetz to the Docket Office, U.S. Dept. of Labor.

Snellings WM, Weil CS, Maronpot RR (1981): Final report on ethylene oxide two-year inhalation study on rats, Project Report 44-20, Bushy Run Research Center, January 28, 1981. Submitted by Union Carbide Corporation to the U.S. Environmental Protection Agency under Section 8(e) of the Toxic Substances Control Act.

Torkelson TR, Sadek SE, Rowe VK, Kodama JK, Anderson HH, Loquvam GS, Hine CH (1961): Toxicologic investigations of 1,2-dibromo-3-chloropropane. Toxicol Appl Pharmacol 3:545–559.

United States Congress (1970): Occupational Safety and Health Act of 1970, Public Law 91-596, 91st Congress, S. 2193, December 29, 1970, 84 STAT. 1590–1620.

Whorton D (1980): Dibromochloropropane (DBCP). In Infante PF, Legator MS (eds): "Proceedings of a Workshop on Methodology for Assessing Reproductive Hazards in the Workplace." DHHS (NIOSH) Pub. No. 81-100. Washington, D.C.: U.S. Government Printing Office, pp 103–105.

American Journal of Industrial Medicine 4:391–393 (1983)

Reproductive Toxicology: State of the Art, 1982

Sergio Fabro, MD, PhD

Key words: reproductive toxicology, chemical environmental hazards, drugs and pregnancy

The twentieth century has witnessed unprecedented, geometrical growth of knowledge in the physical, chemical, and biologic sciences that led to a remarkable increase in our ability to control infections, infestations, and to some degree famine and malnutrition.

Advances in synthetic chemistry have resulted in an increasingly complex chemical environment that includes drugs, pesticides, and food additives. These intentionally synthesized compounds, along with thousands that are inadvertantly generated, are increasingly polluting our air, water, and food. Since many of these chemicals are novel structures not previously existing in nature, their long-term effects in biologic systems in general, and in the human in particular, are not known. Thus, the same technologic advances that have lengthened our life span may now present potential toxicologic threats to human life. Of special concern is the potentially adverse effects of these substances on human fertility, reproduction, and development.

This complex issue is made more difficult because advances in our understanding of the reproductive toxicity of chemicals have not kept pace with the introduction of new chemicals into the environment and because a significant proportion of human reproductive wastage has an unknown etiology. For every 3,000,000 births per year in the United States there are approximately 600,000 spontaneous abortions, and some 24,000 fetuses die in utero. Of those who are born alive, about 10% are premature, approximately 13% have a low birthweight, and another 3–7% possess some type of malformation. In this country, about 30,000 babies per year die during the neonatal period, and an additional 26,000 do not survive into the beginning of their second year.

No one would contest that the incidence of human reproductive wastage is significant. Nevertheless the cause of most of this wastage is unknown: for example, of the approximately 150,000 babies born with a birth defect each year in the United States, it is estimated that 5% are due to a mutant gene; in 20% the defects are attributed to multifactorial causes and in 15% to intrauterine infections. Environmental

Reproductive Toxicology Center, Columbia Hospital for Women, Georgetown University, Washington, DC.
Address reprint requests to Dr. S. Fabro, Reproductive Toxicology Center, Columbia Hospital for Women, Georgetown University, 2425 L Street, N.W., Washington, DC 20037.

0271-3586/83/0401-02-0391$01.50 © 1983 Alan R. Liss, Inc.

causes can be identified with relative certainty in 2–3% of the cases but there is increasing speculation that environmental factors may play a significant part in the etiology of some of the 60% of birth defects for which the cause is presently unknown.

Similarly, a large proportion of the cases of infertility, spontaneous abortion, stillbirth, prematurity, and intrauterine growth retardation have an unknown etiology, and again it has been suggested that environmental factors may play a significant role.

The science of reproductive toxicology has developed in response to the clear need to evaluate what part the environment plays in determining human reproductive failure, fetal wastage, and malformed births. The field is young but growing rapidly. Scientific progress has been difficult to achieve, because of our lack of knowledge of the biology of growth and development at the molecular level and because the mechanisms by which environmental agents may induce their effect is diverse. We must unravel the complexity of the molecular events that form the basis of ovarian maturation, fertilization, cleavage, and morphogenesis. Furthering our knowledge in this difficult area is an important first step toward improving our understanding of the mechanisms by which xenobiotics exert their toxicity on the reproductive process.

Long-term reproductive toxicity is another problem area in which progress is badly needed. Birth defects, intrauterine death, and low birthweight are only the most obvious effects of abnormal human reproduction. We must add to them the offspring who seem perfectly normal at birth but who later in life are affected with mental retardation or with transplacentally induced disease, including cancer, hypertension, and infertility. At present most of the evidence regarding this type of delayed toxicity is derived from experiments in animals, but a few examples in the human have surfaced.

The identification and validation of end points in reproductive toxicity, as well as the development of epidemiologic methods for the long-term monitoring of these effects represent important and attainable goals in this area. Our newly found mobility makes it difficult to follow up populations exposed to potentially harmful environmental factors since a meaningful study would of necessity be not only of a large population sample but would be of a longitudinal nature. Although epidemiologic techniques have significantly improved in recent times, still their sensitivity could be markedly increased by better population sampling methods. Ideally we should develop techniques by which the health and reproductive function of individuals could be followed throughout their lives. This implies that each individual born in this country should be identified with a number (such as a Social Security Number) and all medical records should be filed under that number. A method for retrieving and analyzing this information would also be required. I believe such a system may be difficult to develop in the near future for economic, medical, and legal reasons. For the present, data from epidemiologic studies can only be suggestive of relationships between environment and reproductive toxicity. We must therefore be prepared to deal with the medical, social, and regulatory problems raised by false-positive and false-negative epidemiologic associations.

It is evident that not only are the problems facing reproductive toxicology medically and scientifically complex, but they are also intertwined with problems related to economic and social structure and activities. I anticipate slow progress in areas of reproductive toxicology that require collaborative endeavors from organizations within the scientific, medical, industrial, and regulatory communities, because of their disparate interests.

As I see it, there is a wide communication and/or knowledge gap between different groups that have different interests related to reproductive toxicology. However, reproductive toxicology requires collaboration and input from experts from all of the above sectors. Bridging this gap is probably the most important problem for the development and progress of reproductive toxicology. With these goals in mind the Reproductive Toxicology Center has recently been founded by the author under the auspices of the Research Foundation of Columbia Hospital for Women.

The mandate of the recently established Center is to further understanding of reproductive toxicity through both education and research.* In addition the Center represents a source of information and consultative services for the medical community, industry, and governmental agencies. The initial response to this initiative has been enthusiastic, as we are totally committed to improving communications between various interested groups such as clinicians and basic scientists, workers, employers, and governmental agencies. We are confident that the Center will become a forum for discussion of the problems we are facing today in human reproductive toxicology and that it will contribute to finding approaches to resolving them.

*Further information about goals, services and activities of the Reproductive Toxicology Center can be obtained by writing the Center at: 2425 L Street, N.W., Washington, DC 20037 or by telephoning (202) 293-5137.

Index

PROGRESS IN CLINICAL AND BIOLOGICAL RESEARCH

Series Editors
Nathan Back
George J. Brewer
Vincent P. Eijsvoogel
Robert Grover
Kurt Hirschhorn
Seymour S. Kety
Sidney Udenfriend
Jonathan W. Uhr

Vol 1: **Erythrocyte Structure and Function,** George J. Brewer, *Editor*

Vol 2: **Preventability of Perinatal Injury,** Karlis Adamsons, Howard A. Fox, *Editors*

Vol 3: **Infections of the Fetus and the Newborn Infant,** Saul Krugman, Anne A. Gershon, *Editors*

Vol 4: **Conflicts in Childhood Cancer: An Evaluation of Current Management,** Lucius F. Sinks, John O. Godden, *Editors*

Vol 5: **Trace Components of Plasma: Isolation and Clinical Significance,** G.A. Jamieson, T.J. Greenwalt, *Editors*

Vol 6: **Prostatic Disease,** H. Marberger, H. Haschek, H.K.A. Schirmer, J.A.C. Colston, E. Witkin, *Editors*

Vol 7: **Blood Pressure, Edema and Proteinuria in Pregnancy,** Emanuel A. Friedman, *Editor*

Vol 8: **Cell Surface Receptors,** Garth L. Nicolson, Michael A. Raftery, Martin Rodbell, C. Fred Fox, *Editors*

Vol 9: **Membranes and Neoplasia: New Approaches and Strategies,** Vincent T. Marchesi, *Editor*

Vol 10: **Diabetes and Other Endocrine Disorders During Pregnancy and in the Newborn,** Maria I. New, Robert H. Fiser, *Editors*

Vol 11: **Clinical Uses of Frozen-Thawed Red Blood Cells,** John A. Griep, *Editor*

Vol 12: **Breast Cancer,** Albert C.W. Montague, Geary L. Stonesifer, Jr., Edward F. Lewison, *Editors*

Vol 13: **The Granulocyte: Function and Clinical Utilization,** Tibor J. Greenwalt, G.A. Jamieson, *Editors*

Vol 14: **Zinc Metabolism: Current Aspects in Health and Disease,** George J. Brewer, Ananda S. Prasad, *Editors*

Vol 15: **Cellular Neurobiology,** Zach Hall, Regis Kelly, C. Fred Fox, *Editors*

Vol 16: **HLA and Malignancy,** Gerald P. Murphy, *Editor*

Vol 17: **Cell Shape and Surface Architecture,** Jean Paul Revel, Ulf Henning, C. Fred Fox, *Editors*

Vol 18: **Tay-Sachs Disease: Screening and Prevention,** Michael M. Kaback, *Editor*

Vol 19: **Blood Substitutes and Plasma Expanders,** G.A. Jamieson, T.J. Greenwalt, *Editors*

Vol 20: **Erythrocyte Membranes: Recent Clinical and Experimental Advances,** Walter C. Kruckeberg, John W. Eaton, George J. Brewer, *Editors*

Vol 21: **The Red Cell,** George J. Brewer, *Editor*

Volume 22: **Molecular Aspects of Membrane Transport,** Dale Oxender, C. Fred Fox, *Editors*

Vol 23: **Cell Surface Carbohydrates and Biological Recognition,** Vincent T. Marchesi, Victor Ginsburg, Phillips W. Robbins, C. Fred Fox, *Editors*

Vol 24: **Twin Research, Proceedings of the Second International Congress on Twin Studies,** Walter E. Nance, *Editor* Published in 3 volumes: Part A: **Psychology and Methodology** Part B: **Biology and Epidemiology** Part C: **Clinical Studies**

Vol 25: **Recent Advances in Clinical Oncology,** Tapan A. Hazra, Michael C. Beachley, *Editors*

Vol 26: **Origin and Natural History of Cell Lines,** Claudio Barigozzi, *Editor*

Vol 27: **Membrane Mechanisms of Drugs of Abuse,** Charles W. Sharp, Leo G. Abood, *Editors*

Vol 28: **The Blood Platelet in Transfusion Therapy,** G.A. Jamieson, Tibor J. Greenwalt, *Editors*

Vol 29: **Biomedical Applications of the Horseshoe Crab (Limulidae),** Elias Cohen, *Editor-in-Chief*

Vol 30: **Normal and Abnormal Red Cell Membranes,** Samuel E. Lux, Vincent T. Marchesi, C. Fred Fox, *Editors*

Vol 31: **Transmembrane Signaling,** Mark Bitensky, R. John Collier, Donald F. Steiner, C. Fred Fox, *Editors*

Vol 32: **Genetic Analysis of Common Diseases: Applications to Predictive Factors in Coronary Disease,** Charles F. Sing, Mark Skolnick, *Editors*

Vol 33: **Prostate Cancer and Hormone Receptors,** Gerald P. Murphy, Avery A. Sandberg, *Editors*

Vol 34: **The Management of Genetic Disorders,** Constantine J. Papadatos, Christos S. Bartsocas, *Editors*

Vol 35: **Antibiotics and Hospitals,** Carlo Grassi, Giuseppe Ostino, *Editors*

Vol 36: **Drug and Chemical Risks to the Fetus, Newborn,** Richard H. Schwarz, Sumner J. Yaffe, *Editors*

Vol 37: **Models for Prostate Cancer,** Gerald P. Murphy, *Editor*

Vol 38: **Ethics, Humanism, and Medicine,** Marc D. Basson, *Editor*

Vol 39: **Neurochemistry and Clinical Neurology,** Leontino Battistin, George Hashim, Abel Lajtha, *Editors*

Vol 40: **Biological Recognition and Assembly,** David S. Eisenberg, James A. Lake, C. Fred Fox, *Editors*

Vol 41: **Tumor Cell Surfaces and Malignancy,** Richard O. Hynes, C. Fred Fox, *Editors*

Vol 42: **Membranes, Receptors, and the Immune Response: 80 Years After Ehrlich's Side Chain Theory,** Edward P. Cohen, Heinz Köhler, *Editors*

Vol 43: **Immunobiology of the Erythrocyte,** S. Gerald Sandler, Jacob Nusbacher, Moses S. Schanfield, *Editors*

Vol 44: **Perinatal Medicine Today,** Bruce K. Young, *Editor*

Vol 45: **Mammalian Genetics and Cancer: The Jackson Laboratory Fiftieth Anniversary Symposium,** Elizabeth S. Russell, *Editor*

Vol 46: **Etiology of Cleft Lip and Cleft Palate,** Michael Melnick, David Bixler, Edward D. Shields, *Editors*

Vol 47: **New Developments With Human and Veterinary Vaccines,** A. Mizrahi, I. Hertman, M.A. Klingberg, A. Kohn, *Editors*

Vol 48: **Cloning of Human Tumor Stem Cells,** Sidney E. Salmon, *Editor*

Vol 49: **Myelin: Chemistry and Biology,** George A. Hashim, *Editor*

Vol 50: **Rights and Responsibilities in Modern Medicine: The Second Volume in a Series on Ethics, Humanism, and Medicine,** Marc D. Basson, *Editor*

Vol 51: **The Function of Red Blood Cells: Erythrocyte Pathobiology,** Donald F. H. Wallach, *Editor*

Vol 52: **Conduction Velocity Distributions: A Population Approach to Electrophysiology of Nerve,** Leslie J. Dorfman, Kenneth L. Cummins, Lary J. Leifer, *Editors*

Vol 53: **Cancer Among Black Populations,** Curtis Mettlin, Gerald P. Murphy, *Editors*

Vol 54: **Connective Tissue Research: Chemistry, Biology, and Physiology,** Zdenek Deyl, Milan Adam, *Editors*

Vol 55: **The Red Cell: Fifth Ann Arbor Conference,** George J. Brewer, *Editor*

Vol 56: **Erythrocyte Membranes 2: Recent Clinical and Experimental Advances,** Walter C. Kruckeberg, John W. Eaton, George J. Brewer, *Editors*

Vol 57: **Progress in Cancer Control,** Curtis Mettlin, Gerald P. Murphy, *Editors*

Vol 58: **The Lymphocyte,** Kenneth W. Sell, William V. Miller, *Editors*

Vol 59: **Eleventh International Congress of Anatomy,** Enrique Acosta Vidrio, *Editor-in-Chief.* Published in 3 volumes: Part A: **Glial and Neuronal Cell Biology,** Sergey Fedoroff, *Editor* Part B: **Advances in the Morphology of Cells and Tissues,** Miguel A. Galina, *Editor* Part C: **Biological Rhythms in Structure and Function,** Heinz von Mayersbach, Lawrence E. Scheving, John E. Pauly, *Editors*

Vol 60: **Advances in Hemoglobin Analysis,** Samir M. Hanash, George J. Brewer, *Editors*

Vol 61: **Nutrition and Child Health: Perspectives for the 1980s,** Reginald C. Tsang, Buford Lee Nichols, Jr., *Editors*

Vol 62: **Pathophysiological Effects of Endotoxins at the Cellular Level,** Jeannine A. Majde, Robert J. Person, *Editors*

Vol 63: **Membrane Transport and Neuroreceptors,** Dale Oxender, Arthur Blume, Ivan Diamond, C. Fred Fox, *Editors*

Vol 64: **Bacteriophage Assembly,** Michael S. DuBow, *Editor*

Vol 65: **Apheresis: Development, Applications, and Collection Procedures,** C. Harold Mielke, Jr., *Editor*

Vol 66: **Control of Cellular Division and Development,** Dennis Cunningham, Eugene Goldwasser, James Watson, C. Fred Fox, *Editors.* Published in 2 volumes.

Vol 67: **Nutrition in the 1980s: Contraints on Our Knowledge,** Nancy Selvey, Philip L. White, *Editors*

Vol 68: **The Role of Peptides and Amino Acids as Neurotransmitters,** J. Barry Lombardini, Alexander D. Kenny, *Editors*

Vol 69: **Twin Research 3, Proceedings of the Third International Congress on Twin Studies,** Ligi Gedda, Paolo Parisi, Walter E. Nance, *Editors.* Published in 3 volumes: Part A: **Twin Biology and Multiple Pregnancy** Part B: **Intelligence, Personality, and Development** Part C: **Epidemiological and Clinical Studies**

Vol 70: **Reproductive Immunology,** Norbert Gleicher, *Editor*

Vol 71: **Psychopharmacology of Clonidine,** Harbans Lal, Stuart Fielding, *Editors*

Vol 72: **Hemophilia and Hemostasis,** Doris Ménaché, D. MacN. Surgenor, Harlan D. Anderson, *Editors*

Vol 73: **Membrane Biophysics: Structure and Function in Epithelia,** Mumtaz A. Dinno, Arthur B. Callahan, *Editors*

Vol 74: **Physiopathology of Endocrine Diseases and Mechanisms of Hormone Action,** Roberto J. Soto, Alejandro De Nicola, Jorge Blaquier, *Editors*

Vol 75: **The Prostatic Cell: Structure and Function,** Gerald P. Murphy, Avery A. Sandberg, James P. Karr, *Editors.* Published in 2 volumes: Part A: **Morphologic, Secretory, and Biochemical Aspects** Part B: **Prolactin, Carcinogenesis, and Clinical Aspects**

Vol 76: **Troubling Problems in Medical Ethics: The Third Volume in a Series on Ethics, Humanism, and Medicine,** Marc D. Basson, Rachel E. Lipson, Doreen L. Ganos, *Editors*

Vol 77: **Nutrition in Health and Disease and International Development: Symposia From the XII International Congress of Nutrition,** Alfred E. Harper, George K. Davis *Editors*

Vol 78: **Female Incontinence,** Norman R. Zinner, Arthur M. Sterling, *Editors*

Vol 79: **Proteins in the Nervous System: Structure and Function,** Bernard Haber, Jose Regino Perez-Polo, Joe Dan Coulter, *Editors*

Vol 80: **Mechanism and Control of Ciliary Movement,** Charles J. Brokaw, Pedro Verdugo, *Editors*

Vol 81: **Physiology and Biology of Horseshoe Crabs: Studies on Normal and Environmentally Stressed Animals,** Joseph Bonaventura, Celia Bonaventura, Shirley Tesh, *Editors*

Vol 82: **Clinical, Structural, and Biochemical Advances in Hereditary Eye Disorders,** Donna L. Daentl, *Editor*

Vol 83: **Issues in Cancer Screening and Communications,** Curtis Mettlin, Gerald P. Murphy, *Editors*

Vol 84: **Progress in Dermatoglyphic Research,** Christos S. Bartsocas, *Editor*

Vol 85: **Embryonic Development,** Max M. Burger, Rudolf Weber, *Editors.* Published in 2 volumes: Part A: **Genetic Aspects** Part B: **Cellular Aspects**

Vol 86: **The Interaction of Acoustical and Electromagnetic Fields With Biological Systems,** Shiro Takashima, Elliot Postow, *Editors*

Vol 87: **Physiopathology of Hypophysial Disturbances and Diseases of Reproduction,** Alejandro De Nicola, Jorge Blaquier, Roberto J. Soto, *Editors*

Vol 88: **Cytapheresis and Plasma Exchange: Clinical Indications,** W.R. Vogler, *Editor*

Vol 89: **Interaction of Platelets and Tumor Cells,** G.A. Jamieson, *Editor;* Alice R. Scipio, *Assistant Editor*

Vol 90: **Beta-Carbolines and Tetrahydroisoquinolines,** Floyd Bloom, Jack Barchas, Morton Sandler, Earl Usdin, *Organizers*

Vol 91: **Membranes in Growth and Development,** Joseph F. Hoffman, Gerhard H. Giebisch, Liana Bolis, *Editors*

Vol 92: **The Pineal and Its Hormones,** Russel J. Reiter, *Editor*

Vol 93: **Endotoxins and Their Detection With the Limulus Amebocyte Lysate Test,** Stanley W. Watson, Jack Levin, Thomas J. Novitsky, *Editors*

Vol 94: **Animal Models of Inherited Metabolic Diseases,** Robert J. Desnick, Donald F. Patterson, Dante G. Scarpelli, *Editors*

Vol 95: **Gaucher Disease: A Century of Delineation and Research,** Robert J. Desnick, Shimon Gatt, Gregory A. Grabowski, *Editors*

Vol 96: **Mechanisms of Speciation,** Claudio Barigozzi, *Editor*

Vol 97: **Membranes and Genetic Disease,** John R. Sheppard, V. Elving Anderson, John W. Eaton, *Editors*